THE LOEB CLASSICAL LIBRARY

FOUNDED BY JAMES LOEB, LL.D.

EDITED BY

G. P. GOOLD, PH.D.

FORMER EDITORS

† T. E. PAGE, C.H., LITT.D.   † E. CAPPS, PH.D., LL.D.
† W. H. D. ROUSE, LITT.D.   † L. A. POST, L.H.D.
E. H. WARMINGTON, M.A., F.R.HIST.SOC.

THEOPHRASTUS
ENQUIRY INTO PLANTS

II

THEOPHRASTUS

ENQUIRY INTO PLANTS

AND MINOR WORKS ON ODOURS AND WEATHER SIGNS

WITH AN ENGLISH TRANSLATION BY
SIR ARTHUR HORT, Bart., M.A.
FORMERLY FELLOW OF TRINITY COLLEGE, CAMBRIDGE

IN TWO VOLUMES
II

CAMBRIDGE, MASSACHUSETTS
HARVARD UNIVERSITY PRESS
LONDON
WILLIAM HEINEMANN LTD
MCMLXXVII

American ISBN 0-674-99088-9
British ISBN 0 434 99079 5

First printed 1926
Reprinted 1949, 1961, 1977

Printed in Great Britain

CONTENTS

BOOK VI

OF UNDER-SHRUBS

	PAGE
Of the classification of under-shrubs: the wild kinds: the chief distinction that between spinous and spineless	3
Of spineless under-shrubs and their differences	7
Of certain specially important spineless under-shrubs—silphium and *magydaris*—belonging to ferula-like plants	15
Of spinous under-shrubs and their differences	21
Of cultivated under-shrubs (coronary plants), with which are included those coronary plants which are herbaceous	35
Of the seasons at which coronary plants flower, and of the length of their life	49

BOOK VII

OF HERBACEOUS PLANTS, OTHER THAN CORONARY PLANTS: POT-HERBS AND SIMILAR WILD HERBS

Of the times of sowing and of germination of pot-herbs	59
Of the propagation of pot-herbs, and of differences in their roots	67
Of the flowers and fruits of pot-herbs	75
Of the various forms of some pot-herbs	81
Of the cultivation of pot-herbs; manure and water	93
Of the pests which infest pot-herbs	95
Of the time for which seed of pot-herbs can be kept	97
Of uncultivated herbs: the wild forms of pot-herbs	99

CONTENTS

	PAGE
Of other uncultivated herbs, which may be classed with pot-herbs	103
Of the differences in stem and leaf found in all herbaceous plants	107
Of other differences seen in herbaceous plants in general, as compared with one another and with trees	111
Of the seasons at which herbs grow and flower	115
Of the classes into which herbaceous plants may be divided, as those having a spike and chicory-like plants	119
Of herbs which have fleshy or bulbous roots	125
Of certain properties and habits peculiar to certain herbaceous plants	135

BOOK VIII

OF HERBACEOUS PLANTS: CEREALS, PULSES, AND 'SUMMER CROPS'

Of the three classes and the times of sowing and of germination	143
Of differences in the mode of germination and of subsequent development	149
Of differences in development due to soil or climate	155
Of differences between the parts of cereals, pulses, and summer crops respectively	159
Of the differences between cereals	165
Of the differences between pulses	173
Of sowing, manuring, and watering	177
Of the degeneration of cereals into darnel	183
Of the peculiar character of chick-pea	183
Of special features of 'summer crops'	185
Of treatment of cereals peculiar to special localities	185
Of cereals which grow a second time from the same stock	187
Of the effects of climate, soil, and manuring	189
Of different qualities of seed	191

CONTENTS

	PAGE
Of degeneration of cereals, and of the weeds which infest particular crops	193
Of the conditions in the seeds of pulses known as 'cookable' and 'uncookable,' and their causes	197
Of the grains and pulses which most exhaust the soil, or which improve it	199
Of the diseases of cereals and pulses, and of hurtful winds	201
Of seeds which keep or do not keep well	205
Of the age at which seeds should be sown	209
Of artificial means of preserving seed	211
Of the effect of heat on seeds	211
Of certain peculiarities of the seed of lupin and *aigilops*	213

BOOK IX

OF THE JUICES OF PLANTS, AND OF THE MEDICINAL PROPERTIES OF HERBS

Of the various kinds of plant-juices and the methods of collecting them	217
Of resinous trees and the methods of collecting resin and pitch	223
Of the making of pitch in Macedonia and in Syria	229
Of frankincense and myrrh: various accounts	233
Of cinnamon and cassia: various accounts	243
Of balsam of Mecca	245
Of other aromatic plants—all oriental, except the iris	247
Of the medicinal juices of plants and the collection of them: general account	251
Of the cutting of roots for medicinal purposes, and of certain superstitions connected therewith	255
Of the medicinal uses of divers parts of plants	261
Of hellebores, the white and the black: their uses and distribution	265
Of the various kinds of all-heal	269
Of the various plants called *strykhnos*	271
Of the various kinds of *tithymallos*	275

CONTENTS

	PAGE
Of the two herbs called *libanotis*	277
Of the two kinds of chamaeleon	277
Of the various plants called 'poppy'	279
Of roots possessing remarkable taste or smell	281
Of the time for which roots can be kept without losing their virtue	287
Of the localities which specially produce medicinal herbs	289
Of the medicinal herbs peculiar to Crete	295
Of wolf's-bane and its habitat, and of meadow-saffron	299
Of two famous druggists and of the virtues of hemlock	303
How use diminishes the efficacy of drugs, and how they have not the same effect on all constitutions	305
Of plants that possess properties affecting lifeless objects	309
Of plants whose properties affect animals other than man	309
Of plants possessing properties which affect the mental powers	311
Of plants said to have magical properties	313
A problem as to cause and effect	313
Of certain plants, not yet mentioned, which possess special properties	315

MINOR WORKS

INTRODUCTION TO THE TREATISES CONCERNING ODOURS AND CONCERNING WEATHER SIGNS 324

CONCERNING ODOURS

Introductory: Of odours in general and the classification of them	327
Of natural odours; of those of animals and of the effect of odours on animals	329
Of smell and taste	331
Of odours in plants	333
Of artificial odours in general and their manufacture: especially of the use of perfumes in wine	333

CONTENTS

	PAGE
Of the oils used as the vehicle of perfumes	341
Of the spices used in making perfumes and their treatment	347
Of the various parts of plants used for perfumes, and of the composition of various notable perfumes	351
Of the properties of various spices	355
Of the medicinal properties of certain perfumes	359
Of rules for the mixture of spices, and of the storing of various perfumes	361
Of the properties of certain perfumes	365
Of other properties and peculiarities of perfumes	373
Of the making of perfume-powders and compound perfumes	377
Of the characteristic smells of animals, and of certain curious facts as to the smell of animal and vegetable products	381
Of odours as compared with other sense-impressions	383

CONCERNING WEATHER SIGNS

Introductory: general principles	391
The signs of rain	397
The signs of wind	407
The signs of fair weather	427
Miscellaneous signs	431
INDEX OF PLANTS	435

KEY TO THE INDEX—

I.—List of plants mentioned in the Enquiry under botanical names	487
II.—List of plants mentioned in the Enquiry unde popular names	494

THEOPHRASTUS

ENQUIRY INTO PLANTS

BOOK VI

ΘΕΟΦΡΑΣΤΟΥ
ΠΕΡΙ ΦΥΤΩΝ ΙΣΤΟΡΙΑΣ

Z

I. Περὶ μὲν οὖν δένδρων καὶ θάμνων εἴρηται πρότερον· ἑπόμενον δ' εἰπεῖν περί τε τῶν φρυγανικῶν καὶ ποιωδῶν καὶ εἴ τινες ἐν τούτοις ἕτεραι συμπεριλαμβάνονται φύσεις· οἷον ἥ γε σιτηρὰ ποιώδης ἐστί.

Πρῶτον δὲ λέγωμεν περὶ τῆς φρυγανικῆς· αὕτη γὰρ ἐγγυτέρω τῶν προειρημένων διὰ τὸ ξυλώδης εἶναι. πανταχοῦ μὲν οὖν ἴσως αἰεὶ τὸ ἄγριον τοῦ ἡμέρου πλεῖον, εἰ δὲ μή, περί γε τὴν φρυγανικὴν οὐσίαν· ὀλίγον γὰρ τὸ ἥμερον αὐτῆς, ὅπερ σχεδὸν ἐν τοῖς στεφανωτικοῖς ἐστιν, οἷον ῥοδωνία ἰωνία διόσανθος ἀμάρακος ἡμεροκαλλές, ἔτι δὲ ἕρπυλλος σισύμβριον ἐλένιον ἀβρότονον. ἅπαντα γὰρ ταῦτα ξυλώδη καὶ μικρόφυλλα, δι'
2 ὃ καὶ φρυγανικά. καὶ ἐπὶ τῶν λαχανηρῶν δ'

---

[1] cf. 6. 6. 1.
[2] This hardly corresponds to the definition of φρύγανα

THEOPHRASTUS
ENQUIRY INTO PLANTS

BOOK VI

OF UNDER-SHRUBS.

Of the classification of under-shrubs: the wild kinds: the chief distinction that between spinous and spineless.

I. WE have spoken already of trees and shrubs, and next we must speak of under-shrubs and herbaceous plants and of any other natural classes which are included with these; for instance, cereals come under herbaceous plants.

But first let us tell of under-shrubs, for this class comes near those mentioned above because of its woody character. Now it may be said that with all plants the wild kinds are more abundant than the cultivated, and this is certainly true of the under-shrubs. For the cultivated kinds of this class[1] are not numerous, and consist almost entirely of coronary plants, as rose gilliflower carnation sweet marjoram martagon lily, to which may be added tufted thyme bergamot-mint calamint southernwood. For all these are woody and have small leaves; wherefore[2] they are classed as under-shrubs. This class covers

given in 1, 3. 1, nor do some of the plants here mentioned come under the description. St. considers the text defective.

ὁμοίως, οἷον ῥάφανος πήγανον καὶ ὅσα παραπλήσια τούτοις ἐστίν. ὑπὲρ ὧν οὐχ ἧττον ἴσως ἁρμόττει κατὰ τὴν οἰκείαν προσηγορίαν εἰπεῖν, ὅταν περὶ στεφανωμάτων καὶ λαχάνων ποιώμεθα μνείαν. νῦν δὲ πρῶτον περὶ τῶν ἀγρίων λέγωμεν. ἔστιν αὐτῶν εἴδη καὶ μέρη πλείω, ἃ δεῖ διαιρεῖν καὶ τοῖς καθ' ἕκαστον γένος καὶ τοῖς ὅλοις εἴδεσι.

Μεγίστην δ' ἄν τις λάβοι διαφορὰν τῶν ὅλων γενῶν, ὅτι τὰ μὲν ἀνάκανθα τὰ δὲ ἀκανθώδη τυγχάνει. πάλιν δ' ἐν ἑκατέρῳ τούτων πολλαὶ διαφοραὶ γενῶν καὶ εἰδῶν, ὑπὲρ ὧν καθ' ἑκάτερα πειρατέον εἰπεῖν.

3 Τῶν ἀκανθικῶν δὴ τὰ μὲν ἁπλῶς εἰσιν ἄκανθαι, ὥσπερ ἀσφάραγος καὶ σκορπίος· οὐ γὰρ ἔχουσι φύλλον οὐδὲν παρὰ τὴν ἄκανθαν. τὰ δὲ φυλλάκανθα, καθάπερ ἄκανος ἠρύγγιον κνῆκος· ταῦτα γὰρ καὶ τὰ τοιαῦτα ἐπὶ τῶν φύλλων ἔχει τὴν ἄκανθαν, δι' ὃ καὶ φυλλάκανθα καλεῖται. τὰ δὲ καὶ παρὰ τὴν ἄκανθαν ἕτερον ἔχει φύλλον, ὥσπερ ἡ ὀνωνὶς καὶ ὁ τρίβολος καὶ ὁ φέως, ὃν δή τινες καλοῦσι στοιβήν. ὁ δὲ τρίβολος καὶ περικαρπιάκανθός ἐστιν· ἔχει γὰρ ἀκάνθας ἐν τῷ περικαρπίῳ, δι' ὃ καὶ τοῦτο ἴδιον πρὸς ἅπαντα ὡς εἰπεῖν· ἐπεὶ πτορθάκανθά γε πολλὰ καὶ τῶν δένδρων καὶ τῶν θαμνωδῶν ἐστιν, οἷον ἀχρὰς ῥόα παλίουρος βάτος ῥοδωνία κάππαρις. ἐν μὲν οὖν τοῖς ἀκανθικοῖς ταύτας ἄν τις ὡς εἰπεῖν τύπῳ λάβοι τὰς διαφοράς.

---

[1] ἔστιν . . . εἴδεσι: text needlessly altered by Sch. and W. Sch. himself shews that T. uses εἶδος and γένος almost indiscriminately. Here τῶν ὅλων γενῶν means the same as τοῖς

also pot-herbs, such as cabbage rue and others like them. Of these it is perhaps more appropriate to speak under their proper designation, that is, when we come to make mention of coronary plants and pot-herbs. Now let us first speak of the wild kinds. Of these are several classes and subdivisions, which we must distinguish by the characteristics of each sub-division as well as by those of each class taken as a whole.[1]

The most important difference distinguishing class from class which one could find is that between the spineless and the spinous kinds. Again under each of these two heads there are many differences distinguishing kinds and forms, of which we must endeavour to speak severally.

[2] Of spinous kinds some just consist of spines, as asparagus and *skorpios*; for these have no leaves except their spines. Then there are the spinous-leaved plants, as thistle eryngo safflower; these and the like have their spines on the leaves, whence their name. Others again have leaves as well as their spines, as rest-harrow caltrop and *pheos*,[3] which some call *stoibe*. Caltrop is also[4] spinous-fruited, having spines on the fruit-vessel. Wherefore this peculiarity marks it off from almost all other plants; though many trees and shrubs have spines on the shoots, as wild pear pomegranate Christ's thorn bramble rose caper. Such[5] are the general distinctions which may be made among spinous plants.

ὅλοις εἴδεσι; and below γενῶν and εἰδῶν *both* refer to the smaller divisions called μέρη above.   [2] Plin. 21. 91.
[3] ὁ φέως ὃν conj. Sch.; ὁ φλεὼς ὃ Ald.H.; καὶ ὃ δή τινες καλ. στ. P₂. *cf.* 6. 5. 1 and Index.
[4] καὶ περικαρπάκανθος conj. Sch.; καὶ ἡ περικαρπία φυλλάκανθον UMVAld. *cf.* 6. 5. 3.   [5] οὖν add. Sch.

THEOPHRASTUS

4 Ἐν δὲ τοῖς ἀνακάνθοις οὐκ ἔστιν οὕτως διαλαβεῖν τοῖς γένεσιν· ἡ γὰρ τῶν φύλλων ἀνωμαλία μεγέθει καὶ μικρότητι καὶ σχήμασιν ἄπειρος καὶ ἀσαφής· ἀλλὰ δεῖ πειρᾶσθαι κατ' ἄλλον τρόπον διαιρεῖν. πλείω δέ ἐστι τὰ γένη τὰ τούτων καὶ διαφορὰς ἔχοντα μεγάλας, οἷον κίσθος μήλωθρον ἐρευθέδανον σπειραία κνέωρον ὀρίγανος θύμβρα σφάκος ἐλελίσφακος πράσιον κόνυζα μελισσόφυλλον ἕτερα τοιαῦτα· πρὸς τούτοις ἔτι τὰ ναρθηκώδη καὶ ἐννευρόκαυλα, καθάπερ μάραθον ἱππομάραθον ναρθηκία νάρθηξ καὶ τὸ καλούμενον ὑπό τινων μυοφόνον καὶ ὅσα ὅμοια τούτοις. ἅπαντα γὰρ ἄν τις καὶ ταῦτα καὶ ὅλως εἴ τι ναρθηκῶδές ἐστι τῆς φρυγανικῆς θείη φύσεως.

II. Εἴδη δὲ καὶ διαφοραὶ καθ' ἕκαστον τῶν εἰρημένων εἰσὶ τῶν μὲν φανερώτεραι τῶν δὲ ἀφανέστεραι. καὶ γὰρ κίσθου δύο γένη διαιροῦσι, τὸ μὲν ἄρρεν τὸ δὲ θῆλυ, τῷ τὸ μὲν μεῖζον καὶ σκληρότερον καὶ λιπαρώτερον εἶναι καὶ τὸ ἄνθος ἐπιπορφυρίζον· ἄμφω δὲ ὅμοια τοῖς ἀγρίοις ῥόδοις, πλὴν ἐλάττω καὶ ἄοσμα.

2 Δύο δὲ εἴδη καὶ τοῦ κνεώρου· ὁ μὲν γὰρ λευκὸς ὁ δὲ μέλας. ἔχει δὲ ὁ μὲν λευκὸς τὸ φύλλον

---

[1] *i.e.* there is a gradation.
[2] κίσθος conj. Sch.; κισσὸς Ald.H.
[3] σπειραία conj. Sch. from Plin. 21. 53; σμηρέα Ald.G.
[4] θύμβρα σφάκος conj. Sch.; θύμβρας φακὸς UMVAld.
[5] *cf.* 6. 2. 5.
[6] ναρθηκώδη = hollow-stemmed, ἐννευρόκαυλα = plants with a plain unjointed stem, solid with 'immersed' fibres. In the

ENQUIRY INTO PLANTS, VI. I. 4–II. 2

With spineless plants it is not possible to make such 'generic' distinctions; for the variation of the leaves in size and shape is endless, and the differences are not clearly marked[1]; but we must try to distinguish on another principle. There are many classes of such plants and they differ widely, as rock-rose[2] bryony madder privet[3] *kneoron* marjoram savory *sphakos*[4] (sage) *elelisphakos*[5] (salvia) horehound *konyza* balm, and others like these; and in addition to these we have the plants with a ferula-like stem[6] or with a stem composed of fibre, as fennel horse-fennel[7] *narthekia* (ferula) *narthex* (ferula) and the plant called by some wolf's bane,[8] and others like these. All these, as well as any other ferula-like plants, may be placed in the class of under-shrubs.

Of spineless under-shrubs and their differences.

II. The various forms and the differences between the above mentioned plants are in some cases more, in some less easy to distinguish. Of rock-rose[9] they distinguish two kinds, 'male' and 'female,' in that the one is[10] larger, tougher, more glossy,[11] and has a crimson flower; both however are like the wild rose,[12] save that the flower is smaller and scentless.

There are also two kinds of *kneoron*, one white, the other black. The white has a leathery oblong

examples given here the two classes are taken together, ναρθηκία being ναρθηκώδης, the others ἐννευρόκαυλα; hence the article is not repeated.   [7] *cf.* 6. 2. 7.

[8] Lit. 'mouse-bane': for other Greek names see Index.

[9] κίσθου conj. Sch., *cf.* 6 1. 4; κισσοῦ Ald.H.; Plin. 24. 81; Diosc. 1. 97.

[10] εἶναι conj. W.; ἔχειν UMVAld. (τὸ φύλλον ἔχειν conj. Sch.).

[11] *i.e.* has more glossy leaves.

[12] *cf.* Plin. 21. 55; Theocr. 5. 131. See Index.

δερματῶδες πρόμηκες ὁμοιόσχημον τρόπον τινὰ τῇ ἐλάᾳ, ὁ δὲ μέλας οἷον ἡ μυρίκη σαρκῶδες· ἐπίγειος δὲ μᾶλλον ὁ λευκός· ἔστι δὲ ὀσμώδης, ὁ δὲ μέλας ἄοσμος. τὴν δὲ ῥίζαν τὴν εἰς βάθος ἄμφω μεγάλην ἔχουσι καὶ τοὺς ἀκρεμόνας πολλοὺς καὶ παχεῖς καὶ ξυλώδεις ἀπ' αὐτῆς τῆς γῆς ἢ μικρὸν ἄνω σχιζομένους, ξυλωδεστάτην δέ. γλίσχρον δὲ σφόδρα, δι' ὃ καὶ χρῶνται πρὸς τὸ καταδεῖν καὶ περιλαμβάνειν, ὥσπερ τῷ οἴσῳ. βλαστάνει δὲ καὶ ἀνθεῖ μετ' ἰσημερίαν μετοπωρινὴν καὶ ἀνθεῖ πολὺν χρόνον.

3 Καὶ τῆς ὀριγάνου δὲ ἡ μέλαινα ἄκαρπος ἡ δὲ λευκὴ κάρπιμος. καὶ θύμον τὸ μὲν λευκὸν τὸ δὲ μέλαν εὐανθὲς δὲ σφόδρα· περὶ τροπὰς γὰρ ἀνθεῖ θερινάς. ἀφ' οὗ καὶ ἡ μέλιττα λαμβάνει τὸ μέλι, καὶ τούτῳ φασὶν οἱ μελιττουργοὶ δῆλον εἶναι πότερον εὐμελιτοῦσι ἢ οὔ· καλῶς γὰρ ἀπανθήσαντος εὐμελιτεῖν· βλάπτει δὲ καὶ ἀπόλλυσι τὴν ἄνθησιν ἐὰν ὕδωρ ἐπιγένηται.

Σπέρμα δὲ κάρπιμον ἡ μὲν θύμβρα καὶ ἔτι μᾶλλον ἡ ὀρίγανος ἔχει φανερόν, τοῦ θύμου δ' οὐκ ἔστι λαβεῖν, ἀλλ' ἐν τῷ ἄνθει πως ἀναμέμικται· σπείρουσι γὰρ τοῦτο καὶ ἀναβλαστάνει.
4 ζητοῦσι δὲ καὶ λαμβάνουσιν οἱ ἐξάγειν Ἀθήνησι βουλόμενοι τὸ γένος. ἴδιον δὲ ἔχει καὶ πρὸς ταῦτα καὶ σχεδὸν πρὸς τὰ ἄλλα τὸ κατὰ τοὺς τόπους· οὐ γάρ φασι δύνασθαι φύεσθαι καὶ

---

[1] cf. 1. 10. 4.
[2] Apparently an afterthought, suggested by the mention of the woodiness of the branches.
[3] περιλαμβάνει conj. W. from G; περιλαμβάνειν Ald.
[4] Plin. 21. 55.

leaf, somewhat like that of the olive; the leaf of the black is like that of the tamarisk[1] and fleshy; the white grows more on the ground and is scented, while the black is scentless. In both the root, which runs deep, is large (and the branches which divide at the ground level are numerous thick and woody), and the root is also very woody.[2] It is also very tough, wherefore it is used for binding and to put[3] round things, like the withy. It grows and flowers after the autumnal equinox, and remains in flower a long time.

[4] Of marjoram the black form is barren, the white bears fruit. [5] There is a black and a white thyme, and it flowers very freely: it is in bloom about the summer solstice. It is from this flower that the bee gets the honey, and by it [6] beekeepers say that it is made known whether they have a good yield of honey or not; for, if the thyme flowers abundantly,[7] they have a good yield, but the bloom is injured or even destroyed if it is rained upon.

Savory, and still more marjoram, has a conspicuous fruitful seed, but in thyme it is not easy to find, being somehow mixed up with the flower; for men sow the flower and plants come up from it. [8] This plant is sought and obtained by those in Athens who wish to export such herbs. But it has a peculiarity as compared both with similar plants and with most others, namely the kind of region which it affects[9]; they say that it can not be grown or become

[5] Plin. 21. 56 and 154.
[6] τούτῳ conj. Sch.; τοῦτο Ald.
[7] καλῶς mBod.; ἄλλως UMVAld.H.
[8] Plin. 21. 57.
[9] τὸ κατὰ τοὺς τόπους conj. W.; καὶ κατὰ τοὺς τ. Ald.; καὶ κατὰ τόπους P.

THEOPHRASTUS

λαμβάνειν ὅπου μὴ ἀναπνοὴ διϊκνεῖται ἡ ἀπὸ τῆς θαλάττης· δι' ὃ οὐδ' ἐν Ἀρκαδίᾳ γίνεται· θύμβρα δὲ καὶ ὀρίγανος καὶ τὰ τοιαῦτα πολλὰ καὶ πολλαχοῦ. παραπλήσιον οὖν τὸ συμβαῖνον τοῦτο καὶ ἐπὶ τῆς ἐλάας· οὐδὲ γὰρ οὐδ' ἐκείνη δοκεῖ τριακοσίων σταδίων ἀπὸ θαλάττης ἐπάνω φύεσθαι.

5 Σφάκος δὲ καὶ ἐλελίσφακος διαφέρουσιν ὡσὰν τὸ μὲν ἥμερον τὸ δὲ ἄγριον· λειότερον γὰρ τὸ φύλλον τοῦ σφάκου καὶ ἔλαττον καὶ αὐχμηρότερον, τὸ δὲ τοῦ ἐλελισφάκου τραχύτερον.

Δύο δὲ γένη καὶ τοῦ πρασίου· τὸ μὲν γὰρ ἔχει ποῶδες τὸ φύλλον καὶ μᾶλλον ἐπικεχαραγμένον, ἔτι δὲ τὰς ἐντομὰς ἐνδήλους σφόδρα καὶ βαθείας, ᾧ καὶ οἱ φαρμακοπῶλαι χρῶνται πρὸς ἔνια· τὸ δὲ ἕτερον στρογγυλότερον καὶ αὐχμῶδες σφόδρα, καθάπερ τοῦ σφάκου, καὶ τὰς ἐντομὰς ἀμαυροτέρας ἔχον καὶ ἐπικεχαραγμένον ἧττον.

6 Κονύζης δὲ τὸ μὲν ἄρρεν τὸ δὲ θῆλυ. διαφορὰς δὲ ἔχει καθάπερ τὰ ἄλλα τὰ οὕτω διαιρούμενα· τὸ μὲν γὰρ θῆλυ λεπτοφυλλότερον καὶ ξυνεστηκὸς μᾶλλον καὶ τὸ ὅλον ἔλαττον, τὸ δὲ ἄρρεν μεῖζόν τε καὶ παχυκαυλότερον καὶ πολυκλωνότερον καὶ τὸ φύλλον μεῖζον καὶ λιπαρώτερον ἔχον, ἔτι δὲ τὸ ἄνθος λαμπρότερον. καρποφόρα δὲ ἄμφω· τὸ δὲ ὅλον ὀψιβλαστεῖ καὶ ὀψιανθεῖ περὶ Ἀρκτοῦρον καὶ μετ'

---

[1] λαμβάνειν P₂Ald.: lit. 'take hold,' cf. 6. 2. 6 ; βλαστάνειν conj. W.

[2] σφάκος conj. Sch.: σφάκελος UMVP₂Ald.; Plin. 22, 146 and 147.

established[1] where a breeze from the sea does not reach. This is why it does not grow in Arcadia, while savory marjoram and such plants are common in many parts. (A similar peculiarity is found in the olive; for it appears that it likewise will not grow more than three hundred furlongs from the sea.)

The difference between *sphakos*[2] (sage) and *elelisphakos* (salvia) is like that between cultivated and wild; for the leaf of *sphakos*[3] is smoother smaller and less succulent,[4] while that of *elelisphakos* is rougher.[5]

There are also two kinds of horehound: one has a narrow leaf with a more jagged edge, and the notches are very conspicuous and deep, and this is the plant used by druggists for certain purposes; the other has a rounder[6] leaf, which, like that of *sphakos*, is not at all succulent; the notches are less conspicuous and the edge less jagged.

Of *konyza*[7] there is a 'male' and a 'female' kind, the differences between them being such as are usual between forms so distinguished; the 'female' has slenderer leaves, is more compact, and a smaller plant; the 'male' is larger, has thicker stalks, is more branched, has larger glossier leaves, and moreover the flower is more conspicuous. Both bear fruit; the plant as a whole is late in growing and in blooming; it blooms about the rising of

---

[3] σφάκου conj. Sch.; σφακέλου UMVP₂Ald.

[4] W. omits ἧττον before αὐχμηρότερον.

[5] τραχύτερον conj. Scal. from G; βραχύτερον Ald.H.

[6] στρογγυλότερον: *cf.* 1. 10. 4 n.

[7] See Index. Plin. 26. 58. *cf.* Nic. *Ther.* 875; Diosc. 3. 121.

Ἀρκτοῦρον λαμβάνει. βαρεῖα δὲ ἡ ὀσμὴ τοῦ ἄρρενος, ἡ δὲ τῆς θηλείας δριμυτέρα, δι' ὃ καὶ πρὸς τὰ θηρία χρησίμη.

Ταῦτα μὲν οὖν καὶ τὰ τοιαῦτα ὥσπερ διαφέροντα. πάλιν δὲ ἄλλα μονοειδῆ τυγχάνοντα καὶ τῶν πρότερον εἰρημένων καὶ ἕτερα παρὰ ταῦτα· πλείω γάρ ἐστι.

7 Τὸ δὲ ναρθηκῶδες, καὶ γὰρ καὶ τοῦτο τῶν φρυγανικῶν, πολλὰς περιείληφεν ἰδέας· ἐν οἷς πρῶτον ὑπὲρ τοῦ κοινοῦ πᾶσι λεκτέον, ὑπὲρ νάρθηκός τε καὶ ναρθηκίας, εἴτε τὸ αὐτὸ γένος ἐστὶν ἀμφοῖν διαφέρον δὲ κατὰ μέγεθος, εἴτε καὶ ἕτερον ὥσπερ τινές φασιν. ἡ δ' οὖν φανερὰ φύσις ἀμφοῖν ὁμοία πλὴν κατὰ τὸ μέγεθος· ὁ μὲν γὰρ νάρθηξ γίνεται μέγας σφόδρα ἡ δὲ ναρθηκία μικρά. μονόκαυλα δ' ἄμφω καὶ γονατώδη, ἀφ' ὧν τά τε φύλλα βλαστάνει καὶ καυλοί 8 τινες μικροί· βλαστάνει δὲ παραλλὰξ τὰ φύλλα· λέγω δὲ παραλλὰξ ὅτι οὐκ ἐκ τοῦ αὐτοῦ μέρους τῶν γονάτων ἀλλ' ἐναλλάξ· περιειληφότα δὲ τὸν καυλὸν ἐπὶ πολύ, καθάπερ τὰ τοῦ καλάμου, πλὴν ἀποκεκλιμένα ταῦτα μᾶλλον διὰ τὴν μαλακότητα καὶ τὸ μέγεθος· μέγα γὰρ τὸ φύλλον καὶ μαλακὸν καὶ πολυσχιδές, ὥστε εἶναι σχεδὸν τριχῶδες· ἔχει δὲ μέγιστα τὰ κάτω πρὸς τὴν γῆν καὶ ἀεὶ κατὰ λόγον. ἄνθος δὲ μηλινοειδὲς ἀμαυρόν, καρπὸν δὲ παρόμοιον τῷ ἀνήθῳ πλὴν

---

[1] λαμβάνει Ald.; ἀδρύνει conj. W. But *cf.* the somewhat similar use 6. 2. 4.

[2] *contra bestiarum morsus* Plin. *l.c.*

[3] Plin. 13. 132 and 133.

[4] The form of expression in the repeated ὑπὲρ seems loose,

Arcturus and is full grown[1] after his setting. The smell of the 'male' plant is strong, but that of the 'female' more pungent; wherefore both of them are of use against wild beasts.[2]

These plants then and others like them have, as it were, different forms. Again there are some which have but one form both among those already mentioned and others as well; for there are numerous plants of this class.

[3] The class of ferula-like plants (for this too belongs to the under-shrubs) comprises many kinds: here we must first speak of the characteristic which is common to all, including ferula itself[4] (*narthex*) and *narthekia*, whether they both belong to the same kind and differ only in size, or whether, as some say, they are distinct. The obvious character of both is alike, except as to size; for *narthex* grows very tall, while *narthekia* is a small plant. Each of them has a single stalk, which is jointed; from this spring the leaves and some small stalks; the leaves come alternately—by which I mean that they do not spring from the same part of the joint, but in alternating rows. For a considerable distance they embrace the stalk, like the leaves of the reed, but they turn back from it more owing to their softness and their size; for the leaf is large soft and much divided, so that it is almost hair-like; the largest leaves are the lowest ones next the ground, and so on in proportion. The flower is quince-yellow[5] and inconspicuous, the fruit[6] like dill, but larger.[7] The

and above ἐν οἷς is hardly satisfactory. Sch. suspects corruption.

[5] μηλινοειδὲς: *cf.* 7. 3. 1.
[6] ἀμαυρόν, καρπὸν δὲ conj. Sch.; ἀμαυρόκαρπον Ald.
[7] μείζω conj. Sch.; μεῖζον Ald.

THEOPHRASTUS

μείζω. ἐξ ἄκρου δὲ σχίζεται καὶ ἔχει τινὰς οὐ μεγάλους καυλούς· ἐνταῦθα δὲ τό τε ἄνθος καὶ ὁ καρπός. ἔχει δὲ καὶ ἄνθος καὶ καρπὸν καὶ ἐν τοῖς παρακαυλίζουσι δι' ὅλου, καθάπερ τὸ ἄνηθον. ἐπετειόκαυλον δέ, καὶ ἡ βλάστησις τοῦ ἦρος πρῶτον μὲν τῶν φύλλων ἔπειτα τοῦ καυλοῦ, καθάπερ τῶν ἄλλων. ῥίζαν δὲ ἔχει βαθεῖαν, ἔστι δὲ μονόρριζον. ὁ μὲν οὖν νάρθηξ τοιοῦτος.

9 Τῶν δ' ἄλλων τὰ μὲν ὁμοιότερα τούτῳ τὸν καυλὸν ἔχει <κοῖλον>, καθάπερ ὁ μανδραγόρας καὶ τὸ κώνειον καὶ ὁ ἐλλέβορος καὶ ὁ ἀνθέρικος· τὰ δ' οἷον ἐννευρόκαυλα τυγχάνει, καθάπερ μάραθον μυοφόνον τὰ ὅμοια τούτοις. ἴδιος δὲ ὁ καρπὸς τοῦ μανδραγόρου τῷ μέλας τε καὶ ῥαγώδης καὶ οἰνώδης εἶναι τῷ χυμῷ.

III. Μέγισται δὲ καὶ ἰδιώταται φύσεις ἥ τε τοῦ σιλφίου καὶ ἡ τοῦ παπύρου ἐν Αἰγύπτῳ· ναρθηκώδη γὰρ καὶ ταῦτά ἐστιν· ὑπὲρ ὧν τοῦ μὲν παπύρου πρότερον εἴπομεν ἐν τοῖς ἐνύδροις, ὑπὲρ θατέρου δὲ νῦν λεκτέον.

Τὸ δὲ σίλφιον ἔχει ῥίζαν μὲν πολλὴν καὶ παχεῖαν, τὸν δὲ καυλὸν ἡλίκον νάρθηξ, σχεδὸν δὲ καὶ τῷ πάχει παραπλήσιον, τὸ δὲ φύλλον, ὃ καλοῦσι μάσπετον, ὅμοιον τῷ σελίνῳ· σπέρμα

---

[1] κοῖλον add. W.
[2] See Index: the *stalk* is specially in question here.
[3] *cf.* 6. 1. 4 n. ἐκνευρόκαυλα P_2Ald., *cf.* ἔκλευκος, 'whitish'; ἐννευρόκαυλα conj. Sch. as in 6. 1. 4; but οἷον indicates the coinage of a fresh term. κώνειον seems to be placed in the wrong list.
[4] Plin. 25. 147-150 describes *mandragoras*, but his description is not taken from T. *cf.* Diosc. 4. 75, where three kinds

plant divides at the top and has some small branches, on which grow the flower and the fruit. It also bears flowers and fruit on the side-stalks all the way up, like dill. The stalk only lasts a year, and the growth takes place in spring, the leaves growing first and then the stem, as with other plants. It roots deep and has but a single root. Such is the ferula.

Of the others some to a certain extent resemble ferula, that is, in having a hollow stem [1]; for instance deadly nightshade hemlock hellebore asphodel [2]: while some have a stem more or less, as it were, consisting of fibre,[3] as fennel aconite and others like these. The fruit of deadly nightshade [4] is peculiar in being black and like a grape and like wine in taste.

Of certain specially important spineless under-shrubs—silphium and magydaris—*belonging to ferula-like plants.*

III. Most important and peculiar in their characters are the silphium and papyrus of Egypt. These too come under the class of ferula-like plants; of these we have spoken [5] of the papyrus already under the head of plants living in water; of the other we have now to speak.

[6] The silphium has a great deal of thick root; its stalk is like ferula in size, and is nearly as thick; the leaf, which they call *maspeton*, is like celery: it has a broad fruit, which is leaf-like,

of μανδραγόρας are described: there being only two known species of *mandragora*, the third may be *atropa Belladonna*; and to this plant may also refer an interpolated sentence in Diosc. 4. 73 (ἄνθος . . . σταφυλήν).

[5] 4. 8. 3 and 4. Papyrus is loosely classed with ferula-like plants, as it has not a hollow stem. [6] Plin. 19. 42–45.

THEOPHRASTUS

δ' ἔχει πλατύ, οἷον φυλλῶδες, τὸ λεγόμενον φύλλον. ἐπετειόκαυλον δ' ἐστίν, ὥσπερ ὁ νάρθηξ. ἅμα μὲν οὖν τῷ ἦρι τὸ μάσπετον τοῦτο ἀφίησιν, ὃ καθαίρει τὰ πρόβατα καὶ παχύνει σφόδρα καὶ τὰ κρέα θαυμαστὰ ποιεῖ τῇ ἡδονῇ· μετὰ δὲ ταῦτα καυλόν, <ὃν> ἐσθίεσθαι πάντα τρόπον ἑφθὸν ὀπτόν, καθαίρειν δὲ καὶ τοῦτον φασὶ τὰ σώματα
2 τετταράκοντα ἡμέραις. ὀπὸν δὲ διττὸν ἔχει, τὸν μὲν ἀπὸ τοῦ καυλοῦ τὸν δὲ ἀπὸ τῆς ῥίζης, δι' ὃ καλοῦσι τὸν μὲν καυλίαν τὸν δὲ ῥιζίαν. ἡ δὲ ῥίζα τὸν φλοιὸν ἔχει μέλανα, καὶ τοῦτον περιαιροῦσιν. ἔστι δὲ ὥσπερ μέταλλα τῶν ῥιζοτομιῶν αὐτοῖς, ἐξ ὧν ὁπόσον ἂν δοκῇ συμφέρειν ταμιευόμενοι πρὸς τὰς τομὰς καὶ τὸ προϋπάρχον τέμνουσιν· οὐκ ἔξεστι γὰρ οὔτε παρατέμνειν οὔτε πλεῖον τῶν τεταγμένων· καὶ γὰρ διαφθείρεται καὶ σήπεται τὸ ἀργὸν ἐὰν χρονίζῃ. κατεργάζονται δὲ ἄγοντες εἰς τὸν Πειραιᾶ τόνδε τὸν τρόπον· ὅταν βάλωσι εἰς ἀγγεῖα καὶ ἄλευρα μίξωσι, σείουσι χρόνον συχνόν, ὅθεν καὶ τὸ χρῶμα λαμβάνει καὶ ἐργασθὲν ἄσηπτον ἤδη διαμένει. τὰ μὲν οὖν κατὰ τὴν ἐργασίαν καὶ τομὴν οὕτως ἔχει.

Τόπον δὲ πολὺν ἐπέχει τῆς Λιβύης· πλείω γάρ φασιν ἢ τετρακισχίλια στάδια· πλεῖστα δὲ γίνεσθαι περὶ τὴν σύρτιν ἀπὸ τῶν Εὐεσπερίδων. ἴδιον δὲ τὸ φεύγειν τὴν ἐργαζομένην καὶ ἀεὶ συνεργαζομένης καὶ συνημερουμένης ἐξαναχωρεῖν,

---

[1] οἷον φυλλ. τὸ λεγ. conj. W.; φυλλ. οἷον τὸ λεγ. Ald.H.
[2] I have added ὃν.
[3] μέταλλα U; μέτρα Ald.; ? ἐν μετάλλοις μέτρα.
[4] cf. 9. 1. 7; Diosc. 3. 80.

ENQUIRY INTO PLANTS, VI. III. 1-3

as it were,[1] and is called the *phyllon*. The stalk lasts only a year, like that of ferula. Now in spring it sends up this *maspeton*, which purges sheep and greatly fattens them, and makes their flesh wonderfully delicious; after that it sends up a stalk, which[2] is eaten, it is said, in all ways, boiled and roast, and this too, they say, purges the body in forty days. It has two kinds of juice, one from the stalk and one from the root; wherefore the one is called 'stalk-juice,' the other 'root-juice.' The root has a black bark, which is stripped off. They have regulations, like those in use in mines,[3] for cutting the root, in accordance with which they fix carefully the proper amount to be cut, having regard to previous cuttings and the supply of the plant. For it is not allowed to cut it wrong nor to cut more than the appointed amount; for, if the juice is kept and not used, it goes bad and decays. When they are conveying it to Peiraeus, they deal with it thus[4]:— having put it in vessels and mixed meal with it, they shake it for a considerable time, and from this process it gets its colour, and this treatment[5] makes it thenceforward keep without decaying. Such are the facts in regard to the cutting and treatment.

The plant is found over a wide tract of Libya, for a distance,[6] they say, of more than four thousand furlongs, but it is most abundant[7] near the Syrtis, starting from the Euesperides islands. It is a peculiarity of it that it avoids cultivated ground, and, as the land is brought under cultivation and tamed,

[5] ἐργασθὲν: ἐξοργασθὲν conj. Salm.; from Plin. *l.c.*, *argumentum erat maturitatis color siccitasque sudore finito*.

[6] *cf.* Strabo 2. 5. 20; 17. 3. 20: Scyl. *Periplus*, Libya.

[7] πλεῖστα conj. W.; πλείονα U; τὰ πλείονα MAld.; γίνεσθαι conj. W.; γενέσθαι Ald.

17

ὡς οὐ δεομένου δῆλον ὅτι θεραπείας ἀλλ' ὄντος ἀγρίου. φασὶ δ' οἱ Κυρηναῖοι φανῆναι τὸ σίλφιον ἔτεσι πρότερον ἢ αὐτοὶ τὴν πόλιν ᾤκησαν ἑπτά· οἰκοῦσι δὲ μάλιστα περὶ τριακόσια εἰς Σιμωνίδην ἄρχοντα Ἀθήνησιν.

4 Οἱ μὲν οὖν οὕτω λέγουσιν. οἱ δὲ τοῦ σιλφίου τὴν ῥίζαν φασὶ γίνεσθαι πηχυαίαν ἢ μικρῷ μείζω. ταύτην δὲ ἔχειν ἐπὶ τοῦ μέσου κεφαλήν, ὃ καὶ μετεωρότατόν ἐστι καὶ σχεδὸν ὑπὲρ γῆς, καλεῖσθαι δὲ γάλα· ἐξ ἧς δὴ φύεσθαι μετὰ ταῦτα καὶ τὸν καυλόν, ἐκ δὲ τούτου μαγύδαριν τὸ καὶ καλούμενον φύλλον· τοῦτο δ' εἶναι σπέρμα· καὶ ὅταν νότος λαμπρὸς πνεύσῃ μετὰ Κύνα διαρρίπτεσθαι, ἐξ οὗ φύεσθαι τὸ σίλφιον. τῷ αὐτῷ δὲ ἔτει τήν τε ῥίζαν γίνεσθαι καὶ τὸν καυλόν· οὐθὲν δὲ τοῦτο ἴδιον, καὶ γὰρ ἐπ' ἄλλων, εἰ μὴ τοῦτο λέγουσιν ὅτι εὐθὺς φύεται μετὰ τὴν διάρριψιν.

5 Καὶ τοῦτο ἴδιον καὶ διάφορον τοῖς πρότερον, ὅτι φασὶ δεῖν ὀρύττειν ἐπέτειον· ἐὰν δὲ ἐαθῇ, φέρειν μὲν τὸ σπέρμα καὶ τὸν καυλόν, χείρω δὲ γίνεσθαι καὶ ταῦτα καὶ τὴν ῥίζαν, ὀρυττομένας δὲ βελτίους γίνεσθαι διὰ τὸ μεταβάλλεσθαι τὴν γῆν. ἐναντίον δὲ τοῦτο τῷ φεύγειν τὴν ἐργάσιμον. ἐσθίεσθαι καὶ τὰς ῥίζας προσφάτους κατατεμνομένας εἰς ὄξος. τὸ δὲ φύλλον τῇ χροιᾷ χρυσοειδὲς

---

[1] cf. Hdt. 4. 158.   [2] B.C. 310.
[3] ταύτην δὲ ... τὸ σίλφιον: text as restored conjecturally by W., chiefly by alteration in the order of the words in Ald.
[4] καλεῖσθαι δὲ γάλα after φύεσθαι in Ald.
[5] In 6. 3. 7 this name is applied to a distinct plant. μαγ. τὸ καὶ conj. Salm.; μαγ. καὶ τὸ P₂Ald.H.   [6] cf. 6. 3. 2.
[7] τοῦτο conj. Salm.; τούτου UMVAld.; τούτων P₂.

ENQUIRY INTO PLANTS, VI. III. 3-5

it retires, plainly shewing that it needs no tendance but is a wild thing. The people of Cyrene say that the silphium appeared[1] seven years before they founded their city; now they had lived there for about three hundred years before the archonship at Athens of Simonides.[2]

Such is their account. Others however say that the root of the silphium grows to the length of a cubit or a little longer, and in the middle of this is a head,[3] which is the highest part and almost comes above ground, and is called the 'milk',[4] from this then presently grows the stalk, and from that the *magydaris*,[5] which is also called the *phyllon*[6]; but it[7] is really the seed, and, when a strong south wind blows after the setting of the dog-star, it is scattered[8] abroad and the silphium grows from it. The root and the stalk grow in the same year; nor is this a singular feature—unless they mean that it grows immediately after the dispersal[9] of the seed—since the same thing occurs with other[10] plants also.

There is this singular statement, which is inconsistent with what was said above, that, it is said, it is necessary to dig the ground every year, and that, if it be left alone, it bears[11] the seed and the stalk, but these are inferior and so is the root; on the other hand, that with digging they are improved because the soil is changed. (This is inconsistent with the statement that silphium avoids cultivated land.) They add that the roots are cut up into vinegar and eaten fresh, and that the leaf is of a golden

---

[8] διαρρίπτεσθαι conj. Sch.; διαρρίπτεται Ald.; διαρριπτεῖται U; διαρρίπτηται M.

[9] διάρριψιν conj. Sch.; δίριψιν UM; ἔκριψιν Ald.

[10] ἐπ' ἄλλων conj. W.; τῶν ἄλλων Ald.H.

[11] μὲν conj. Sch.; πᾶν Ald.

6 εἶναι. ἐναντίον δὲ καὶ τὸ μὴ καθαίρεσθαι τὰ πρόβατα τὸ φύλλον ἐσθίοντα· φασὶ γὰρ καὶ τοῦ ἦρος καὶ τοῦ χειμῶνος εἰς ὅρος ἀφιέναι, νέμεσθαι δὲ τοῦτό τε καὶ ἕτερον ὅμοιον ἀβροτόνῳ· θερμαντικὰ δ' ἄμφω δοκεῖ εἶναι καὶ κάθαρσιν μὲν οὐ ποιεῖν, ἀναξηραίνειν δὲ καὶ συμπέττειν· ἐὰν δέ τι νοσοῦν ἢ κακῶς ἔχον εἰσέλθῃ πρόβατον, ὑγιάζεσθαι ταχέως ἢ ἀποθνήσκειν, ὡς δ' ἐπὶ τὸ πολὺ σώζεσθαι μᾶλλον. ταῦτα μὲν ὁποτέρως ἔχει σκεπτέον.

7 Ἡ δὲ καλουμένη μαγύδαρις ἕτερόν ἐστι τοῦ σιλφίου μανότερόν τε καὶ ἧττον δριμὺ καὶ τὸν ὀπὸν οὐκ ἔχει· διάδηλος δέ ἐστι καὶ τῇ ὄψει τοῖς ἐμπείροις. γίνεται δὲ περὶ Συρίαν καὶ οὐκ ἐν Κυρήνῃ· φασὶ δὲ καὶ ἐν τῷ Παρνασίῳ ὄρει πολλήν· ἔνιοι δὲ σίλφιον τοῦτο καλοῦσιν. εἰ δὲ φεύγει τὴν ἐργάσιμον ὥσπερ τὸ σίλφιον σκεπτέον, ὡσαύτως δὲ καὶ εἴ τι ὅμοιον ἢ παραπλήσιον ἔχει φύλλου τε πέρι καὶ καυλοῦ, καὶ εἰ ὅλως ἀφίησί τι δάκρυον. τὴν μὲν οὖν ναρθηκώδη [καὶ ὅλως τὴν ἀκανθώδη] φύσιν ἐν τοῖς τοιούτοις θεωρητέον.

IV. Τῆς δ' ἀκανθικῆς, ἑπόμενον γὰρ τοῦτο εἰπεῖν, ἐπειδὴ διῄρηται τὸ μὲν ἀκανθῶδες ὅλως τὸ δὲ φυλλάκανθον, ὑπὲρ ἑκατέρου χωρὶς λεκτέον,

---

[1] cf. Arr. Anab. 3. 28. 6 and 7.
[2] Artemisia camphorata : Index App. (24).
[3] Plin. 19. 46; Diosc. 3. 94; Hesych. s.v.; Photius, Gloss. s.v.; cf. 6. 3. 4 n.

colour. We have also the inconsistent statement that sheep are not purged by eating the leaves; for they say that in spring and in winter they are driven into the hill-country, where [1] they feed on this and on another plant [2] which is like southernwood; both these plants appear to be heating and not to cause purging, but, on the contrary, to have a drying effect and promote digestion. It is also said that, if a sheep which is sick or in bad condition comes to that district, it is quickly cured or else dies, but usually it recovers. Which of these accounts is true is matter for enquiry.

[3] The plant called *magydaris* is distinct from silphium, being of later growth and less pungent, and it does not produce the characteristic juice; experts can also easily distinguish it by its appearance. It grows in Syria and not in Cyrene, and they say that it is also abundant on Mount Parnassus, and some call it silphium. Whether however, like silphium, it avoids cultivated ground is matter for enquiry, as also whether it has any resemblance or likeness in leaf and stalk, and, in general, whether it produces a juice. In these examples we may consider the class of ferula-like plants [and, in general, that of spinous plants.[4]]

Of spinous under-shrubs and their differences.

IV. Taking next the class of spinous plants (for we must next speak of them), we have already distinguished [5] those which are altogether spinous and those which have spinous leaves, and now we must

---

[4] καὶ ... ἀκανθώδη. These words occur only in U: they cannot belong here. Note that τὸ μὲν ἀκανθῶδες ὅλως occurs just below.   [5] 6. 1. 3.

καὶ τρίτου δὴ περὶ τοῦ καὶ παρὰ τὴν ἄκανθαν ἔχοντος φύλλον, ὥσπερ ὅ τε φέως¹ καὶ ὁ τρίβολος. καὶ ἡ κάππαρις ἴδιον ἔχει τὸ μὴ μόνον τὴν ἐκ τῶν καυλῶν ἄκανθαν ἔχειν ἀλλὰ καὶ τὸ φύλλον ἐπακανθίζον. τῶν δὲ διῃρημένων εἰδῶν πλεῖστον μέν ἐστι τὸ φυλλάκανθον, ἐλάχιστον δὲ ὡς εἰπεῖν τὸ ἀκανθῶδες ὅλως. βραχὺ γάρ τι πάμπαν ἐστίν, ὥσπερ ἐλέχθη, καὶ σχεδὸν οὐ ῥᾴδιον λαβεῖν παρά τε τὸν ἀσφάραγον καὶ τὸν σκορπίον.

2 Ἀμφότερα δὲ ταῦτα ἀνθεῖ μετὰ ἰσημερίαν φθινοπωρινήν. ὁ μὲν σκορπίος² ἐν τῷ σαρκώδει τῷ ἐποιδοῦντι τῷ ὑπὸ τὸ ἄκρον τῆς ἀκάνθης ἔχων τὸ ἄνθος ἐξ ἀρχῆς μὲν λευκὸν ὕστερον δ' ἐπιπορφυρίζον. ὁ δὲ ἀσφάραγος ἐκφύων παρὰ τὰς ἀκάνθας κορυνῶδες μικρόν, ἐκ τούτου δέ ἐστι τὸ ἄνθος μικρόν. ὁ δὲ σκορπίος μονόρριζον καὶ βαθύρριζον, ὁ δὲ ἀσφάραγος βαθύρριζόν τε εὖ μάλα καὶ πολύρριζον πυκναῖς ταῖς ῥίζαις, ὥστε τὸ ἄνω συνεχὲς εἶναι αὐτῶν, ἀφ' οὗ καὶ αἱ βλαστήσεις αὐτῶν τῶν καυλῶν· ἀναβλαστάνει δὲ ὁ καυλὸς ἐκ τῆς ἀσφαραγιᾶς τοῦ ἦρος καὶ ἐδώδιμός ἐστιν· εἶθ' οὕτως ἀποτραχύνεται καὶ ἐξακανθοῦται προϊούσης τῆς ὥρας· ἡ δὲ ἄνθησις οὐκ ἐκ τούτων μόνον ἀλλὰ καὶ ἐκ τῶν πρότερον· οὐ γὰρ ἐπετειόκαυλόν ἐστι. τὰ μὲν οὖν ὅλως ἀκανθώδη τοιαύτην τινὰ ἔχει φύσιν.

3 Τῶν δὲ φυλλακάνθων³ τὸ πλεῖστον γένος ὡς

---

¹ φέως conj. St.; φλεὼς Ald. cf. 6. 1. 3.
² cf. 6. 1. 3. ³ Plin. 21. 91; 22. 39.

ENQUIRY INTO PLANTS, VI. iv. 1-3

speak of each of these classes separately, and also, in the third place, of those which have leaves as well as their spines, such as *pheos*[1] and caltrop. Moreover caper has the peculiarity of possessing not only spines on its stems but also a spinous leaf. Of the classes thus distinguished that with spinous leaves is the largest, while that which is altogether spinous is about the smallest. It is indeed, as was said, a very small class, and it would not be easy to find examples of such plants besides asparagus and *skorpios*.[2]

[3] Both of these flower after the autumnal equinox. *Skorpios* produces its flower in the fleshy swelling[4] below the top[5] of the spinous twig; at first it is white, but afterwards it becomes purplish. Asparagus produces alongside of the spines a small knob, and from this grows the flower, which is of small size. *Skorpios* has a single root which runs deep; asparagus roots very deep and its roots are numerous and matted, the upper part of them being in one piece,[6] and from this the actual shoots spring. The stalk comes up from the plant in spring and is edible; afterwards, as the season advances, it acquires its rough and spinous character[7]; the bloom appears not only on this stalk, but on those of previous years, for the stalk is not annual. Such is the character of plants which are altogether spinous.

[8] Of those which have spinous leaves the largest class, one may say, consists of those plants which

[4] ἐποιδοῦντι conj. Scal.; ἐπωδοῦντι U; ἐσποδοῦντι MAld.
[5] τὸ ἄκρον conj. Scal.; τὸ ἄκρατον UMAld.; τῆς ἀκάνθης om. Scal.
[6] *i.e.* tuberous. *cf.* Col. 11. 3. 43; Pall. 3. 24. 8; 4. 9. 11.
[7] ἐξακανθοῦται conj. Link. *ap.* Sch.; ἐξανθοῦται UM; ἐξανθεῖται Ald.   [8] Plin. 21. 94.

THEOPHRASTUS

ἁπλῶς εἰπεῖν ἀκανῶδες τυγχάνει· λέγω δὲ τὸ ἀκανῶδες, ὅτι τὸ κύημα καὶ ἐν ᾧ τὸ ἄνθος ἢ καὶ ὁ καρπὸς ἄκανος ἢ ἀκανῶδες πάντων ἐστί. διαφορὰν δὲ ἔχει ἐν ἑαυτῷ καὶ μεγέθει καὶ σχήματι καὶ χρώματι καὶ πλήθει καὶ ὀλιγότητι τῶν ἀκανθῶν καὶ τῶν ἄλλων. ἔξω γὰρ ὀλίγων πάνυ, καθάπερ τοῦ στρουθίου τε καὶ τοῦ σόγκου καὶ εἴ τινων ἑτέρων, τὰ λοιπὰ πάντα ὡς εἰπεῖν τοιαύτην ἔχει τὴν φύσιν· ἐπεὶ καὶ ὁ σόγκος τήν γε φύσιν ἀκανθώδη ἔχει, τὸ δὲ σπερματικὸν οὐχ ὅμοιον· ἀλλὰ τά γε τοιαῦτα πάντα οἷον ἄκορνα λευκάκανθα χάλκειος κνῆκος πολυάκανθος ἀτρακτυλὶς ὀνόπυξος ἰξίνη χαμαιλέων· πλὴν οὗτος οὐ φυλλάκανθος, σκόλυμος δέ, ὃς καὶ λειμωνία, φυλλ-
4 άκανθος· καὶ τἆλλα, πλείω γάρ ἐστι. διαφέρουσι δ' ἀλλήλων πρὸς τοῖς εἰρημένοις τῷ τὰ μὲν πολύκαυλα εἶναι καὶ ἀποφύσεις ἔχειν, ὥσπερ ὁ ἄκανος, τὰ δὲ μονόκαυλα καὶ μὴ ἔχειν, ὥσπερ ὁ κνῆκος, ἔνια δ' ἄνωθεν ἔχειν ἐξ ἄκρου, καθάπερ τὸ ῥύτρος. καὶ τὰ μὲν εὐθὺς τοῖς πρώτοις ὑετοῖς βλαστάνειν τὰ δ' ὕστερον, ἔνια δὲ καὶ τοῦ θέρους, ὥσπερ καὶ ἡ τετράλιξ ὑπό τινων καλουμένη καὶ ἡ ἰξίνη· καὶ ἐπὶ τῶν ἀνθῶν ὁμοίως· ὀψιανθὴς γὰρ ὁ σκόλυμος καὶ ἐπὶ πολὺν χρόνον.

---

[1] ἀκανῶδες mBas.; ἀκανθῶδες Ald. cf. 1. 13. 3, where ἀκανῶδες is restored by W.'s certain conj.

[2] ἀκανῶδες conj. Sch.; ἀκανθῶδες Ald.H.; acanaceum G.

[3] ἄκανος ἢ ἀκανῶδες Ald.; ἄκανθος ἢ ἀκανθῶδες mBas. v. supra.

[4] σόγκος conj. Sch.; κνῆκος Ald. The correction seems necessary in view of 6. 4. 8.

[5] ἄκορνα conj. Sch.; ἄκαρνα Ald. cf. Plin. l.c.

[6] ὃς καὶ λειμωνία I conj.; ἢ καὶ λειμωνία conj. Scal. from

are thistle-like,[1] by which [2] I mean that the swollen part, that part which contains the flower, or, it may be, the fruit, is in all cases a thistle-head,[3] or has that appearance. However there are differences in the 'head' itself, in size shape colour number of spines and in other respects. For, apart from quite a few plants, such as soap-wort sow-thistle and possibly some others, nearly all the rest have this character (even sow-thistle [4] has a spinous character, but its seed-process is different). The list includes all the following: *akorna* [5] milk-thistle *khalkeios* safflower *polyakanthos* distaff-thistle *onopyxos ixine* chamaeleon (the last-named, however, has not spinous leaves, though golden thistle, which is also called 'meadow-thistle,' [6] has[7]), and so on, for there are many more. These differ from one another not only in the aforesaid ways, but in that some of them have many stalks and side-growths, like the pine-thistle, while some have a single stalk and no side-growths, like the safflower, and some again have out-growths above from the top of the plant, like the globe-thistle.[8] Again some grow directly the first rains come, others at a later time, some again in summer, as the plant which some call yellow star-thistle, and *ixine*.[9] So too [10] the flowering-time differs: golden thistle blooms late [11] and is in bloom for a long time.

Plin. 22. 86; ἡλυλειμωνία UM_1; ἡλυλειμωνία M_2Ald. καὶ λειμωνία conj. W. But λειμωνία is not mentioned again in the following description, which is against its being a distinct plant from σκόλυμος.

[7] φυλλάκανθος I conj.; φυλλάκανθα MSS.
[8] ῥύτρος: *rhutrum* G; but Plin. *l.c.* has *eryngen*.
[9] Plin. 22. 23. [10] καὶ ἐπὶ conj. Sch.; καὶ ἡ ἐπὶ Ald.H.
[11] ὀψιανθής conj. Bod. from Plin. *l.c.* floret sero et diu; εὐανθής Ald.

THEOPHRASTUS

5 Διαφοραὶ δὲ τῶν μὲν ἀκάνων οὐκ εἰσί, τῆς κνήκου δ' εἰσίν· ἡ μὲν γὰρ ἀγρία ἡ δ' ἥμερος. τῆς δ' ἀγρίας δύο εἴδη, τὸ μὲν προσεμφερὲς σφόδρα τῷ ἡμέρῳ πλὴν εὐθυκαυλότερον, δι' ὃ καὶ πηνίοις ἔνιαι τῶν ἀρχαίων ἐχρῶντο γυναικῶν. καρπὸν δὲ ἔχει μέλανα καὶ μέγαν καὶ πικρόν. ἡ δ' ἑτέρα δασεῖα καὶ τοὺς καυλοὺς ἔχει σογκώδεις, ὥστε τρόπον τινὰ ἐπιγειόκαυλος γίνεται· διὰ γὰρ μαλακότητα τῶν καυλῶν κατακλίνεται πρὸς τὰς ἀρούρας· καρπὸν δ' ἔχει μικρὸν πώγονος· σπερματώδεις πᾶσαι, πλὴν μείζοσι καὶ πυκνοτέροις αἱ ἄγριαι. ἴδιον δὲ ἔχει πρὸς τὰ ἄλλα ἄγρια· τὰ μὲν γὰρ σκληρότερα καὶ ἀκανθωδέστερα τῶν ἡμέρων, αὕτη δὲ μαλακωτέρα καὶ λειοτέρα.

6 Ἡ δ' ἄκορνα προσεμφερὴς ὡς ἁπλῶς εἰπεῖν κατὰ τὴν πρόσοψιν τῇ κνήκῳ τῇ ἡμέρῳ, χρῶμα δ' ἐπίξανθον ἔχει καὶ χυλὸν λιπαρόν. ἀτρακτυλὶς δέ τις καλεῖται καὶ λευκοτέρα τούτων· ἴδιον δὲ ἔχει τὸ περὶ τὸ φύλλον, ὅτι ἀφαιρούμενον καὶ τῇ σαρκὶ προσφερόμενον αἱματώδη ποιεῖ τὸν χυλόν, δι' ὃ καὶ φόνον ἔνιοι καλοῦσι τὴν ἄκανθαν ταύτην· ἔχει δὲ καὶ τὴν ὀσμὴν δεινὴν καὶ φονώδη· ὀψὲ δὲ καὶ τελειοῖ τὸν καρπὸν πρὸς τὸ μετό-

---

[1] ἀκάνων Ald.; ἀκαρνῶν mBas.; ἀκάνου or ἀκόρνης conj. Sch., the plural being awkward.
[2] πηνίοις conj. R. Const.; σπληνείοις U; σπληνίοις MAld; colu G and Plin. l.c.; cf. Diosc. 3. 107.
[3] σογκώδεις : Plin. l.c. seems to have read ὀγκώδεις (torosiore caule).
[4] καυλῶν conj. Scal. from Plin. l.c.; φύλλων Ald.
[5] μικρὸν conj. Spr. from Plin. l.c. (minutum semen); πικρὸν Ald.
[6] πώγονος· σπερματώδεις Ald.: so U, but πωγωνος, and M, but

Pine-thistle[1] has but one kind, but there are different kinds of safflower, the wild and the cultivated. Again of the wild kind there are two forms, one very like the cultivated except that the stalk is straighter; wherefore in ancient times women sometimes used it to make distaffs.[2] It has a fruit which is black large and bitter. The other is leafy, and its stalks are like those of the sow-thistle,[3] so that to some extent it comes to have a prostrate stem; for on account of the softness of the stalks[4] it bends down towards the ground; and it has a small[5] fruit, which is bearded. All the forms produce abundant seed,[6] but it is larger[7] and more crowded in the wild forms. This kind has also a peculiarity as compared with other wild plants; these are usually coarser and more spinous than the cultivated forms, but in this plant the wild form is softer and smoother.

The *akorna* resembles in a general way in appearance the cultivated safflower, but has a yellowish colour and a sticky juice.[8] There is also a plant called distaff-thistle, which is whiter than these. A peculiarity of the leaf of this is that, if it is stripped off and applied to the flesh, the contact makes the juice blood-coloured,[9] wherefore some call this kind of spinous plant 'blood-wort'; also it has an abominable smell, like that of blood; it matures its fruit late,

---

πωγωνὸs : G. has *fructum amarum* (see last note) *frequentem barbaeque modo hirsutum gignit: sunt ambo seminosa.* W. conj. πωγωνοσπέρματα δ' εἰσὶ πᾶσαι, which is not convincing. I have retained the corrupt text and translate in the light of G.

[7] μείζοσι : sc. σπέρμασι, but σπερματώδεις cannot be right.

[8] χυλὸν add. St.; om. Ald.; *succo pingui* G ; *pinguiore succo* Plin. *l.c.*   [9] *cf.* 9. 1. 1. Plin. 21. 95.

πωρον. τὸ δ' ὅλον ὡς ἁπλῶς εἰπεῖν ἅπασα ἡ ἀκανικὴ φύσις ὀψίκαρπος. ἅπαντα δὲ ταῦτα φύεται καὶ ἀπὸ τοῦ σπέρματος καὶ ἀπὸ τῆς ῥίζης, ὥστε βραχύν τινα γίνεσθαι τὸν ἀνὰ μέσον χρόνον τῆς ἐκφύσεως τε καὶ τῆς τοῦ σπέρματος τελειώσεως.

7 Τοῦ σκολύμου δὲ οὐχ ὅτι τοῦτο μόνον ἴδιον, ὅτι τὴν ῥίζαν ἐδώδιμον ἔχει καὶ ἐφθὴν καὶ ὠμήν, ἀλλὰ καὶ ὅτι τότε ἀρίστην ὅταν ἀνθῇ καὶ ὅτι σκληρυνομένη ἀφίησιν ὀπόν. ἴδιον δὲ καὶ τὸ τῆς ἀνθήσεως· ἐπεὶ περὶ τροπάς.

8 Σαρκώδης δὲ καὶ ἐδώδιμος ἡ τοῦ σόγκου· ἡ δὲ κύησις οὐκ ἀκανώδης ἀλλὰ προμήκης αὐτοῦ· καὶ τοῦτ' ἴδιον μόνον ἔχει τῶν φυλλακάνθων ἀντεστραμμένως ἢ ὁ χαμαιλέων· ὁ μὲν γὰρ ἀφυλλάκανθος ὢν ἀκανίζει. γηράσκον δὲ τὸ ἄνθος ἐκπαπποῦται, καθάπερ τὸ τῆς ἀπάπης καὶ τὸ τῆς μυρίκης καὶ ὅσα παραπλήσια τούτοις. παρακολουθεῖ δὲ μέχρι τοῦ θέρους τὸ μὲν κυοῦν τὸ δὲ ἀνθοῦν τὸ δὲ σπέρμα τίκτον, μικρὰν ἰκμάδα καὶ κέντρον ἔχον· ξηραινόμενον δὲ τὸ φύλλον διαχεῖται καὶ οὐκέτι κεντεῖ.

9 Ἡ ἰξίνη δὲ φύεται μὲν οὐ πολλαχοῦ, ῥιζόφυλλον δέ ἐστιν. ἀπὸ δὲ τῆς ῥίζης μέσης ὁ σπερματικὸς ἄκανος ἐπιπέφυκεν, ὥσπερ μῆλον εὖ

---

[1] ἀκανικὴ conj. Bod., cf. 6. 4. 4 nn.; ἀκανθικὴ Ald.
[2] cf. Hes. Op. 582.
[3] σόγκου conj. C. Hoffmann; ὕγκου Ald.
[4] κύησις: i.e. flower-head. cf. κύημα 6. 4. 3; Plin. 21. 94.
[5] ἀκανώδης conj. Scal.; ἀκανθώδης Ald. cf. 6. 4. 3 nn.
[6] cf. 6. 4. 3. T.'s information seems to be incorrect, as

ENQUIRY INTO PLANTS, VI. iv. 6–9

towards autumn. Indeed, generally speaking, all plants like the thistle-tribe [1] are late fruiting. All these plants grow both from seed and from the root, so that there is but a short period between the beginning of growth and the maturing of the seed.

Golden thistle has not only this peculiarity, that it has a root which is edible, whether boiled or raw, but the root is best when the plant is in flower, and, as it becomes hard, it produces a juice. The flowering time [2] is also peculiar, about the solstice.

The root of the sow-thistle [3] is also fleshy and edible; but the swollen part [4] is elongated and not thistle-like [5]; and, alone of the spinous-leaved plants, it has this peculiarity, in which it is the reverse of the chamaeleon,[6] (for that plant, though it has not spinous leaves, has a thistle-like flower-head). The flower of the sow-thistle, as it ages, turns into down, as do that of the dandelion [7] the tamarisk [8] and other plants like these. In its growth [9] there is a succession up to the summer, part forming flowers, part flowering, and part producing seed [10]; this [11] has little moisture in it and has a sharp point. The leaf, as it dries, becomes flaccid and no longer pricks.

Ixine does not grow in many places, and it has leaves on the root. From the middle of the root grows the seed-bearing thistle-head, which is like

both of the plants which he calls χαμαιλέων (see Index) have spinous leaves.

[7] ἀπάπης conj. Sch., cf. 7. 8. 3; πάπνης U; δαπάνης P; δάφνης Ald.

[8] μυρίκης conj. Sch.; μυρίνης M; μυρρίνης Ald.

[9] cf. Plin. l.c.

[10] σπέρμα τίκτον I conj.; σπέρματος μὲν Ald.H.; σπερμοτόκουν conj. Sch.

[11] Text perhaps defective.

μάλα ἐπικεκρυμμένον ὑπὸ τῶν φύλλων· οὗτος δὲ ἐπὶ τοῦ ἄκρου φέρει τὸ δάκρυον εὔστομον, καὶ τοῦτό ἐστιν ἡ ἀκανθικὴ μαστίχη. ταῦτα μὲν οὖν καὶ τὰ τοιαῦτα πανταχοῦ σχεδόν ἐστιν.

10 Ἡ δὲ κάκτος καλουμένη περὶ Σικελίαν μόνον, ἐν τῇ Ἑλλάδι δὲ οὐκ ἔστιν. ἴδιον δὲ παρὰ τἆλλα τὸ φυτόν· ἀφίησι γὰρ εὐθὺς ἀπὸ τῆς ῥίζης καυλοὺς ἐπιγείους, τὸ δὲ φύλλον ἔχει πλατὺ καὶ ἀκανθῶδες· καλοῦσι δὲ τοὺς καυλοὺς τούτους κάκτους· ἐδώδιμοι δέ εἰσι περιλεπόμενοι μικρὸν ἐπίπικροι, καὶ θησαυρίζουσιν αὐτοὺς ἐν ἅλμῃ.

11 Ἕτερον δὲ καυλὸν ὀρθὸν ἀφίησιν, ὃν καλοῦσι πτέρνικα· γίνεται δὲ καὶ οὗτος ἐδώδιμος πλὴν ἀθησαύριστος. τὸ δὲ περικάρπιον, ἐν ᾧ τὸ σπέρμα, τὴν μὲν μορφὴν ἀκανῶδες, ἀφαιρεθέντων δὲ τῶν παππωδῶν σπερμάτων ἐδώδιμον καὶ τοῦτο καὶ ἐμφερὲς τῷ τοῦ φοίνικος ἐγκεφάλῳ· καλοῦσι δὲ αὐτὸ σκαλίαν. τὰ μὲν οὖν φυλλάκανθα σκεπτέον ἐν τοιαύταις διαφοραῖς.

V. Τὰ δὲ καὶ παρὰ τὴν ἄκανθαν ἔχοντα φύλλον, οἷον τὰ τοιαῦτα φέως ὄνωνις παντάδουσα τρίβολος ἱππόφεως μυάκανθος . . . . τε σφόδρα καὶ τὸ φύλλον ἔχει σαρκῶδες· πολυ-

---

[1] ὑπὸ conj. Sch.; ἐπὶ Ald.; Plin. *l.c. malum contectum sua fronde.*   [2] *cf.* 9. 1. 3.   [3] *cf.* Plin. 12. 72.
[4] Plin. 21. 97; Athen. 2. 83.
[5] πλατὺ add. Scal. from Athen. *l.c.*, *cf.* Plin. *l.c.*; om. Ald. H. The 'stems' are the petioles of the leaves.
[6] ἀκανῶδες conj. Sch.; ἀκανθῶδες Ald.

ENQUIRY INTO PLANTS, VI. iv. 9–v. 1

an apple and well hidden by[1] the leaves; this on its head produces its gum,[2] which is pleasant to the taste, and this is the 'thorn-mastich.'[3] These plants and others like them are found almost everywhere.

[4] But the plant called *kaktos* (cardoon) grows only in Sicily, and not in Hellas. It is a plant quite different from any other; for it sends up straight from the root stems which creep on the ground, and its leaf is broad[5] and spinous: these stems are called *kaktoi*; they are edible, if peeled, and are slightly bitter, and men preserve them in brine.

There is another kind which sends up an erect stem, called the *pternix*. This too is edible, but cannot be preserved. The fruit-vessel, which contains the seed, is in shape like a thistle-head[6]: and when the downy seeds are taken off, this too is edible and resembles the 'brain'[7] of the palm; and it is called *skalias*.[8] Such are the different characteristics in the light of which we may observe the spinous-leaved plants.

V. Examples of plants which have leaves as well as spines are *pheos*[9] rest-harrow star-thistle caltrop 'horse-*pheos*'[10] (spurge) butcher's broom[11] . . . ,[12] and it has a fleshy leaf: it is much divided and has

[7] *i.e.* 'cabbage.' *cf.* 2. 6. 2.
[8] *ascaliam* Plin. *l.c.*; ἀσκάληρον Athen. *l.c.* Modern Greek σκάληρα. English 'bottom.' See Index κάκτος (2).
[9] φέως conj. St,; φλέως Ald. *cf.* 6. 1. 3.
[10] ἱπποφέως conj. Salm., *cf.* 6. 5. 2; ἱππόφυον Ald. *cf.* Plin. 21. 91.
[11] Diosc. 2. 125; Plin. 19. 151.
[12] Text defective: the end of one sentence is missing and the beginning of the next, containing the name of a plant. G attaches the following description to φέως. The plants presently described do not correspond to this list.

σχιδὲς δὲ καὶ πολύρριζον, οὐ μὴν κατὰ βάθους γε τὰς ῥίζας ἔχον. βλαστάνει δὲ ἅμα Πλειάδι καὶ τοῖς πρώτοις ἀρότοις καὶ ἀφίησι τότε τὸ φύλλον· οὐ γάρ ἐστιν ἐπέτειον ἀλλὰ χρονιώτερον.

2 Τὸ δὲ τῆς καππάριος ἴδιον, ὥσπερ ἐλέχθη, παρὰ ταῦτα· καὶ γὰρ τὸ φύλλον ἐπακανθίζον ἔχει καὶ τὸν καυλόν, οὐχ ὥσπερ ὁ φέως καὶ ἱππόφεως ἀνάκανθα τοῖς φύλλοις· μονόρριζον δὲ καὶ ἐπίγειον καὶ χαμαίκαυλον· βλαστάνει δὲ καὶ ἀνθεῖ τοῦ θέρους καὶ διαμένει τὸ φύλλον χλωρὸν ἄχρι Πλειάδος. χαίρει δὲ ὑφάμμοις καὶ λεπτογείοις χωρίοις· λέγεται δὲ ὡς ἐν τοῖς ἐργασίμοις οὐ θέλει φύεσθαι, καὶ ταῦτα περὶ τὰ ἄστη καὶ ἐν εὐγείοις τόποις φυομένη καὶ οὐχ ὥσπερ σίλφιον ἐν ὀρεινοῖς· τοῦτο μὲν <οὖν> οὐ πάντως ἀληθές.

3 Ὁ δὲ τρίβολος ἴδιον ἔχει, διότι περικαρπιάκανθός ἐστι. δύο δ' αὐτοῦ γένη· τὸ μὲν γὰρ ἔχει φύλλον ἐρεβινθῶδες, ἕτερος δὲ φυλλάκανθος· ἐπίγειοι δὲ ἄμφω καὶ πολλαχῇ σχιζόμενοι· ὀψιβλαστὴς δὲ μᾶλλον ὁ φυλλάκανθος καὶ φύεται περὶ τὰς αὐλάς. τὸ δὲ σπέρμα τοῦ μὲν πρωΐου σησαμῶδες, τοῦ δὲ ὀψίου στρογγύλον ἐπίμελαν ἐν λοβῷ. καὶ τὰ μὲν οὖν παρὰ τὰ φύλλα καὶ ἄκανθαν ἔχοντα σχεδὸν ἐν τούτοις.

Ἡ δ' ὀνωνίς ἐστι πτορθάκανθον· ἐπέτειον δὲ τὸ φύλλον ἔχει πηγανῶδες παραπεφυκὸς παρ'

---
[1] ἀρότοις conj. Bod.; ἀρότροις Ald. *cf.* 8. 1. 2.
[2] τότε conj. St.; τοῦτο Ald.   [3] *cf.* Pall. 10. 13. 2.
[4] ὁ φέως conj. St.; ὄφεως Ald. Bas. Cam. H.; ὁ φλεὼς mBas.
[5] Plin. 21. 91.

ENQUIRY INTO PLANTS, VI. v. 1-3

many roots, but is not deep-rooting. It grows at the rising of the Pleiad, the first seed-time,[1] and then [2] puts forth its leaf; for it is not annual, but lives longer than one year.

[3] Caper, as was said, is quite distinct from these; it has a spinous leaf and a spinous stem, whereas *pheos* [4] and 'horse-*pheos*' have no spines on their leaves [5]; it has a single [6] root, is low-growing,[7] and has a creeping stem; it grows and flowers in summer, and the leaf remains green till the rising of the Pleiad. It rejoices in sandy light soils, and it is said that it is unwilling to grow on cultivated land, and that though it grows near towns and in good soil, and not, like silphium, in mountain country. This account however [8] is not altogether accurate.

[9] A peculiarity of caltrop is that it is spinous-fruited.[10] There are two kinds; one has a leaf like that of chick-pea, the other has spinous leaves. Both are low-growing and much divided, but the spinous-leaved form grows later and is found near enclosures. The seed of the early kind is like that of sesame, that of the late kind is round and blackish and enclosed in a pod. These may serve as examples of plants which have spines as well as leaves.[11]

[12] Rest-harrow has spines on the shoots; the leaf, which is annual,[13] is like that of rue, and grows right along the stem, so that the general appearance is

---

[6] Diosc. 2. 173 gives a different account.
[7] *cf.* 7. 8. 1.
[8] οὖν add. W. (in comm.) from G.
[9] Plin. 21. 98.   [10] *cf.* 6. 1. 3.
[11] τὰ μὲν οὖν παρὰ τὰ φύλλα conj. Sch. (οὖν add. W.); τὰ μὲν οὖν ὥσπερ ἀνάφυλλα Ald.H.   [12] Plin. 21. 98.
[13] I have altered the punctuation; πτορθάκανθον, ἐπέτειον δὲ τὸ φ. κτλ. W. after UMP.

33

THEOPHRASTUS

ὅλον τὸν καυλόν, ὥστε καθάπερ στεφάνου τὴν ὅλην εἶναι μορφήν, διαλαμβανομένων ἐπαλλήλων· κολοβοανθὴς δὲ καὶ ἐλλοβόκαρπος ἀδια-
4 φράκτως· φύεται δ' ἐν τῇ γλίσχρᾳ καὶ γανώδει καὶ μάλιστα ἐν τῇ σπορίμῳ καὶ γεωργουμένῃ, δι' ὃ καὶ πολέμιον τοῖς γεωργοῖς· καὶ ἔστι δυσώλεθρος· ὅταν γὰρ λάβῃ χώρας βάθος, ὠθεῖται κάτω εὐθὺς καὶ καθ' ἕκαστον ἔτος ἀποφύσεις ἀφεμένη εἰς τὰ πλάγια πάλιν εἰς τὸ ἕτερον ὠθεῖται κάτω· σπαστέα μὲν οὖν ὅλη· τοῦτο δὲ βραχείσης γίνεται τῆς γῆς καὶ ἀπόλλυται ῥᾷον· ἐὰν δὲ καὶ μικρὸν ἀπολειφθῇ, ἀπὸ τούτου πάλιν βλαστάνει· ἄρχεται δὲ τῆς βλαστήσεως θέρους τελειοῦται δὲ μετοπώρου. τὰ μὲν οὖν ἄγρια τῶν φρυγανικῶν ἐκ τούτων θεωρείσθω.

VI. Τὰ δὲ ἥμερα βραχεῖάν τινα ἔχει θεωρίαν, ἅπερ ἐν τοῖς στεφανωματικοῖς ἐστι.

Τὰ δὲ καθ' ὅλου πειρατέον περὶ στεφανωμάτων εἰπεῖν, ὅπως ἅπαν περιληφθῇ τὸ γένος. ἡ γὰρ στεφανωματικὴ φύσις ἰδίαν τινὰ ἔχει τάξιν, ἐπιμιγνυμένη δὲ τὰ μὲν τοῖς φρυγανικοῖς τὰ δὲ τοῖς ποιώδεσι· δι' ὃ κἀκεῖνα συμπεριληπτέον ἐπιμιμνησκομένους ὡς ἂν ᾖ ὁ καιρός, ἀρξαμένους
2 πρῶτον ἀπὸ τῶν φρυγανικῶν. διχῇ δὲ ἡ τούτων

---

[1] Evidently some conventional way of making a wreath.
[2] διαλαμβανομένων ἐπαλλήλων conj. W.; διαλαμβανομένην ὑπ' ἀλλήλων Ald. *cf.* Plat. *Prot.* 346 E, where the verb means 'to punctuate.'
[3] κολοβοανθὴς; *cf.* 8. 3. 3.

that of a garland,[1] the leaves being set at intervals alternately along it[2]; the flower is irregular,[3] and the fruit contained in a pod,[4] which is not divided into compartments.[5] It grows in sticky rich soil and especially in sown and cultivated land; wherefore it is an enemy to husbandmen, and it is hard to kill; for, when it gets hold of a piece of ground, it immediately pushes its roots down deep,[6] and every year it sends up new growths at the sides and the next year[7] it roots these again. Wherefore it has to be dragged up entire[8]; this is done when the ground has been moistened, and then it is easier to destroy. But, if but a small piece is left, it shoots again from this. It begins to grow in summer and completes its growth in autumn. Let these examples serve for a survey of the wild forms of under-shrubs.

Of cultivated under-shrubs (coronary plants), with which are included those coronary plants which are herbaceous.

VI. The cultivated kinds need but a brief survey; these[9] come under the class of coronary plants.

Of coronary plants we must endeavour to give a general account, so that the whole class may be included. This group has a somewhat peculiar position, since it overlaps partly the under-shrubs, partly the herbaceous plants; wherefore the latter must also be included and we must mention them as occasion serves, taking first the under-shrubs.

[4] ἐλλοβόκαρπος conj. Sch.; ἐλλοβοάνθης Ald.
[5] *cf.* 8. 5. 2.
[6] ὠθεῖται κάτω conj. Sch.; ὠθεῖ τὰ κάτω Ald.
[7] εἰς τὸ ἕτερον, sc. ἔτος; τῷ ἑτέρῳ conj. Sch.
[8] σπαστέα μὲν οὖν ὅλη conj. W.; σταθεῖσα μὲν οὕτως ὅλη Ald.
[9] ἅπερ conj. Sch.; εἴπερ UMAld.G.

διαίρεσις ἡ κατὰ τὴν χρείαν. τῶν μὲν γὰρ τὸ ἄνθος μόνον χρήσιμον· καὶ τούτων τὸ μὲν εὔοσμον, ὥσπερ ἴον, τὸ δ᾽ ἄνοσμον, ὥσπερ διόσανθος φλόξ. τῶν δὲ καὶ οἱ κλῶνες καὶ τὰ φύλλα καὶ ὅλως ἡ πᾶσα φύσις εὔοσμος, οἷον ἑρπύλλου ἑλενίου σισυμβρίου τῶν ἄλλων. ἄμφω δὲ φρυγανικά. κἀκείνων τῶν ἀνθικῶν πολλῶν ἡ φύσις φρυγανώδης, ἡ μὲν ἐπέτειος οὖσα μόνον, ἡ δὲ πολυχρονιωτέρα, πλὴν ἰωνίας τῆς μελαίνης· αὕτη γὰρ ἄκλων ὅλως ἀλλὰ προσριζόφυλλος καὶ ἀείφυλλος, ὡς δέ τινές φασι καὶ δυναμένη δι᾽ ὅλου φέρειν τὸ ἄνθος, ἐὰν τρόπῳ τινὶ θεραπεύηται. τοῦτο μὲν ἴδιον ἂν ἔχοι.

3 Τῶν δὲ ἄλλων μᾶλλον δὲ τῶν πάντων αἱ μὲν ὅλαι μορφαὶ σχεδὸν πᾶσι φανεραί· εἰ δέ τινας ἄλλας ἰδιότητας ἔχουσι, ταύτας λεκτέον, οἷον εἰ τὰ μὲν ἁπλᾶ δοκεῖ τοῖς εἴδεσιν εἶναι τὰ δὲ ἔχειν διαφοράς.

Ἁπλᾶ μὲν οὖν τὰ ξυλώδη, καθάπερ ἕρπυλλος σισύμβριον ἑλένιον· πλὴν εἰ τὰ μὲν ἄγρια τὰ δὲ ἥμερα καὶ <τὰ μὲν> εὔοσμα τὰ δὲ ἀοσμότερά ἐστι· τούτων δὲ καὶ αἱ θεραπεῖαι καὶ αἱ χῶραι διάφοροι καὶ οἱ ἀέρες. ἔνια δὲ καὶ τῶν ἀνθῶν, οἷον τὸ μέλαν ἴον· οὐ γὰρ ἔχειν δοκεῖ τοῦτο διαφορὰν

---

[1] Plin. 21. 59.
[2] So Plin. *l.c.*; but Nic. *ap.* Athen. 15. 31 calls this flower fragrant.
[3] πολλῶν conj. W.; πολλὰ UMAld.
[4] οἷον εἰ conj. W.; ὅτι Ald.   [5] οὖν conj. W.; οἷον Ald.

[1] These may be divided into two groups according to their uses. Of some only the flower is serviceable; and of these some are sweet-scented, as gilliflower, some scentless, as carnation [2] and wallflower. Of others again the branches leaves and in fact the whole growth are sweet-scented, as with tufted thyme calamint bergamot-mint and the rest. Both groups however belong to the under-shrubs. And of the first-mentioned, those valued for their flowers, the growth is in most [3] cases that of undershrubs, (in some annual merely, in others of longer duration) except in the violet; for this is altogether without branches, its leaves grow close to the root, and it is always in leaf; while, as some say, it is able to bear flowers continuously, if it is tended in a certain way. This may be considered a peculiar characteristic of this plant.

Of the others, or rather of all the group, the general appearance is in each case plain to all; any peculiarities that they may exhibit we must mention, for instance, if [4] some appear to have but a single form, while others have various forms.

Thus [5] those of woody character, as tufted thyme bergamot-mint calamint, have but one form, unless one counts wild and cultivated, scented and scentless plants, as belonging to distinct forms; and again there are with these plants differences of culture of position and of climate. Some also [6] of the group valued for their flowers [7] have each but one form, for instance, the black *ion* (violet); for this does not appear to have different forms

[6] ἔνια δὲ καὶ conj. W.; ἔνιοι δὲ UMAld.
[7] ἀνθῶν in the sense of ἀνθικῶν § 2, which perhaps should be read here.

ὥσπερ τὸ λευκόν· ἐμφανὴς γὰρ ἡ τούτων χροιὰ διαλλάττουσα, καὶ ἔτι δὴ μᾶλλον ἡ τῶν κρίνων, εἴπερ δή, καθάπερ φασίν, ἔνια καὶ πορφυρᾶ ἐστι.

4 Τῶν δὲ ῥόδων πολλαὶ διαφοραὶ πλήθει τε φύλλων καὶ ὀλιγότητι καὶ τραχύτητι καὶ λειότητι καὶ εὐχροίᾳ καὶ εὐοσμίᾳ. τὰ μὲν γὰρ πλεῖστα πεντάφυλλα, τὰ δὲ δωδεκάφυλλα καὶ εἰκοσίφυλλα, τὰ δ' ἔτι πολλῷ πλεῖον ὑπεραίροντα τούτων· ἔνια γὰρ εἶναί φασιν ἃ καὶ καλοῦσιν ἑκατοντάφυλλα· πλεῖστα δὲ τὰ τοιαῦτά ἐστι περὶ Φιλίππους· οὗτοι γὰρ λαμβάνοντες ἐκ τοῦ Παγγαίου φυτεύουσιν· ἐκεῖ γὰρ γίνεται πολλά· σμικρὰ δὲ σφόδρα τὰ ἐντὸς φύλλα· ἡ γὰρ ἔκφυσις αὐτῶν οὕτως ὥστε εἶναι τὰ μὲν ἐκτὸς τὰ δ' ἐντός· οὐκ εὔοσμα δὲ οὐδὲ μεγάλα τοῖς μεγέθεσιν. ἐν δὲ τοῖς μεγά-
5 λοις εὐώδη μᾶλλον ὧν τραχὺ τὸ κάτω. τὸ δὲ ὅλον, ὥσπερ ἐλέχθη, καὶ ἡ εὔχροια καὶ ἡ εὐοσμία παρὰ τοὺς τόπους ἐστίν· ἐπεὶ καὶ τὰ ἐν γῇ τῇ αὐτῇ γινόμενα ποιεῖ τινα παραλλαγὴν εὐοσμίας καὶ ἀοσμίας. εὐοσμότατα δὲ τὰ ἐν Κυρήνῃ, δι' ὃ καὶ τὸ μύρον ἥδιστον. ἁπλῶς δὲ καὶ τῶν ἴων καὶ τῶν ἄλλων ἀνθῶν ἄκρατοι μάλιστα ἐκεῖθι

---

[1] *cf.* 6. 8. 1 n.; Diosc. 3. 102.
[2] Plin. 21. 14–21; Athen. 15. 29.
[3] *i.e.* of the bark. *cf.* Plin. 21. 17, *scabritia corticis*.
[4] Sc. in 'double' roses.
[5] *i.e.* the hip; called ὀμφαλός Arist. *Probl.* 12. 8, where the same statement is made; called μῆλον below, §6.

like the white *ion* (gilliflower) in which the colour evidently varies; as does still more that of the lilies, if it be true, as some say, that there is a crimson kind.[1]

[2] Among roses there are many differences, in the number of petals, in roughness,[3] in beauty of colour, and in sweetness of scent. Most have five petals, but some have twelve or twenty, and some a great many more than these; for there are some, they say, which are even called 'hundred-petalled.' Most of such roses grow near Philippi; for the people of that place get them on Mount Pangaeus, where they are abundant, and plant them. However the inner petals[4] are very small, (the way in which they are produced being such that some are outside, some inside). Some kinds are not fragrant nor of large size. Among those which have large flowers those in which the part[5] below the flower is rough are the more fragrant. In general, as has been said, good colour and scent depend upon locality; for even bushes which are growing in the same[6] soil shew some variation in the presence or absence of a sweet scent. Sweetest-scented of all are the roses of Cyrene, wherefore the perfume made from these is the sweetest. (Indeed it may be said generally that the scents[7] of the gilliflowers[8] also and of the other flowers of that place are the purest, and especially the

[6] τῇ αὐτῇ conj. Sch.; τοιαύτῃ U; τοιαῦτα M.
[7] ἄκρατοι μάλιστα ἐκεῖθι αἱ ὀσμαί conj. Sch. after Saracenus on Diosc. 1. 25; Athen. *l.c.* (ἄκρατοι μάλιστα καὶ θεῖαι αἱ ὀσμαί); ἄκρατοι· μάλιστα δ' ἐκείνου αἱ ὀσμαί Ald.; ἐκεῖ αἱ ὀσμαί (rest uncertain) U. *cf. C.P.* 6. 18. 3.
[8] ? violets and gilliflowers: see Index.

THEOPHRASTUS

αἱ ὀσμαί, διαφερόντως δὲ τοῦ κρόκου· πλεῖστον
6 γὰρ οὗτος δοκεῖ παραλλάττειν. φύεται μὲν οὖν
ἡ ῥοδωνία καὶ ἐκ τοῦ σπέρματος· ἔχει δὲ ὑπὸ τὸ
ἄνθος ἐν τῷ μήλῳ κνηκῶδες ἢ ἀκανῶδες, ἔχον δέ
τινα χνοῦν ὥστε ἐγγὺς εἶναι τῶν παππωδῶν
σπερμάτων· οὐ μὴν ἀλλὰ διὰ τὸ βραδέως παρα-
γίνεσθαι κατακόπτοντες, ὡς ἐλέχθη, τὸν καυλὸν
φυτεύουσιν. ἐπικαιομένη δὲ καὶ ἐπιτεμνομένη
βέλτιον φέρει τὸ ἄνθος· ἐωσμένη γὰρ ἐξαύξεται
καὶ ἀπολοχμοῦται. δεῖ δὲ καὶ μεταφυτεύειν
πολλάκις· καὶ γὰρ οὕτω φασὶ κάλλιον γίνεσθαι
τὸ ῥόδον. αἱ δ' ἄγριαι τραχύτεραι καὶ ταῖς
ῥάβδοις καὶ τοῖς φύλλοις, ἔτι δὲ ἄνθος ἀχρού-
στερον ἔχουσι καὶ ἔλαττον.

7 Τὸ δὲ ἴον τὸ μέλαν τοῦ λευκοῦ διαφέρει κατά
τε ἄλλα καὶ κατ' αὐτὴν τὴν ἰωνίαν, ὅτι πλατύ-
φυλλός τε καὶ ἐγγειόφυλλος καὶ σαρκόφυλλός
ἐστι, πολλὴν ἔχουσα ῥίζαν.

8 Τὰ δὲ κρίνα τῇ μὲν χροιᾷ τὴν εἰρημένην ἔχει
διαφοράν. μονόκαυλα δέ ἐστιν ὡς ἐπὶ πᾶν,
δικαυλεῖ δὲ σπανίως· τάχα δὲ τοῦτο χώρας καὶ
ἀέρος διαφορᾶς. καθ' ἕκαστον δὲ καυλὸν ὁτὲ μὲν
ἓν κρίνον ὁτὲ δὲ πλείω γίνεται· βλαστάνει γὰρ
τὸ ἄκρον· σπανιώτερα δὲ ταῦτα· ῥίζαν δὲ ἔχει
πολλὴν σαρκώδη καὶ στρογγύλην· ὁ δὲ καρπὸς

---

[1] διαφερόντως δὲ τοῦ κρόκου conj. Saracenus from Athen. *l.c.*; διαφερόντως δὴ τοῦ χρόνου Ald. *cf* Callim. *Hymn to Apollo* 83, whence it appears that an autumnal crocus (*crocus sativus*) is meant. See below § 10.
[2] ἀκανῶδες conj. Sch. from G, *acanaceum*; ἀνθῶδες UMAld.
[3] παππωδῶν conj. Sch.; πρώτων Ald.
[4] Plin. 21. 27.

ENQUIRY INTO PLANTS, VI. vi. 5-8

scent of the saffron-crocus,[1] a plant which seems to vary in this respect more than any other). Roses can be grown from seed, which is to be found below the flower in the 'apple,' and is like that of safflower or pine-thistle,[2] but it has a sort of fluff, so that it is not unlike the seeds which have a pappus.[3] As however the plant comes slowly from seed, they make cuttings of the stem, as has been said, and plant them. If the bush is burnt or cut over, it bears better flowers; for, if left to itself, it grows luxuriantly and makes too much wood. Also it has to be often transplanted; for then, they say, the roses are improved. The wild kinds are rougher both in stem and in leaf, and have also smaller flowers of a duller colour.

[4] The black *ion* (violet) differs from the white *ion* (gilliflower) not only in other respects but in the plant itself, in that in the former the leaves are broad, lie close to the ground, and are fleshy, and there is much root.

[5] *Krina* (lilies) shew the variation in colour which has been already mentioned.[6] The plant has in general a single stem, but occasionally divides into two, which may be due to differences[7] in position and climate. On each stem grows sometimes one flower, but sometimes more; (for it is the top of the stem which produces the flower[8]) but this sort is less common. There is an ample root, which is fleshy and round. If the fruit is taken off, it

---

[5] Plin. 21. 25. The account of *herbaceous* coronary plants seems to begin here. *cf.* 6. 6. 10.   [6] 6. 6. 3.
[7] διαφορᾶς U; διαφορᾷ W. after Sch.
[8] βλαστάνει. But this word in T. has usually a more general sense. ? 'for in that case the top of the stem branches' (lit. 'makes fresh growth').

ἀφαιρούμενος ἐκβλαστάνει καὶ ἀποδίδωσι τὸ κρίνον πλὴν ἔλαττον. ποιεῖ δέ τινα καὶ δακρυώδη συρροήν, ἣν καὶ φυτεύουσιν, ὥσπερ εἴπομεν.

9 Ὁ δὲ νάρκισσος ἢ τὸ λείριον, οἱ μὲν γὰρ τοῦτο οἱ δ' ἐκεῖνο καλοῦσι, τὸ μὲν ἐπὶ τῇ γῇ φύλλον ἀσφοδελῶδες ἔχει, πλατύτερον δὲ πολύ, καθάπερ ἡ κρινωνία, τὸν δὲ καυλὸν ἄφυλλον μὲν ποώδη δὲ καὶ ἐξ ἄκρου τὸ ἄνθος, καὶ ἐν ὑμένι τινὶ καθάπερ ἐν ἀγγείῳ <καρπὸν> μέγαν εὖ μάλα καὶ μέλανα τῇ χροιᾷ σχήματι δὲ προμήκη. οὗτος δ' ἐκπίπτων ποιεῖ βλάστησιν αὐτόματον· οὐ μὴν ἀλλὰ καὶ συλλέγοντες πηγνύουσι καὶ τὴν ῥίζαν φυτεύουσιν. ἔχει ῥίζαν σαρκώδη στρογγύλην μεγάλην. ὄψιον δὲ σφόδρα· μετὰ γὰρ Ἀρκτοῦρον ἡ ἄνθησις καὶ περὶ ἰσημερίαν.

10 Ὁ δὲ κρόκος ποώδης μὲν τῇ φύσει, καθάπερ καὶ ταῦτα, πλὴν φύλλῳ στενῷ, σχεδὸν γὰρ ὥσπερ τριχόφυλλόν ἐστιν· ὀψιανθὲς δὲ σφόδρα καὶ ὀψιβλαστὲς ἢ πρωϊανθές, ὁποτέρως τις λαμβάνοι τὴν ὥραν. <μετὰ> Πλειάδα γὰρ ἀνθεῖ καὶ ὀλίγας ἡμέρας· εὐθὺς δ' ἅμα τῷ φύλλῳ καὶ τὸ ἄνθος ὠθεῖ· δοκεῖ δὲ καὶ πρότερον· ῥίζα δὲ πολλὴ καὶ σαρκώδης, καὶ τὸ ὅλον εὔζωον· φιλεῖ δὲ καὶ πατεῖσθαι καὶ γίνεται καλλίων κατατριβομένης

---

[1] cf. 2. 2. 1 n., 9. 14 ; C.P. 1. 4. 4–6. Plin. 21. 26 describes a method of artificially producing crimson lilies from the bulbils of a white lily. cf. Geop. 11. 20.
[2] cf. 6. 8. 1 and 3. See Index. [3] cf. 7. 13. 1.
[4] ποώδη :·cf. 4. 10. 3.

germinates and produces a fresh plant, but of smaller size; the plant also produces a sort of tear-like exudation, which men also plant, as we have said.[1]

The narcissus[2] or *leirion* (for some call it by the one name, some by the other) has its ground-leaves like those of the asphodel,[3] but much broader, like those of the *krinon* (lily); its stem is leafless and grass-green[4] and bears the flower at the top; the fruit[5] is in a kind of membrane-like vessel, and is very large, black in colour, and oblong in shape. This as it falls germinates of its own accord; however men collect and set[6] the seed, and also plant the root, which is fleshy round and large. The plant blooms very late,[7] after the setting of Arcturus about the equinox.

[8] The saffron-crocus is herbaceous in character, like the above-mentioned plants,[9] but has a narrow leaf; indeed the leaves are, as it were, hair-like; it blooms very late, and grows either late or early, according as one looks at the season[10]; for it blooms after[11] the rising of the Pleiad and only for a few days. It pushes up the flower at once with the leaf, or even seems to do so earlier. The root[12] is large and fleshy, and the whole plant vigorous; it loves even to be trodden on and grows fairer when the root is crushed into the ground by the

---

[5] καρπὸν omitted in MSS.; add. Dalec. from Diosc. 4. 158.
[6] πηγνύουσι: cf. 7. 4. 3 n.
[7] cf. C.P. 1. 10. 5; Plin. *l.c.* (a much confused passage).
[8] Plin. 21. 31–34.
[9] Sc. κρίνον and νάρκισσος; cf. 6. 6. 8 n.
[10] *i.e.* whether at the end of one season or the beginning of the next. cf. C.P. 1. 10. 5. λαμβάνοι U; λαμβάνει Ald.
[11] μετὰ add. W.   [12] cf. 7. 9. 4.

πάτῳ τῆς ῥίζης· δι' ὃ καὶ παρὰ τὰς ὁδοὺς καὶ ἐν τοῖς κροτητοῖς κάλλιστος. ἡ δὲ φυτεία ἀπὸ ῥίζης.

11 Ταῦτα μὲν οὖν οὕτω γεννᾶται. τὰ δ' ἄλλα ἄνθη τὰ προειρημένα πάντα σπείρεται, οἷον ἰωνία διόσανθος ἴφυον φλὸξ ἡμεροκαλλές· καὶ γὰρ αὐτὰ καὶ αἱ ῥίζαι ξυλώδεις· σπείρεται δὲ καὶ ἡ οἰνάνθη· καὶ γὰρ καὶ τοῦτο ἀνθῶδες. τὰ μὲν οὖν ἀνθικὰ σχεδὸν ἐν τούτοις καὶ τοῖς ὁμοίοις ληπτέον.

VII. Τὰ δὲ ἕτερα πάντα μὲν ἀνθεῖ καὶ σπερμοφορεῖ, δοκεῖ δὲ οὐ πάντα διὰ τὸ μὴ φανερὸν εἶναί τινων τὸν καρπόν· ἐπεὶ καὶ τὸ ἄνθος ἐνίων ἀμαυρόν· ἀλλ' ὅτι βραδέως καὶ χαλεπωτέρως παραγίνεται, τῇ φυτείᾳ χρῶνται μᾶλλον, 2 ὥσπερ ἐλέχθη καὶ κατ' ἀρχάς. καίτοι διατείνονταί τινες ὡς οὐκ ἐχόντων καρπόν· οἵ τε πεπειρᾶσθαι φάσκοντες καὶ τούτων εἰσίν, αὐτοὶ γὰρ ξηρᾶναι πολλάκις καὶ ἀποτρῖψαι καὶ σπεῖραι, καὶ οὐδεπώποτε βλαστεῖν οὔτε ἕρπυλλον οὔτε ἐλένιον οὔτε σισύμβριον οὔτε μίνθαν· πεπειρᾶσθαι γὰρ καὶ ταύτης. ἀλλ' ὅμως ἐκεῖνο ἀληθέστερον, ἥ τε τῶν ἀγρίων φύσις ἐπιμαρτυρεῖ· καὶ γὰρ ἕρπυλλός ἐστιν ἄγριος, ὃν κομίζοντες ἐκ τῶν ὁρῶν φυτεύουσι καὶ ἐν Σικυῶνι καὶ Ἀθήνησιν ἐκ τοῦ Ὑμηττοῦ· παρ' ἄλλοις δὲ ὅλως ὅρη πλήρη καὶ λόφοι, καθάπερ ἐν τῇ Θρᾴκῃ· καὶ σισύμβριον

---

[1] πάτῳ conj. Turneb. and others ; κάτω Ald,
[2] κροτητοῖς : Plin. *l.c. iuxta semitas ac fontes.* Did he read κρουνοῖς?
[3] ἀνθικὰ conj. Scal.; ἀκανθικὰ Ald. *cf.* 6. 6. 2.
[4] ἀλλ' ὅτι conj. W. from G ; ἄλλα δὲ UMPAld.

ENQUIRY INTO PLANTS, VI. vi. 10–vii. 2

foot[1]: wherefore it is fairest along the roads and in well-worn places.[2] It is propagated from the root.

These are the ways then in which the above plants are grown. All the above-mentioned flowers are grown from seed, as gilliflower carnation spike-lavender wall-flower martagon-lily; these plants themselves, as well as their roots, are woody. Drop-wort is also grown from seed; for that too is a plant grown for its flower. These and other plants like them may serve as examples of plants grown for their flowers.[3]

VII. All the others flower and bear seed, though they do not all appear to do so, since in some cases the fruit is not obvious. Indeed in some the flower too is inconspicuous, but, because[4] these grow slowly and with some difficulty, men propagate them rather by off-shoots, as was said at the beginning. However some contend that they have no fruit: and there are men who have actually tried with the following plants[5]; they have, they say, themselves often dried and rubbed out and sown the apparent fruit of thyme calamint bergamot-mint and green mint (for even that they have tried) and there was no germination from such sowing. However, the account given above is the truer, and the character of the wild forms testifies to this; for there is also a wild thyme (Attic thyme[6]), which they bring from the mountains and plant at Sicyon, or from Hymettus and plant at Athens; and in other districts the mountains and hills[7] are quite covered with it, for instance in Thrace. There is also a

[5] οἵ τε ... εἰσίν transposed by Sch.; in MSS. after ἀληθέστερον.
[6] Plin. 19. 172; Athen. 15. 28.
[7] λόφοι conj. W.; τόποι Ald.

δὲ καὶ τᾶλλα δριμυτέραν ἔχοντα τὴν ὀσμήν· ἕρπυλλος δ' ἐνίοτε καὶ παντελῶς θυμώδης· ἃ δῆλον ὅτι ταύτην τὴν γένεσιν λαμβάνει.

3 Ἀβρότονον δὲ μᾶλλον ἀπὸ σπέρματος βλαστάνει ἢ ἀπὸ ῥίζης καὶ παρασπάδος· χαλεπῶς δὲ καὶ ἀπὸ σπέρματος· προμοσχευόμενον <δὲ> ἐν ὀστράκοις, ὥσπερ οἱ Ἀδώνιδος κῆποι, τοῦ θέρους· δύσριγον γὰρ σφόδρα καὶ ὅλως ἐπίκηρον καὶ ὅποι ὁ ἥλιος σφόδρα λάμπει· ἐμβιῶσαν δὲ καὶ αὐξηθὲν μέγα καὶ ἰσχυρὸν καὶ δενδρῶδες ὥσπερ τὸ πήγανον, πλὴν ξυλωδέστερον πολὺ τοῦτο καὶ ξηρότερον καὶ αὐχμωδέστερον.

4 Ὁ δὲ ἀμάρακος ἀμφοτέρως φύεται, καὶ ἀπὸ παρασπάδος καὶ ἀπὸ σπέρματος· πολύσπερμον δέ, καὶ τὸ σπέρμα εὔοσμον ὀσμῇ μαλακωτέρᾳ· δύναται δὲ καὶ μεταφυτεύεσθαι. πολύσπερμον δὲ καὶ τὸ ἀβρότονον καὶ οὐκ ἄοσμον. τοῦτο δὲ ῥίζας μὲν ἔχει ὀρθὰς καὶ κατὰ βάθους. ἔστι γὰρ ὥσπερ μονόρριζον τῇ παχείᾳ τὰς δ' ἄλλας <ἀφίησιν> ἀπ' αὐτῆς· ὁ δ' ἀμάρακος καὶ ὁ ἕρπυλλος καὶ τὸ σισύμβριον καὶ τὸ ἐλένιον ἐπιπολαίους καὶ πολυσχιδεῖς καὶ ταρρώδεις· ξυλώδεις δὲ πᾶσαι, πολὺ δὲ μᾶλλον ἡ τοῦ ἀβροτόνου καὶ διὰ τὸ μέγεθος καὶ τῇ ξηρότητι.

---

[1] ἐνίοτε conj. W.; ἐνίοις Ald.
[2] Plin. 21. 57. Description of various forms of ἕρπυλλος has perhaps dropped out after this word: cf. § 5, καθάπερ ἐλέχθη.
[3] *i.e.* from seed. ταύτην conj. W.; πάντα UMAld.; ? πάντα ταύτην.
[4] Plin. 21. 34: cf. *C.P.* 1. 4. 2. ἀβρότονον ... θέρους, text nearly as given by Ald. and by UM (?)—supported by Plin.

wild bergamot-mint, and wild forms of the other plants mentioned, having a more pungent smell. Thyme is sometimes [1] quite like cultivated thyme.[2] Now it is plain that these wild forms possess this means of reproducing themselves.[3]

Southernwood actually grows more readily from seed than from a root [4] or a piece torn off (though it grows even from seed with difficulty); however it can be propagated by layering in pots in summertime, like the 'gardens of Adonis' [5]; it is indeed very sensitive [6] to cold and generally delicate even where the sun shines brightly; but, when it is established and has grown, it becomes tall and strong and tree-like, like rue, except that the latter is much more woody drier and less succulent.

[7] Sweet marjoram grows in either way, from pieces torn off or from seed; it produces a quantity of seed, which is fragrant with a delicate scent; it can also be transplanted.[8] Southernwood also produces much seed, which has some scent. This plant has straight roots which run deep; it has, as it were, its single stout root, from which the others spring; [9] while sweet marjoram thyme bergamot-mint and calamint have surface [10] roots which are much divided and matted; in all these plants the roots are woody, but especially in southernwood, because of its size and because it is so dry.

l.c. so far as that passage is intelligible—but δὲ before ἐν ὀστράκοις add. W.; after ἐν ὀστρ. supply βλαστάνει.

[5] *cf.* Plat. *Phaedo* 276 B and Thompson's n. Sir W. Thiselton-Dyer in *Companion to Greek Studies*, § 99, p. 65.

[6] *cf* C.P. 4. 3. 2.   [7] Plin. 21. 61.

[8] μεταφυτεύεσθαι conj. Sch. from G; μεταφύεσθαι Ald.

[9] ἀφίησιν add. W.

[10] ἐπιπολαίους conj. Scal.; ἐπὶ πολλοὺς MAld. *cf. C.P.* 2. 16. 5.

5   Τοῦ δὲ ἑρπύλλου ἴδιος ἡ αὔξησις ἡ τῶν βλαστῶν· δύναται γὰρ ἐφ' ὁσονοῦν προϊέναι κατὰ μῆκος χάρακα λαβὼν ἢ πρὸς αἱμασιὰν φυτευθεὶς ἢ κάτω καθιέμενος· εὐαυξέστατος δὲ εἰς φρέαρ. εἴδη δὲ τοῦ μὲν ἡμέρου λαβεῖν οὐκ ἔστι, καθάπερ ἐλέχθη. τοῦ δὲ ἀγρίου φασὶν εἶναι. τοῦ γὰρ ἐν τοῖς ὄρεσιν τὸν μὲν θυμβρώδη τινὰ καὶ δριμύτατον τὸν δ' εὔοσμον εἶναι καὶ μαλακώτερον.

6   Ὥρα δὲ τῆς φυτείας πλείστων μετόπωρον, ἐν ᾧ σπεύδουσιν ὡς πρῶτα φυτεύειν· οὐ μὴν ἀλλ' ἔνια καὶ τοῦ ἦρος φυτεύουσιν. ἅπαντα φιλόσκια καὶ φίλυδρα καὶ φιλόκοπρα μάλιστα· αὐχμὸν δὲ δέχεται καὶ ὅλως ὀλιγοϋδρότατος ὁ ἕρπυλλος. κόπρῳ δὲ χαίρει, μάλιστα δὲ καὶ τῇ τῶν λοφούρων· φασὶ δὲ καὶ μεταφυτεύειν δεῖν πολλάκις· καλλίω γάρ. τὸ δὲ σισύμβριον, ὥσπερ ἐλέχθη, καὶ ἐξίσταται μὴ μεταφυτευόμενον.

VIII. Τῶν δ' ἀνθῶν τὸ μὲν πρῶτον ἐκφαίνεται τὸ λευκόϊον, ὅπου μὲν ὁ ἀὴρ μαλακώτερος εὐθὺς τοῦ χειμῶνος, ὅπου δὲ σκληρότερος ὕστερον, ἐνιαχοῦ τοῦ ἦρος. ἅμα δὲ τῷ ἴῳ ἢ μικρόν τι ὕστερον καὶ τὸ φλόγινον καλούμενον τὸ ἄγριον·

---

[1] *cf.* Plin. 20. 245 and 246 (not from T.); *C.P.* 2. 18. 2; Diosc. 3. 38; Index ἕρπυλλος.
[2] *cf.* Plin. 19. 172, which refers however to σισύμβριον; Nic. *ap.* Athen. 15. 31.
[3] Plin. 21. 61.

[1] The growth of the shoots of thyme is peculiar. If it has a stake, or is planted against a wall, it can send them out to any length; so also if it is let grow downwards; indeed it is most vigorous when grown into a pit. [2] It is not possible to distinguish different forms of the cultivated kind, as has been said, but they say that of the wild kind (Attic thyme) there is more than one form; for that of the kind which grows on the mountains one form is like savory and very pungent, while the other is fragrant and more delicate.

[3] The season for planting most of these is autumn, and then men hasten to plant them as early as possible; however some are planted also in spring. All of them love shade,[4] water, and especially dung; however thyme is patient of drought and, in general, needs moisture less than the others. These plants especially delight in the dung of beasts of burden; and it is said that they should often be transplanted, for that it improves them, while bergamot-mint, as has been said, actually degenerates [5] if it is not transplanted.

Of the seasons at which coronary plants flower, and of the length of their life.

VIII. [6] Of the flowers the [7] first to appear is the gilliflower; where the air is mild, it appears as soon as winter comes, but, where it is more severe, later, sometimes in spring. Along with the gilliflower, or a little later, appears the flower called the wild wall-

[4] φιλόσκια conj. Scal. from G; φιλοίκια UMAld. *cf.* Plin. *l.c.*
[5] ἐξίσταται conj. Scal. from G, *degenerat*; ἐξήτασται MAld.
[6] Plin. 21. 64–66; Athen. 15. 26 and 27. ἀνθῶν: ? in the sense of ἀνθικῶν, as in 6. 6. 3.
[7] τὸ conj. Scal.; τοῦ Ald.

ταῦτα γὰρ ὧν οἱ στεφανήπλοκοι χρῶνται πολὺ
ἐκτρέχει τῶν ἄλλων. μετὰ δὲ ταῦτα ὁ νάρκισσος καὶ τὸ λείριον, <καὶ τῶν ἀγρίων ἀνεμώνης γένος τὸ καλούμενον ὄρειον,> καὶ τὸ τοῦ
βολβοῦ κώδυον· ἐμπλέκουσι γὰρ ἔνιοι καὶ τοῦτο
εἰς τοὺς στεφάνους. ἐπὶ δὲ τούτοις ἡ οἰνάνθη
καὶ τὸ μέλαν ἴον καὶ τῶν ἀγρίων ὅ τε ἐλειόχρυσος καὶ τῆς ἀνεμώνης ἡ λειμωνία καλουμένη καὶ τὸ ξίφιον καὶ ὑάκινθος καὶ σχεδὸν
2 ὅσοις ἄλλοις χρῶνται τῶν ὀρείων. τὸ δὲ ῥόδον
ὑστερεῖ τούτων καὶ τελευταῖον μὲν φαίνεται,
πρῶτον δ' ἀπολείπει τῶν ἐαρινῶν· ὀλιγοχρονία
γὰρ ἡ ἄνθησις. ὀλιγοχρόνια δὲ καὶ τῶν ἀγρίων τὰ λοιπὰ πλὴν τῆς ὑακίνθου καὶ τῆς ἀγρίας
καὶ τῆς σπαρτῆς· αὕτη δὲ διαμένει καὶ τὸ λευκὸν
ἴον καὶ ἔτι πλείω τὸ φλόγινον· τὸ δὲ δὴ μέλαν ἴον,
ὥσπερ εἴρηται, δι' ἐνιαυτοῦ θεραπείας τυγχάνον.
ὡσαύτως δὲ καὶ ἡ οἰνάνθη, καὶ γὰρ τοῦτο ἀνθι-
κὸν μὲν ποῶδες δὲ τὴν φύσιν, ἐάν τις ἀποκνίζῃ
καὶ ἀφαιρῇ τὸ ἄνθος καὶ μὴ ἐᾷ σπερματοῦσθαι
καὶ ἔτι τόπον εὔειλον ἔχῃ· τὸ δὲ ἄνθος βοτρυῶδες
καὶ λευκὸν **καθάπερ τῶν ἀγρίων** ... ταῦτα μὲν
οὖν ὥσπερ ἐαρινὰ φαίνεται.

3 Τὰ δὲ θερινὰ μᾶλλον ἥ τε λυχνὶς καὶ τὸ
διόσανθος καὶ τὸ κρίνον καὶ τὸ ἴφυον καὶ ὁ

---

[1] Evidently both distinct from the νάρκισσος ἢ λείριον of 6. 6. 9 ; 6. 8. 3. See Index.

[2] καὶ τῶν ... ὄρειον ins Sch. from Athen. *l.c.* with alteration of ὀρείων to ἀγρίων. *cf.* Plin. *l.c.*

[3] *i.e.* the *flower* of muscari, mentioned in this way because elsewhere (*e.g.* 7. 12. 1) the edible *root* is in question, which was properly called βολβός.

[4] *cf.* 9. 19. 3.   [5] See Index.

ENQUIRY INTO PLANTS, VI. VIII. 1-3

flower. These, of all the flowers that the garland-makers use, far outrun the others. After these come pheasant's eye[1] and polyanthus[1] narcissus (and, among wild plants, the kind of anemone which is called the 'mountain anemone')[2] and the 'head'[3] of purse-tassels; for this too some interweave in their garlands. After these come dropwort violet, and of wild plants, gold-flower,[4] the meadow kind of anemone corn-flag *hyakinthos* (squill), and pretty well all the mountain flowers that are used. The rose comes last of these, and is the first of the spring flowers to come to an end, as it is the first to appear, for its time of blooming is short. So too is that of the rest of the wild plants mentioned, except *hyakinthos*,[5] the wild kind (squill), and also the cultivated (larkspur); this lasts on, and so does the gilliflower, and for a still longer time the wallflower, while the violet, as has been said,[6] blooms throughout the year, if it receives tendance. So too dropwort[7] (for that too is one of the plants valued for their flowers, though it is herbaceous[8] in character) if one pinches off and removes the flower instead of letting it go to seed, and if, further,[9] it has a sunny position. The flower is clustering and white, like that of the wild . . . .[10] These then are, we may say, the plants of spring.

[11] The following belong rather to summer: rose-campion carnation *krinon*[12] (lily) spike-lavender and

---

[6] 6. 6. 2; *cf. C.P.* 1. 13. 12.   [7] *cf.* 6. 6. 11.

[8] ποῶδες: sense not obvious; εὐῶδες conj. Dalec. *cf. C.P.* 1. 13. 12.   [9] ἔτι conj. W.; ὅτι UMAld.

[10] *Ut labruscae* G, perhaps a guess: see οἰνάνθη in Index.

[11] Plin. 21, 67 and 68.

[12] κρίνον Sch. from Athen. *l.c.*; so also Plin, *l.c.*; κήρινθον Ald.

51

THEOPHRASTUS

ἀμάρακος ὁ Φρύγιος· ἔτι δὲ ὁ πόθος καλούμενος· οὗτος δ᾽ ἐστὶ διττός, ὁ μὲν ἔχων τὸ ἄνθος ὅμοιον τῇ ὑακίνθῳ, ὁ δὲ ἕτερος ἄχρους λευκός, ᾧ χρῶνται περὶ τοὺς τάφους· καὶ χρονιώτερος οὗτος. ἀνθεῖ δὲ καὶ ἡ ἶρις τοῦ θέρους καὶ τὸ στρούθιον καλούμενον· τῇ μὲν ὄψει καλὸν τὸ ἄνθος ἄοσμον δέ. μετοπώρου δὲ τὸ λείριον τὸ ἕτερον καὶ ὁ κρόκος, ὅ τε ὀρεινὸς ἄοσμος καὶ ὁ ἥμερος· εὐθὺς γὰρ ἀνθοῦσι τοῖς πρώτοις ὕδασι. χρῶνται δὲ καὶ τῶν ἀγρίων τῷ τῆς ὀξυακάνθου καρπῷ καὶ τῷ ἄνθει τῷ τῆς μίλακος.

4 Καὶ ταῖς μὲν ὥραις οὕτως ἑκάστων ἡ γένεσις. ὡς δὲ ἁπλῶς εἰπεῖν οὐδεὶς διαλείπεται χρόνος οὐδ᾽ ἔστιν ἀνανθής, ἀλλὰ καὶ ὁ χειμὼν ἔχει καίπερ ἄγονος δοκῶν εἶναι διὰ τὴν κατάψυξιν τῶν μετοπωρινῶν μεταλαμβανόντων, ἐὰν δὲ δὴ καὶ μαλακὸς ᾖ, πολλῷ μᾶλλον. ἁπλῶς γὰρ πάντ᾽ ἢ τὰ πολλὰ καὶ ἐπεκτείνεται τῆς οἰκείας ὥρας, καὶ ἐὰν ὁ τόπος εὔειλος ᾖ μᾶλλον· δι᾽ ὃ καὶ συνέχεια γίνεται. χρόνοι μὲν οὖν οὗτοι καὶ ὧραι κατὰ τὰς γενέσεις.

5 Βίος δὲ ἰωνίας μὲν τῆς λευκῆς ἔτη μάλιστα τρία· γηράσκουσα δὲ ἐλαττοῦται καὶ ἴα λευκότερα φέρει. ῥοδωνίας δὲ πέντε τὰ πρὸς τὴν ἀκμὴν μὴ ἐπικαομένης· χείρω δὲ καὶ ταύτης τὰ ῥόδα γηρασκούσης. πρὸς εὐοσμίαν δὲ καὶ ῥόδων καὶ ἴων καὶ τῶν ἄλλων ἀνθῶν μέγιστον ὁ τόπος

---

[1] cf. C.P. 1. 4. 1.
[2] cf. the Eng. plant-name 'love-in-absence'; see πόθος in Index.
[3] λευκὸς. ἔκλευκος, 'whitish,' Athen. l.c.
[4] Evidently the νάρκισσος ἢ λείριον of 6. 6. 9; cf. 6. 8. 1 n.

ENQUIRY INTO PLANTS, VI. viii. 3–5

the Phrygian sweet marjoram[1]; also the plant called 'regret,'[2] of which there are two kinds, one with a flower like that of larkspur, the other not coloured but white,[3] which is used at funerals; and this one lasts longer. The iris also blooms in summer, and the plant called soap-wort, which has a beautiful flower but is scentless. In autumn bloom the other kind of narcissus,[4] the crocus, both the scentless mountain form and the cultivated one (saffron-crocus); for these bloom directly the first rains come. The fruit[5] of the cotoneaster and the flower of the smilax, both of them wild plants, are also used in garlands.

Such are the seasons at which each appears; and, to speak generally, there is no interval of time nor flowerless period, but even winter produces flowers, for all that it seems to be unproductive by reason of the cold, since the autumn flowers continue into winter, and to a much greater extent if the season be mild. For all things,[6] one may say, or at least most of them, extend beyond their proper season, and all the more if the place be sunny; so that there is a continuous succession. These then are the periods and seasons at which the various flowers are produced.

[7] The life of the gilliflower is at most three years; as it ages it degenerates and produces paler flowers.[8] A rose-bush lives five years, after which its prime[9] is past, unless it is pruned by burning[10]; with this plant too the flowers become inferior as it ages. Position and a suitable climate contribute most to

---

[5] καρπῷ : Plin. l.c. apparently read ἄνθει.
[6] πάντ' ἢ conj. St.; πάντη Ald. H.   [7] Plin. 21. 69.
[8] ἴα conj. St.; ἀεὶ Ald.
[9] ἀκμὴν conj. Scal.; ἀκτὴν Ald.   [10] cf. 6. 6. 6.

53

συμβάλλεται καὶ ὁ ἀὴρ πρὸς ἕκαστον οἰκεῖος· ἐν Αἰγύπτῳ γὰρ τὰ μὲν ἄλλα πάντ' ἄοσμα καὶ ἄνθη καὶ ἀρώματα, αἱ δὲ μυρρίναι θαυμασταὶ τῇ εὐοσμίᾳ. προτερεῖν δέ φασι τῶν ἐνταῦθα καὶ ῥόδα καὶ ἴα καὶ τὰ ἄλλα ἄνθη καὶ διμήνῳ, καὶ διαμένειν πλείω τῶν παρ' ἡμῖν ἢ οὐκ ἐλάττω χρόνον ταῦτα.

6 Δοκεῖ δὲ πολὺ πρὸς εὐοσμίαν διαφέρειν, ὥσπερ ἐλέχθη, καὶ ὁ ἐνιαυτὸς τοῖος ἢ τοῖος γενόμενος, οὐ μόνον ἐπομβρίαις καὶ αὐχμοῖς ἀλλὰ καὶ τῷ κατὰ καιρὸν γίνεσθαι καὶ ὕδατα καὶ πνεύματα καὶ ἁπλῶς τὰς τοῦ ἀέρος μεταβολάς. τὰ δὲ ἐν τοῖς ὄρεσιν ὡς ἁπλῶς εἰπεῖν καὶ ῥόδα καὶ ἴα καὶ τὰ ἄλλα <καλῶς μὲν ἀνθεῖν> τῇ δὲ ὀσμῇ πολλὰ χείρω γίνεσθαι. καὶ περὶ μὲν τῶν στεφανωματικῶν καὶ ἁπλῶς τῶν φρυγανικῶν σχεδὸν ἐν τούτοις καὶ τοῖς ὁμοίοις ἐστὶν ἡ ἱστορία.

---

[1] ?'violets and gilliflowers; so also below.
[2] Plin. *l.c.*; *cf. C.P.* 6. 18. 3.
[3] ἄνθη conj. St. from G; ἀνανθὴ Ald. *cf. C.P.* 6. 19. 4.
[4] Plin. 15. 37.

the fragrance of roses gilliflowers[1] and other flowers. Thus in Egypt,[2] while all other flowers[3] and sweet herbs are scentless, the myrtles[4] are marvellously fragrant. In that country it is said that the roses gilliflowers and other flowers are as much as two months ahead of those in our country, and also that they[5] last a longer, or at least not a shorter, time than those of our country.

And, as has been said, the particular season according to its character, makes a great difference to the fragrance, not only by reason of rains and droughts, but also according as rain, wind, and in general, the changes of climate occur or do not occur at the fitting moment. Also it appears that in general roses gilliflowers and the rest bloom well on the mountains, but many of them have there an inferior scent.[6] Concerning coronary plants and under-shrubs in general these examples and others like them suffice for our enquiry.

[5] ταῦτα conj. W.; τούτου Ald.
[6] ἄνθη τῇ ὀσμῇ πολλῷ Ald.; ἄνθη τῇ δὲ ὀσμῇ πολλὰ UM, whence Sch. and W. conj. that some such words as καλῶς μέν have dropped out and ἀνθεῖν has been altered to ἄνθη. cf. C.P. 6. 20. 1.

BOOK VII

Η

I. Ἑπόμενον δὲ τοῖς εἰρημένοις περὶ τῶν ποιωδῶν εἰπεῖν· τοῦτο γάρ ἐστι λοιπὸν τῶν ἐξ ἀρχῆς διαιρεθέντων γενῶν, ἐν ᾧ συμπεριλαμβάνονταί πως τὸ λαχανηρὸν καὶ τὸ σιτῶδες. καὶ πρῶτον περὶ τοῦ λαχανώδους λεκτέον ἀρξαμένους ἀπὸ τῶν ἡμέρων, ἐπεὶ γνώριμα μᾶλλον τυγχάνει τῶν ἀγρίων.

Εἰσὶ δὴ τρεῖς ἄροτοι πάντων τῶν κηπευομένων, ἐν οἷς ἕκαστα σπείρουσι διαιροῦντες ταῖς ὥραις. εἷς μὲν οὖν ὁ χειμερινός, ἄλλος δὲ ὁ θερινός, τρίτος δὲ ὁ μεταξὺ τούτων μεθ' ἡλίου τροπὰς 2 χειμερινάς. καλοῦσι δ' οὕτως οὐ πρὸς τὴν σπορὰν βλέποντες ἀλλὰ πρὸς τὴν γένεσιν καὶ τὴν χρείαν ἑκάστου· ἐπεὶ ἥ γε σπορὰ σχεδὸν ἐν τοῖς ἐναντίοις γίνεται. τοῦ χειμερινοῦ μὲν γὰρ ἀρχὴ μετὰ τροπὰς θερινὰς τοῦ Μεταγειτνιῶνος μηνός, ἐν ᾧ σπείρουσι ῥάφανον ῥαφανίδα γογγυλίδα καὶ τὰ καλούμενα ἐπίσπορα· ταῦτα δ' ἐστὶ τεύτλιον θριδακίνη εὔζωμον λάπαθον νᾶπυ κορίαννον ἄνηθον κάρδαμον· καλοῦσι δὲ

[1] cf. C.P. 3. 20. 7 and 8.

BOOK VII

Of Herbaceous Plants, other than Coronary Plants:
Pot-herbs and similar Wild Herbs.

Of the times of sowing and of germination of pot-herbs.

I. Next we have to tell of herbaceous plants: for this class remains of those which we distinguished at the outset, and it includes to some extent the classes of pot-herbs and of cereals. And first we must speak of the class of pot-herbs, beginning with the cultivated kinds, since it happens that these are better known than the wild kinds.

[1] There are three seed-times for all things grown in gardens, at which men sow the various herbs, distinguishing by the season. One is the 'winter' seed-time, another the 'summer,' and the third is that which falls between these, coming after the winter solstice. These terms however are given in regard not to the sowing, but to the growth and use of each kind; for the actual sowing takes place, one might almost say, at the opposite seasons. Thus, the 'winter' period begins after the summer[2] solstice in the month Metageitnion,[3] in which they sow cabbage radish turnip, and what are called 'secondary crops,' that is to say, beet lettuce rocket monk's rhubarb mustard coriander dill cress; and

[2] θερινὰς conj. Scal.; χειμερινὰς U(?)MP₂Ald.G (ed. Bas. and Par. but not ed. Tarv.).

[3] July. δὲ before M. om. Sch.

καὶ πρῶτον τοῦτον τῶν ἀρότων. τοῦ δὲ δευτέρου πάλιν μεθ' ἡλίου τροπὰς τοῦ Γαμηλιῶνος μηνός, ἐν ᾧ σπείρουσι καὶ πηγνύουσι πράσον σέλινον γήθυον ἀδράφαξυν. τοῦ τρίτου δέ, ὃν καλοῦσι θερινόν, τοῦ Μουνυχιῶνος· ἐν τούτῳ δὲ σπείρεται σίκυος κολοκύντη βλίτον ὤκιμον ἀνδράχνη θύμβρον. ποιοῦνται δὲ πλείους ἀρότους τῶν ὁμοίων καθ' ἑκάστην ὥραν, οἷον ῥαφανίδος ὠκίμου τῶν ἄλλων. πᾶσι δὲ σπείρεται τοῖς ἀρότοις τὰ ἐπίσπορα.

3 Διαφύεται δ' οὐκ ἐν ἴσοις πάντα χρόνοις, ἀλλὰ τὰ μὲν θᾶττον τὰ δὲ βραδύτερον ὅσα δυσφυῆ. τάχιστα μὲν οὖν ὤκιμον καὶ βλίτον καὶ εὔζωμον καὶ τῶν χειμερινῶν ῥαφανίς· τριταῖα γὰρ ὡς εἰπεῖν. θριδακίναι δὲ τεταρταῖαι ἢ πεμπταῖαι. σίκυος δὲ καὶ κολοκύντη περὶ τὰς πέντε ἢ ἕξ, οἱ δέ φασιν ἑπτά· πρότερον δὲ καὶ θᾶττον ὁ σίκυος. ἀνδράχνη δ' ἐν πλείοσι τούτων. ἄνηθον δὲ τεταρταῖον. κάρδαμον δὲ καὶ νᾶπυ πεμπταῖα. τεύτλιον δὲ θέρους μὲν ἑκταῖον χειμῶνος δὲ δεκαταῖον. ἀδράφαξυς δὲ ὀγδοαία. ῥάφανος δὲ δεκαταία. πράσον δὲ καὶ γήθυον οὐκ ἐν ἴσοις, ἀλλὰ τὸ μὲν ἐννεακαιδεκαταῖον ἐνιαχοῦ δὲ εἰκοσταῖον, γήθυον δὲ δεκαταῖον ἢ δωδεκαταῖον. κορίαννον δὲ δυσφυές· οὐδὲ γὰρ ἐθέλει βλαστάνειν τὸ νέον ἐὰν μὴ βρεχθῇ. θύμβρα δὲ καὶ ὀρίγανος ἐν πλείοσιν ἢ τριάκοντα. δυσφυέστατον δὲ πάντων τὸ σέλινον· τεσσαρακοσταῖον γάρ φασιν οἱ τὰ συντομώτερα λέγοντες,

---

[1] January. [2] April. [3] Plin. 19. 117.
[4] τῶν χειμερινῶν: cf. 7. 1. 1.

ENQUIRY INTO PLANTS, VII. I. 2–3

this is also called the 'first' period of cultivation. The second period begins after the winter solstice in the month Gamelion,[1] in which they scatter or plant the seed of leeks celery long onion orach. The third period, which is called the 'summer' period, begins in the month Munychion[2]: in this are sown cucumber gourd blite basil purslane savory. Moreover they make several sowings of the same herb at each season, as of radish basil and the others. And at all the periods are sown the 'secondary crops.'

[3] Not all herbs germinate within the same time, but some are quicker, others slower, namely those which germinate with difficulty. The speediest are basil blite rocket, and of those sown for winter[4] use, radish; for these germinate in about three days. Lettuce takes four or five, cucumber and gourd about five or six, or, as some say, seven; however, cucumber is earlier and quicker than the others. Purslane takes a longer time, dill four days, cress and mustard five. Beet in summer takes six days, in winter ten, orach takes eight, and turnip ten. Leek[5] and long onion do not take the same time, but the former nineteen to twenty days, the latter ten to twelve. Coriander germinates with difficulty; indeed fresh seed will not come up at all unless it is moistened.[6] Savory[7] and marjoram take more than thirty days; but celery germinates with the greatest difficulty of all; for those who make the time comparatively short say forty days, and others fifty, and

---

[5] πράσον conj. Bod.; πράσιον P₂Ald.H.
[6] βρεχθῇ conj. Bod. cf. C.P. 4. 3. 1; ἐλιχθῇ Ald.; ἐλιχθῇ P₂Bas.; so also G.
[7] cf. C.P. 4. 3. 1; Plin. 19. 7.

61

οἱ δὲ πεντηκοσταῖον, καὶ τοῦτο κατὰ πάντας τοὺς ἀρότους· ἐπισπείρουσι γάρ τινες ἐπὶ πᾶσιν.

4 Ὅλως δὲ ὅσα κατὰ πλείους ὥρας σπείρεται, ταῦτ' οὐδὲν θᾶττον τέλεια γίνεται τοῦ θέρους. καὶ θαυμαστὸν εἰ καὶ μηθὲν ἡ ὥρα συμβάλλεται καὶ ὁ ἀὴρ πρὸς τὸ θᾶττον, ἐὰν δὲ μοχθηρὰ καὶ ψυχρὰ καὶ τῷ ἀέρι περισκεπὴς βραδύτερον· ἐπεὶ καὶ χειμώνων ἢ εὐδιῶν ἐπιγινομένων τοῖς ἀρότοις ὁτὲ μὲν βραδύτερον ὁτὲ δὲ θᾶττον ἡ βλάστησις· διαφέρει δὲ ταῦτα κατὰ τοὺς ἀρότους ἑκάστων· πρωϊαίτατον γὰρ ἐν τοῖς εὐείλοις καὶ εὐκράσιν.

5 Ὡς γὰρ ἁπλῶς εἰπεῖν ἐν πλείοσι δεῖ τὰς αἰτίας ὑπολαβεῖν τῶν τοιούτων, ἔν τε τοῖς σπέρμασιν αὐτοῖς καὶ ἐν τῇ χώρᾳ καὶ τῷ ἀέρι καὶ ταῖς ὥραις αἷς ἕκαστα σπείρουσι καὶ χειμώνων καὶ εὐδιῶν. ἀλλὰ τοῦτο μὲν σκεπτέον, ἐφ' ὧν τε παραλλάττουσιν οἱ χρόνοι καὶ ἐφ' ὧν οὔ· καὶ γὰρ τὴν ῥαφανίδα φασί τινες τριταίαν καὶ θέρους καὶ χειμῶνος, τὸ δὲ τεύτλιον, ὥσπερ εἴρηται, παραλλάττει κατὰ τὰς ὥρας. χρόνοι δ' οὖν οὗτοι τῆς βλαστήσεώς εἰσι καὶ λέγονται καθ' ἕκαστον.

6 Διαφέρει δὲ πρὸς τὸ θᾶττον καὶ βραδύτερον καὶ ἡ τῶν σπερμάτων παλαιότης. τὰ μὲν γὰρ ἀπὸ νέων παραγίνεται θᾶττον, οἷον πράσον γήθυον

---

[1] ὥρας Vo.H.; χώρας UM; so also G.
[2] τέλεια conj. W. (comm.); γε πολλὰ MSS.; τὰ πολλὰ Vo.Sch.W. (text); γίνεται conj. Sch. from G; γίνεσθαι Ald.
[3] καὶ τῇ ἀέρι ... βραδύτερον: grammar doubtful and text perhaps defective: so given in UM; καὶ ὁ ἀὴρ περισκεπὴς

that too, at whichever period it is sown, for some sow it as a 'secondary crop' at all the periods.

Generally speaking, those herbs which are sown at more than one season[1] do not mature[2] faster in the summer. Howbeit it is strange if the season and the state of the atmosphere do not contribute at all to quicker growth, and if, when there is an unfavourable cold season and the atmosphere is cloudy, these conditions do not tend to make growth slower,[3] seeing that, when stormy or fair weather follows the sowing, germination is slower or quicker accordingly. And there is another thing which makes a difference as to the raising of the various herbs; germination begins earlier in sunny places which have an even temperature.

As a matter of fact, to speak roundly, the causes of such differences must be found in several different circumstances, in the seeds themselves, in the ground, in the state of the atmosphere, and in the season at which each is sown, according as it is stormy or fair. However it is a point for consideration with which herbs the time of sowing makes a difference and with which it makes none; thus it is said that radish germinates on the third day whether it be sown in summer or in winter, while beet, as has been said, behaves differently according to the season. Anyway such are and are said to be the seasons of germination in each case.

[4]Another thing which makes a difference as to the rapidity with which the seeds germinate is their age; for some herbs come up quicker from fresh seed, as

πρὸς τὸ βραδύτερον conj. Sch. (with μοχθ. κ. ψυχρὰ supply ἡ ὥρα ᾖ).
[4] Plin. 19. 118. δὲ conj. Scal.; γὰρ Ald.H.

THEOPHRASTUS

σίκυος κολοκύντη· ἔνιοι δὲ καὶ προβρέχουσι τὸν σίκυον πρὸς τὸ θᾶττον ἢ ἐν γάλακτι ἢ ἐν ὕδατι. τὰ δ' ἀπὸ παλαιῶν, οἷον σέλινον τεύτλιον κάρδαμον θύμβρα κορίαννον ὀρίγανον· εἴπερ μὴ <φυτεύεται> αὐτὰ ἀπὸ τοῦ νέου, καθάπερ εἴπομεν. ἴδιον δέ φασιν ἐπὶ τοῦ τευτλίου συμβαίνειν· οὐ γὰρ διαφύεσθαι πᾶν εὐθὺς ἀλλ' ὕστερον πολλῷ, τὸ δὲ καὶ τῷ ἐχομένῳ ἔτει καὶ τῷ τρίτῳ, δι' ὃ καὶ ἐκ πολλοῦ σπέρματος ὀλίγον βλαστάνειν.

7 Ἕκαστον δὲ τῶν σπερμάτων, ἐὰν ἁδρυνθέντα ἀποπέσῃ, διαμένει πρὸς τὴν ὥραν τὴν ἑαυτοῦ καὶ οὐ πρότερον ἐκβλαστάνει· καὶ κατὰ λόγον ἐστί· καὶ γὰρ ἐπὶ τῶν ἀγρίων ὁρῶμεν συμβαῖνον, ἐὰν μὴ φθαρῇ. αἱ δὲ τελειώσεις τῶν καρπῶν ἁπάντων γίνονται τοῦ θέρους, πρότερον δὲ καὶ θᾶττον ὡς ἁπλῶς εἰπεῖν τῶν πρότερον σπαρέντων. διαφέρει δὲ καὶ ἡ ὥρα· τὰ γὰρ ἐν ταῖς θερμημερίαις σπαρέντα θᾶττον ἐκκαυλεῖ καὶ ἐκσπερματοῦται, καθάπερ ῥαφανὶς γογγυλίς. ἔνια δὲ οὐκ ἐνιαύσια φέρει τὸν καρπὸν ἀλλὰ δίενα, καθάπερ σέλινον πράσον γήθυον, ἃ καὶ διαμένει χρόνον πλείονα καὶ οὐκ ἔστιν ἐπέτεια· τὰ γὰρ πολλὰ τούτων ἅμα τῇ τελειώσει τῶν σπερμάτων αὐαίνεται.

8 Πάντα δὲ ὡς εἰπεῖν ὅσα ἐκκαυλεῖ καὶ τελειοῖ τὸν καρπὸν ἀποτελειοῦται κατὰ τὸ σχῆμα τοῦ παραβλαστήσεις ἐκ τῶν καυλῶν ἔχειν ἀκρεμονικάς, πλὴν ὅσα μονόκαυλα, καθάπερ πράσον καὶ γήθυον καὶ κρόμυον καὶ σκόροδον.

Φίλυδρα δὲ καὶ φιλόκοπρα πάντα, μᾶλλον δὲ

---

[1] φυτεύεται αὐτὰ conj. W.; οὐ τὸ UMAld.
[2] Sc. soaking.

ENQUIRY INTO PLANTS, VII. I. 6-8

leek long onion cucumber gourd; (some even soak the seed of cucumber first in milk or water, to make it germinate quicker). Some come up quicker from old seed, as celery beet cress savory coriander marjoram (unless indeed they are raised[1] from fresh seed in the manner[2] which we have mentioned). There is, they say, a singular feature about beet[3]; the seed does not all germinate at once, but some of it not for some time, some even in the next or in the third year; wherefore it is said that little comes up from much seed.

Any of the seeds, if they are ripe when they fall, last till their own proper season and do not sprout till then. And in this they are consistent; for we note that the same thing happens with the seed of wild plants, unless it is destroyed. However all mature their fruits in the summer, though sooner and quicker, generally speaking, when they are sown earlier. The season also[4] makes a difference; things sown in the hot season push up their shoots and go to seed sooner, as radish and turnip. Some however bear their fruit not in the same year but in the next, as celery[5] leek long onion, which plants also last a longer time, and are not annual; for most herbs wither with the ripening of their seed.

Generally speaking, all those that push up shoots and mature their fruit reach their perfection of form in having side-shoots branching from the main stem —except those which have but a single stem, as leek long onion onion garlic.

All these herbs are lovers of water and of dung,

---

[3] *cf. C.P.* 4. 3. 2; Plin. *l.c.*
[4] δὲ conj. W.; γὰρ Ald.H.
[5] Plin. *l.c.*

65

τὰ ἀσθενέστερα καὶ πλείονος ἐπιμελείας δεόμενα, τὰ δὲ καὶ τροφῆς.

II. Φύεται δὲ πάντα ἀπὸ τοῦ σπέρματος, ἔνια δὲ καὶ ἀπὸ παρασπάδος καὶ κλωνὸς καὶ ῥίζης. ἀπὸ μὲν παρασπάδος ἡ ῥάφανος· δεῖ γάρ τι καὶ ῥιζῶδες προσλαβεῖν. ἀπὸ δὲ τῶν βλαστῶν πήγανον ὀρίγανος ὤκιμον· ἀποφυτεύουσι γὰρ καὶ τοῦτο ὅταν σπιθαμιαῖον ᾖ μεῖζον γένηται τεμόντες εἰς τὸ ἥμισυ. ἀπὸ ῥίζης δὲ σκόροδον καὶ κρόμυον καὶ βολβὸς καὶ ἄρον καὶ ἁπλῶς τὰ τοιαῦτα τῶν κεφαλορρίζων. φύεται δὲ καὶ εἴ τινων αἱ ῥίζαι διαμένουσιν ἐπὶ πλείονα χρόνον ἐπετειοκαύλων ὄντων. ὅτι δὲ ἀπὸ σπέρματος πάντα βλαστάνει φανερόν· καὶ γὰρ τὸ πήγανον, ὅπερ οὔ φασί τινες, ἀλλὰ βραδέως, δι' ὃ καὶ ἀποφυτεύουσιν.

2 Ὅσα δὲ ἀπὸ ῥίζης φύεται, τούτων ἡ μὲν ῥίζα χρόνιος αὐτὰ δὲ ἐπετειόκαυλα, δι' ὃ καὶ παραβλαστάνουσιν αἱ ῥίζαι τῶν τοιούτων καὶ γίνονται πλείους οὐ μόνον ἐν τοῖς ἡμέροις καὶ κηπευομένοις ἀλλὰ καὶ ἐν τοῖς ἀγρίοις, ὥσπερ εἴπομεν, οἷον βολβοῖς γηθύοις σκίλλαις καὶ τοῖς ἄλλοις. παραβλαστάνει δ' ἔνια καὶ τῶν μὴ κεφαλορρίζων χρονιωτέρων δέ, οἷον σέλινον καὶ τεύτλιον· ἀφιᾶσι γὰρ ῥίζας ἀφ' ὧν φύονται φύλλα καὶ καυλοί.

---

[1] Plin. 19. 121.   [2] cf. C.P. 1. 4. 2.
[3] δεῖ γάρ τι UP₂; ἀεὶ γάρ τι Ald. H. G; Sch. suggests δὲ for γὰρ, missing the sense.
[4] βλαστῶν corresponds to κλωνὸς above.

and especially the weaker ones, which require more attention or in some cases more feeding.

Of the propagation of pot-herbs, and of differences in their roots.

II. [1]All these herbs are propagated from seed, and some also by a piece torn off, a shoot, or a piece of root. Cabbage is propagated by a piece torn off,[2] since it is essential[3] in this case to take a piece which has root attached to it. From cuttings[4] are grown rue marjoram basil; for slips of this too men plant when it has grown to the height of a span or more, cutting off half the plant.[5] By root[6] are planted garlic onion purse-tassels cuckoo-pint and in general such bulbous plants. Such propagation is also possible in cases where the roots persist for more than a year, though the shoots last but for a year. And it is plain that all these herbs can be grown from seed; for even rue can (which some deny), though the process is slow, and so cuttings are also taken.

Of those which are propagated by a piece of root the root is long-lived, though the plant itself may be annual; wherefore the roots of such plants make offsets and so increase; and this is true not only of plants cultivated in the garden, but also of wild plants, as we have said, for instance of purse-tassels long onion[7] squill and so forth. Some plants even which are not bulbous[8] but longer-lived make offsets, as celery and beet; for these send out roots from which grow leaves and stems. Long onion and

---

[5] *cf. C.P.* 1. 4. 3. [6] *i.e.* offsets.
[7] γηθύοις om. some editors, as not being wild.
[8] *i.e.* and so annual.

THEOPHRASTUS

παραβλαστάνει δὲ καὶ γήθυον καὶ πράσον καὶ παραφύει κάτωθεν οἷον βολβώδη τινὰ κεφαλήν, ἐξ ἧς ἡ βλάστησις γίνεται τῶν φύλλων, αὐανθέντος δὲ τοῦ καυλοῦ καὶ τοῦ σπέρματος ἀφαιρεθέντος· ἀλλὰ διὰ τὸ μὴ χρησίμας εἶναι τὰς τούτων κεφαλὰς οὐ συλλέγουσιν εἰς ξηρασίαν, δι' 3 ὃ καὶ οὐ φυτεύουσι. τάχα δὲ ταῦτα καὶ ὁμογενῆ καὶ σύνεγγύς πως τῇ τοῦ κρομύου φύσει, δι' ὃ καὶ οὐ θαυμαστόν. ἀλλ' ὁμοίως [καὶ] ἐπὶ πάντων καὶ ἡμέρων καὶ ἀγρίων, ὅσα χρονιώτερα μέν ἐστιν ἐπετειόκαυλα δέ, τούτων καὶ αἱ ῥίζαι ἐπιβλαστάνουσιν, ὥσπερ καὶ ἐπὶ τῶν φρυγανικῶν καὶ τῶν θαμνωδῶν· ἀλλ' ἐπὶ τῶν κρομύων καὶ σκορόδων καὶ βολβῶν καὶ ὥσπερ ἀριθμός τις γίνεται τούτων. ἡ δὴ γένεσις, ὥσπερ εἴρηται, τριχῶς ἐστιν, ἀπὸ σπέρματος μὲν πάντων, ἀπὸ δὲ καυλοῦ καὶ ῥίζης τῶν εἰρημένων.

4 Τῶν δὲ καυλῶν κολουσθέντων πάντα μὲν ὡς εἰπεῖν βλαστάνει πλὴν τῶν ἀποκαύλων, ἐμφανέστατα δ' ὥσπερ καὶ εἰς χρείαν ὤκιμον θρῖδαξ ῥάφανος. καὶ τῆς μὲν θρίδακος ἡδίους φασὶ τοὺς παλιμβλαστεῖς εἶναι καυλούς· τὸν γὰρ πρῶτον ὀπώδη καὶ πικρὸν εἶναι ὡς ἄπεπτον· οἱ δὲ τὸ ἐναντίον ὀπωδεστέρους τούτους ἀλλ' ἕως ἂν ὦσιν ἁπαλοὶ φαίνεσθαι γλυκυτέρους. ἀλλ' ἐπὶ τῆς

---

[1] πράσον conj. St.; πράσιον Ald.H.
[2] διὰ τὸ μὴ conj. W.; μὴ διὰ τὸ UM(?)Ald.
[3] i.e. offset bulbs.
[4] W. omits μὲν (Ald.UM(?)) after συλλέγουσι.
[5] i.e. the plant is increased by seed only and not by offsets. cf. 7. 4. 10; Plin. 19. 103.
[6] ὁμοίως conj. Sch.; ὅμως PAld.H.(UM?).
[7] ἐπετειόκαυλα conj. Sch.; ἐπιγειότερα PAld.H.

leek[1] also make offsets, sending out a 'head' below, like the bulb of purse-tassels, from which the leaves spring; but this only takes place when the stem has withered and the seed has been removed. But, as[2] the 'heads'[3] of such plants are not useful, they do not collect them[4] for storing dry; wherefore also they do not plant these.[5] It may be that somehow these are akin and closely allied to onion, wherefore what has been said is not surprising. However in all those plants, both wild and cultivated alike,[6] which have an annual stem,[7] but yet live longer than a year, there is an outgrowth of the roots, just as there is in under-shrubs and shrubby plants: while in onions garlic and purse-tassels even a number,[8] as it were, of such roots is formed. In fact,[9] they are reproduced in three ways, as has been said; from seed in all cases and from the stem[10] and root in those specified.

[11] Almost all shoot again if the stem is broken (except those which are stemless), but most obviously basil lettuce cabbage, which are, as it were, broken for a practical reason. Indeed they say that the stems of lettuce which thus grow again are sweeter,[12] for that the original stem has a taste like fig-juice and is bitter, as being not properly ripened. Some however say that the later stems have the taste of fig-juice more than the original one, but that, so long as they are tender, they appear sweeter. Be that as

[8] ἀριθμὸς is clearly corrupt, and has displaced an unusual word for which ὥσπερ apologises.
[9] δὴ conj. Sch.; δὲ Ald.
[10] καυλοῦ is here that part of the plant which is above ground.   [11] Plin. 19. 122.
[12] ἡδίους Vo.mBas.H., so too G, Plin. *l.c.*, Athen. 2. 69; ἰδίους UAld.   *cf. C.P.* 2. 15. 6.

ῥαφάνου τοῦτο ὁμολογούμενον, ὡς εἰ πάλιν βλα-
στήσειεν¹ ἡδίων ἀφαιρεθέντων γε τῶν φύλλων πρὸ
τοῦ διακαυλίσαι.

5 Διαμένουσι δὲ αἱ ῥίζαι πλειόνων, ἀλλ' αἱ μὲν
βλαστάνουσι πάλιν αἱ δὲ οὔ. ῥαφανὶς γοῦν καὶ
γογγυλὶς διαμένουσι γῆς ἐπιβληθείσης ἄχρι
θέρους καὶ αὔξησιν λαμβάνουσιν, ὅπερ ποιοῦσί
τινες ἐξεπίτηδες τῶν κηπουρῶν· οὐ βλαστάνουσι
δὲ οὐδ' ἀφιᾶσι φύλλον οὐδ'² εἴ τις ἀφέλοι τὴν
ἐπισεσαγμένην γῆν. ἰδεῖν δὲ τοῦτο καὶ ἐπὶ τῶν
ἄλλων ἐστί. τὰ δὲ πλεῖστα τῶν λαχάνων μονόρριζα τῇ παχείᾳ κατὰ βάθους ῥίζῃ· καὶ γὰρ ὅσα
παραφύει τὰς ἰσοπαχεῖς ταύτας, ὥσπερ σέλινον
καὶ τεύτλιον, ἀπὸ τῆς μέσης πως ἡ παράφυσίς
ἐστι καὶ οὐκ εὐθὺς ἀπὸ τῆς ἀρχῆς ἡ σχίσις· ἐκ³
δὲ ταύτης τῆς μιᾶς ἀπήρτηνται αἱ ἀποφυάδες αἱ
μικραὶ καὶ τῆς ῥαφανίδος καὶ τῆς γογγυλίδος.
καὶ αὗται μὲν δὴ πᾶσι φανεραὶ διὰ τὴν χρείαν.

6 Ἡ δὲ τοῦ τευτλίου μία μὲν μακρὰ καὶ παχεῖα
καὶ ὀρθή, καθάπερ ἡ τῶν ῥαφανίδων, ἀποφύσεις δὲ
ἔχει παχείας ὁτὲ μὲν δύο ὁτὲ δὲ καὶ τρεῖς ὁτὲ δὲ
καὶ μίαν, τὰς δὲ μικρὰς ἐκ τούτων. σαρκώδης
δὲ ἡ ῥίζα καὶ τῇ γεύσει γλυκεῖα καὶ ἡδεῖα, δι' ὃ
καὶ ὠμὴν ἐσθίουσί τινες· ὁ δὲ φλοιὸς οὐ παχὺς
οὐδὲ ἀφαιρετός, ὥσπερ ὁ τῶν ῥαφανίδων, ἀλλὰ
μᾶλλον οἷος ὁ τῶν ἱπποσελίνων. ὡσαύτως δὲ
καὶ ἡ τῆς ἀδραφάξυος μία μὲν εἰς βάθος ἐκ
ταύτης δὲ ἄλλαι.

---

¹ βλαστήσειεν conj. Sch.; βλαστήσει Ald.
² οὐδ' εἴ τις Ald.H.; εἰ μή τις conj. Scal. supported by G.
³ ἐκ ... μικραὶ conj. W.; εἰς δὲ ταύτην τὴν μίαν ἢ ἀπ' αὐτῆς

ENQUIRY INTO PLANTS, VII. II. 4–6

it may, it is admitted that in the case of cabbage the stem is sweeter if it should have grown[1] again after being broken, provided that the leaves are stripped off before the plant runs to stalk.

In most cases the roots persist, but they do not in all cases produce fresh growth. Thus radish and turnip persist till summer, if earth is thrown on them, and they increase in size; and some gardeners do this deliberately; but they do not make fresh growth nor send out leaves, even if one[2] removes the earth heaped over them. And this may also be observed in other plants. However, most pot-herbs have the single stout root which runs deep; for even in those which produce these side-roots of equal stoutness, as celery and beet, the side-growth comes, as it were, from the middle root and it is not separate to start with; but to this single root are attached the small out-growths,[3] both in radish and in turnip. These instances are familiar to all because of the use[4] which is made of these plants.

The beet has a single long stout straight root like that of the radish, and has stout out-growths, sometimes two, sometimes three, sometimes only one, and the small ones are attached to these. The root is fleshy and sweet and pleasant to the taste, wherefore some even eat it raw. The 'bark' is not thick and cannot be detached, like that of the radish, but rather resembles that of alexanders. In like manner the root of orach is single and runs deep, and other roots are attached to it.

τε καὶ τῆς ἀποφυάδος καὶ μικρὰ Ald.H.; so also M, omitting τε. W.'s restoration of a very corrupt text is at least consistent with what follows in § 6.
[4] *i.e.* for food.

7 Μονορριζότατον δὲ τούτων πάντων τὸ λάπαθον· οὐ γὰρ ἔχει παχείας ἀποφύσεις ἀλλὰ τὰς λεπτάς· βαθυρριζότατον δὲ πάντων, ἔχει γὰρ μείζω τριῶν ἡμιποδίων· τὸ δ' ἄγριον βραχυτέραν, πολύκαυλον δὲ καὶ πολύκλαδον καὶ ἡ ὅλη μορφὴ τελειωθεῖσα παραπλησία τῇ τοῦ τευτλίου· πολυχρονιώτερον δὲ καὶ τοῦ ἀγρίου καὶ ὅλως δὲ πάντων τῶν λαχάνων ὡς εἰπεῖν· διαμένει γὰρ ὁποσονοῦν χρόνον ὥς φασιν. ἔχει δὲ σαρκώδη τὴν ῥίζαν καὶ ἔνικμον, δι' ὃ καὶ ἐξαιρεθεῖσα ζῇ πολὺν χρόνον.

Τὸ δ' ὤκιμον μίαν μὲν τὴν παχεῖαν τὴν κατὰ βάθους τὰς δ' ἄλλας τὰς ἐκ πλαγίου λεπτὰς ἐπιεικῶς εὐμήκεις.

Ἔνια δ' οὐκ ἔχει τὴν μίαν τὴν ὀρθήν, οἷον τὸ βλίτον, ἀλλ' εὐθὺ πολλὰς ἐξ ἄκρου καὶ εὐπαχεῖς καὶ μακροτέρας τῆς ἀδραφάξυος.

8 Τῶν δὲ ῥιζῶν ξυλωδέσταται πασῶν αἱ τοῦ ὠκίμου, καθάπερ καὶ ὁ καυλός. ἡ γὰρ τοῦ βλίτου καὶ τῆς ἀδραφάξυος καὶ τῶν τοιούτων ἧττον ξυλώδης. εἰσὶ γὰρ ὡς ἁπλῶς εἰπεῖν πασῶν αἱ μὲν σαρκώδεις αἱ δὲ ξυλώδεις. <σαρκώδεις>, οἷον ἡ τοῦ τευτλίου καὶ τοῦ σελίνου καὶ ἱπποσελίνου καὶ λαπάθου καὶ ῥαφανίδος καὶ γογγυλίδος καὶ πάντων μάλιστα τῶν κεφαλοβαρῶν· οὐδὲ γὰρ ἀναξηραινόμεναι σκληρύνονται τελείως. ξυλώδεις δέ,

---

[1] cf. 1. 6. 6.
[2] τὰς Ald., cf. τὰς δὲ μικρὰς § 6 ; τινας conj. W. cf. Plin. 19. 98 (who mistranslates).
[3] cf. 7. 6. 1 ; C.P. 3. 1. 4.   [4] See Index.

ENQUIRY INTO PLANTS, VII. ii. 7–8

Monk's rhubarb [1] however has a single root in a truer sense than any of the others, for it has no stout out-growths of root, but only the [2] slender ones; its root also runs deeper than that of the others, being more than a foot and a half long. The wild sort [3] however has a shorter root, and has several stems and branches, and its shape, as a whole, when fully grown resembles that of beet. Cultivated monk's rhubarb moreover is longer lived than the wild form,[4] and, in general, we may say, than any other pot-herb, for, they say, it may live any time. It has a fleshy root,[5] full of moisture, wherefore, if pulled up, it will live some time.

Basil has the single stout root, the one which runs deep, and the others at the sides are slender and fairly long.

Some herbs, as blite, have not the single straight root, but a number of roots which start directly from the top and are of a good stoutness [6] and longer than those of orach.

The roots of basil are woodier than those of any of the other herbs, as also is its stem; for those of blite orach and the like are less woody. In general we may say that the roots of any [7] of these herbs are either woody or fleshy. Examples of fleshy [8] roots are beet celery alexanders monk's rhubarb radish turnip, and especially all 'heavy-headed' [9] kinds, for the roots of these do not wither up altogether even when they are dried. Examples of those with woody roots

[5] ῥίζαν conj. Sch.; σάρκα Ald.
[6] Plin. *l.c.* seems to have read a different word from εὐπαχεῖς, or to have misunderstood it.
[7] πασῶν conj. W.; παρ' ὧν UMP; also Ald.H., omitting αἱ.
[8] σαρκώδεις add. Scal. from G.
[9] *i.e.* bulbous; *cf.* 1. 6. 8.

ὥσπερ αἱ τοῦ ὠκίμου καὶ βλίτου καὶ ἀδραφάξυος καὶ εὐζώμου καὶ ἀνήθου [καὶ λαπάθου] καὶ κοριάννου καὶ ἁπλῶς τῶν νευροκαύλων· ἔχει γὰρ δὴ καὶ τὸ ἄνηθον καὶ τὸ κορίαννον ὄντα μονόρριζα ξυλώδη τε τὴν ῥίζαν καὶ οὐ μακρὰν οὐδὲ τὰς λεπτὰς ἀποφυάδας ἔχουσαν πολλάς· πολύκαυλα δὲ ἄμφω καὶ πολύοζα, δι' ὃ καὶ οὐ κατὰ λόγον οὐδενὶ τούτων τὸ ἄνω πρὸς τὸ κάτω.

9 Βραχύρριζα δὲ ταῦτά ἐστιν, οἷον θρῖδαξ ἀνδράχνη, τῇ ὀρθῇ καὶ ταῖς εἰς τὰ πλάγια. ἡ δὲ θρῖδαξ, ὥσπερ οὐκ ἔχει τὰς τοιαύτας ἀποφύσεις ἀλλὰ μόνον τὰς λεπτάς, καὶ μάλιστα δὴ μονόρριζον ὡς εἰπεῖν. ἁπλῶς δὴ πάντα τὰ θερινὰ βραχύρριζα· καὶ γὰρ ὁ σίκυος καὶ ἡ κολοκύντη καὶ ἡ σικύα καὶ διὰ τὴν ὥραν καὶ ἴσως ἔτι μᾶλλον διὰ τὴν φύσιν, ἥπερ συνηκολούθηκε τῇ ὥρᾳ. ἡ δὲ μεταφυτευομένη θρῖδαξ βραχυτέραν ἔχει τὴν ῥίζαν τῆς σπαρείσης· παραβλαστάνει γὰρ ἐκ τῶν πλαγίων μᾶλλον· βραχυτέραν δὲ καὶ ἡ ἀγρία τῆς ἡμέρου, καὶ ἐκ τῶν ἄνωθεν πολυκαυλοτερα.

III. Ἀνθεῖ δὲ τῶν μὲν ἄλλων ἕκαστον ἀθρόον, τὸ δὲ ὤκιμον κατὰ μέρος, τὰ κάτω πρῶτον εἶτ' ὅταν ταῦτα ἀπανθήσῃ τὰ ἄνω, δι' ὃ καὶ πολυ-

---

[1] After ἀνήθου Ald. H. have καὶ λαπάθου: bracketed by W. after Sch.

[2] ἀποφυάδας conj. Scal.; ἀποφυλλάδας Ald.

[3] ταῦτα conj. Sch.; τὰ τοιαῦτα UM; τοιαῦτα Ald.

[4] Athen. 2. 79. Sch. suggests that the name of a plant has dropped out after ὥσπερ: ? ἡ ἀνδράχνη.

are basil blite orach rocket dill[1] coriander, and in general, those with fibrous stems; for in dill and coriander, which have a single root, the root is woody and not long, and the slender side-roots[2] from it are not numerous; but both plants have several stems and branches; wherefore in neither of these plants does the part above ground correspond to the part which is below.

The following[3] have short roots: lettuce and purslane, in which both the straight main root and the side ones are short. [4] Lettuce may be said to have no such side-roots, but only the slender ones, and may be called in the strictest sense a plant of a single root. In general all summer herbs have short roots: we may include cucumber gourd and bottle-gourd, both because of the season to which they belong and perhaps still more because of their character, which corresponds to the season. However the transplanted lettuce has a shorter root than one that is raised from seed, since it is more apt to send out side-growths; also the wild kind has a shorter[5] root than the cultivated, and the part above ground has more stems.[6]

Of the flowers and fruits of pot-herbs.

III. [7]All, except one, of these herbs produce all their bloom at once, but basil has a succession of flowers, the lower part of the plant flowering first, and then, when that bloom is over, the upper part. Wherefore its season of bloom is a long one, like that of the

---

[5] βραχυτέραν conj. Sch.; βραχυτέρα Ald.
[6] ἄνωθεν πολυκαυλοτέρα conj. Sch. from G; ἄνω· τὰ δὲ πολυκ. Ald. *cf.* Diosc. 2. 136. [7] Plin. 19. 100.

THEOPHRASTUS

χρόνιον ἐν τῷ ἀνθεῖν, καθάπερ κύαμος καὶ τῆς πόας τὸ ἡλιοτρόπιον καλούμενον καὶ ἄλλα δὲ τῶν ἀγρίων. ἀνθεῖ δὲ καὶ ὁ σίκυος πολὺν χρόνον· καὶ γὰρ ἐπιβλαστάνειν τούτῳ γε συμβαίνει. τὰ δὲ ἄνθη τῶν μὲν ἔκλευκα τῶν δὲ μηλινοειδῆ τῶν δὲ μικρὸν ἐπιπορφυρίζοντα, εὔχρουν δ' οὐθέν.

2 Τὰ δὲ σπέρματα διαφέρει καὶ τοῖς σχήμασι· τὰ μὲν γὰρ πλεῖστα στρογγύλα τὰ δὲ προμήκη τὰ δ' αὖ πλατέα καὶ φυλλώδη, καθάπερ τὰ τῆς ἀδραφάξυος· ὅμοιον γὰρ τῷ τοῦ σιλφίου· τὰ δὲ στενὰ καὶ γραμμώδη, καθάπερ τοῦ κυμίνου. καὶ τοῖς χρώμασιν ὁμοίως, τὰ μὲν μέλανα τὰ δὲ ξυλώδη τὰ δὲ λευκότερα. πάντα δὴ ἐλλοβοσπέρματα ἢ γυμνοσπέρματα ἢ ἐμφλοιοσπέρματα ἢ παππόσπερματα· ῥαφανὶς μὲν γὰρ καὶ νᾶπυ καὶ γογγυλὶς ἐλλοβοσπέρματα, κορίαννον δὲ καὶ μάραθον καὶ ἄνηθον καὶ κύμινον γυμνοσπέρματα, βλίτον δὲ καὶ τεύτλιον καὶ ἀδράφαξυς καὶ ὤκιμον ἐμφλοιοσπέρματα, θριδακίνη δὲ παππόσπερματον.

3 Πάντα δὲ πολύκαρπα καὶ πολυβλαστῆ, πολυκαρπότατον δὲ τὸ κύμινον. ἴδιον δὲ καὶ ὃ λέγουσι κατὰ τούτου· φασὶ γὰρ δεῖν καταρᾶσθαί τε καὶ βλασφημεῖν σπείροντας, εἰ μέλλει καλὸν ἔσεσθαι καὶ πολύ.

Δυσξήραντα δὲ πάντα μὲν ὡς εἰπεῖν πλὴν τοῦ κυμίνου, οὐχ ὡς ὁ σῖτος· οὗτος γὰρ κἂν ἅπαξ

---

[1] For the collective sense of πόα (= τὰ ποώδη) cf. 1. 3. 1.

[2] πολὺν χρόνον conj. W., which at least gives the required sense ; καλούμενος Ald.

[3] μηλινοειδὲς : cf. 6. 2. 8.

? 'orange.'   [5] Plin. 19. 119.

bean, and among herbaceous plants [1] that of the plant called *heliotropion*, and also other wild plants. Cucumber also has a long period [2] of bloom, for this plant has a second growth. The flowers are in some cases whitish, in others quince-yellow,[3] in others somewhat reddish [4]; but the flower is never of a bright colour.

[5] The seeds too differ in shape; most are round, but some are oblong; some again are broad and leaf-like, as those of orach, for the seed of this is like that of silphium; others again are narrow or marked in lines,[6] as those of cummin. They also vary in colour, some being black, some the colour of wood,[7] some paler. The seeds of all are either in pods or naked, or have an integument or have a pappus. Radish mustard and turnip have their seeds in pods; coriander fennel dill and cummin have naked seeds; those of blite beet orach and basil are enclosed in an integument; those of lettuce have a pappus on them.

All have numerous fruits and numerous shoots, but cummin has the most [8] fruits of all. [9] And there is another peculiarity told of this plant: they say that one must curse and abuse it, while sowing, if the crop is to be fair and abundant.

Nearly all of these, except cummin, are hard to dry for keeping,—unlike corn [10]; for this, when once

[6] γραμμώδη: cf 4. 12. 2.; *canaliculata* Plin. *l.c.*
[7] ? 'brown' *cf.* 7 9. 3.
[8] *cf.* 8. 3. 5; *C P.* 4. 15. 2.
[9] *cf* 9. 8. 8; Plin. *l.c.* applies this to ὤκιμον, Pall. 4. 9. 5 to πήγανον.
[10] σῖτος· οὗτος γὰρ I conj.; σῖτος γάρ UMH.; P omits γὰρ; σῖτος ὃς W. after Sch.; *nec modo frumenti consistunt, quod* G.

THEOPHRASTUS

ἁδρυνθῇ ταχὺ ξηραίνεται καὶ ἀποπίπτει· δυσξηραντότερα δὲ τὰ ἐμφλοιοσπέρματα καὶ τούτων
4 <μάλιστα τὸ ὤκιμον. ἅπαντα δὲ ξηρανθέντα πολυκαρπότερα γίνεται, δι' ὃ> καὶ προαφαιροῦντες αὐτὰ ξηραίνουσιν. ἅπαντα δὲ πολύχοα καὶ πολυσπέρματα, πολυκαρπότατον δὲ τὸ ὤκιμον.

Ἔστι δὲ τὰ μὲν ἀκρόκαρπα, καθάπερ ὤκιμον πράσον κρόμυον· τὰ δὲ πλαγιόκαρπα μᾶλλον, οἷον ῥαφανὶς γογγυλὶς καὶ τὰ τοιαῦτα· τὰ δ' ἀμφοτέρως, οἷον βλίτον ἀδράφαξυς· ἀμφότερα γὰρ ταῦτα καὶ ἐκ τοῦ πλαγίου, καὶ τό γε βλίτον εὐθὺς παρ' ἕκαστον ὄζον προσκαθήμενον ἔχει τὸ σπέρμα βοτρυῶδες. τὰ δ' ἐκ παλαιοτέρων σπερμάτων θᾶττον ἐκκαυλεῖ, τάχιστα δὲ τὰ ἐκ τῶν ἀκμαζόντων· ἔστι γάρ τις ἀκμὴ καὶ τούτων. ἀνὰ λόγον δὲ καὶ τὸ κάλλος ἀκολουθεῖ τῶν ... ἐὰν τὰ ἄλλα τὴν αὐτὴν ἔχωσι θεραπείαν.

Δοκεῖ δὲ καὶ εἰς τὸ αὐτὸ ἀθρόα θεμένων καλλίω γίνεσθαι καὶ βλαστάνειν· οὕτω γὰρ τὸ τοῦ πράσου καὶ τὸ τοῦ σελίνου τιθέασιν ἀποδήσαντες εἰς ὀθόνιον καὶ γίνονται μεγάλα.

5 Συμβάλλεται δέ τι καὶ ὁ τόπος πρὸς αὔξησιν· κελεύουσι γοῦν, ὅταν τις μεταφυτεύῃ τὰ σέλινα, πάτταλον κατακρούειν ἡλίκον ἂν βούληται ποιεῖν τὸ σέλινον· τιθέναι δὲ καὶ ἐν ὀθονίῳ πάτταλον κατακρούσαντα καὶ πλήσαντα κόπρου καὶ γῆς.

---

[1] μάλιστα ... δι' ὃ missing in UMAld.Bas.; text as restored by Sch. from Cam., G and Plin. *l.c.*

[2] τό γε βλίτον conj. W.; τό γε πλεῖστον U; τό τε πλεῖστον Ald.H.

[3] ἐκκαυλεῖ: *cf.* 7. 1. 7; 7. 4. 3, and esp. *C.P.* 4. 3. 5.

[4] After ἀκολουθεῖ τῶν follows a lacuna of one and a half lines

it is ripened, quickly dries and is shed, and the herbs whose seed have an integument are harder still to dry, especially basil. All however, when dried, produce more fruit: wherefore [1] it is the custom to gather the seed early and dry it. All of them are prolific and produce many seeds, but basil produces most of all.

Examples of those which produce their fruit at the top of the stem are basil leek onion: of those which produce it rather at the sides, radish turnip and the like; of those which produce it in both ways, blite and orach; both of these produce it at the side as well as at the top; in fact blite [2] has its seed in clusters, closely attached to each branch. Some push up their shoots [3] fairly soon from old seed, but seed from plants in their prime is the most rapid; for these plants too have a time when they are at their best. The beauty of the plant also corresponds [4] in proportion, provided that equal care in [5] other respects is shewn in cultivation.

[6] It likewise appears that, if a quantity of seed is sown in the same place, the resulting crop comes up and germinates better; thus they tie up seed of leek and celery in a piece of cloth [7] before sowing, and then there is a large [8] crop.

The position also contributes to growth; at least, when celery is transplanted, they suggest that one should hammer [9] in a peg of whatever size one wishes to make the celery; and also that one should sow the seed in a piece of cloth [10] after hammering in a peg and filling the hole with dung and soil.

in UMAld.; text as given by Cam., which however omits τῶν; τῶν σπειρομένων H.; τῶν τοιούτων Vo.Vin.
[5] *cf.* 7. 4. 7.   [6] Plin. 19. 120.   [7] *cf. C.P.* 5. 6. 9.
[8] μεγάλα conj. St.; μεγάλαι Ald.H.
[9] Made clearer *C.P.* 5. 6. 7.   [10] *cf. C.P.* 5. 6. 9.

THEOPHRASTUS

Ἔνια δὲ καὶ τοῖς σχήμασιν ἐξομοιοῦται καὶ τοῖς τόποις· ἡ γὰρ σικύα ὁμοιοσχήμων γίνεται ἐν ᾧ ἂν τεθῇ ἀγγείῳ.

Καὶ διαφορὰν λαμβάνει κατὰ τοὺς χυμοὺς ἔνια προθεραπευθέντα τῶν σπερμάτων, οἷον τὸ τοῦ σικύου ἐὰν ἐν γάλακτι βρέξαντες σπείρωσιν. ἀλλὰ τὰ μὲν τοιαῦτα ἴσως οἰκειότερα τῆς θεραπείας.

IV. Γένη δὲ τῶν μέν ἐστι πλείω τῶν δ' οὐκ ἔστιν, οἷον ὠκίμου λαπάθου βλίτου καρδάμου εὐζώμου ἀδραφάξυος κοριάννου ἀνήθου πηγάνου· τούτων γὰρ οὔ φασιν εἶναι ⟨γένους διαφοράν.⟩ τῶν δὲ ἔστι, ῥαφανῖδος ῥαφάνου τευτλίου σικύου κολοκύντης κυμίνου σκορόδου θριδακίνης. διαιροῦσι δὲ τοῖς τε φύλλοις καὶ ταῖς ῥίζαις καὶ τοῖς χρώμασι καὶ τοῖς χυλοῖς καὶ τοῖς ἄλλοις τοῖς τοιούτοις.

2 Οἷον τῆς ῥαφανῖδος ⟨γένη Κορινθίαν Κλεωναίαν Λειοθασίαν⟩ ἀμωρέαν Βοιωτίαν· εὐαυξεστάτην δὲ τὴν Κορινθίαν, ἣ καὶ τὴν ῥίζαν ἔχει γυμνήν· ὠθεῖται γὰρ εἰς τὸ ἄνω καὶ οὐχ ὡς αἱ ἄλλαι κάτω. τὴν δὲ Λειοθασίαν, ἣν ἔνιοι καλοῦσι Θρᾳκίαν,

---

[1] καὶ τοῖς τόποις Ald ; κατὰ τοὺς τόπους conj W. *cf. C.P.* 5. 6. 7.
[2] ἀγγείῳ ... λαμβάνει om. UMPAld.; διαφορὰν δὲ καὶ Cam.; τόπῳ· διαφέρειν δὲ καὶ H.; ἀγγείῳ conj. W. from *C.P.* 5. 6. 7; καὶ διαφορὰν conj. Sch. *cf. Geop.* 12. 19. 6.
[3] *cf.* 7. 1. 6 ; *Geop.* 12. 20. 3.
After εἶναι there is a lacuna in UMAld.; Cam. supplies γένους διαφοράν· τῶν δὲ ἀνάπαλιν πλείω γένη ; H. has πλείω γένη

Some things again come to resemble in their shape even the position[1] in which they grow: thus the bottle-gourd becomes like in shape to the vessel[2] in which it has been placed.

Moreover differences in taste are acquired in some cases when the seed has been treated specially beforehand; for instance, the seed of the cucumber produces a fruit with different taste if it is soaked[3] in milk before sowing. But such matters belong perhaps more properly to the subject of cultivation.

Of the various forms of some pot-herbs.

IV. Of some herbs there are several kinds, but of others only one, as basil monk's rhubarb blite cress rocket orach coriander dill rue; of each of these they say that there is[4] but one kind. But of others there is more than one, as radish cabbage beet cucumber gourd cummin garlic lettuce. Differences are marked in the leaves, the root, the colour, the taste, and so forth.

Thus of radish they recognise these various kinds[5] —the Corinthian, that of Cleonae, the Leiothasian, *amorea*, the Boeotian. The Corinthian is said to be the strongest in growth, and it has an exposed root; for it pushes upwards, and not downwards like the others. The Leiothasian[6] is called by some the

οὐδὲ γένους διαφοράν· τῶν δὲ ἀνάπαλιν πλείω γένη; Plin. 19. 123 rather supports H. ? read as in H.: τῶν δὲ ἐστι is perhaps an attempt to fill the lacuna.

[5] *cf.* Plin. 19. 75 and 76, who gives a kind called *viride* in place of T.'s ἀμωρέα: see below. After ῥαφανῖδος there is a lacuna in UMAld. (but U has τὴν δὲ μόραν Βοιωτίαν). Text restored from Athen. 2. 48 (*cf.* Plin. *l.c.*). Cam.H.Bas. (also Vo.Vin.(?)) give substantially the same.

[6] The name suggests Thasos, off the Thracian coast.

ἰσχυροτάτην πρὸς τοὺς χειμῶνας. τὴν δὲ Βοι
ωτίαν γλυκυτάτην καὶ τῷ σχήματι στρογγύλην,
οὐχ ὥσπερ τὴν Κλεωναίαν μακράν. ὅσων δ' ἂν ᾖ
λεῖα τὰ φύλλα, γλυκύτεραι καὶ ἡδίους, ὅσων δ' ἂν
τραχέα, δριμύτεραι. γένος δέ τι παρὰ ταῦτα
ἔστιν ὃ ἔχει τὸ φύλλον εὐζώμῳ ὅμοιον. ῥαφανῖδος μὲν οὖν ταῦτα.

3 Γογγυλίδος δὲ οἱ μέν φασιν εἶναι οἱ δ' οὔ
φασιν, ἀλλὰ τῷ ἄρρενι καὶ τῇ θηλείᾳ διαφέρειν,
γίνεσθαι δὲ ἐκ τοῦ αὐτοῦ σπέρματος ἄμφω.
πρὸς δὲ τὸ ἀποθηλύνεσθαι πηγνύναι δεῖν μανάς·
ἐὰν γὰρ πυκνάς, πάσας ἀπαρρενοῦσθαι, τὸν αὐτὸν
δὲ τρόπον κἂν ἐν γῇ μοχθηρᾷ σπαρῶσι· δι' ὃ καὶ
πρὸς σπερματισμὸν μεταφέροντες φυτεύουσι τὰς
ἐκφύσεις καὶ πλατείας. ἔστι δὲ καὶ τὸ σπέρμα
τῇ ὄψει τὸ χεῖρον καὶ βέλτιον φανερόν· τῆς μὲν
γὰρ χρηστῆς λεπτὸν τῆς δὲ μοχθηρᾶς ἁδρόν.
χειμαζομένη δὲ χαίρει καὶ αὕτη καὶ ἡ ῥαφανίς·
οἴονται γὰρ ἅμα γλυκαίνεσθαί τε καὶ τὴν αὔξησιν
εἰς τὴν ῥίζαν τρέπεσθαι καὶ οὐκ εἰς τὰ φύλλα.
τοῖς δὲ νοτίοις καὶ ταῖς εὐδίαις ἐκκαυλεῖ ταχύ.
τοῦτο μὲν οὖν λόγου δεῖται τῆς ὁμοιώσεως ἐν
ἀμφοῖν εἶναι τὰς διαφοράς.

---

[1] Diosc. 2. 112 mentions a kind called by the Romans ἀρμοράκιον. Plin. 19. 82 has *armoracia* and says that this was called *armon* in Pontus; Sch. suggests that the latter name may have given rise to both *armoracia* and ἀμωρέα.

[2] Plin. 18. 129, *cf.* 19. 75; Athen. 9. 7.

[3] πηγνύναι. The verb is used of planting seeds singly; *cf.* 6. 6. 9; 7. 1. 2; 7. 5. 3.

Thracian radish, and it stands the winter best. The Boeotian is said to be the sweetest and to be round in shape, not of a long shape like that of Cleonae. Those kinds whose leaves are smooth are sweeter and pleasanter to the taste, those whose leaves are rough have a somewhat sharp taste. Besides the above-mentioned kinds[1] there is yet another, whose leaves resemble those of rocket. These then are the different kinds of radish.

Of the turnip[2] all do not agree that there are several kinds, but some say that the only difference is between the 'male' and the 'female,' and that both forms come from the same seed. In order to produce 'female' plants it is said that the seed should be sown[3] thinly, for that, if it is sown thick, the result is all 'male' plants; and that the same result follows if the seed is sown in poor soil. Wherefore, when they are shifting plants for seeding,[4] they plant the seedlings[5] wide apart.[6] Good and inferior seed can be easily distinguished by their appearance; the seed of a good plant is fine, that of a poor one coarse. Both this plant and radish like exposure to winter; for it is supposed that this makes them sweeter and that they are thus made to grow roots rather than leaves. With a south wind and warm weather they run up quickly. It needs explanation that both plants should thus adapt[7] themselves in special ways.

[4] πρὸς σπερματισμὸν conj. W.; τοὺς σπερματισμοὺς Ald H. cf. 7. 5. 3.   [5] ἐκφύσεις : cf. 3. 3. 7.

[6] καὶ πλατείας corrupt. διεστηκυίας (W.) gives the required sense; but there may be a loss of some words, πλατείας indicating that the object is to produce *broader* plants. cf. C.P. 5. 6. 9 and Sch.'s note.

[7] τῆς ὁμοιώσεως probably corrupt: no correction suggests itself.

4 Τῆς δὲ ῥαφάνου τριχῇ διαιρουμένης, οὐλοφύλλου τε καὶ λειοφύλλου καὶ τρίτης τῆς ἀγρίας, <ἡ ἀγρία> τὸ μὲν φύλλον ἔχει λεῖον μικρὸν δὲ καὶ περιφερές, πολύκλαδος καὶ πολύφυλλος, ἔτι δὲ χυλὸν ἔχουσα δριμὺν καὶ φαρμακώδη, δι' ὃ καὶ πρὸς τὰς κοιλίας αὐτῷ χρῶνται οἱ ἰατροί. ὁμοίως δὲ καὶ ἐν ἐκείναις δοκοῦσι διαφοραὶ καθ' ἑκατέραν· ἐπεὶ ἄσπερμόν τι γένος αὐτῶν ἐστιν ἢ κακόσπερμον. τὸ δ' ὅλον ἡ οὔλη τῆς λείας εὐχυλοτέρα καὶ μεγαλοφυλλοτέρα.

Εὐχυλότερον δὲ καὶ τῶν τευτλίων τὸ λευκὸν τοῦ μέλανος καὶ ὀλιγοσπερμότερον, ὃ καλοῦσί τινες Σικελικόν.

5 Ὡσαύτως δὲ καὶ τῆς θριδακίνης· ἡ γὰρ λευκὴ γλυκυτέρα καὶ ἁπαλωτέρα. γένη δὲ αὐτῆς ἐστιν ἄλλα τρία, τό τε πλατύκαυλον καὶ στρογγυλόκαυλον καὶ τρίτον τὸ Λακωνικόν· αὕτη δὲ τὸ μὲν φύλλον ἔχει σκολυμῶδες, ὀρθὴ δὲ καὶ εὐαυξὴς καὶ ἀπαράβλαστος ἐκ τοῦ καυλοῦ. τῶν δὲ πλατειῶν οὕτω τινὲς πλατύκαυλοι γίνονται ὥστ' ἐνίους φασὶ καὶ θύραις χρῆσθαι κηπουρικαῖς. τὸ δὲ ὀπῶδες σφόδρα καὶ μικρόφυλλον καὶ λευκοκαυλότερον ἔοικεν ἀγρία.

6 Τῶν δὲ σελίνων καὶ ἐν τοῖς φύλλοις καὶ ἐν τοῖς καυλοῖς αἱ διαφοραί· τὸ μὲν γὰρ πυκνὸν καὶ οὖλον καὶ δασὺ τὸ φύλλον ἔχει, τὸ δὲ μανότερον καὶ πλατύτερον καυλὸν δὲ μείζω. τούτων δὲ πάλιν τὰ μὲν λευκόκαυλα τὰ δὲ πορφυρόκαυλα ἢ ποικιλόκαυλα· τὸ δ' ὅλον ἅπαν τὸ τοιοῦτον ἐμφερέστερον τῷ ἀγρίῳ.

---

[1] Athen. 9. 9; Plin. 19. 80. [2] Wild radish. See Index.

ENQUIRY INTO PLANTS, VII. iv. 4–6

[1] Of cabbage three kinds are distinguished, the curly-leaved, the smooth-leaved, and thirdly, the wild form.[2] The wild form [3] has a small round leaf, it has many branches and many leaves, and further a sharp medicinal taste ; wherefore physicians use it for the stomach. Between the other two kinds [4] there seem also to be differences, inasmuch as one of them bears no seed or only inferior seed. In general the curly-leaved kind has a better flavour than the smooth and it has larger leaves.

[5] So too with beet; the white kind has a better flavour than the black and produces fewer seeds ; some call it ' Sicilian ' beet.

So too with lettuce; the white kind is sweeter and tenderer. Of this plant there are three other kinds,[6] the flat-stalked, the round-stalked, and the Laconian; the last-named has a leaf like the golden thistle,[7] but is erect and strong-growing and has no side-shoots [8] from the main stem. Of the ' flat ' kinds some have such flat stalks that some, they say, use them to make a garden trellis.[9] The third kind, which has much milky juice and small leaves and a whiter stem, is like a wild plant.

[10] In celery the differences between the various kinds lie in the leaves and stem ; one kind is close and curly and has rough leaves, the other is more open in growth and flatter, but has a larger stalk. Again there are kinds with stems white, red or particoloured ; and in general all such forms resemble more the wild kind.

[3] ἡ ἀγρία add. W.
[4] ἐκείναις conj. Sch. from Plin. *l.c.*; ἐκείνῳ Ald.H.
[5] Athen. 9. 11 ; Plin. 19. 132.
[6] Plin. 19. 125.   [7] Athen. 2. 79.   [8] *cf.* 7. 2. 4.
[9] *ostiola olitoria* Plin. 19. 125.   [10] Plin. 19. 124.

THEOPHRASTUS

Σικύου δὲ καὶ κολοκύντης τοῦ μὲν εἶναί φασι γένη τῆς δ' οὐκ εἶναι, καθάπερ τῆς ῥαφανῖδος καὶ τῆς γογγυλίδος, ἀλλ' ἐν τῷ αὐτῷ γένει τὰς μὲν βελτίους τὰς δὲ χείρους. τοῦ δὲ σικύου τρία, Λακωνικὸν σκυταλίαν Βοιώτιον· τούτων δὲ ὁ μὲν Λακωνικὸς ὑδρευόμενος βελτίων, οἱ δ' ἕτεροι ἀνύδρευτοι.

7 Διαφέρει δὲ γένει καὶ τὰ κρόμυα καὶ τὰ σκόροδα. πλείω δὲ τοῦ κρομύου τὰ γένη, οἷον τὰ κατὰ τὰς χώρας ἐπικαλούμενα Σάρδια Κνίδια Σαμοθράκια, καὶ πάλιν τὰ σητάνια καὶ σχιστὰ καὶ Ἀσκαλώνια. τούτων δὲ τὰ μὲν σητάνια μικρὰ γλυκέα δὲ εὖ μάλα, τὰ δὲ σχιστὰ καὶ ἀσκαλώνια καὶ ταῖς θεραπείαις διαφέροντα καὶ δῆλον ὅτι τῇ φύσει· τὸ γὰρ σχιστὸν τῷ μὲν χειμῶνι μετὰ τῆς κόμης ἐῶσιν ἀργόν, ἅμα δὲ τῷ ἦρι τὰ φύλλα περιαιροῦσι τὰ ἔξω καὶ τὰ ἄλλα θεραπεύουσι· περιαιρεθέντων δὲ τῶν φύλλων ἕτερα βλαστάνει καὶ ἅμα κάτω σχίζεται, δι' ὃ καλοῦσι σχιστά. οἱ δὲ καὶ ὅλως φασὶ πάντων δεῖν, ὅπως ἡ δύναμις εἰς τὸ κάτω καὶ μὴ σπερ-
8 μοφυῇ. τῶν δὲ Ἀσκαλωνίων ἰδία τις ἡ φύσις· μόνα γὰρ <οὐ> σχιστὰ καὶ ὥσπερ ἄγονα ἀπὸ τῆς ῥίζης, ἔτι δὲ ἐν αὐτοῖς ἀναυξῆ καὶ ἀνεπίδοτα·

---

[1] Athen. 3. 4; Plin. 19. 68.
[2] Plin. 19. 101–104.
[3] Σάρδια conj. Meurs. from Plin. *l.c.*; γάρδια Ald. H.
[4] *i.e.* making offsets.
[5] Ἀσκαλώνια, whence Eng. shallot; though this name is applied to κ. σχιστόν. [6] τὸ add. W.

As to cucumber and gourd, it is said that there are various forms of the former, but of the latter, just as in radish and turnip, the differences are only between better and inferior individuals. [1] Of the cucumber there are three forms, the Laconian the cudgel-shaped and the Boeotian. Of these the Laconian is better with moisture, the others without it.

[2] There are also various kinds of onion and of garlic; those of the onion are the more numerous, for instance, those called after their localities Sardian,[3] Cnidian, Samothracian; and again the 'annual' the 'divided'[4] (shallot) and that of Ascalon.[5] Of these the annual kind is small but very sweet, while the divided and the Ascalonian differ plainly as to their character as well as in respect of their cultivation. For the 'divided'[6] kind they leave untended in winter with its foliage,[7] but in spring they strip off[8] the outside leaves and tend the plant in other ways; when the leaves are stripped off, others grow, and at the same time division takes place under ground, which is the reason of the name 'divided.'[9] Some indeed say that all kinds should be thus treated, in order that the force of the plant may be directed downwards and it may not go to seed. The Ascalonian kind has a somewhat peculiar character; it is the only kind which does not[10] divide and which does not, as it were, reproduce itself from the root; moreover in the plant[11] itself there is no power of increasing and multiplying; wherefore

κόμης ἐῶσιν conj. Scal.; κοιμησέως UMP₂Ald.
περιαιροῦσι conj Scal. from Plin. *l.c.* and G; περιάγουσι P₂Ald. H.   [9] *cf.* Pall. 3. 24. 3.
[10] οὐ add. Scal.   [11] *i.e.* the part above ground.

δι' ὃ καὶ οὐ πηγνύουσιν ἀλλὰ σπείρουσιν αὐτὰ καὶ σπείρουσιν ὀψὲ πρὸς τὸ ἔαρ, εἶθ' ὅταν βλαστήσῃ μεταφυτεύουσι· τελειοῦται δὲ οὕτω ταχέως ὥσθ' ἅμα τοῖς ἄλλοις ἢ καὶ πρότερον ἐξαιρεῖσθαι· πλέονα δὲ χρόνον ἐαθέντα ἐν τῇ γῇ σήπεται· φυτευθέντα δὲ καυλὸν ἀφίησι καὶ σπέρμα φύει μόνον, εἶτα κενοῦται καὶ αὐαίνεται. τούτων μὲν οὖν τοιαύτη τις ἡ φύσις.

9 Διαφέρει δ' ἔνια καὶ τοῖς χρώμασιν· ἐν Ἰσσῷ γὰρ τὰ μὲν ἄλλα ὅμοια τοῖς ἄλλοις, λευκὰ δὲ σφόδρα τῇ χροιᾷ· φέρειν δέ φασιν ὅμοια τοῖς Σαρδιανοῖς. ἰδιωτάτη δὲ ἡ φύσις ἡ τῶν Κρητικῶν, παραπλησία δὲ τρόπον τινὰ τοῖς Ἀσκαλωνίοις, εἰ μὴ ἄρα καὶ ἡ αὐτή. ἐν Κρήτῃ γάρ ἐστί τι γένος ὃ σπειρόμενον μὲν ῥίζαν ποιεῖ φυτευόμενον δὲ καυλὸν καὶ σπέρμα, κεφαλὴν δὲ οὐκ ἴσχει, γλυκὺ δὲ τῷ χυμῷ· τοῦτο γὰρ οἷον ἀνάπαλιν ἔχει τοῖς 10 ἄλλοις. ἅπαντα γὰρ πηγνύμενα καὶ βελτίω καὶ θᾶττον παραγίνεται. πάντα δὲ φυτεύεται μετ' Ἀρκτοῦρον ἔτι θερμῆς οὔσης τῆς γῆς, ὅπως τὰ ὕδατα πεφυτευμένα καταλαμβάνῃ. καὶ ὅλα δὲ φυτεύεται καὶ διατεμνόμενα παρὰ τὴν κεφαλήν. οὐχ ὅμοιαι δὲ αἱ ἐκβλαστήσεις, ἀλλ' ἐκ μὲν τοῦ κάτω γίνεται κρόμυον, ἐκ δὲ τοῦ ἄνω

---

[1] πηγνύουσι: cf. 7. 4. 3 n. The word evidently has a different sense here; cf. § 10, where πηγνύω and φυτεύω seem to be synonymous.

[2] οὕτω conj. Sch. from G; τοῖς ἄλλοις Ald.

[3] i.e. instead of being raised from seed. cf. what is said 7. 2. 2 of the offsets of γήθυον.

ENQUIRY INTO PLANTS, VII. iv. 8–10

many do not plant[1] these, but raise them from seed; and the sowing is made late, towards the spring; and then, when the seed has germinated, they transplant. And the plant arrives at maturity so[2] fast that it is taken up with the others or even earlier; whereas, if it is left a longer time in the ground, it rots. If planted on the other hand,[3] it sends up a stem and merely produces seed, and then shrivels up[4] and withers. Such then is the character of these.

Some also shew differences in colour; thus at Issus[5] are found plants which in other respects resemble the others,[6] but which are extremely white in colour; and they bear, it is said, onions like those of Sardis. Most distinct however is the character of the Cretan kind, which resembles to some extent that of Ascalon, if indeed it be not the same. For in Crete there is a kind which when sown produces a root, but when planted produces a stem and seed but has no 'head';[7] and it is sweet in flavour. This kind in fact has just the contrary character to the others; for they all grow better and faster when they are planted. All are planted[8] after the rising of Arcturus while the earth is still warm, so that the rains may come upon them after planting. They are planted[9] either entire or else in sections made by cutting at the 'head.' The growth which results is not uniform; from the lower part comes an onion,

[4] κενοῦται conj. St. from G *exinaniuntur*; καινοῦται Ald.

[5] Ἴσσῳ conj. Sch. from G and Plin. *l.c.*; Ἴσῳ UM; νήσῳ Ald.H.

[6] ἄλλοις conj. Sch; λευκοῖς Ald.   [7] Sc. bulb.

[8] φυτεύεται conj. Sch; φύεται Ald. See next note.

[9] φυτεύεται M; φύεται Ald. cf. *C.P.* 1. 4. 5.

χλόη μόνον· ὀρθὸν δὲ διατμηθὲν ὅλως ἀβλαστές ἐστι. τὸ δὲ γήτειον καλούμενον ἀκέφαλόν τι καὶ ὥσπερ αὐχένα μακρὸν ἔχον, ὅθεν καὶ ἡ βλάστησις ἄκρα· καὶ ἐπικείρεται πολλάκις, ὥσπερ τὸ πράσον, δι' ὃ καὶ σπείρουσιν αὐτὸ καὶ οὐ φυτεύουσι. τὰ μὲν οὖν κρόμυα σχεδὸν ταύτας ἔχει τὰς ἰδέας.

11 Τὸ δὲ σκόροδον φυτεύεται μὲν μικρὸν πρὸ τροπῶν ἢ μετὰ τροπὰς διαιρούμενον κατὰ γέλγεις. διαφορὰ δέ ἐστιν αὐτῶν ἥ τε τῶν ὀψίων πρὸς τὰ πρώϊα· γένος γάρ τι τυγχάνει τοιοῦτον ὃ ἐν ἑξήκοντα ἡμέραις τελειοῦται, καὶ μεγέθει καὶ μικρότητι. καὶ τῷ μεγέθει γένος τι διάφορόν ἐστι, μάλιστα δὲ τὸ Κύπριον καλούμενον τοιοῦτον, ὅπερ οὐχ ἑψοῦσιν ἀλλὰ πρὸς τοὺς μυττωτοὺς χρῶνται, καὶ ἐν τῇ τρίψει θαυμαστὸν ποιεῖ τὸν ὄγκον ἐκπνευματούμενον. καὶ ἔτι τῷ μὴ ἔχειν ἔνια τὰς γέλγεις. ἡ δὲ γλυκύτης καὶ ἡ εὐωδία καὶ ἡ ἁδρότης σχεδὸν παρὰ τὰς χώρας γίνεται καὶ τὰς θεραπείας, ὥσπερ καὶ τῶν ἄλλων. τελειοῦται δὲ καὶ ἀπὸ σπέρματος ἀλλὰ βραδέως· τῷ πρώτῳ γὰρ ἔτει κεφαλὴν ἡλίκην πράσου λαμβάνει, τῷ δ' ὕστερον γελγιδοῦται, καὶ τῷ τρίτῳ τέλειον γίνεται, καὶ οὐδὲν χεῖρον ἀλλ' ἔνιοί γε
12 καὶ κάλλιόν φασι τοῦ πηκτοῦ. τῆς δὲ ῥίζης ἡ γένεσις οὐχ ὁμοία τοῦ τε σκορόδου καὶ τοῦ κρομύου· ἀλλὰ τοῦ μὲν σκορόδου ὅταν ἀνοιδήσῃ ἡ γελγὶς κυρτοῦται πᾶσα καὶ ἐνταῦθα αὐξηθεῖσα διαιρεῖται πάλιν εἰς τὰς γέλγεις καὶ ἐξ ἑνὸς πολλὰ γίνεται τῷ τελειοῦσθαι τὴν κεφαλήν, τὸ δὲ

---

[1] *i.e.* bulb; *cf.* 9. 11. 6.   [2] *cf.* 7. 2. 2.
[3] Plin. 19. 111 and 112.

ENQUIRY INTO PLANTS, VII. iv. 10–12

from the upper only foliage; while, if the plant is divided vertically, no growth at all takes place. The kind called horn-onion has no 'head,'[1] but has as it were a long neck, at the top of which comes the new growth; it is often cut, like the leek; wherefore it is raised from seed and not planted.[2] Such then, one may say, are the forms of the onion.

[3] Garlic is planted a little before or after the solstice, when it divides into cloves.[4] There are different kinds distinguished as late or early, for there is one kind which matures in[5] sixty days. There are also differences as to size. There is one kind which excels in size, especially that variety which is called Cyprian, which is not cooked but used for salads, and, when it is pounded up, it increases wondrously in bulk, making a foaming dressing. There is a further difference, in that some kinds cannot be divided into cloves. The sweetness of taste and smell and the vigour depend on the position[6] and on cultivation, as with other herbs. Garlic reaches maturity from seed, but slowly, for in the first year it acquires a 'head' which is only as large as that of the leek, but in the next year it divides into cloves, and in the third is fully grown, and is not inferior, indeed some say it is superior, to the garlic which has been planted.[7] The growth of the root in garlic and onion is not the same; in garlic, when the clove has swollen, the whole of it becomes convex[8]; then it increases and divides again into the cloves, and becomes several plants instead of one by the maturing of the 'head,'

---

[4] γέλγεις conj. Scal. from G (*nucleatim divisum*); γένη Ald.
[5] δ ἐν conj. Sch.; ὅθεν UMAld.
[6] χώρας conj. Dalec.; ὥρας UMP₂Ald.
[7] Sc. not raised from seed. [8] So W. renders.

κρόμυον εὐθὺς ἐκ τῆς ῥίζης ἄλλο καὶ ἄλλο παραφίησι, καθάπερ καὶ βολβοὶ καὶ σκίλλα καὶ πάντα τὰ τοιαῦτα. καὶ γὰρ τὰ κρόμυα καὶ τὰ σκόροδα μὴ ἀναιρούντων ἀλλ' ἐώντων πολλὰ γίνεται. φέρειν δέ φασι καὶ τὸ σκόροδον ἐπὶ τῆς φύσιγγος σκόροδα καὶ τὸ κρόμυον κρόμυα· περὶ μὲν οὖν τῶν γενέσεων ἱκανῶς εἰρήσθω.

V. Φίλυδρα δὲ πάντα τὰ ἄλλα λάχανα καὶ φιλόκοπρα πλὴν πηγάνου, τοῦτο δὲ ἥκιστα φιλόκοπρον. τὰ χειμερινὰ δὲ οὐχ ἧττον τῶν θερινῶν καὶ τὰ ἐπίκηρα τῶν ἰσχυρῶν. κόπρον δὲ μάλιστα ἐπαινοῦσι τὴν συρματῖτιν, τὴν δὲ τῶν ὑποζυγίων μοχθηρὰν διὰ τὸ μάλιστα ἐξικμάζεσθαι· ζητοῦσι δὲ τὴν κόπρον ἅμα τῷ σπόρῳ μάλιστα συναναμιχθεῖσαν· οἱ δὲ καὶ σπείροντες ἐπιβάλλουσι· χρῶνται δὲ καὶ τῇ ἀνθρωπίνῃ ὠμῇ πρὸς τὴν χύλωσιν. φιλυδρότερα δὲ τὰ χειμερινὰ τῶν θερινῶν καὶ τὰ ἀσθενῆ τῶν ἰσχυρῶν, ἔτι δὲ τὰ πλείστης δεόμενα τροφῆς. φίλυδρα καὶ τὸ κρόμυον καὶ τὸ γήθυον· καίτοι φασί τινες οὐ ζητεῖν, ἐὰν τὸ πρῶτον ἐπιγένηται δὶς ἢ τρίς.

2 τῶν δὲ ὑδάτων ἄριστα τὰ πότιμα καὶ τὰ ψυχρά, χείριστα δὲ τὰ ἁλυκὰ καὶ δυσμανῆ, δι' ὃ καὶ ἐκ τῶν ὀχετῶν οὐ χρηστά· συμπεριφέρει γὰρ σπέρματα πόας. ἀγαθὰ δὲ τὰ ἐκ διός· ταῦτα

---

[1] cf. 7. 2. 2 and 3.
[2] φύσιγγος conj. Casaub. on Athen. 2. 78; σφύριγγος UM Ald. See LS φύσιγξ.
[3] καὶ τὸ κρόμυον κρόμυα conj. Sch.; καὶ τὰ κρόμμυα UMAld.
[4] Plin. 19. 156.

while the onion puts out another and another growth straight from the root, as do purse-tassels[1] and squill and all such plants. For both onions and garlic multiply if they are not removed but left alone. They say also that garlic produces garlic heads on the stalk,[2] and that the onion in like manner produces onions.[3] Let this suffice for an account of their ways of growth.

Of the cultivation of pot-herbs; manure and water.

V. [4] All the pot-herbs are lovers of water and of dung, except rue, which does not at all like dung; this is true of the winter no less than of the summer herbs, and of the tender no less than of the strong ones. The dung which is most commended is that which is mixed with litter, while that of beasts of burden is held to be bad, because it is most apt to lose its moisture. Dung which is mixed with the seed is most in request, but some cast the manure on while they are sowing, and they also use fresh human dung as a liquid manure.[5] The winter crops like moisture more than the summer ones, and the weak more than the strong, as well as those which specially need feeding. Onion and long onion also love moisture, though some say that they do not require it, if at the outset it has been applied twice or thrice. [6] Fresh cold water is the best, and the worst is that which is brackish and thick:[7] wherefore the water from irrigation ditches is not good, for it brings with it seeds of weeds. Rain

[5] Lit. 'for their liquid-manuring.' *cf. C.P.* 3. 9. 2, where χύλωσις must have the same sense.
[6] Plin. 12. 182 and 183.
[7] δυσμανῆ UMAld.; δυσμενῆ H.

γὰρ δοκεῖ καὶ φθείρειν τὰ θηρία [γινόμενα] τὰ γόνιμα κατεσθίοντα. φασὶ δέ τινες οὔτε τοῖς σικύοις συμφέρειν οὔτε τοῖς κρομύοις. ἀρδεύουσι δὲ τὰ μὲν ἄλλα πρωὶ ἢ πρὸς ἑσπέραν, ὅπως μὴ καθέψηται, τὸ δὲ ὤκιμον καὶ μεσημβρίας· καὶ γὰρ διαβλαστάνειν θᾶττόν φασι θερμῷ τὸ πρῶτον ἀρδευόμενον. τὸ δὲ πολὺ λίαν ὕδωρ δοκεῖ συμφέρειν ἄλλως τε καὶ ἐὰν [μὴ] ἔχῃ κόπρον· πολλάκις γὰρ πεινῆν τὰ λάχανά φασι, καὶ ταῦτα γνωρίζειν τοὺς ἐμπείρους τῶν κηπουρῶν.

3 Μεταφυτευόμενα δὲ πάντα καλλίω καὶ μείζω γίνεται· καὶ γὰρ τὰ τῶν πράσων μεγέθη καὶ τὰ τῶν ῥαφανίδων ἐκ μεταφυτείας. μάλιστα δὲ μεταφυτεύουσι πρὸς τοὺς σπερματισμούς· καὶ τὰ μὲν ἄλλα ὑπομένει, οἷον γήθυον πράσον ῥάφανος σίκυος σέλινον γογγυλὶς θρῖδαξ, <τὰ δὲ> γλίσχρως. ἅπαντα δ' εὐαυξέστερα καὶ μείζω πηγνυμένων τῶν σπερμάτων ἢ σπειρομένων.

4 Θηρία δὲ γίνεται ταῖς μὲν ῥαφανῖσι ψύλλαι, τῇ δὲ ῥαφάνῳ κάμπαι καὶ σκώληκες, καὶ ἐν τῇ θριδακίνῃ καὶ ἐν τοῖς πράσοις καὶ ἐν ἄλλοις δὲ πλείοσιν αἱ πρασοκουρίδες. ταύτας μὲν οὖν ἡ κράστις ἀθροισθεῖσα ἀπόλλυσι καὶ ὅταν κόπρος

---

[1] γινόμενα τὰ γόνιμα H.; γινόμενα γόνιμα UMAld.; ? τὰ τὰ γόνιμα. Either γινόμενα or γόνιμα seems to be due to dittography. For γόνιμα cf. C.P. 1. 15. 1: τὰς γονίμους ἀρχάς.
[2] καθέψηται conj. Sch. after Plin. l.c.; καθάψηται P₂Ald.
[3] ἔχῃ κόπρον conj. Dalec.; μὴ ἔχῃ κ. Ald.; μετέχῃ κόπρου conj. W. cf. 7. 5. 1, χύλωσιν; C.P. 3. 9. 2.
[4] Plin. 19. 183.

water is good, for it also appears to destroy the pests which devour the young plants.[1] Some however say that rain-water is not good for melons nor for onions. Most herbs are watered in early morning or at evening, so that they may not be dried up[2]; but basil is watered even at noon, for it is said that it grows more quickly if it is watered at first with warm water. In general water seems to be extremely beneficial, especially if it is mixed with dung[3]; for, they say, pot-herbs often are hungry, and experienced gardeners can recognise when this is so.

[4] All herbs grow finer and larger if transplanted; for even the size of leeks and radishes depends on transplantation. Transplanting is done especially in view of collecting seed[5]: and, while most herbs bear it well, as long onion leek cabbage cucumber celery turnip lettuce, others bear it less well.[6] All however make better growth and are larger if the seed is planted[7] rather than scattered.

Of the pests which infest pot-herbs.

[8] As for pests,—radish is attacked by spiders,[9] cabbage by caterpillars and grubs, while in lettuce, leek, and many other herbs occur 'leek-cutters.'[10] These are destroyed by collecting green fodder,[11] or when they have been caught somewhere in a mass

---

[5] σπερματισμοὺς conj. Scal.; σπερματικοὺς UMAld. *cf.* 7. 4. 3.

[6] τὰ δὲ γλίσχρως conj. Sch., adding τὰ δὲ; γλίσχρως U; γλίσχροι M; γλίσχρος Ald.; γλήχων conj. Sch. Sch. also conjectures τὰ λίσχρα: see LS. *s.v.*

[7] πηγνυμένων: *cf.* 6. 6. 9; 7. 4. 3.   [8] Plin. 19. 177.

[9] ψύλλαι: *cf.* Arist. *H.A.* 9. 39. 1.

[10] πρασοκουρίδες: ? leaf-maggots. *cf.* Arist. *H.A.* 5. 19. 20; *Geop.* 12. 9.

[11] κράστις conj. R. Const.; κρᾶσις Ald.

ἀθρόα που καταλάβῃ· φιλόκοπρον δ' ὂν τὸ θηρίον ἀναδύεται καὶ ἐνδῦσα κοιμᾶται ἐν τῇ κόπρῳ, δι' ὃ δὴ ῥᾴδιον θηρεύειν· ἄλλως δ' οὐκ ἔστι. ταῖς δὲ ῥαφανῖσι πρὸς τὰς ψύλλας πρόσφορον τὸ ἐπισπείρειν ὀρόβους. πρὸς δὲ τὸ μὴ γίνεσθαι ψύλλας οὔ φασιν εἶναι φάρμακον οὐδέν. ὑπὸ δὲ τὸ ἄστρον ὤκιμον μὲν λευκαίνεται κορίαννον δὲ ἁλμᾷ. τὰ μὲν οὖν συμβαίνοντα διὰ τούτων θεωρητέον.

5 Τῶν δὲ σπερμάτων τὰ μέν ἐστιν ἰσχυρότερα τὰ δὲ ἀσθενέστερα πρὸς διαμονήν· ἰσχυρότερα μὲν οἷον κορίαννον τεύτλιον πράσον κάρδαμον νᾶπυ εὔζωμον θύμβρα, ἁπλῶς τὰ δριμέα πάντα· ἀσθενέστερα δὲ γήθυον, τοῦτο γὰρ οὐκ ἐθέλει μένειν, ἀδράφαξυς ὤκιμον κολοκύντη σίκυος, ἁπλῶς τὰ θερινὰ τῶν χειμερινῶν μᾶλλον. διαμένει δὲ οὐδὲν πλέον τεττάρων ἐτῶν ὥστε ἔτι χρήσιμον εἶναι πρὸς τοὺς σπόρους· ἀλλὰ διένα μὲν βελτίω, τὰ δὲ τριένα οὐδὲν χείρω, τὸ δ' ὑπερτεῖνον ἤδη χεῖρον.

6 Πρὸς δὲ τὴν μαγειρικὴν χρείαν ἐπὶ πλείω δια-

---

[1] κόπρος ἀθρόα που καταλάβῃ Ald.; κόπρον ἀθρόαν πού τις καταβάλῃ conj. W. after S·h·; κόπρον ἀθρόαν conj. Scal.
[2] φιλόκοπρον δ' ὂν τὸ θηρίον ἀναδύεται καὶ ἐνδῦσα conj. W.; φιλόπονον τὸ θηρίον ἀναδέυεται καὶ ἐν αἷς κοιμᾶται UMAld.; φίλυπνον conj. R. Const., but W.'s conj. is confirmed by *Geop. l.c.* The change of gender in ἐνδῦσα is strange.
[3] πρὸς τὰς ψύλλας πρόσφορον τὸ mBas.; ψύλλας πρὸς τὸ Ald. H.; πρὸς τὰς ψύλλας ἀρκεῖ τὸ conj. W.
[4] ψύλλας Ald.; καμπὰς conj. Sch. followed by W.
[5] *cf. Geop.* 12. 7; Pall. 1. 35. 8; Plin. *l.c.*

ENQUIRY INTO PLANTS, VII. v. 4–6

of dung,[1] the pest being fond of dung emerges, and, having entered the heap, remains dormant there[2]; wherefore it is then easy to catch, which otherwise it is not. To protect[3] radishes against spiders[4] it is of use to sow vetch[5] among the crop; to prevent the spiders from being engendered they say that there is no specific. [6] Basil turns pale about the rising of the dog-star, and coriander becomes mildewed.[7] In these instances we may observe the accidents which occur to pot-herbs.

Of the time for which seed of pot-herbs can be kept.

[8] Of seeds some have more vitality than others as to keeping; among the more vigorous ones are coriander beet leek cress mustard rocket savory, and in general[9] those of pungent taste; among the less vigorous are long onion—which will not keep —orach basil gourd cucumber; and in general the summer herbs keep less well than the winter ones. No seed will keep more than four years so as still to be of use for sowing; though it is better in the second year,[10] in some cases it does not deteriorate in three years,[11] but after that time[12] deterioration begins.

However for cooking purposes seed will keep a

---

[6] Plin. 19. 176.

[7] ἀλμᾷ conj. W.; ἄλμαι MAld.; ἀλμαίνεται Vo.Vin.; ἀλμᾶται mBas. *cf.* 8. 10. 1; *C.P.* 6. 10. 5. In all three places W. introduces this word, comparing ψωριᾶν ἐρυσιβᾶν, etc.

[8] Plin. 19. 181.

[9] ἁπλῶς conj. St. from G; ἄλλως Ald.; ἅλως U.

[10] διένα conj. Scal.; δι' ἕνα UMAld.H.

[11] τὰ δὲ τριένα conj.W.; διὰ δὲ τρεῖς UMAld.H.

[12] ὑπερτείνον conj. Scal.: *cf.* 8. 11. 5; ὑπὲρ γαῖον UMAld.; ὑπερβαῖνον H.

μένει, πλὴν ἀσθενέστερα ταῦτα ἀναγκαῖον εἶναι διὰ τὴν ἀναπνοὴν καὶ τὴν σκωλήκωσιν. φθορὰ δὲ μάλιστα μὲν ὑπὸ τῶν θηρίων· γίγνεται γὰρ ἐν ἅπασι καὶ τοῖς δριμέσιν, ἥκιστα δὲ ἐν τῷ σικυωνί· οὐ μὴν ἀλλὰ καὶ ἐξικμαζόμενα πικρὰ γίνεται τῇ γεύσει, δι' ὃ καὶ πρὸς τὴν χρείαν χείρω. καὶ περὶ μὲν τῶν σπερμάτων καὶ ἁπλῶς τῶν κηπευομένων ἱκανῶς εἰρήσθω.

VI. Περὶ δὲ τῶν ἀγρίων καὶ τῶν καλουμένων ἀρουραίων πειρατέον ὁμοίως εἰπεῖν. τυγχάνει δὲ τὰ μὲν ὁμώνυμα τοῖς ἡμέροις· ἅπαντα γάρ ἐστι τὰ γένη ταῦτα καὶ ἄγρια, καὶ σχεδὸν τά γε πολλὰ παραπλησίαν ἔχοντα τὴν ὄψιν τοῖς ἡμέροις, πλὴν τοῖς γε φύλλοις ἐλάττω ταῦτα καὶ τραχύτερα καὶ τοῖς καυλοῖς καὶ μάλιστα τοῖς χυλοῖς δριμύτερα καὶ ἰσχυρότερα, καθάπερ ἥ τε θύμβρα καὶ ἡ ὀρίγανος ἥ τε ῥάφανος καὶ τὸ πήγανον· ἐπεὶ καὶ τὸ λάπαθον ἄγριον, καίπερ εὐστομώτερον τοῦ ἡμέρου ὄν, τὸν χυλὸν ὅμως ὀξύτερον ἔχει καὶ τούτῳ μάλιστα διαφέρει. πάντα δὲ καὶ ξηρότερα τῶν ἡμέρων, καὶ ἴσως αὐτῷ τούτῳ τά γε πολλὰ καὶ δριμύτερα καὶ ἰσχυρότερα.

2 Ἰδίως δὲ ἡ ῥάφανος ἔχει παρὰ τὰ ἄλλα τοὺς καυλοὺς περιφερεστέρους καὶ λειοτέρους τῆς ἡμέ-

---

[1] i.e. drying-up; cf. Plat. *Tim.* 85 A.
[2] σκωλήκωσιν conj. Sch.; κώλυσιν Ald.; σκωλήκησιν conj. R. Const.
[3] σικυῶνι Ald.: perhaps here a general term for cucumbers, gourds, etc.; σικυῶν M; σικύῳ conj. W.
[4] Plin. 19. 185.

longer time, except that such seed must necessarily become less vigorous by reason of 'evaporation'[1] and destruction by worms.[2] The chief cause of loss is vermin; for vermin occur in all the seeds, even those which are pungent, though least in the gourd[3] tribe; such seeds however, as they lose their moisture, become bitter in taste and inferior for use. Let this suffice for an account of the seeds and in general of herbs cultivated in gardens.

Of uncultivated herbs: the wild forms of pot-herbs.

VI. [4] We must now endeavour to speak in the same way of the wild kinds and of those which are called uncultivated herbs. Some of these have the same names as the cultivated[5] kinds; for all these kinds exist also in a wild form, and most of them resemble the cultivated kinds in appearance, except that in the wild forms the leaves and also the stalks are smaller and rougher, and in particular these forms are more pungent and stronger in taste, for instance, savory[6] marjoram cabbage and rue; the wild monk's rhubarb (dock) indeed, though it has a pleasanter taste than the cultivated, yet has[7] a sharper flavour; and this is the chief difference. Moreover all the wild kinds are less juicy than the cultivated, and perhaps this is the very reason why most[8] of them are more pungent and stronger.

A peculiarity of 'wild cabbage' as compared with the others is that its stems are rounder and smoother

---

[5] ἡμέροις conj. Sch.; εἰρημένοις Ald. The correction would seem unnecessary but that Ald. gives εἰρημένοις in §4 where ἡμέρους is required.   [6] *cf.* Diosc. 3. 37.

[7] ὄν, τὸν I conj.; τὸν δὲ MSS.W.

[8] γε conj. Sch.; τε UMAld.

ρου, καὶ τὴν τοῦ φύλλου πρόσθεσιν ἐκείνη μὲν ἔχει πλατεῖαν αὕτη δὲ περιφερεστέραν, καὶ αὐτὸ δὲ τὸ φύλλον ἀγωνότερον· ἐπεὶ τά γε ἄλλα τραχύτερα¹ καὶ τοῖς καυλοῖς καὶ τοῖς φύλλοις.

Ἡ δὲ γογγυλὶς καὶ τὴν ῥίζαν ἔχει μακρὰν καὶ ῥαφανιδώδη καὶ τὸν καυλὸν βραχύν.

Θριδακίνη² δὲ τό τε φύλλον βραχύτερον τῆς ἡμέρου, καὶ τελεουμένης³ ἀκανθοῦται, καὶ τὸν καυλὸν ὁμοίως, τὸν ὀπὸν δὲ δριμὺν καὶ φαρμακώδη. φύεται δ' ἐν ταῖς ἀρούραις· ὀπίζουσι δ' αὐτὴν ὑπὸ πυραμητόν, καί φασι καθαίρειν ὕδρωπα καὶ ἀχλὺν ἀπ' ὀφθαλμῶν ἀπάγειν καὶ ἄργεμα ἀφαιρεῖν ἐν γάλακτι γυναικείῳ.

3 Τὸ δ' ἱπποσέλινον καὶ ἑλειοσέλινον καὶ ὀρεοσέλινον καὶ πρὸς ἑαυτὰ διαφορὰν ἔχει καὶ πρὸς τὸ ἥμερον· τὸ μὲν γὰρ ἑλειοσέλινον τὸ παρὰ τοὺς ὀχετοὺς καὶ ἐν τοῖς ἕλεσι φυόμενον μανόφυλλόν τε καὶ οὐ δασὺ γίνεται, προσεμφερὲς δέ πως τῷ σελίνῳ καὶ τῇ ὀσμῇ καὶ τῷ χυλῷ καὶ τῷ σχήματι. τὸ δ' ἱπποσέλινον φύλλον μὲν ἐμφερὲς τῷ ἑλειοσελίνῳ, δασὺ δὲ καὶ μεγαλόκαυλον καὶ τὴν ῥίζαν ὥσπερ ῥαφανὶς ἔχει τὸ πάχος μέλαιναν· μέλας δὲ καὶ ὁ καρπός, μέγεθος δὲ μεῖζον ὀρόβου. χρήσιμα δ' ἄμφω φασὶ πρὸς στραγγουρίαν εἶναι ἐν οἴνῳ γλυκεῖ λευκῷ καὶ τοῖς λιθιῶσι· φύεται δὲ

---

¹ τραχύτερα conj. Sch.; τραχύτερον Ald., which contradicts what has just been said.
² Plin. 20. 20 ; Diosc. 2. 110.
³ τελεουμένης conj. W.; τελεούμενος U; τελειούμενον P₂Ald. cf. C.P. 4. 3. 5.

ENQUIRY INTO PLANTS, VII. vi. 2–3

than in the cultivated kind, and, while in the latter the attachment of the leaf is flat, in the wild kind it is rounder, and the leaf itself has less angles; in other cases the wild form is the rougher[1] both in stem and leaf.

[2] The wild turnip has a long root, like that of the radish, and a short stem.

The wild lettuce has a shorter leaf than the cultivated kind, and, as the plant matures,[3] it becomes spinous; the stem is also shorter, while the juice is pungent and medicinal. It grows in fields; they extract its juice at the time of wheat-harvest, and it is said that it purges away dropsy and takes away dimness of sight and removes ulcers[4] on the eye; for which purpose it is administered in human milk.

[5] 'Horse-celery' (alexanders) 'marsh-celery' and 'mountain-celery' (parsley) differ both from one another and from the cultivated kind; 'marsh-celery,' which grows by irrigation-ditches and in marshes, has scanty leaves,[6] and is not of close habit, [7]yet it somewhat resembles the cultivated kind in smell taste and appearance. 'Horse-celery' has a leaf like that of the marsh kind, but is of close habit and has a big stalk, and its root is as thick as a radish and black; [8] the fruit is also black, and in size is larger than the seed of a vetch. They say that both kinds are serviceable in cases of strangury and for those suffering from stone, being administered in sweet white wine. Both kinds grow equally [9]

[4] 9. 9. 5; Plin. 20. 58; Diosc. 2. 136.
[5] Plin. 19. 124.
[6] μανόφυλλον: Plin. *l.c.* seems to have read μονόφυλλον.
[7] Diosc. 3. 64.   [8] Diosc. 3. 67.
[9] ὁμοίως conj. Sch.; ὅμως Ald.

ὁμοίως πανταχοῦ· γίνεται δὲ καί τι δάκρυον ἐξ αὐτοῦ ὅμοιον τῇ μύρρᾳ· οἱ δέ φασιν ὅλως μύρραν.

4 Τὸ δὲ ὀρεοσέλινον μείζους ἔτι διαφορὰς ἔχει· τὸ μὲν γὰρ φύλλον ἔοικε κωνείῳ, ῥίζα δὲ λεπτή, τὸν δὲ καρπὸν ἔχει καθάπερ ἄνηθον πλὴν ἐλάττω· διδόασι δὲ τοῦτον ἐν οἴνῳ αὐστηρῷ τῶν γυναικείων χάριν.

Ἔνια δὲ ὅλως ἀσύμβλητα τοῖς ἡμέροις ἐστὶ κατά γε τοὺς χυλοὺς καὶ τὰς δυνάμεις, ὥσπερ σίκυος ὅ τε ἄγριος καὶ ὁ ἥμερος, ἀλλ' ἐκ τῆς προσόψεως ἔχει τὴν ὁμοιότητα, καθάπερ καὶ ἐν τοῖς στεφανώμασιν ἡ ἰωνία· τὸ γὰρ φύλλον ἔχει παρόμοιον. τούτων μὲν οὖν ἐν τοῖς εἰρημένοις αἱ διαφοραί.

VII. Τῶν δὲ ἀρουραίων λεγομένων μετὰ ταῦτα ῥητέον, καὶ ὅλως εἴ τι ποιῶδές ἐστιν ὃ μὴ τυγχάνει βρωτόν. καλοῦμεν γὰρ λάχανα τὰ πρὸς τὴν ἡμετέραν χρείαν· ἐν δὲ τῷ καθ' ὅλου κἀκεῖνα περιέχεται, δι' ὃ καὶ περὶ ἐκείνων λεκτέον. λάχανα μὲν δὴ καὶ τὰ τοιαῦτα καλεῖται, κιχόρη ἀπάπη χόνδρυλλα ὑποχοιρὶς ἠριγέρων, καὶ ὅλως

---

[1] τι conj. Sch.; τὸ Ald. cf. 9. 1. 4.
[2] ὅλως P₂Bas ; ὅλως Ald ; ? ἁπλῶς W.
[3] κωνείῳ conj. Sch.; κονίῳ Ald. cf 1. 5. 3 n.
[4] καρπὸν conj. Cornarius on Diosc. 3. 67. and Dalec.; καυλὸν UMAld. cf. Diosc. l.c.
[5] ἡμέροις H.; εἰρημένοις UMAld. cf. 7. 6. 1 n.
[6] See Index, σίκυος.
[7] i.e. which gives them a common name.
[8] εἴ τι ποιῶδές ἐστιν H.; ἐγγειποιώδες U; ἐγγειποιῶδές ἐστι MAld.

ENQUIRY INTO PLANTS, VII. vi. 3–vii. 1

everywhere. There is also a sort[1] of gum which exudes from the plant, like myrrh, and some say that it *is*[2] myrrh.

'Mountain-celery' (parsley) exhibits even greater differences; its leaf is like that of hemlock,[3] the root is slender, and the fruit[4] like that of dill, but smaller; it is given in dry wine for diseases of women.

In some cases however the wild kinds are not in the least like the cultivated[5] in taste and properties; thus the wild and the cultivated cucumber[6] are quite different, and their resemblance[7] is due only to their general look, as, among coronary plants, there is resemblance between the wild and the cultivated kinds of gilliflower; for the leaves are alike. We have then described the differences which these plants present.

Of other uncultivated herbs, which may be classed with pot-herbs.

VII. Next we must speak of the differences found in the herbs called 'uncultivated,' and in general in any herbaceous plants[8] which are not edible. For we give the name of 'pot-herbs' to those which are cultivated for our own use, but in a wider sense the term includes these also; wherefore we must speak of them too. [9] Under the name 'pot-herbs' are included also[10] such plants as chicory dandelion[11] *khondrylla*[12] cat's ear groundsel, and in general all

---

[9] Plin. 21. 89. [10] καὶ add. Scal.

[11] ἀπάπη (or ἀπάτη) conj. Sch.; ἀφάκη Ald. The latter is a leguminous plant mentioned 8. 5. 3, etc.: for ἀπάπη *cf.* 6. 4. 8; 7. 8. 3; 7. 11. 3; for spelling see notes on the last two passages.

[12] χόνδρυλλα conj. Salm. from Plin. *l.c., cf.* 7. 11. 4 n.; ἀνδρύαλα Ald.G. *cf.* Plin. 21. 105; Diosc. 2. 133.

THEOPHRASTUS

ὅσα κιχοριώδη καλεῖται¹ διὰ τὴν ὁμοιότητα τῶν φύλλων· πάντα γάρ πως ἐμφερῆ ἔχει τῷ κιχορίῳ· πάλιν καυκαλὶς ἔνθρυσκον² ἡδύοσμον. οἱ δὲ μυρία ἄλλα καλοῦσιν, σκάνδιξ καὶ ὅσα ἄλλα τοιαῦτα σκανδικώδη, τραγοπώγων, οἱ δὲ κόμην καλοῦσιν, ὃ τὴν μὲν ῥίζαν ἔχει μακρὰν καὶ γλυκεῖαν τὰ δὲ φύλλα τῷ κρόκῳ ὅμοια πλὴν μακρότερα,³ τὸν καυλὸν δὲ βραχύν, ἐφ' οὗ τὴν κάλυκα⁴ μεγάλην καὶ ἐξ ἄκρου μέγαν τὸν πάππον⁵ πολιόν, ἀφ' οὗ καλεῖται τραγοπώγων.

2 Ὁμοίως δὲ καὶ ὅσα ἄλλα τοιαύτας⁶ μὲν ἰδέας ἔχει τοὺς δὲ χυλοὺς ἐδωδίμους ἢ ὠμοὺς ἢ ἑφθούς· ἔνια γὰρ δεῖται πυρώσεως, ὥσπερ μαλάχη καὶ τευτλὶς καὶ τὸ λάπαθον καὶ ἡ ἀκαλύφη καὶ τὸ παρθένιον· τὸν δὲ στρύχνον καὶ ὠμὸν ἐσθίουσιν, ὃν καὶ εὐκήπευτόν τινες πρότερον .... καὶ ἕτερα δὲ πλείω τούτων, ἐν οἷς καὶ ὁ παροιμιαζόμενός ἐστι διὰ πικρότητα κόρχορος ἔχων τὸ φύλλον ὠκιμῶδες. πάντα δὲ τὰ μὲν ἐπέτεια τὰ δὲ ἐπετειόκαυλα τυγχάνει· τὰ μὲν γὰρ ἐξαυαίνονται τῶν δὲ διαμένουσιν εἰς πλείω χρόνον αἱ ῥίζαι· σχεδὸν δὲ οὐκ ἐλάττω τὰ τοιαῦτά ἐστι.

3 Φύεται δὲ τὰ μὲν καὶ ἀπὸ τῶν ῥιζῶν καὶ ἀπὸ τῶν σπερμάτων, τὰ δὲ ἕτερα μόνον ἀπὸ σπέρ-

---

¹ καλεῖται conj. Sch.; ταῦτα Ald.
² ἔνθρυσκον: Sch. conjectured ἔνθρυσκος, form corrected by L.Dindorf; ἐνθουσικόν Ald.G. cf. Plin. 22. 81.
³ Plin. 27. 142; Diosc. 2. 138.
⁴ κάλυξ: cf. 8. 2. 4; 8. 4. 3.
⁵ πάππον conj. W.; παγητὸν UMAld.; πώγωνα H. cf. Diosc. l.c., where Saracenus corrects καρπὸς to πάππος.
⁶ τοιαύτας (sc. herbaceous) PmBas.; τοιαῦτα τὰς Ald.; τὰς αὐτὰς conj. W.

those that are called [1] 'chicory-like' because of the resemblance in the leaves; for to a certain extent the leaves of all these are like those of chicory; and we may add *kaukalis* chervil [2] green mint. Some include under the name countless others, as wild chervil and all plants that resemble it, and goat's beard,[3] which some call *kome* ('hair'), which has a long sweet root and leaves like those of the crocus, but longer, and a short stem, on which is set the sheath [4]; this is large, and on the top is the large mass of grey pappus,[5] from which it gets its name of 'goat's beard.'

In like manner all those may be included which have a similar [6] appearance, but juices suitable for food whether raw or cooked; for some need the action of fire, as *malakhe* (cheese-flower) beet monk's rhubarb nettle and bachelor's buttons; while garden nightshade [7] is also eaten raw, and some in former times [8] considered it worth growing in gardens. There are also many more, including the plant which has become proverbial [9] for its bitterness, blue pimpernel, which has a leaf like basil. All these are either annual or have annual stems; for some of them wither away altogether in one season, while of others the roots persist for a longer time, and to this class belong the majority.

Some of these plants grow from roots and also from seed—unless in some cases they come up

---

[7] *i.e.* στρύχνος ὁ ἐδώδιμος: *cf.* 7. 15. 4. The American 'wonder-berry.'

[8] πρότερον Ald.; *πρότερον Bas.; ὠνόμασαν conj. W. Text probably defective.

[9] κόρχορος ἐν λαχάνοις is the proverb. *cf.* Ar. *Vesp.* 239, Schol.; Plin. 21. 183. (= 'Is Saul also among the prophets?')

ματος, εἰ μή τι καὶ αὐτόματον. ἡ δὲ βλάστησις καὶ τούτων καὶ τῶν ἄλλων τῶν μὲν ἅμα τοῖς πρώτοις ὑετοῖς ἐστι μετ' ἰσημερίαν, οἷον ἀπάπης καὶ τοῦ κύνωπος καὶ ἣν καλοῦσί τινες βούπρηστιν, τῶν δὲ μετὰ Πλειάδα, καθάπερ καὶ κιχορίου καὶ σχεδὸν τῶν ἄλλων τῶν κιχοριωδῶν. καὶ τὰ μὲν εὐθὺς ἅμα τῇ βλαστήσει τὸ ἄνθος ἀφίησι, καθάπερ ἡ ἀφία, τὰ δὲ ὕστερον οὐ πολλῷ, καθάπερ ἡ ἀνεμώνη, τὰ δὲ ἅμα τῷ ἦρι καὶ ἐκκαυλεῖ καὶ ἀνθεῖ, καθάπερ τὸ κιχόριον καὶ τὰ κιχοριώδη καὶ τῶν ἀκανθικῶν ὅσα λαχανώδη.

4 Διαφορὰ δὲ τῶν ἀνθῶν πολλή, περὶ ἧς ἐν τοῖς πρότερον εἴρηται· σχεδὸν γάρ ἐστι κοινὸν ἁπάντων· ἔνια δὲ καὶ ὅλως ἀνανθῆ, καθάπερ καὶ τὸ ἐπίπετρον. συμβαίνει δὲ τοῖς ἅμα τῷ καυλῷ τὸ ἄνθος ἀφιεῖσι ταχεῖαν εἶναι τὴν ἀπάνθησιν· πλὴν ἡ μὲν ἀπάπη γηράσαντος τοῦ πρώτου πάλιν ἄλλο καὶ ἄλλο παραφύει, καὶ τοῦτο ποιεῖ παρ' ὅλον τὸν χειμῶνα καὶ τὸ ἔαρ ἄχρι τοῦ θέρους. πολὺν δὲ χρόνον καὶ ὁ ἠριγέρων. τὰ δὲ ἄλλα οὐ ποιεῖ τοῦτο, καθάπερ οὐδὲ ὁ κρόκος οὔτε ὁ εὔοσμος οὔθ' ὁ λευκὸς οὔθ' ὁ ἀκανθώδης· οὗτοι δὲ ἄοσμοι.

VIII. Κοινὴ δὲ διαφορὰ πάντων τῶν ποιωδῶν ἡ τοιάδε· τὰ μὲν γάρ ἐστιν ὀρθόκαυλα καὶ νευρό-

---

[1] δὲ after τούτων om. W.
[2] ἀπάπης (or ἀπάτης) conj. Sch.; ἀφάκης U; ἀφάκεις MAld. cf. 7. 7. 1 n. Plin. l.c., however, has aphace.
[3] ἐπίπετρον conj. Scal. from Plin. l.c.; ἐπίμετρον UMAld.G. cf. Hesych. [4] καυλῷ conj. Sch.; καρπῷ UMAld.G.

spontaneously. The growth alike of these[1] and of others takes place in some cases with the first rains after the equinox, for instance, dandelion[2] ribgrass and the plant which some call *buprestis*; in other cases after the rising of the Pleiad, for instance, chicory and most of the plants of that class. Some produce their flower immediately at the time of making growth, as lesser celandine, some not long after, as anemone, while some as soon as spring comes send up both their stems and flower, as chicory and the plants which resemble it, and those spinous plants which come under the head of pot-herbs.

There is much difference in the flowers, of which we have spoken already; for such difference is a thing common to all; and some are altogether flowerless, as stonecrop.[3] Those which produce their flower with the stem[4] quickly shed the flower; except that dandelion,[5] when the first flower is past its prime, produces another and yet another, and continues to do so right through the winter and spring up to the summer. Groundsel[6] also blooms for a long time; the others however do not do this; for instance the crocus does not, neither the scented (saffron crocus) nor the white nor the spinous kind,[7]—which last are scentless.

Of the differences in stem and leaf found in all herbaceous plants.

VIII. A distinction which is found in all herbaceous plants alike is the following:—some have straight

[5] ἀπάπη γηρήσαντος conj. W.; ἀπηγηράσαντος U; ἀπογηράσαντος MAld.; ἀφάκη ἀπογηράσαντος H. *cf.* Plin. *l.c.*; 7. 7. 1 n.
[6] *cf. C.P.* 1. 22. 4; Plin. 25. 106.
[7] See Index. This plant can only have been called κρόκος because it produced a yellow dye.

καυλα, τὰ δὲ ἐπιγειόκαυλα, καθάπερ μαλάχη σκάνδιξ σίκυος ἄγριος· τὸ δὲ ἡλιοτρόπιον ἔτι μᾶλλον ὡς εἰπεῖν τοιοῦτον, ὥσπερ καὶ ἐν τοῖς ἀκανθώδεσιν οὖσιν τρίβολος καὶ ἡ κάππαρις καὶ ἄλλα πλείω· καὶ γὰρ ἐκείνων ἡ διαφορὰ πλείων. ἔνια δὲ περιαλλόκαυλα, μὴ ἔχοντα δὲ ποῦ προσπέσωσιν ἐπιγειόκαυλα,[1] καθάπερ ἐπετίνη καὶ ἀπαρίνη καὶ ἁπλῶς ὧν ὁ καυλὸς λεπτὸς καὶ μαλακὸς καὶ μακρός, δι' ὃ καὶ φύονται ταῦτα ὡς ἐπὶ τὸ πᾶν ἐν ἄλλοις· κοινὴ δὴ καὶ αὕτη ἡ διαφορὰ πάντων οὐ μόνον τῶν ποιωδῶν καὶ φρυγανικῶν ἀλλὰ καὶ τῶν θαμνωδῶν· καὶ γὰρ ἡ ἕλιξ καὶ ἔτι μᾶλλον ἡ σμῖλαξ περιαλλόκαυλον.

2 Ἔτι δὲ καὶ τῶν ποιωδῶν τὰ μὲν πολύκαυλα τὰ δὲ μονόκαυλα· καὶ τῶν μονοκαύλων τὰ μὲν ἀπαράβλαστα κατὰ τὸν καυλὸν τὰ δὲ παραβλαστικά, καθάπερ καὶ ἐν τοῖς ἡμέροις ἥ τε ῥαφανὶς καὶ ἄλλ' ἄττα. πολύκαυλα δὲ ὡς ἁπλῶς εἰπεῖν τὰ ἐπιγειόκαυλα, μονόκαυλα δὲ καὶ ὀλιγόκαυλα τὰ ὀρθόκαυλα. τούτων δὲ ἀπαράβλαστα τὰ λειόκαυλα κρόμυον πράσον σκόροδον,[2] ὥσπερ καὶ ἐν τοῖς ἡμέροις καὶ τὰ μὲν εὐθύκαυλα τὰ δὲ σκολιόκαυλα καὶ τούτων [τοῖς ἡμέροις] ὑπάρχει.

3 Διαφορὰ δέ τις καὶ τοιάδε τῶν ποιωδῶν ἐστι· τὰ μὲν γὰρ ἐπιγειόφυλλα τὰ δ' ἐπικαυλόφυλλα τυγχάνει τὰ δ' ἀμφοτέρως. ἐπιγειόφυλλα μὲν

---

[1] ἐπιγειόκαυλα conj. Cornarius; ἐπετειόκαυλα Ald H.
[2] cf. 7. 15. 1; Diosc. 4. 190 and 191; Plin. 22. 57.

and fibrous stems, some prostrate stems,[1] as *malakhe* (cheese-flower) wild chervil 'wild cucumber' (squirting cucumber); while *heliotropion*[2] has this character[3] to an even greater extent, and so, among spinous plants, have caltrop caper and several others; for in these too the above-mentioned distinction is even more marked. Some again have clasping stems, but if they have nothing on which to throw themselves, their stems become prostrate, as *epetine* bedstraw and in general those which have a slender soft long stem; wherefore these in general grow in the midst of[4] other plants. This point of difference too is common not only to all herbaceous plants and under-shrubs, but also to shrubby ones; for *helix* (ivy) has a clasping stem, and, still more, smilax.

Again of herbaceous plants too some have several stems, some only one; and of the latter some have no side-shoots along the stem, while others have side-shoots, for instance, among cultivated plants radish and some others. Those with prostrate stems have generally more than one, while those with erect stems have but one or a few. Of these those with smooth stems have no side-shoots, as onion leek garlic—the wild, as well as the cultivated forms; and of these[5] again some have straight, some crooked stems.

There is also the following point of difference in herbaceous plants:—some have their leaves on the ground, some on the stem, some have both characters. The following have ground leaves—crowsfoot[6] the

[3] τοιοῦτον conj. Sch. from G; τούτων Ald.
[4] ἐν; G seems to have read ἐπ'.
[5] τοῖς ἡμέροις probably repeated by mistake from above.
[6] *cf. C.P.* 2. 5. 4; Plin. 22. 48; Diosc. 2. 130.

THEOPHRASTUS

κορωνόπους ἄνθεμον ἀφύλλανθες ἄγχουσα πόα ἀνεμώνη ἀπαργία ἀρνόγλωσσον ἀπάπη· ἐπικαυλόφυλλα δὲ κρηπὶς ἄνθεμον τὸ φυλλῶδες λωτὸς λευκόϊον· ἀμφοτέρως δὲ τὸ κιχόριον· καὶ γὰρ ἐπὶ τῶν καυλῶν ἅμα ταῖς ἐκφύσεσι ταῖς ἀκρεμονικαῖς ἐκφύει τι καὶ ἄνθος· καὶ τῶν φυλλακάνθων ἔνια, πλὴν ἀκανθώδεσι κομιδῇ, καθάπερ ὁ σόγκος.

IX. Ἔστι δὲ καὶ τὰ μὲν ἄκαρπα τὰ δὲ κάρπιμα. καὶ ὅλως τῶν ποιωδῶν τὰ μὲν ἄχρι τῶν φύλλων ἀφικνεῖται, τὰ δὲ καυλὸν ἔχει καὶ ἄνθος καρπὸν δὲ οὔ. τὰ δὲ καὶ καρπὸν ὥσπερ τελειοτάτην φύσιν, εἰ μή τι καὶ ἄνευ τοῦ ἄνθους καρποφόρον, ὥσπερ ἐπὶ τῶν δένδρων.

Διαφέρει δὲ καὶ τὰ φύλλα σχεδὸν οὐκ ἐλάττοσιν ἀλλὰ πλείοσι διαφοραῖς ἢ τὰ τῶν δένδρων· καὶ πρὸς αὐτὰ δὲ ἐκεῖνα διαφορὰς ἔχει· μεγίστην μὲν ὡς εἰπεῖν ὅτι τὰ μὲν ἀπὸ μίσχου προσπέφυκε, τὰ δὲ αὐτὰ μὲν ὡς ἁπλῶς, τὰ δὲ καυλικῇ τινι προσφύσει. καὶ τῶν μὲν ἐν τῇ βλαστήσει προτερεῖ <ὁ καυλός>, τῶν δὲ πλείστων τὰ φύλλα, καὶ σχεδὸν ἐν τῇ ἀρχῇ μέγιστα γίνονται καὶ μάλιστα ἐδώδιμα· τὰ δὲ ἐκ τῶν δένδρων προωθεῖ τινα καυλόν.

---

[1] ἀφύλλανθες placed after ἄνθεμον by Sch.; in Ald. placed after ἀνεμώνη. cf. 7. 14. 2; Plin. 21. 56.
[2] ἀπάπη U; ἀπάτη Ald. cf. 7. 7. 1 n.
[3] cf. 1. 13. 1. and Index.
[4] ἐκφύει τι καὶ MSS.; ? ἐκφύει φύλλον τε καὶ W.

anthemon whose flowers have no petals[1] (wild camomile) alkanet grass anemone hawk's beard plantain dandelion[2]; the following have leaves on the stem— ox-tongue the *anthemon* which has petalled flowers[3] trefoil gilliflower; while chicory has both kinds of leaves; for this plant produces,[4] as well as leaves, a certain number of flowers on the stems at the points where the side-shoots are attached. Similar too are some of the plants with spinous leaves, but not those that are altogether spinous, as sow-thistle.

Of other differences seen in herbaceous plants in general, as compared with one another and with trees.

IX. Again some are barren, while others bear fruit, and, speaking generally, of herbaceous plants some get as far as producing leaves only, others have a stem and flower, but no fruit; some again have fruit as the completion of their development, while some bear fruit even though they have no flower, as is the case with some trees.

[5] The leaves of herbaceous plants again differ in hardly fewer, nay, even in more, ways than those of trees, and further, they present differences as compared with these, the chief being perhaps that some are attached by a leaf-stalk, some are attached directly, some attached with cauline appendages.[6] And in some herbaceous plants the stalk[7] is the first part to grow, but in most the leaves, which almost at the outset grow to their largest and are best for eating; whereas the leaves of trees always push out first a sort of stalk.

[5] Plin. 21. 100.
[6] *i.e.* petiolate, sessile, and decurrent respectively.
[7] ὁ καυλὸς add. Sch. from G.

THEOPHRASTUS

2 Διαφέρουσι δὲ καὶ τοῖς ἄνθεσι πολύ· ἐν μὲν γὰρ τοῖς δένδρεσι τά γε πλεῖστα λευκά, τὰ δὲ μικρὸν ἐπιπορφυρίζοντα, τὰ δὲ ποώδη καὶ χλοώδη, κεχρωσμένον δὲ ἀνθινῷ <οὐδέν· ἐν δὲ τοῖς ποιώδεσι τῶν ἀνθῶν> πολλαὶ καὶ παντοδαπαὶ χροιαὶ καὶ ἄκρατοι καὶ μεμιγμέναι καὶ εὔοσμοι δὴ καὶ ἄοσμοί εἰσιν. καὶ τὰ μὲν δένδρα τὴν ἄνθησιν ἀθρόαν ποιεῖται, τούτων δ' ἔνια κατὰ μέρος, ὥσπερ ἐλέχθη καὶ περὶ τοῦ ὠκίμου, δι' ὃ καὶ πολὺν χρόνον ἀνθεῖ, καθάπερ ἄλλα τε πολλὰ καὶ τὸ ἡλιοτρόπιον καὶ τὸ κιχόριον.

3 Πολλαὶ δὲ καὶ τῶν ῥιζῶν διαφοραὶ καὶ τρόπον τινὰ αἱ τούτων φανερώτεραι· εἰσὶ γὰρ αἱ μὲν ξυλώδεις αἱ δὲ σαρκώδεις καὶ ἰνώδεις, ὥσπερ καὶ τῶν ἡμέρων, καθάπερ αἵ τε τοῦ σίτου καὶ τῆς πόας τῆς πλείστης. αὐτῶν δὲ τούτων ἕκασται πλείστας ἔχουσι διαφορὰς χρώμασιν ὀσμαῖς χυμοῖς μεγέθεσιν· αἱ μὲν γὰρ λευκαὶ αἱ δὲ μέλαιναι αἱ δ' ἐρυθραί, καθάπερ ἥ τε τῆς ἀγχούσης καὶ τοῦ ἐρευθεδάνου· αἱ δ' ὥσπερ ξανθαὶ καὶ ξυλοειδεῖς· καὶ γλυκεῖαι δὲ καὶ πικραὶ καὶ δριμεῖαι καὶ εὐώδεις καὶ κακώδεις, καὶ ἔνιαι φαρμακώδεις, ὡς ἐν ἄλλοις εἴρηται.

4 Διαφοραὶ δὲ καὶ τῶν σαρκωδῶν· αἱ μὲν γὰρ στρογγύλαι αἱ δὲ προμήκεις καὶ βαλανώδεις, ὥσπερ ἀσφοδέλου καὶ κρόκου· καὶ αἱ μὲν λεπυριώδεις, ὥσπερ ἡ τοῦ βολβοῦ καὶ τῆς σκίλλης καὶ ὅσαι βολβώδεις καὶ κρομύου δὲ καὶ γηθύου καὶ

---

[1] cf. 1. 13. 1.
[2] οὐδέν ... ἀνθῶν add. Scal. from G (κεχρωσμένων δὲ ἀνθικῷ πολλαὶ UMAld.); ἀνθινῷ for ἀνθικῷ conj. W., who also added τῶν ἀνθῶν. See LS. ἀνθινός.

ENQUIRY INTO PLANTS, VII. ix. 2-4

There is also much difference as to the flowers between herbaceous plants and trees; for in trees[1] most of the flowers are white, while some are slightly reddish, others are greenish or greenish-yellow, but none of them[2] have distinct gay colours; while in herbaceous plants the flowers shew many and various colours, both simple and in combination, and further, some of them are scented, others not. Again[3] trees produce all their bloom at once, while some herbaceous plants have a succession of flowers, as we said[4] of basil; wherefore it is in flower for a long period, as are many other herbs, such as *heliotropion* and chicory.

There are also many differences in the roots, and, in a way, the differences in these are more obvious; some are woody, some fleshy and fibrous, just as in the cultivated kinds, as are those of corn and most kinds of grass. Again the roots themselves exhibit in each case very many differences in colour smell taste and size; some are white, some black, some red, as those of alkanet and madder; some are yellowish,[5] or the colour of wood.[6] Again there are roots which are sweet, bitter, pungent, fragrant, evil-smelling; and some are medicinal, as has been said elsewhere.

There are also differences between those with fleshy roots; the roots of some are round, of some oblong and acorn-shaped, as those of asphodel and crocus;[7] some consist of several layers, as those of purse-tassels squill and others which belong to that class, onion long onion and others like

---

[3] ἄοσμοι εἰσιν. καὶ τὰ μὲν conj. W.; ἄοσμοι. καὶ ἐνίων τὰ μὲν Ald.H.   [4] 7. 3. 1.
[5] ὥσπερ MSS.; πλεῖσται conj. W.
[6] ξυλοειδεῖς: *cf.* 7. 3. 2.   [7] *cf.* 1. 6. 7; 6. 6. 10.

ὅσα τούτοις ὅμοια. αἱ δὲ ὁμαλεῖς καὶ ψαθυραὶ καὶ μαλακαὶ δι' ὅλου καὶ ὥσπερ ἄφλοιοι, καθάπερ τοῦ ἄρου· αἱ δὲ φλοιὸν ἔχουσι πρὸς τῇ σαρκί, καθάπερ ἡ τοῦ κυκλαμίνου καὶ τῆς γογγυλίδος. οὐχ ἅπασαι δ' αἱ εὐώδεις ἢ γλυκεῖαι ἢ εὔστομοι καὶ ἐδώδιμοι, οὐδ' αἱ πικραὶ ἄβρωτοι· ἀλλ' ὅσαι ἀβλαβεῖς εἰσι τῷ σώματι μετὰ τὴν προσφοράν· ἔνιαι γὰρ γλυκεῖαι μὲν θανάσιμοι δὲ καὶ νοσώδεις, αἱ δὲ πικραὶ μὲν ἢ κακώδεις ὠφέλιμοι δέ. τὸν αὐτὸν δὲ τρόπον καὶ φύλλα καὶ καυλοί, καθάπερ τοῦ ἀψινθίου καὶ τοῦ κενταυρίου. διαφορὰ δὲ καὶ κατὰ τὴν βλάστησιν καὶ κατὰ τὴν ἄνθησιν, οἷον ἀρχομένου χειμῶνος καὶ μεσοῦντος καὶ πάλιν ἦρος ἢ θέρους ἢ μετοπώρου. καὶ ἐπὶ τῶν καρπῶν δὲ ὁμοίως τῷ βρωτοὺς εἶναι καὶ ἐγχύλους ἐνίοις καὶ φύλλα καὶ σπέρματα καὶ ῥίζας· καὶ ἐν αὐτοῖς τούτοις κατὰ τοὺς χυλούς, οἷον ὀξύτητι καὶ δριμύτητι καὶ γλυκύτητι καὶ αὐστηρότητι καὶ ταῖς ἄλλαις ταῖς τοιαύταις ἁπλῶς τε καὶ κατὰ τὸ μᾶλλον. τὰς μὲν οὖν διαφορὰς ἐν τούτοις ληπτέον.

X. Διῃρημένων δὲ κατὰ τὰς ὥρας ἑκάστων πρός τε τὰς βλαστήσεις καὶ κατὰ τὰς ἀνθήσεις καὶ τελειώσεις τῶν καρπῶν, οὐδὲν ἀναβλαστάνει πρὸ τῆς οἰκείας ὥρας οὔτε τῶν ῥιζοφυῶν οὔτε τῶν σπερμοφυῶν, ἀλλ' ἕκαστον ἀναμένει τὴν

---

[1] καὶ conj. W.; ἢ Ald.
[2] τῷ βρωτοὺς εἶναι καὶ ἐγχύλους ἐνίοις conj. W.; τὸ βρωτὰ εἶναι καὶ καυλοὺς εἶναι Ald.H. Text probably defective.

these. Some are smooth loose and soft throughout, and, as it were, without 'bark,' as those of cuckoo-pint, while some have a 'bark' attached to the fleshy part, as those of cyclamen and turnip. And not all those that are fragrant or sweet or pleasant to the taste are also [1] edible, any more than all those that are bitter are uneatable; any (whether sweet or bitter) that are harmless to the body after being eaten are edible; for some that are sweet are deadly and dangerous to health, while some are beneficial even if they are bitter or have an evil smell. The same may be said of the leaves and stalks, as in the case of wormwood and centaury. There are also differences in the time of growth and of flowering, the season being variously the beginning or middle of winter, or again spring, summer, or autumn. So too is there in like manner a difference in the fruits, which [2] in some of these plants are edible and juicy, as well as [3] the leaves seeds and roots. And in these cases there are further differences in the taste (of those which are edible and juicy), which may be sharp, pungent, sweet, dry, or exhibit other similar differences, either altogether or in degree. These are examples of the differences which we find.

Of the seasons at which herbs grow and flower.

X. Each plant having its proper season for growth, flowering and maturing of the fruit, nothing grows [4] before its proper season either of those grown from a root or of those grown from seed, but each awaits

[3] καὶ ῥίζας seems irrelevant.
[4] ἀναβλαστάνει conj. W.; ἀναβλαστεῖ καὶ Ald.H.

οἰκείαν οὐδ᾽ ὑπὸ τῶν ὑδάτων οὐδὲν πάσχον· ἔνια γὰρ θερινὰ κομιδῇ καὶ τῇ βλαστήσει καὶ τῇ ἀνθήσει, καθάπερ ὅ τε σκόλυμος καὶ ὁ σίκυος ἄγριος, ὥσπερ καὶ περὶ τῶν φρυγανικῶν ἐλέχθη περὶ κονύζης τε καὶ καππάριδος καὶ τῶν ἄλλων· οὐδὲν γὰρ οὐδὲ ἐκείνων ἀνθεῖ καὶ βλαστάνει πρὸ
2 τῆς οἰκείας ὥρας. δι᾽ ὃ κἂν ταύτῃ δόξαιεν ἂν διαφέρειν τῶν δένδρων. τῶν μὲν γὰρ ἅμα πως πάντων ἢ ἐγγὺς ἡ βλάστησις, εἰ δὲ μὴ κατὰ μίαν γε ὥραν ὡς εἰπεῖν· τούτων δὲ ἐν πολλαῖς μᾶλλον δὲ ἐν ἁπάσαις ἡ βλάστησις καὶ ἔτι μᾶλλον ἡ ἄνθησις, ὥστε εἴ τις ἐθέλει κατανοεῖν σχεδὸν συνεχὲς γίνεται καθ᾽ ὅλον ἐνιαυτὸν καὶ ἡ βλάστησις καὶ ἡ ἄνθησις· αἰεὶ γὰρ ἕτερον ἐξ ἑτέρου διαδεχόμενον πάσας καταλαμβάνει τὰς ὥρας· οἷον μετὰ τὴν ἀπάπην κρόκος ἔσται καὶ ἀνεμώνη καὶ ὁ ἠριγέρων καὶ τὰ ἄλλα χειμερινά, μετὰ δὲ ταῦτα τὰ ἠρινὰ ⟨καὶ θερινὰ⟩ καὶ μετοπωρινά.
3 πολλὰ δέ, ὥσπερ ἐλέχθη, διὰ τὸ κατὰ μέρος ἀνθεῖν ἐπιτείνεται ταῖς ὥραις· ἔνια γὰρ οὕτως ἀνθεῖ, καθάπερ ἥ τε ἀπάπη καὶ τὸ ὀνοχειλὲς καὶ τὸ κιχόριον καὶ τὸ ἀρνόγλωσσον καὶ ἄλλα· διὰ δὲ τὴν συνέχειαν καὶ τὴν περικατάληψιν τὴν ὑπ᾽ ἀλλήλων οὐ φαίνεται ῥᾴδιον ἐν ἐνίοις οὐδ᾽ ὁρίσαι ποῖα πρῶτα βλαστάνει καὶ ποῖα ὀψιβλαστῆ· πλὴν εἴ τις ὑποθοῖτο τοῦ ἔτους τὴν ἀρχήν τινα

---

[1] οὐδ᾽ conj. W.; οὔθ᾽ Ald.   [2] Reference not discoverable.
[3] ἀπάπην conj. W.; ἀφάκην UMAld. *cf.* 7. 7. 1 n.
[4] κρόκος conj. Sch. (adding ὁ); ἦρος U (corrected); κρος M; om. Ald.; τὸ κρόκον mBas.
[5] καὶ θερινὰ seems to have dropped out.
[6] διὰ τὸ κατὰ μέρος ἀνθεῖν conj. W.; καὶ τῶν κατὰ μέρος ἀνθέων Ald.

its proper season and is not[1] affected even by rain. For some are plants which belong properly to summer as to their growth and their flowering, as golden thistle and squirting cucumber, as was said[2] of shrubby plants and of *konyza* caper and the rest; for of these too none blooms or grows before its proper season. Wherefore in this respect too these plants would seem to differ from trees. For trees make their growth all at once or nearly so, or at all events we may say that they do so all at one season; but the plants of which we are now speaking have their times of growing and still more of flowering at many or rather at all seasons; so that, if one will consider it, both the growing and the flowering are almost continuous throughout the year; for one continually succeeds to another, so that all seasons are covered; thus after the dandelion[3] will come the crocus[4] anemone groundsel and the other plants of winter, and after these those of spring summer[5] and autumn. Some again, as was said, because they do not produce all their bloom at once,[6] cover a longer season; for there are some that thus bloom, for instance dandelion[7] bugloss[8] chicory plantain, and others; but because of this continuity and overlapping it does not seem easy in some cases to define which first make growth and which are late in growing,— unless[9] one were to lay down that the 'year'

[7] ἀπάπη conj. W.; ἀφάκη Ald. *cf.* 7. 7. 1 n.

[8] ὀνοχειλὲς conj. Sch. from Plin. 21. 100; Diosc. 4. 24; ὀνοκίχλης UMAld.

[9] *i.e.* unless one has a fixed starting-point. τινα ὥρας τινος ἀρχῇ conj. W.; τινα πρὸς τῇ ἵνα ζῇ(?) U; text defective in MAld., but both give ἵνα ζῇ; W. conjectures also τροπὰς τὰς χειμερινάς. ? εἶναι ὥρας τινος ἀρχήν or εἶναι ὥραν τινα (omitting εἰ ζῇ as a trace of a lost sentence).

4 ὥρας τινὸς ἀρχῇ. καὶ αὐτῶν δὲ τούτων τὰς γενέσεις καὶ τὰς ὥρας ὅταν τελειωθέντων τῶν καρπῶν πάλιν ἄλλας ἀρχὰς ἐνίστωνται τῆς γενέσεως· ὅπερ μάλιστα δοκεῖ συμβαίνειν μετ' ἰσημερίαν μετοπωρινήν· τότε γὰρ ἤδη τὰ σπέρματα πλεῖστα τετελείωται καὶ τῶν δενδρικῶν καρπῶν οἱ πολλοί, καὶ ἅμα μεταβολή τις αὐτοῦ τούτου προσγίνεται καὶ τῆς ὥρας· ὅσα δὲ ἀτελῆ καὶ ἄπεπτα περικαταλαμβάνεται, τούτοις κατὰ λόγον ἐκ περιόδου καὶ ἡ βλάστησις γίνεται καὶ ἡ ἄνθησις καὶ ἡ τελείωσις· δι' ὃ συμβαίνει τὰ μὲν ὑπὸ τροπὰς ἀνθεῖν τὰ δ' ὑπὸ Κύνα τὰ δὲ καὶ μετὰ Ἀρκτοῦρον καὶ ἰσημερίαν μετοπωρινήν.

5 Ἀλλὰ ταῦτα μὲν ἔοικε κοινοτέραν ἔχειν σκέψιν εἰς ἀφορισμὸν ἀρχῆς. ὅτι δὲ αἱ διαφοραὶ πλείους ἢ οὐκ ἐλάττους ἐν τούτοις φανερόν. ἐπεὶ καὶ ἀείφυλλα τῶν τοιούτων ἐστὶν ἔνια, καθάπερ τὸ πόλιον καὶ ἡλιοτρόπιον καὶ τὸ ἀδίαντον.

XI. Ἀφωρισμένων οὖν τούτων περὶ τὰς διαφορὰς ἐν οἷς γίνονται καὶ πῶς λεκτέον ἤδη τὰς καθ' ἕκαστον ἱστορίας . . . ὅσα μὴ κατὰ τὴν

---

[1] *i.e.* to fix the date of the beginning and end of the cycle of the plant's life.
[2] αὐτοῦ τούτου: ? the plant itself. αὐτοῦ τοῦ ἔτους conj. Sch.
[3] *i.e.* according as the seed ripened last year or this year. In this rather obscure section I follow W.'s explanations.
[4] Plin. 21. 100.

begins when a certain season begins. Further in these plants it is not easy to define[1] in each case the time of first growth and the season when, the fruits being matured, it makes a fresh start in reproduction. This seems chiefly to occur after the autumnal equinox; for by that time most of the seeds are matured, like most of the fruits of trees: moreover a change then takes place in the seed[2] itself as well as in the season. But in the case of any seeds which are still immature and unripe and so are overtaken by winter, the period of first growth, the flowering of the new plant, and the period of maturity are proportionally later. Wherefore it comes to pass that some bloom at the solstice, some at the rising of the dog-star, and some after the rising of Arcturus and the autumnal equinox.[3]

But these matters seem to require a wider investigation in order to determine when the process begins. However it is clear from what has been said that these plants present at least as many differences as trees. [4] For some again of this class are evergreen, as hulwort *heliotropion* and maidenhair.

Of the classes into which herbaceous plants may be divided, as those having a spike and chicory-like plants.

XI. Having then made these explanations we must now give a separate account of each plant, discussing the differences (in those plants in which they occur) and saying how they arise . . . .[5] except those peculiarities which belong to the

[5] Lacuna in UMAld.Cam.Bas., leaving the connexion of the next clause obscure.

ἰδίαν ἑκάστου φύσιν. λέγω δὲ οἷον τὰ σταχυώδη καὶ τὰ σκανδικώδη καὶ μονοφυῆ, κἂν εἴ τι ἕτερόν ἐστι τοιοῦτον κοινὸν ἐπὶ πάντων λαβεῖν, ὃ τῇ αἰσθήσει γνώριμον ἢ φύλλοις ἢ ἄνθεσιν ἢ ῥίζαις ἢ καρποῖς· ἐκ γὰρ τῶν φανερῶν ὁ μερισμὸς ὥσπερ καὶ ἐκ τῶν ῥιζῶν.

2 Σταχυώδη μὲν οὖν ἐστιν ὅ τε κύνωψ ὑπό τινων καλούμενος πλείους ἔχων ἰδέας ἐν ἑαυτῷ· καὶ ὁ ἀλωπέκουρος καὶ ὁ στελέφουρος καὶ ὑπ' ἐνίων δὲ ἀρνόγλωσσον τῶν δὲ ὄρτυξ καλούμενος· παρόμοιον δὲ τούτῳ τρόπον τινὰ καὶ ἡ θρυαλλίς. ἁπλᾶ δὲ καὶ μονοειδῆ τρόπον τινὰ ταῦτα καὶ στάχυν οὐκ ὀξὺν οὐδ' ἀθερώδη ἔχοντα· ὁ δ ἀλωπέκουρος μαλακὸν καὶ χνοωδέστερον, ὅτι καὶ ὅμοιον ταῖς τῶν ἀλωπέκων οὐραῖς, ὅθεν καὶ τοὔνομα μετείληφεν. ὅμοιος δὲ τούτῳ καὶ ὁ στελέφουρος, πλὴν οὐχ ὥσπερ ἐκεῖνος ἀνθεῖ κατὰ μέρος ἀλλὰ δι' ὅλου τοῦ στάχυος ὥσπερ ὁ πυρός. ἡ δὲ ἄνθησις ἀμφοῖν χνοώδης, καθάπερ καὶ τοῦ σίτου· παρόμοιον δὲ τῇ ὅλῃ μορφῇ τῷ πυρῷ πλὴν πλατυφυλλότερον. ὡσαύτως δὲ τούτοις καὶ τὰ ἄλλα τὰ σταχυώδη λεκτέον.

3 Τὰ δὲ κιχοριώδη πάντα μὲν ἐπετειόφυλλα καὶ ῥιζόφυλλα, βλαστάνει δὲ μετὰ Πλειάδα πλὴν τῆς ἀπάπης, τοῖς δὲ καυλοῖς καὶ ταῖς ῥίζαις

---

[1] *i.e.* spicate.

[2] σκανδικώδη: *i.e.* umbellate. One would expect κιχοριώδη, to correspond with § 3; but the three classes mentioned seem to be merely 'samples' of classification: of the three only one (τὰ σταχυώδη) is described below, and other classes are added.

[3] μονοφυῆ I conj.: *i.e.* those which have a scape: *cf.* 2. 6. 9; *C.P.* 1. 1. 3. μυόφαα U; μυοφῶα MAld.; om. G.

ENQUIRY INTO PLANTS, VII. xi. 1-3

character of individual kinds. I mean for instance the plants which have a spike,[1] those which may be classed with wild chervil,[2] and those which have a single stem,[3] . . . . or any other such class in which one can find some such general characteristics obvious to the senses either in leaves flowers roots or fruits; (for the classification is to be made by the visible parts, as well as by the roots).[4]

[5] An example of the plants which have a spike is the plant which some call 'dog's eye[6]' (rib-grass), which comprises several forms; we have also 'fox-brush,' *stelephuros* (plantain), which some call 'lamb's tongue' and some 'quail-plant'; and somewhat similar to this is *thryallis*. These are simple plants and uniform in character, having a spike which is not pointed nor bearded; while in 'fox-brush' it is soft and somewhat downy, in that it actually resembles the brush of a fox, whence also it has obtained its name. Similar to this is *stelephuros* (plantain), except that it does not, like that plant, flower here and there, but all up the spike like wheat. The bloom of both is downy like that of corn, and the plants in their general appearance resemble wheat, but have broader leaves. Of the other plants which have a spike a similar account may be given.

[7] The chicory-like plants all have annual leaves and have root-leaves, and they begin to grow after the Pleiad, except dandelion[8]; but in their stems

[4] Roots being the basis of classification in xii. below.
[5] Plin. 21. 101.
[6] κύνωψ conj. Sch.; ἀχύνωψ UAld.; Plin. *l.c.* has *cynops* (*cf.* 7. 7. 3); *oculus caninus* G.
[7] *i.e.* composites. Plin. *l.c.*
[8] ἀπάπης U; ἀπάτης MAld. *cf.* 7. 7. 1 n.; 7. 8. 3 n.

μεγάλας ἔχουσι διαφοράς· οἱ μὲν γὰρ τῶν ἄλλων ἁπλούστεροι καὶ ἐλάττους,¹ ὁ δὲ τοῦ κιχορίου μέγας καὶ ἀποφύσεις ἔχων πολλάς, ἔτι δὲ γλίσχρος καὶ δυσδιαίρετος, δι' ὃ καὶ δεσμῷ χρῶνται· παραβλαστητικὸν δὲ καὶ τῇ ῥίζῃ καὶ ἄλλως μακρόρριζον, δι' ὃ καὶ δυσώλεθρον· ὅταν γὰρ ἐκλαχανίσωνται, πάλιν τὸ³ ὑπόλοιπον ἀρχὴν λαμβάνει γενέσεως. συμβαίνει δὲ καὶ παρανθεῖν αὐτοῦ μέρος ἄλλο καὶ ἄλλο, καὶ τοῦτο ἄχρι τοῦ μετοπώρου, σκληροῦ⁴ δοκοῦντος εἶναι τοῦ καυλοῦ. φέρει δὲ καὶ λοβὸν ἐν ᾧ τὸ σπέρμα περὶ τὰ ἄκρα τῶν καυλῶν.

4 Ἡ δὲ ὑποχοιρὶς⁵ λειοτέρα καὶ ἡμερωτέρα τῇ προσόψει καὶ γλυκυτέρα καὶ οὐχ ὥσπερ⁶ ἡ χόνδρυλλα· τὸ γὰρ ὅλον οὐκ ἐδώδιμος αὐτὴ καὶ ἄβρωτος καὶ ἐν τῇ ῥίζῃ δριμὺν ὀπὸν ἔχει καὶ πολύν.

Ἄβρωτος δὲ καὶ πικρὰ ἡ ἀπάπη· πρωϊανθὴς δὲ καὶ ταχὺ γηράσκει καὶ ἀποπαπποῦται, εἶτ' ἄλλο φύεται πάλιν καὶ ἄλλο καὶ τοῦτο παρ' ὅλον ποιεῖ τὸν χειμῶνα καὶ τὸ ἔαρ ἄχρι τοῦ θέρους· τὸ δ' ἄνθος μηλινοειδές.

Ὡσαύτως δὲ καὶ ἡ πικρίς· καὶ γὰρ αὕτη τῷ ἦρι ἀνθεῖ, καὶ παραπλησία δι' ὅλου τοῦ χειμῶνος καὶ τοῦ θέρους παρανθεῖ· τῇ γεύσει δὲ πικρά, δι' ὃ καὶ τοὔνομα εἴληφε. ταῦτα μὲν οὖν ἐν ταύ-

---

¹ ἐλάττους conj. Scal.; θάττους UMAld.
² Plin. 21. 88.   ³ τὸ conj. Sch.; τὴν Ald.
⁴ Meaning not obvious ; σκληροῦ is perhaps corrupt.
⁵ ὑποχοιρὶς conj. Scal.; ὑποχώρησις UMAld. cf. 7. 7. 1 n.
⁶ οὐχ ὥσπερ : an adjective has perhaps dropped out between these words ; ? πικρὰ (amara Plin. l.c.).

and roots they exhibit great differences; for in some these are simpler and fewer,[1] but the stem of chicory is large and has many side-shoots; also it is tough and hard to break, wherefore it is used for withes[2]; it makes side-growths from the root, and also has long roots, wherefore it is hard to kill; for, when the top is taken off to use as a vegetable, what remains[3] starts growing again. Moreover different parts of it flower at different times, and the flowering goes on till autumn, since the stem appears to be hard.[4] Also it bears a pod, which contains the seed, at the top of the stem.

Cat's ear[5] is smoother and has a more cultivated appearance, and is also sweeter and not like[6] *khondrylla*[7]; for the latter is altogether uneatable and[8] unfit for food, and its root contains a quantity of pungent juice.

Dandelion[9] is also unfit for food and bitter: it flowers early and quickly waxes old and the flower turns to pappus; but then another flower forms, and yet another, and this goes on right through the winter and spring up to the summer; and the flower is yellow.

[10] The like may be said of *pikris* : for this plant too blooms in spring, and like dandelion it flowers throughout the winter, and it flowers also to some extent in summer; in taste it is bitter, whence its name. These are the special points of difference

[7] χόνδρυλλα conj. St.; χανδρυ αλλα U; χανδρὺ αλλα M; χανδρὰς ἀλλὰ Ald.H.; *cadryalia* G (Tarv.); *candralia* G (Bas. Par.).
[8] τὸ γὰρ ὅλον οὐκ ἐδώδιμος αὕτη καὶ conj. W.; τὸ ὅλον οὐκ ἐδ. αὕτη γὰρ Ald.; αὕτη γὰρ ἄβρωτος conj. Sch. from Plin. 21. 105.
[9] ἀπάπη conj. W.; ἀφάκη Ald. *cf.* 7. 7. 1 n.
[10] Plin. 22. 66.

ταῖς ταῖς διαφοραῖς. πειρατέον δέ, ὡς ἐλέχθη, καὶ τῶν ἄλλων λαμβάνειν ὁμοίως.

XII. Πολὺ δέ τι γένος ἐστὶ καὶ τῶν σαρκορρίζων ἢ κεφαλορρίζων, ἃ καὶ πρὸς τὰ ἄλλα καὶ καθ' αὑτὰ τὰς διαφορὰς ἔχει ῥίζαις τε καὶ φύλλοις καὶ καυλοῖς καὶ ταῖς ἄλλαις μορφαῖς. τῶν γὰρ ῥιζῶν, ὥσπερ εἴρηται πρότερον, αἱ μὲν λεπυριώδεις αἱ δὲ σαρκώδεις, καὶ αἱ μὲν ἔχουσαι φλοιὸν αἱ δ' ἄφλοιοι, ἔτι δὲ αἱ μὲν στρογγύλαι αἱ δὲ προμήκεις καὶ αἱ μὲν ἐδώδιμοι αἱ δ' ἄβρωτοι. ἐδώδιμοι μὲν γὰρ οὐ μόνον βολβοὶ καὶ τὰ ὅμοια τούτοις, ἀλλὰ καὶ ἡ τοῦ ἀσφοδέλου ῥίζα καὶ ἡ τῆς σκίλλης, πλὴν οὐ πάσης ἀλλὰ τῆς Ἐπιμενιδείου καλουμένης, ἣ ἀπὸ τῆς χρήσεως ἔχει τὴν προσηγορίαν· αὕτη δὲ στενοφυλλοτέρα τε καὶ λειοτέρα τῶν λοιπῶν ἐστιν.

2 Ἐδώδιμος δὲ καὶ ἡ τοῦ ἄρου καὶ αὐτὴ καὶ τὰ φύλλα προαφεψηθέντα ἐν ὄξει καί ἐστιν ἡδεῖά τε καὶ πρὸς τὰ ῥήγματα ἀγαθή. πρὸς δὲ τὴν αὔξησιν αὐτῆς, ὅταν ἀποφυλλίσωσιν, ἔχει δὲ μέγα σφόδρα τὸ φύλλον, ἀνορύξαντες στρέφουσιν, ὅπως ἂν μὴ διαβλαστάνῃ ἀλλὰ πᾶσαν ἕλκῃ τὴν τροφὴν εἰς ἑαυτήν, ὃ καὶ ἐπὶ τῶν βολβῶν τινες

---

[1] cf. 7. 11. 2 ad fin.   [2] Plin. 19. 93 and 94.
[3] μορφαῖς: cf. 8. 4. 2.   [4] 7. 9. 4.
[5] The legends about Epimenides suggest that the 'use' was possibly in magic: cf. what is said of σκίλλα 7. 13. 4. cf. Plin. l.c.

about these plants; now we must endeavour, as was said,[1] to set forth the special points of the other classes in like manner.

Of herbs which have fleshy or bulbous roots.

XII. [2] There is a large class of these which have fleshy or bulbous roots: these exhibit differences both as compared with other plants and with one another both in roots leaves stems and their other prominent features.[3] Of the roots, as has been said[4] already, some are in layers, some fleshy, some have a 'bark,' some not; and again some are round, some oblong, some edible and some not fit for food. Among edible roots are not only purse-tassels and others which resemble them, but also the roots of asphodel and squill, though not of all kinds of the latter, but only of the kind called 'Epimenides' squill (French sparrow-grass) which gets its name from its use[5]; this kind has narrower leaves and is smoother than the others.

[6] The root of cuckoo-pint is also edible, and so are the leaves, if they are first boiled down in vinegar; they are sweet, and are good for fractures. To increase the root, having first stripped[7] off the leaves (and the leaf is very large), they dig[8] it up and invert[9] it in order that it may not shoot,[10] but may draw all the nourishment into itself. This some

[6] Plin. 19. 96; 24. 162.
[7] ἀποφυλλίσωσιν conj. Sch.; ἀποφυλάσσωσιν U; ἀποφυλλάσ- σωσιν MAld.
[8] ἀνορύξαντες conj. St.; ὃ ἀνορύξαντες Ald.
[9] *cf.* 1. 6. 10; Plin. 19. 94 and 97, who seems to have read κατορύξαντες: so also G. ? 'they plunge it in a pit.'
[10] διαβλαστάνῃ: *cf. C.P.* 4. 8. 1.

ποιοῦσι συντιθέντες·[1] ἡ δὲ τοῦ δρακοντίου, καλοῦσι γάρ τι δρακόντιον ἄρον διὰ τὸ τὸν καυλὸν ἔχειν τινὰ ποικιλίαν, ἄβρωτος καὶ φαρμακώδης.

3 Ἀλλὰ ἡ τοῦ φασγάνου καλουμένου γλυκεῖά τε ἑψηθεῖσα, καὶ τριφθεῖσα μιγνυμένη τῷ ἀλεύρῳ ποιεῖ τὸν ἄρτον γλυκὺν καὶ ἀσινῆ· στρογγύλη δέ ἐστι καὶ ἄφλοιος καὶ ἀποφύσεις ἔχουσα μικράς, ὥσπερ τὸ γήθυον· πολλὰς δὲ εὑρίσκουσιν ἐν ταῖς σκαλοπιαῖς·[2] χαίρει γὰρ καὶ συλλέγει τὸ ζῶον.

Ἡ δὲ τοῦ θησείου τῇ μὲν γεύσει πικρά, τριβομένη δὲ κοιλίαν ὑποκαθαίρει. φαρμακώδεις δέ τινές εἰσι καὶ ἕτεραι, πολλῶν δὲ οὔτε φαρμακώδεις οὔτε ἐδώδιμοι. καὶ αὗται μὲν ἐν ταῖς ῥίζαις αἱ διαφοραί.

XIII. Κατὰ δὲ τὰ φύλλα τοῖς τε μεγέθεσιν καὶ τοῖς σχήμασιν. ὁ μὲν ἀσφόδελος μακρὸν καὶ στενότερον καὶ ὑπόγλισχρον ἔχει τὸ φύλλον, ἡ δὲ σκίλλα πλατὺ καὶ εὐδιαίρετον, τὸ δὲ φάσγανον ὑπό τινων δὲ καλούμενον ξίφος ξιφοειδές, ὅθεν ἔσχε καὶ τοὔνομα, ἡ δὲ ἶρις καλαμωδέστερον· τὸ δὲ τοῦ ἄρου πρὸς τῇ πλατύτητι καὶ ἔγκοιλον καὶ σικυῶδές ἐστιν· ὁ δὲ νάρκισσος στενὸν καὶ πολὺ καὶ λιπαρόν· βολβὸς δὲ καὶ τὰ βολβώδη παντελῶς στενὰ καὶ τοῦ κρόκου δ' ἔτι στενότερον.

2 Καυλὸν δὲ τὰ μὲν οὐκ ἔχει τὸ ὅλον οὐδ' ἄνθος, ὥσπερ τὸ ἄρον τὸ ἐδώδιμον· τὰ δὲ τὸν τοῦ ἄνθους μόνον, ὥσπερ ὁ νάρκισσος καὶ ὁ κρόκος· ἔνια δὲ ἔχει, καθάπερ ἡ σκίλλα καὶ ὁ βολβὸς καὶ ἡ ἶρις καὶ τὸ ξίφιον· μέγιστον δὲ πάντων ἀσφό-

---

[1] συντιθέντες: sense doubtful. Sch. and W. mark the word as corrupt.
[2] cf 9. 20. 3; Plin. 24. 142; Diosc. 2. 166.

do also with purse-tassels, when they lay them by.[1] [2] However the root of edder-wort (for a kind of cuckoo-pint is so called because of its variegated stem) is not good for food, but is used for drugs.

But the root of the plant called corn-flag is sweet, and, if cooked and pounded up and mixed with the flour, makes the bread sweet and wholesome. It is round and without 'bark,' and has small offsets like the long onion. Many of them are found in moles' runs[3]; for this animal likes them and collects them.

[4] The root of *theseion* is bitter to the taste, but when pounded purges the bowels. There are also certain others of these roots which have medicinal properties, but of many the roots are neither medicinal nor edible. Such are the differences in the roots.

XIII. [5] In the leaves the differences are in size and shape. Asphodel has a long leaf, which is somewhat narrow and tough, while that of squill is broad and tears easily; corn-flag, which is called by some *xiphos* ('sword'), has a sword-like leaf, whence its name, and iris one more like a reed. That of cuckoo-pint, in addition to being broad, is concave and like that of cucumber; that of the narcissus is narrow substantial and glossy, those of purse-tassels and plants of that character are quite narrow, and that of crocus narrower still.

[6] Some have not a stem at all, nor a flower, as the edible cuckoo-pint; some have only the flower-stem, as narcissus and crocus; some however have a stem, as squill purse-tassels iris and corn-flag; but asphodel

---

[3] παρὰ ταῖς σκαλοπιαῖς conj. Sch.; ἐν ταῖς σκ. conj. W.; ταῖς σκολοπίαις UMAld.   [4] Plin. 22. 66.   [5] Plin. 21. 108.
[6] Plin. 21. 108 and 109.

δελος· ὁ γὰρ ἀνθέρικος μέγιστος· ὁ δὲ τῆς ἴριδος ἐλάττων μὲν σκληρότερος δὲ τὸ δὲ ὅλον ἀνθερικώδης. ἔστι δὲ καὶ πολύκαρπος ὁ ἀσφόδελος, καὶ ὁ καρπὸς αὐτοῦ ξυλώδης τῇ μὲν μορφῇ τρίγωνος τῷ δὲ χρώματι μέλας· γίνεται δὲ ἐν τῷ στρογγύλῳ τῷ ὑποκάτω τοῦ ἄνθους, ἐκπίπτει δὲ τοῦ
3 θέρους, ὅταν τοῦτο διαχάνῃ. τὴν ἄνθησιν ποιεῖται κατὰ μέρος, ὥσπερ καὶ ἐπὶ τῆς σκίλλης, ἄρχεται δὲ πρῶτον ἀπὸ τῶν κάτωθεν. ἐν δὲ τῷ ἀνθερίκῳ συνίσταται σκώληξ, ὃς εἰς ἄλλο μεταβάλλει ζῷον ἀνθρηνοειδές, εἶθ' ὅταν ὁ ἀνθέρικος αὐανθῇ διεσθίον ἐκπέταται. δοκεῖ δὲ ἴδιον ἔχειν πρὸς τὰ ἄλλα τὰ λειόκαυλα, διότι στενὸς ὢν ἀποφύσεις ἄνωθεν ἔχει. πολλὰ δὲ εἰς τροφὴν παρέχεται χρήσιμα· καὶ γὰρ ὁ ἀνθέρικος ἐδώδιμος σταθευόμενος καὶ τὸ σπέρμα φρυγόμενον καὶ πάντων δὲ μάλιστα ἡ ῥίζα κοπτομένη μετὰ σύκου καὶ πλείστην ὄνησιν ἔχει καθ' Ἡσίοδον.

4 Ἅπαντα μὲν οὖν φιλόζωα τὰ κεφαλόρριζα μάλιστα δ' ἡ σκίλλα· καὶ γὰρ κρεμαννυμένη ζῇ καὶ πλεῖστόν γε χρόνον διαμένει· δύναται δὲ καὶ τὰ θησαυριζόμενα σῴζειν, ὥσπερ τὴν ῥόαν ἐμπηγνυμένου τοῦ μίσχου, καὶ τῶν φυτευομένων δ' ἔνια βλαστάνει θᾶττον ἐν αὐτῇ· λέγεται δὲ καὶ πρὸ τῶν θυρῶν τῆς εἰσόδου φυτευθεῖσαν ἀλεξητήριον εἶναι τῆς ἐπιφερομένης δηλήσεως. πάντα δὲ ταῦτα ἀθρόα φύεται, καθάπερ καὶ τὰ κρόμυα καὶ τὰ σκόροδα· παραβλαστάνουσι γὰρ ἀπὸ τῆς

---

[1] ἐκπίπτει conj. W.; ἐκπίπτων Ald.
[2] ἄνθησιν conj. Scal.; ἄναυσιν corr. to αὔανσιν U; ἄναυσιν M; αὔανσιν Ald.; *floret per partes* G.

has the tallest of all—for the *antherikos* (asphodel-stalk) is very tall : that of iris is smaller, but tougher, though in general it is like the asphodel-stalk. Asphodel also produces much fruit, and its fruit is woody : in shape it is triangular and in colour black ; it is found in the round vessel which is below the flower, and it falls out [1] in summer when this splits open. It does not produce all its flowers [2] at once ; in which respect it resembles squill, but the flowering begins at the bottom. In the stalk of asphodel forms a grub which changes into another creature like a hornet, and then, when the stem withers, eats its way out and flies away. A peculiarity of the plant as compared with others which have a smooth stem appears to be that, though it is slender, it has outgrowths at the top. It provides many things useful for food : the stalk is edible when fried, the seed when roasted, and above all the root [3] when cut up with figs; in fact, as Hesiod says,[4] the plant is extremely profitable.

Now all bulbous plants are tenacious of life, but especially squill ; for this even lives when hung up and continues to do so for a very long time ; it is even able to keep other things that are stored, for instance the pomegranate, if the stalk of the fruit is set in it ; and some cuttings [5] strike more quickly if set in it ; and it is said that, if planted before the entrance door of a house, it wards off mischief [6] which threatens it. All these bulbs grow in masses, as do onions and garlic ; for they make offsets from the root, and some

[3] *cf.* 7. 9. 4 ; 9. 9. 6.
[4] Hes. *Op.* 41.
[5] *cf* 2. 5. 5 ; *C.P.* 5. 6. 10.
[6] Sc. witchcraft. *veneficiorum noxam* Plin. 21. 108.

ρίζης· ἔνια δὲ καὶ ἀπὸ τοῦ σπέρματος φανερῶς, οἷον ὅ τε ἀνθέρικος καὶ τὸ λείριον καὶ τὸ φάσγανον καὶ ὁ βολβός.

5 Ἀλλ' ἴδιον τοῦτο τοῦ βολβοῦ λέγεται, τὸ μὴ ἀπὸ πάντων βλαστάνειν ἅμα τῶν σπερμάτων, ἀλλὰ τοῦ μὲν αὐτοετὲς τοῦ δ' εἰς νέωτα, καθάπερ τὸν αἰγίλωπά¹ φασι καὶ τὸν λωτόν. τοῦτο μὲν οὖν εἴπερ ἀληθὲς κοινὸν ἑτέρων. κοινὸν δὲ ἴσως καὶ τὸ μέλλον λέγεσθαι, πλὴν οὐ πολλῶν, θαυμαστὸν δὲ ἐπὶ πάντων, ὅπερ ἐπί τε τῆς σκίλλης καὶ τοῦ ναρκίσσου συμβαίνει· τῶν μὲν γὰρ ἄλλων καὶ τῶν ἐξ ἀρχῆς φυτευομένων καὶ τῶν βλαστανόντων καθ' ὥραν ἔτους τὸ φύλλον ἀνατέλλει πρῶτον, εἶθ' ὕστερον ὁ καυλός· ἐπὶ δὲ τούτων ὁ καυλὸς πρότερον.

6 Τοῦ ναρκίσσου δὲ ὁ τοῦ ἄνθους μόνον εὐθὺ προωθῶν² τὸ ἄνθος· τῆς δὲ σκίλλης καθ' αὑτόν, εἰς ὕστερον ἐπὶ τούτῳ τὸ ἄνθος ἀνίσχον προσκαθήμενον· ποιεῖται δὲ τὰς ἀνθήσεις τρεῖς, ὧν ἡ μὲν πρώτη δοκεῖ σημαίνειν τὸν πρῶτον ἄροτον, ἡ δὲ δευτέρα τὸν μέσον, ἡ δὲ τρίτη τὸν ἔσχατον· ὡς γὰρ ἂν αὗται γένωνται καὶ οἱ ἄροτοι σχεδὸν οὕτως ἐκβαίνουσιν· ὅταν δὲ οὗτος ἀπογηράσῃ, τότε ἡ τῶν φύλλων βλάστησις πολλαῖς ἡμέραις ὕστερον· ὡσαύτως δὲ καὶ ἐπὶ τοῦ ναρκίσσου, πλὴν οὔτε καυλὸν ἕτερον ἔχει παρὰ τὸν τοῦ ἄνθους, ὥσπερ εἴπομεν, οὔτε καρπὸν φανερόν, ἀλλ'

---

¹ αἰγίλωπα conj. Sch. from Plin. 21. 103; γίλωπα UMAld.
² εὐθὺ προωθῶν conj. W.; εὐθυπρόωρον Ald.H. cf. Plin. 21. 66, where however the statement is transferred to the crocus.

plainly are also increased by seed, as the asphodel polyanthus narcissus corn-flag and purse-tassels.

However it is said to be a peculiarity of purse-tassels that all the seeds do not germinate at once, but some in the same year, some the next year; a like account is given of *aigilops*[1] and trefoil. If then this is true, it is not peculiar to this plant. Nor perhaps is the following characteristic, which is not found in many plants and is marvellous wherever it does occur—and it is found in squill and narcissus: namely that, whereas in most plants, whether those originally planted or those which are produced from them in season, the leaf comes up first and then presently the stem, in these plants the stem comes up first.

In the case of narcissus it is only the flower-stem which comes up, and it immediately pushes up[2] the flower. But in squill it is the stem[3] proper which thus appears, and presently the flower appears emerging[4] from and sitting on it. And it makes three flowerings,[5] of which the first appears to mark the first seed-time, the second the middle one, and the third the last one; for, according as these flowerings have occurred,[6] so the crops usually turn out. But, when the flower-stem[7] has waxed old, then the growth of the leaves follows many days later. So also is it with narcissus, except that it has no second stem besides the flower-stem, as we said, nor any visible fruit; but the flower itself

[3] *i.e.* the whole 'bud.'

[4] ἀνίσχον Ald.; ἀνίσχων conj. Sch. followed by W.

[5] Plin. *l.c.*; *cf.* 18. 237.

[6] *i.e.* the flowering is the sign when to sow. The same is said of the fruiting of σχῖνος *de signis* 55.

[7] οὗτος conj. Sch.; οὕτως Ald.

αὐτὸ τὸ ἄνθος ἅμα τῷ καυλῷ καταφθίνει καὶ ὅταν αὐανθῇ τότε τὰ φύλλα ἀνατέλλει.

7 Πρὸς μὲν οὖν τὰ ἄλλα τὰ συνάμφω ταῦτα ἴδια· πρὸς δὲ τὰ προανθοῦντα τῶν φύλλων καὶ τῶν καυλῶν, ὅπερ δοκεῖ ποιεῖν τὸ τίφυον καὶ ἕτερα τῶν ἀνθικῶν, ἔτι τε τῶν δένδρων ἡ ἀμυγδαλῆ μάλιστα ἢ μόνον, ὅτι ταῦτα μὲν ἅμα τῷ ἄνθει προφαίνει τὸ φύλλον ἢ εὐθὺς κατόπιν, ὥστε καὶ διαζητεῖσθαι περί τινων, ἐπὶ δὲ τούτων οἷον ἀφ' ἑτέρας ἀρχῆς φαίνεται καὶ διὰ τὸ πλῆθος τῶν ἡμερῶν καὶ διὰ τὸ μὴ πρότερον βλαστάνειν πρὶν τοῦ μὲν τὸ ἄνθος τοῦ δὲ καὶ ὁ καυλὸς ὅλος ἀπογηράσῃ. ἡ δὲ βλάστησις προτέρα μὲν τῆς σκίλλης, ὑστέρα δὲ τοῦ ναρκίσσου· πολὺ δὲ πλέον τὸ φύλλον οὗτος ἀφίησι, καί ἐστιν ἡ ῥίζα αὐτὴ μικρὰ καὶ οὐ μεγάλη, προσεμφερὴς δὲ κατὰ τὸ σχῆμα τῷ βολβῷ, πλὴν ⟨οὐ⟩ λεπυριώδης. ταῦτα μὲν οὖν ἔχει σκέψιν.

8 Τῶν δὲ βολβῶν ὅτι πλείω γένη φανερόν, καὶ γὰρ τῷ μεγέθει καὶ τῇ χρόᾳ καὶ τοῖς σχήμασι διαφέρουσι καὶ τοῖς χυλοῖς· ἐνιαχοῦ γὰρ οὕτω γλυκεῖς ὥστε καὶ ὠμοὺς ἐσθίεσθαι, καθάπερ ἐν Χερρονήσῳ τῇ Ταυρικῇ. μεγίστη δὲ καὶ ἰδιωτάτη διαφορὰ τῶν ἐριοφόρων· ἔστι γάρ τι γένος τοιοῦτον, ὃ φύεται μὲν ἐν αἰγιαλοῖς ἔχει δὲ τὸ ἔριον ὑπὸ τοὺς πρώτους χιτῶνας, ὥστε ἀνὰ μέσον εἶναι

---

[1] τίφυον Ald., *cf. C.P.* 1. 10. 5; τ' ἴφυον conj. W.; *iphyum* GBas.Par. *cf.* 6. 6. 11. [2] ἢ add. Sch.
[3] ἐπὶ conj. H.; περὶ UMAld. [4] W. adds ἂν.
[5] αὐτὴ: sc. apart from offsets.
[6] μικρὰ conj. Sch.; οὐ μικρὰ Ald.

ENQUIRY INTO PLANTS, VII. xiii. 6–8

perishes with the stem, and when it has withered, then the plant puts up its leaves.

These two plants then, as compared with the other bulbous plants are peculiar; and, as compared with those which bloom before the leaves and stems appear (as the autumn squill[1] seems to do, and other plants with conspicuous flowers, as well as, among trees, the almond especially, if not alone), there is the distinction that, while these two put forth their leaves along with the flowers or[2] immediately afterwards (so that about some the matter is uncertain) in[3] the case of these two the flower appears, as it were, from a different starting-point, there being a considerable number of days in between, and the growth of the leaves not beginning till,[4] in the case of one of them, the flower, and in the case of the other, the whole stem has withered. Squill produces its leaves before the flower, narcissus afterwards; but the latter produces much more abundant foliage, and the individual[5] root is small[6] rather than large, resembling purse-tassels in shape, except that it is not formed of scales.[7] About these matters then there is doubt.

Of purse-tassels it is plain that there are several kinds; for they differ in size colour shape and taste. [8] In some places they are so sweet as to be eaten raw, as in the Tauric Chersonese. But the greatest and most distinct difference is shown by the 'wool-bearing[9]' purse-tassels; for there is such a kind, and it grows on[10] the sea-shore, and has the wool beneath the outer tunic, so that it is between

---

[7] οὐ λεπυριώδης conj. Sch. from G, *non squamata*; οὐδὲ πυρώδη UMAld.; οὐ λεπυρώδης H.
[8] Plin. 19. 95; Athen. 2. 64.
[9] Plin. 19. 32. See Index.   [10] ἐν after μὲν add. W.

τοῦ τε ἐδωδίμου τοῦ ἐντὸς καὶ τοῦ ἔξω· ὑφαίνεται δὲ ἐξ αὐτοῦ καὶ πόδεια καὶ ἄλλα ἱμάτια· δι' ὃ καὶ ἐριῶδες τοῦτο καὶ οὐχ ὥσπερ τὸ ἐν Ἰνδοῖς τριχῶδες.

9 Πλείω δὲ καὶ τὰ βολβώδη καὶ ἐλάττω. ταῦτα δὲ ... καθάπερ τὸ λευκόϊον καὶ βολβίνη καὶ ὀπιτίων καὶ κύϊξ καὶ τρόπον τινὰ τὸ σισυρίγχιον. βολβώδη δὲ ταῦτα ὅτι στρογγύλα ταῖς ῥίζαις· ἐπεὶ τοῖς γε χρώμασι λευκὰ καὶ οὐ λεπυριώδη. ἴδιον δὲ τοῦ σισυριγχίου τὸ τῆς ῥίζης αὐξάνεσθαι τὸ κάτω πρῶτον, ὃ καλοῦσι ... χειμῶνα, τοῦ δ' ἦρος ὑποφαίνοντος τοῦτο μὲν ταπεινοῦσθαι τὸ δ' ἄνω τὸ ἐδώδιμον αὐξάνεσθαι. καὶ τὰ μὲν τοιαύτας ἔχει τὰς διαφοράς.

XIV. Ἴδια δὲ καὶ ταῦτα ἐν τοῖς ποιώδεσιν, οἷον τό [τε] ἐπὶ τοῦ ἀδιάντου συμβαῖνον· οὐδὲ γὰρ ὑγραίνεται τὸ φύλλον βρεχόμενον οὐδ' ἐπίδροσόν ἐστι διὰ τὸ μὴ τὴν νοτίαν ἐπιμένειν, ὅθεν καὶ ἡ προσηγορία. γένη δὲ αὐτοῦ δύο, τὸ μὲν λευκὸν τὸ δὲ μέλαν, χρήσιμα δ' ἀμφότερα πρὸς ἔκρυσιν κεφαλῆς τριχῶν ἐν ἐλαίῳ τριβόμενα. φύεται δὲ

---

[1] δι' ὃ καὶ ἐρ. τοῦτο: text probably defective. ? δι' ὃ καὶ <χρήσιμον τὸ> ἐριῶδες τοῦτο: 'wherefore this woolly kind is serviceable, which the Indian hairy kind is not.'

[2] Plin. 19. 95.

[3] καὶ ἐλάττω· ταῦτα δὲ: text corrupt and defective.

[4] ὀπιτίων H.; ὁ πιτίων Ald.; *pithyon* Plin. *l.c.*; ὀπιτίων and κύϊξ were possibly earth-nuts.

[5] γε conj. Sch ; τε Ald.

the edible inside and the outside : of it are woven felt shoes and other articles of apparel. Wherefore[1] this kind is woolly and distinct from the Indian kind, which is hairy.

[2] There are also several kinds of plants of the same class as purse-tassels . . . .[3] such as snowdrop star-flower *opition*[4] *kyix*, and to a certain extent Barbary nut. These belong to this class only in having round roots ; for in colour[5] they are white, and the bulbs are not formed of scales. A peculiarity of Barbary nut is that the lower end of the root grows first, and this is called . . . . ; it grows[6] during winter, but, when spring appears, it decreases, while the upper part, which is edible, grows. Such are the differences in these plants.

Of certain properties and habits peculiar to certain herbaceous plants.

XIV. There are also the following peculiarities in herbaceous plants, for instance that[7] which we find in ' wet-proof' (maidenhair); [8] the leaf does not even get wet when it is watered, nor does it catch the dew,[9] because the dew does not[10] rest on it; whence its name. [11] There are two kinds, the white ' wet-proof' (English maidenhair), and the black (maidenhair); and both are useful to prevent the falling off of the hair of the head, for which purpose they are pounded up and mixed with olive-oil. They grow

[6] 'Grows' supplied from G and Plin. *l.c.*, who have no trace of ὃ καλοῦσι.

[7] I have bracketed τε.   [8] Plin. 22. 62–65.

[9] ἐπίδροσον conj. W.; ἐπίδηλον UP₂MAld.; *nec quicquam adhaesisse humoris constat* G.

[10] μὴ before τὴν add. W.

[11] Plin. *l.c.*; 27. 138 ; 25. 132.

μάλιστα πρὸς τὰ ὑδρηλά. ὡς δὲ οἴονταί τινες, καὶ πρὸς στραγγουρίαν τὸ τριχομανὲς ποιεῖ· ἔχει δὲ τὸν καυλὸν ὅμοιον τῷ ἀδιάντῳ τῷ μέλανι, φύλλα δὲ μικρὰ σφόδρα καὶ πυκνὰ καὶ πεφυκότα καταντικρὺ ἀλλήλων, ῥίζα δὲ οὐχ ὕπεστι· χωρία δὲ φιλεῖ σκιερά.

2 Τῶν δὲ κατὰ μέρος ἀνθούντων ἴδιον τὸ περὶ τὸ ἄνθεμον, ὅτι τῶν μὲν ἄλλων πάντων τὰ κάτω πρῶτον ἀπανθεῖ τούτου δὲ τὰ ἄνω· τυγχάνει δ' αὐτοῦ τὸ μὲν κύκλῳ τὸ λευκὸν ἄνθος τὸ δὲ ἐν τῷ μέσῳ τὸ χλωρόν· καὶ καρπὸς ὃς ἐκπίπτει, καθάπερ τοῖς ἀκανθώδεσι, καταλιπὼν τὴν πρόσφυσιν κενήν· εἴδη δ' αὐτοῦ πλείω.

3 Ἴδιον δὲ καὶ τὸ περὶ τὴν ἀπαρίνην, ἣ καὶ τῶν ἱματίων ἀντέχεται διὰ τὴν τραχύτητα καί ἐστι δυσαφαίρετον· ἐν τούτῳ γὰρ ἐγγίνεται τῷ τραχεῖ τὸ ἄνθος οὐ προϊὸν οὐδὲ ἐκφαῖνον ἀλλ' ἐν ἑαυτῷ πεττόμενον καὶ σπερμογονοῦν· ὥστε παρόμοιον εἶναι τὸ συμβαῖνον ὥσπερ ἐπὶ τῶν γαλεῶν καὶ ῥινῶν· ἐκεῖνά τε γὰρ ἐν ἑαυτοῖς ᾠοτοκήσαντα ζῳογονεῖ, καὶ αὕτη τὸ ἄνθος ἐν ἑαυτῇ κατέχουσα καὶ πέττουσα καρποτοκεῖ.

XV. Ὅσα δὲ τὰς ἀνθήσεις λαμβάνουσι ἀκολουθοῦντα τοῖς ἄστροις, οἷον τὸ ἡλιοτρόπιον καλούμενον καὶ ὁ σκόλυμος, ἅμα γὰρ ταῖς τροπαῖς καὶ οὗτος, ἔτι δὲ τὸ χελιδόνιον, καὶ γὰρ τοῦτο ἅμα τῷ χελιδονίᾳ ἀνθεῖ, ταῦτα δὲ δόξειεν ἂν τῇ μὲν φυσικὴν ἔχειν τὴν αἰτίαν τῇ δὲ συμπτωματικήν.

---

[1] *i.e.* the white kind. Sch. followed by G adds τὸ καὶ τριχομανὲς καλούμενον after τὸ μὲν λευκὸν above.

especially in damp places. Some think that *trikhomanes*[1] (English maidenhair) is also useful in cases of strangury. Its stem is like that of the black kind, but it has small leaves, which are close set and grow in opposite pairs; there is no root below, and the plant loves shady places.

Of those plants which do not flower all at once *anthemon* has the peculiarity that, while in all others[2] the lower part flowers first, in this plant it is the upper part which does so; the outer circle of the flower is white,[3] and the centre green[4]; and the fruit falls off, as in spinous plants, leaving the attachment bare. There are several forms of it.

[5] Bedstraw has the peculiarity that it sticks to clothes owing to its roughness, and it is hard to pull away; indeed it is in this rough part that the flower is contained: it does not project nor show, but matures within itself and produces seed; so that its habit is like that of weasels and sharks; for, as these animals[6] likewise produce eggs in themselves and then bear their young alive, so this plant keeps its flower within itself, matures it and produces fruit.

XV. [7] As to these plants whose flowering time is dependent on the heavenly bodies,[8] as the plant called *heliotropion*, golden thistle (for this also blooms at the solstice), and also 'swallow-plant' (greater celandine)—for this blooms when the[9] Swallow-wind blows—the reason in these cases would seem to be partly in their nature and partly accidental.

---

[2] ἴδιον after πάντων om. W. after Sch.
[3] τὸ λευκὸν: ? λευκὸν τὸ.   [4] ? om. τὸ before χλωρόν.
[5] Plin. 21. 104.   [6] cf. Arist. H.A. 6. 11.
[7] Athen. 15. 32.   [8] ἄστροις conj. St.; ἀγρίοις Ald.
[9] τῷ conj. Sch.; τῇ MAld. cf. Plin. 2. 122.

2 Πολλὰ δὲ τοιαῦτά ἐστι καὶ ἐν ἑτέροις ἴδια· οἷον καὶ ἡ τοῦ ἀειζώου φύσις τὸ διαμένειν ὑγρὸν ἀεὶ καὶ χλωρόν, φύλλον σαρκῶδες ἔχον καὶ λεῖον καὶ πρόμηκες. φύεται δὲ ἔν τε τοῖς ἀλιπέδοις τοῖς τε ἐπὶ τῶν τειχῶν ἀνδήροις καὶ οὐχ ἥκιστα ἐπὶ τῶν κεράμων, ὅταν ἐπιγένηται γῆς τις ἀμμώδης συρροή.

3 Πολλὰ δ' ἄν τις ἴσως λάβοι καὶ ἕτερα περιττά. χρὴ δέ, ὥσπερ πολλάκις εἴρηται, τὰς ἰδιότητας θεωρεῖν καὶ τὰς διαφορὰς πρὸς τὰ ἄλλα. τὰ μὲν ἐν πλείοσιν ἰδέαις ἐστὶ καὶ σχεδὸν οἷον ὁμωνύμοις, ὥσπερ ὁ λωτός· τούτου γὰρ εἴδη πολλὰ διαφέροντα καὶ φύλλοις καὶ καυλοῖς καὶ ἄνθεσι καὶ καρποῖς, ἐν οἷς καὶ ὁ μελίλωτος καλούμενος· καὶ δυνάμει δὲ τῇ κατὰ τὴν προσφοράν, ἔτι τε τῷ μὴ τοὺς αὐτοὺς τόπους ζητεῖν. ὁμοίως δὲ καὶ ἕτερα πλείω.

4 Τὰ δὲ ἐν ἐλάττοσιν, ὥσπερ ὁ στρύχνος ὁμωνυμίᾳ τινὶ παντελῶς εἰλημμένος· ὁ μὲν γὰρ ἐδώδιμος καὶ ὥσπερ ἥμερος, καρπὸν ἔχων ῥαγώδη, ἕτεροι δὲ δύο εἰσίν, ὧν ὁ μὲν ὕπνον ὁ δὲ μανίαν ἐμποιεῖν δύναται, πλείων δ' ἔτι δοθεὶς καὶ κτείνει. ὁμοίως δὲ τοῦτο καὶ ἐφ' ἑτέρων ἐστὶ λαβεῖν, ἃ πολλὴν ἔχει διάστασιν. περὶ μὲν οὖν τῶν ἄλλων τῶν ποιωδῶν ἱκανῶς εἴρηται. περὶ δὲ τοῦ σίτου καὶ τῶν σιτωδῶν μετὰ ταῦτα λεκτέον· τοῦτο γὰρ ἔτι κατάλοιπον ἦν.

---

[1] οἷον conj. W.; διὸ Ald.
[2] ἀλιπέδοις conj. Sch.; ἀληπέδοις U; ἀληπέδοις M; ἀλοπέδοις Ald. cf. Xen. Hell. 2. 4. 30.
[3] τε after τοῖς add. W. after Lobeck.
[4] e.g. λωτός and μελίλωτος. See Index, λωτός.
[5] μελίλωτος conj. Bod.; μελίας σῖτος Ald.

Such peculiarities are common in other plants also; thus[1] it is the nature of the house-leek to remain always moist and green, its leaf being fleshy smooth and oblong. It grows on flat shores,[2] on the[3] earthy tops of walls, and especially on tiled roofs, when there is on them a sandy accumulation of earth.

Possibly one might mention many other eccentricities. But, as has been repeatedly said, we must only observe the peculiarities and differences which one plant has as compared with others. Some plants are found in several forms which have almost[4] the same name, for instance the *lotos*; for of this there are many forms differing in leaves stems flowers and fruit, including the plant called *melilotos*[5]; there are also forms differing in the virtues for which[6] they are used as food, and again in their fondness for different localities. So too is it with many other plants.

Others are found in fewer forms, as *strykhnos*,[7] which is a general name covering plants that are quite distinct; one is edible and like a cultivated plant, having a berry-like fruit, and there are two others,[8] of which the one is said to induce sleep, the other to cause madness, or, if it is administered in a larger dose, death. The same thing may be observed in other plants which are widely different. Now about the other herbaceous plants enough has been said; but concerning corn and corn-like plants we must speak next; for this subject still lies before us.

[6] δὲ τῇ conj. Sch.; διτταί UM; διτταῖς Ald.
[7] *cf.* 7. 7. 2; Plin 21. 177–179; Diosc. 4. 70–73; Index.
[8] In 9. 11. 5 these two plants are said to be συνώνυμοι, *i.e.* different forms of the same plant, whereas the 'edible' στρύχνος is the same only in name (ὁμωνυμίᾳ). *cf.* 9. 12. 5.

BOOK VIII

Θ

I. Περὶ μὲν οὖν τῶν ἄλλων ποιωδῶν ἱκανῶς εἰρήσθω· περὶ δὲ σίτου καὶ τῶν σιτωδῶν λέγωμεν ὁμοίως τοῖς πρότερον· τοῦτο γὰρ κατάλοιπον ἦν τῶν ποιωδῶν.

Δύο δὲ αὐτοῦ γένη τὰ μέγιστα τυγχάνει· τὰ μὲν γὰρ σιτώδη, οἷον πυροὶ κριθαὶ τίφαι ζειαὶ τὰ ἄλλα τὰ ὁμοιόπυρα ἢ ὁμοιόκριθα· τὰ δὲ χεδροπά, οἷον κύαμος ἐρέβινθος πισὸς καὶ ὅλως τὰ ὄσπρια προσαγορευόμενα· τρίτον δὲ παρ' αὐτὰ κέγχρος ἔλυμος σήσαμον καὶ ἁπλῶς τὰ ἐν τοῖς θερινοῖς ἀρότοις ἀνώνυμα κοινῇ προσηγορίᾳ.

2 Ἔστι δὲ ἡ μὲν γένεσις αὐτῶν μία καὶ ἁπλῆ· φύεται γὰρ ἀπὸ σπέρματος, ἐὰν μή τι σπάνιον καὶ ὀλίγον ἀπὸ τῆς ῥίζης. ὧραι δὲ τοῦ σπόρου τῶν πλείστων δύο· πρώτη μὲν καὶ μάλιστα ἡ περὶ Πλειάδος δύσιν, ᾗ καὶ Ἡσίοδος ἠκολούθηκε

---

[1] τῶν before ποιωδῶν om. Sch.
[2] Plin. 18. 48–80.
[3] ἔλυμος: μελίνη appears to be the Attic name for this plant. Sch. would restore it for ἔλυμος here and 4. 4. 10; 8. 11. 1.

BOOK VIII

OF HERBACEOUS PLANTS: CEREALS, PULSES, AND 'SUMMER CROPS.'

Of the three classes and the times of sowing and of germination.

I. Let the above suffice for an account of the other herbaceous[1] plants; let us now discuss corn and corn-like plants in the same manner as those already treated; for this class of herbaceous plants we reserved.

[2] There are two principal classes; there are the corn-like plants such as wheat barley one-seeded wheat rice-wheat and the others which resemble either of the first two; and again there are the leguminous plants, as bean chick-pea pea, and in general those to which the name of pulses is given. Besides these there is a third class, which includes millet[3] Italian millet, sesame and in general the plants which belong to the summer seed-time,[4] which lack any common designation.

There is only one single way of propagating these; they grow from seed, except that some may grow rarely and scantily from a root. There are two seasons for sowing most of them; the first and most important is about the setting of the Pleiad[5]; this rule we find even Hesiod[6] following with

[4] *cf.* 8. 7. 3.
[5] Πλειάδος conj. Sch.; πλειάδας U; πλειάδων Ald.
[6] Hes. *Op.* 383.

143

καὶ σχεδὸν οἱ πλεῖστοι, δι' ὃ καὶ καλοῦσί τινες αὐτὴν ἄροτον· ἄλλη δ' ἀρχομένου τοῦ ἦρος μετὰ τὰς τροπὰς τοῦ χειμῶνος. οὐ τῶν αὐτῶν δὲ ἑκατέρα. τὰ μὲν γὰρ αὐτῶν φιλεῖ πρωϊσπορεῖσθαι, τὰ δὲ ὀψὲ διὰ τὸ μὴ δύνασθαι φέρειν τοὺς χειμῶνας, τὰ δὲ πρὸς ἀμφοτέρας τὰς ὥρας οὐ κακῶς ἔχει, καὶ πρὸς χειμῶνα καὶ πρὸς ἔαρ.

3 Πρωΐσπορα μὲν οὖν ἐστι πυρὸς κριθή, καὶ τούτων ἡ κριθὴ πρωϊσπορώτερον· ἔτι δὲ ζειά τίφη ὀλύρα καὶ εἴ τι ἕτερον ὁμοιόπυρον· ἁπάντων γὰρ σχεδὸν ὁ αὐτὸς χρόνος τῆς σπορᾶς· τῶν δὲ χεδροπῶν μάλιστα ὡς εἰπεῖν κύαμος καὶ ὦχρος· ταῦτα γὰρ διὰ τὴν ἀσθένειαν προλαβεῖν τῇ ῥιζώσει βούλεται τοὺς χειμῶνας· πρωΐσπορον δὲ καὶ ὁ θέρμος· ἀπὸ τῆς ἅλω γάρ φασι δεῖν καταβάλλειν εὐθύς.

4 Ὀψίσπορα δὲ τούτων γε αὐτῶν ὅσα διαφέρει τοῖς γένεσιν, οἷον πυρῶν τέ τι γένος καὶ κριθῶν ὃ καλοῦσι τρίμηνον διὰ τὸ ἐν τοσούτῳ τελειοῦσθαι· καὶ τῶν χεδροπῶν τὰ τοιάδε, φακὸς ἀφάκη πισός. ἐν ἀμφοτέραις δὲ ταῖς ὥραις τῶν χεδροπῶν, καθάπερ ὄροβος ἐρέβινθος· οἱ δὲ καὶ τὸν κύαμον ὀψὲ σπείρουσιν, ἐὰν ὑστερήσωσι τῶν πρώτων ἀρότων. ἁπλῶς δὲ πρωϊσποροῦσι τὰ μὲν δι' ἰσχὺν ὡς δυνάμενα φέρειν τοὺς χειμῶνας, τὰ δὲ δι' ἀσθένειαν, ὅπως προλάβωσι ταῖς εὐδίαις τὴν αὔξησιν. δύο μὲν οὖν αὗται. τρίτη δὲ τῶν θερινῶν ἦν

---

[1] A cultural variety of ζειά. cf. 8. 9. 2.
[2] τῶν δέ γε Ald.; γε om. Sch.
[3] ὦχρος conj. W.; κέγχρος Ald.; om. G. cf. 8. 3. 1 and 2.
[4] cf. 8. 11. 8.

most authorities; wherefore some call it simply
'the seed-time.' Another time is at the beginning
of spring after the winter equinox. However
different crops are sown at the two seasons. For
some of them love to be sown early, some late
because they cannot bear the winters, while some
will do not amiss at either season, both towards
winter and towards spring.

Crops sown early are wheat and barley, and of
these the latter is sown the earlier; also rice-wheat
one-seeded wheat *olyra*,[1] and others which resemble
wheat. For all of these the time of sowing is about
the same. Of leguminous plants[2] bean and *okhros*,[3]
it may be said, are specially sown at this time; for
these on account of their weakness like to be well
rooted before the winter. Lupin is also sown early;
in fact they say it should be sown straight from the
threshing-floor.[4]

Those which are sown late are certain special
varieties[5] of these very kinds, as a certain kind of
wheat, and of barley the kind which is called 'three
months barley' because it takes that time to mature;
and among leguminous plants lentil tare pea.
However some of these plants are sown at both
seasons, as vetch and chick-pea; some also sow
beans late, if they have missed the first seed-time.
To speak generally, some crops are sown early
because of their robustness, since they can stand
the winters, some because of their weakness, so
that their growth may be secured[6] in the fine
weather. These then are the two seasons; the
third is that of the summer crops of which we

[5] τοῖς γένεσιν: τῇ γενέσει W. *i.e.* 'certain kinds which differ as to their germination.'   [6] *cf. C.P.* 4. 7. 2.

εἴπομεν, ἐν ᾗ κέγχρος σπείρεται καὶ μέλινος καὶ σήσαμον, ἔτι δ' ἐρύσιμον καὶ ὅρμινον. χρόνοι μὲν οὖν ἑκάστων οὗτοι.

5 Βλαστάνει δὲ τὸ μὲν θᾶττον τὸ δὲ βραδύτερον· καὶ κριθὴ μὲν καὶ πυρὸς ἑβδομαῖα μάλιστα· προτερεῖ δὲ ἡ κριθὴ μᾶλλον· τὰ δ' ὄσπρια τεταρταῖα ἢ πεμπταῖα πλὴν κυάμων· κύαμος δὲ καὶ τῶν σιτωδῶν ἔνια πλείοσιν· ἐνιαχοῦ γὰρ καὶ πεντεκαιδεκαταῖος, ὁτὲ δὲ καὶ εἰκοσταῖος· δυσφυέστατον γὰρ τοῦτο πάντων, ἐὰν δὲ δὴ καὶ σπαρέντος ἐπὶ πλέον ὕδωρ ἐπιγένηται, καὶ παντελῶς. εἰ δὲ τῶν ἐν τοῖς ἠρινοῖς ἀρότοις θᾶττον ἡ ἔκφυσις διὰ τὴν ὥραν σκεπτέον.

6 Χρὴ δὲ τὰς ἀναβλαστήσεις καὶ τὰς διαφύσεις ταύτας ὡς ἐπὶ τὸ πᾶν διαλαβεῖν· ἐνίοτε γὰρ ἐνιαχοῦ καὶ ἐν ἐλάττοσιν ἡμέραις, καθάπερ ἐν Αἰγύπτῳ <κριθή>· τριταίαν γάρ φασι καὶ τεταρταίαν ἀνατέλλειν· παρ' ἄλλοις δὲ ἐν πλείοσι τῶν εἰρημένων, ὅπερ καὶ οὐκ ἄλογον, ὅταν καὶ χώρα καὶ ἀὴρ διαφέρῃ καὶ πρωϊαίτερον ἢ ὀψιαίτερον ἀρόσῃ καὶ τὰ ἐπιγινόμενα ἀνόμοια τυγχάνῃ. ἡ μὲν γὰρ μανὴ καὶ κούφη καὶ εὐκράτῳ ἀέρι ταχὺ καὶ ῥᾳδίως ἀναδίδωσιν, ἡ δὲ γλίσχρα καὶ βαρεῖα βραδέως, ἡ δὲ τοῖς τόποις αὐχμωδεστέρα βραδύτερον.

7 Ἔτι δὲ ἂν χειμῶνες ἐπιγένωνται καὶ αὐχμοὶ καὶ εὐδίαι καὶ πάλιν ὕδατα· καὶ γὰρ ἐν τούτοις πολὺ παραλλάττουσιν. ὡσαύτως δὲ καὶ ἐὰν ἡ

---

[1] ἔνια conj. W.; ἐν Ald.
[2] The reason is given *C.P.* 4. 8. 2.
[3] *cf.* 7. 1. 4.

ENQUIRY INTO PLANTS, VIII. I. 4–7

spoke, in which are sown millet Italian millet sesame, and also *erysimon* and *horminon*. Such then are the times for each.

Some are quicker in coming up, some slower. Barley and wheat generally come up on the seventh day, but barley is the earlier. Pulses take four or five days, except beans; for they, like some kinds[1] of corn, require a longer time; in some places they take as much as fifteen days, or even twenty. This crop indeed is the slowest to start of all, and if after the sowing there is a long spell of wet weather, it is extremely slow.[2] Whether the sprouting[3] of crops sown at the spring seed-time is quicker because of the season is matter for enquiry.

These times of sprouting or germination must be taken generally; for at some times and places germination takes fewer days, as with barley[4] in Egypt, where it is said to come up on the third or fourth day; while elsewhere it takes longer than the period mentioned, which is not surprising when both soil and climate are different, when one makes the sowing earlier or later, and when the crop is subjected to different influences afterwards. For open light soil with a favourable[5] climate produces quick and easy growth, while soil that is sticky and heavy tends to slow growth, and that of a specially dry district to slower growth still.

Moreover the time of growth is affected, according as storms supervene, or droughts, or fine weather or again rain; for these conditions make wide differences. So too it makes a difference if the

---

[4] κριθή add. W.
[5] εὐκράτῳ conj. Scal. from G (*benigno caelo*); εὐκάρπῳ Ald.

γῇ προειργασμένῃ καὶ κόπρον ἔχουσα τυγχάνῃ, καὶ ἐὰν μηδὲν τούτων· ἐπεὶ καὶ περὶ τὸ πρωϊσπορεῖν ἕκαστα καὶ ὀψισπορεῖν αἱ χῶραι διαφέρουσιν. ἔνιοι δὲ καὶ περὶ τὴν Ἑλλάδα πάντα πρωϊσπορεῖν εἰώθασι διὰ ψυχρότητα τῆς χώρας, ὥσπερ οἱ Φωκεῖς, ὅπως ἂν οἱ χειμῶνες μὴ νήπια καταλαμβάνωσιν.

II. Βλαστάνει δὲ τὰ μὲν ἐκ τοῦ αὐτοῦ τὴν ῥίζαν ἀφιέντα καὶ τὸ φύλλον, τὰ δὲ ἑκάτερον ἐξ ἑκατέρου τοῦ ἄκρου. πυρὸς μὲν οὖν καὶ κριθὴ καὶ τίφη καὶ ὅλως ὅσα σιτώδη πάντα ἐξ ἑκατέρου ὥσπερ ἐν τῷ στάχυϊ πέφυκεν, ἀπὸ μὲν τοῦ κάτω τοῦ παχέος τὴν ῥίζαν ἀπὸ δὲ τοῦ ἄνω τὸν βλαστόν· ἐν δέ τι καὶ συνεχὲς γίνεται τὸ ἀμφοῖν τῆς τε ῥίζης καὶ τοῦ καυλοῦ. κύαμος δὲ καὶ τὰ ἄλλα χεδροπὰ οὐχ ὁμοίως, ἀλλ᾽ ἐκ τοῦ αὐτοῦ τὴν ῥίζαν καὶ τὸν καυλόν, καθ᾽ ὃ καὶ ἡ πρόσφυσις αὐτῶν ἐστι πρὸς τὸν λοβόν, ἐν ᾧ καὶ ἔχουσιν οἷον ἀρχήν τινα φανεράν· ἐπ᾽ ἐνίων δὲ καὶ αἰδοιῶδες φαίνεται, καθάπερ ἐπὶ τῶν κυάμων καὶ τῶν ἐρεβίνθων καὶ μάλιστα τῶν θέρμων· ἐκ τούτου γὰρ ἡ μὲν ῥίζα κάτω τὸ δὲ φύλλον καὶ ὁ καυλὸς ἄνω χωρεῖ.

2 Ταύτῃ μὲν οὖν πῃ διαφέρει. τῇ δὲ ὁμοίως ἔχει τῷ πάντα κατὰ τὴν πρόσφυσιν τοῦ λοβοῦ καὶ τοῦ στάχυος ἀφιέναι τὴν ῥίζαν καὶ μὴ

---

[1] ὡς προειργ. Ald. H.; ὡς om. Sch. from G.   [2] cf. 8. 8. 2.
[3] ὥσπερ conj. Scal.; πάντα Ald. (? repeated by mistake). cf. C.P. 4. 7. 4.

ground has been well tilled[1] and given dung, or if neither of these things has been done: for the soil makes a difference even as to the early or late germination of each crop. In Hellas some are used to sow everything earlier because of the coldness of the soil, for instance the Phocians;[2] the object being that the winter may not overtake the crop while it is still tender.

Of differences in the mode of germination and of subsequent development.

II. In germinating some of these plants produce their root and their leaves from the same point, some separately, from either end of the seed. Wheat barley one-seeded wheat, and in general all the cereals produce them from either end, in a manner corresponding to[3] the position of the seed in the ear, the root growing from the stout lower part, the shoot from the upper part; but the part corresponding to the root and that corresponding to the stem form a single continuous whole. Beans and other leguminous plants do not grow in the same manner, but they produce the root and the stem from the same point, namely the point at which the seed is attached to the pod, which, it is plain, is a sort of starting point of fresh growth. In some cases there is also a formation resembling the *penis*, as in beans chick-peas and especially in lupins; from this[4] the root grows downwards, the leaf and the stem upwards.

There are then these different ways of germinating; but a point[5] in which all these plants agree is that they all send out their roots at the place where

[4] τούτου conj. Sch.; τούτων Ald. *cf. C.P.* 4. 7. 4.
[5] *cf. C.P.* 4. 7. 7.

THEOPHRASTUS

καθάπερ ἐν τοῖς δενδρικοῖς τισιν ἀνάπαλιν, οἷον ἀμυγδαλῇ καρύῳ βαλάνῳ τοῖς τοιούτοις. ἐν ἅπασι δὲ ἡ ῥίζα μικρῷ πρότερον ἐκφύεται τοῦ καυλοῦ· συμβαίνει δὲ ἔν γέ τισι τῶν δένδρων ὥστε τὸν μὲν βλαστὸν ἐν αὐτῷ τῷ σπέρματι βλαστάνειν πρῶτον, αὐξανομένου δὲ διΐστασθαι τὰ σπέρματα—πάντα γάρ πως καὶ ταῦτα διμερῆ, τὰ δὲ δὴ χεδροπὰ φανερῶς πάντα δίθυρα καὶ σύνθετα—τὴν δὲ ῥίζαν εὐθὺς ἔξω προωθεῖσθαι· ἐν δὲ τοῖς σιτηροῖς διὰ τὸ καθ' ἓν αὐτὰ εἶναι τοῦτο μὲν οὐ συμβαίνει, προτερεῖ δὲ ἡ ῥίζα μικρόν.

3 Ἀναφύεται δὲ ἡ μὲν κριθὴ καὶ ὁ πυρὸς μονόφυλλα, ὁ δὲ πισὸς καὶ ὁ κύαμος καὶ ὁ ἐρέβινθος πολύφυλλα. ῥίζαν δὲ ἔχει τὰ μὲν χεδροπὰ πάντα ξυλώδη καὶ μίαν ἀπὸ δὲ ταύτης καὶ ἀποφύσεις λεπτάς. βαθυρριζότατον δὲ ὡς εἰπεῖν τούτων ὁ ἐρέβινθος, ἐνίοτε δὲ καὶ παρακαθίησιν· ἀλλ' ὁ πυρὸς καὶ ἡ κριθὴ καὶ τὰ ἄλλα τὰ σιτώδη πολύρριζα καὶ λεπτόρριζα, δι' ὃ καὶ ταρρώδη. καὶ πολύκλαδα καὶ πολύκαυλα πάντα τὰ τοιαῦτα. σχεδὸν δὲ καὶ ἐναντίωσίς τις ἑκατέρων ἐστί· τὰ μὲν γὰρ χεδροπὰ μονόρριζα ὄντα πολλὰς ἄνωθεν ἀπὸ τῶν καυλῶν ἀποφύσεις ἔχει πλὴν κυάμου· τὰ δὲ σιτηρὰ πολύρριζα πολλοὺς μὲν ἀνίησι

---

[1] βαλάνῳ: διοσβαλάνῳ Sch. from mBod.
[2] τισι τῶν δένδρων conj. W.; σιτώδεσιν UMAld.; τοῖς δενδρικοῖς conj. Sch. This and W.'s other conjectures in this section are rather desperate, but are accepted provisionally as at least restoring a satisfactory sense. The passage looks as if it had been deliberately tampered with by someone who misunderstood it.

ENQUIRY INTO PLANTS, VIII. II. 2-3

the seed is attached to the pod or ear, whereas the contrary is the case with the seeds of certain trees, as almond hazel acorn[1] and the like. And in all these plants the root begins to grow a little before the stem; whereas in certain trees[2] the bud first begins to grow within the seed itself, and, as it increases in size, the seeds split—for all such seeds are in a manner in two halves, and those of leguminous plants again all plainly have two valves and are double—and then the root is immediately thrust out; but in cereals,[3] since the seeds are in one piece,[4] this does not[5] occur, but the root grows a little before the bud.

Barley and wheat come up with a single leaf, but peas beans and chick-peas with several. [6] All the leguminous plants have a single woody root, and also slender[7] side-roots springing from this. The chick-pea is about the deepest rooting of these, and sometimes it has side-roots; but wheat barley and the other cereals have a number of fine roots, wherefore they are matted together.[8] Again all such plants have many branches and many stems. And there is a sort of contrast between these two classes; the leguminous plants, which have a single root, have many side-growths above from the stem— all except beans; while the cereals, which have many roots, send up many shoots,[9] but these have

[3] σιτηροῖς conj. W.; χεδροποῖς UMAld.
[4] καθ' ἓν αὐτὰ conj. W.; κατὰ τὸ αὐτὸ UMAld.
[5] οὐ conj. Scal. from G; οὖν UMAld.
[6] Plin. 18. 51.
[7] λεπτάς conj. St.; λεπταί Ald.H.
[8] ταρρώδη: cf. 6. 7. 4.
[9] μὲν conj. Sch.; γὰρ Ald.H.

151

βλαστούς, ἀπαράβλαστοι δὲ οὗτοι, πλὴν εἴ τι γένος πυρῶν τοιοῦτον, οὓς καλοῦσι σιτανίας καὶ κριθανίας.

4 Τὸν μὲν οὖν χειμῶνα ἐν τῇ χλόῃ μένει τὰ σιτώδη, διαγελώσης δὲ τῆς ὥρας καυλὸν ἀφίησιν ἐκ τοῦ μέσου καὶ γονατοῦται. συμβαίνει δ' εὐθὺς ἐν τῷ τρίτῳ γόνατι, τοῖς δὲ ἐν τῷ τετάρτῳ, καὶ τὸν στάχυν ἔχειν ἀλλ' οὐ φανερὸν ἐν τῷ ὄγκῳ·—γίνεται δὲ ἐν τῷ ὅλῳ καλάμῳ πλείω τούτων·—ὥστε σχεδὸν ἅμα τῷ καλαμοῦσθαι συνίστασθαι ⟨ἢ⟩ μικρὸν ὕστερον· ἀλλ' οὐ πρότερον φανερὸς γίνεται πρὶν ἂν προαυξηθεὶς ἐν τῇ κάλυκι γένηται, τότε δὲ ἡ κύησις φανερὰ διὰ τὸν ὄγκον.

5 Ἀπολυθεὶς δ' εὐθὺς ἀνθεῖ μεθ' ἡμέρας τέτταρας ἢ πέντε καὶ πυρὸς καὶ κριθὴ καὶ ἀνθεῖ σχεδὸν τὰς ἴσας, οἱ δὲ τὰς πλείστας λέγοντες ἐν ταῖς ἑπτά φασιν ἀπανθεῖν. ἀλλὰ τῶν χεδροπῶν χρόνιος ἡ ἄνθησις· χρονιωτάτη δὲ τῶν μὲν ἄλλων ὀρόβου καὶ ἐρεβίνθου, τούτων δ' ἁπάντων τοῦ κυάμου καὶ ἐν μεγίστῃ διαφορᾷ· τετταράκοντα γὰρ ἡμερῶν ἀνθεῖν λέγουσι· πλὴν οἱ μὲν ἀεὶ παρανθοῦντος ἑτέρου καὶ ἑτέρου λέγουσι, κατὰ μέρος γὰρ ἀνθεῖν, οἱ δὲ ἁπλῶς. ἡ γὰρ ἄνθησις τῶν μὲν σταχυηρῶν ἀθρόως τῶν δὲ ἐλλοβωδῶν καὶ χεδροπῶν πάντων κατὰ μέρος· πρῶτα γὰρ ἀνθεῖ τὰ κάτω, καὶ ὅταν ταῦτα ἀπανθήσῃ τὰ ἐχόμενα, καὶ οὕτως αἰεὶ βαδίζει πρὸς τὰ ἄνω.

---

[1] Plin. 18. 52.   [2] Plin. 18. 56.
[3] cf. 7. 7. 1 ; 8. 4. 3.

no side-shoots—except such sorts of wheat as are called *sitanias* and *krithanias* ('barley-wheat').

[1] During winter cereals remain in the blade, but, as the season begins to smile, they send up a stem from the midst and it becomes jointed. [2] And it comes to pass that the ear also at once appears in the third, or in some cases in the fourth joint, though it is not distinctly seen in the mass of growth (the whole stem contains more joints than three or four), so that it must be formed at the same time that the straw grows or but a little later; though it does not become conspicuous till it has first swollen and formed in the sheath,[3] and by that time its size makes its development visible.

Four or five days after being set free [4] wheat and barley flower and remain in bloom for a like number of days; those who put the period at the longest say that the bloom is shed in seven days. [5] On the other hand the flowering period of leguminous plants lasts a long time; that of vetch and chick-pea is longer than that of most, but that of the bean is far longer than that of any of them; they say that it is in bloom for forty days; some however give this period absolutely, others say that at different times different parts are in flower,[6] since the whole plant does not flower at once. For plants with an ear bloom all at once, but plants with pods and all leguminous plants bloom part at a time; the lower part blooms first, and, when this bloom has fallen, the part next above it, and so on up to the top.

---

[4] Sc. from the sheath. ἀπολυθείς Ald.H.; ἀποχυθείς conj. Sch. followed by W. *cf.* ἀπόχυσις 8. 3. 4.

[5] Plin. 18. 59.

[6] παρανθοῦντος conj. H.; παρανθοῦντες Ald.

δι' ὃ πολλὰ τῶν ὀρόβων τίλλεται τὰ μὲν κάτω κατερρυηκότα τὰ δ' ἄνω χλωρὰ πάμπαν.

6 Μετὰ δὲ τὴν ἀπάνθησιν ἁδρύνονται καὶ τελειοῦνται πυρὸς μὲν καὶ κριθὴ τετταρακοσταῖα μάλιστα· παραπλησίως δὲ καὶ τίφη καὶ τἆλλα τὰ τοιαῦτα. τετταρακοσταῖον δέ φασι καὶ τὸν κύαμον, ὥστε ἐν ἴσαις ἀνθεῖν καὶ τελειοῦσθαι· τὰ δ' ἄλλα ἐν ἐλάττοσιν· ἐλαχίσταις δὲ ὁ ἐρέβινθος, εἴπερ ἀπὸ τῆς σπορᾶς ἐν τετταράκοντα τελειοῦται ταῖς ἁπάσαις ὥσπερ τινές φασιν· ἐπεὶ τό γ' ὅλον ὅτι τάχιστα φανερόν. οἱ δὲ κέγχροι καὶ τὰ σήσαμα καὶ οἱ μέλινοι¹ καὶ ὅλως τὰ θερινὰ σχεδὸν ὁμολογεῖται τὰς τετταράκονθ' ἡμέρας λαμβάνειν· οἱ δέ φασι καὶ ἐλάττους.

7 Διαφέρει δὲ καὶ πρὸς τὴν τελείωσιν χώρα τε χώρας καὶ ἀὴρ ἀέρος· ἐν ἐλάττοσι γὰρ ἔνιαι δοκοῦσιν ἐκφέρειν, ὥσπερ ἄλλαι τε καὶ μάλιστα ἐπιδήλως Αἴγυπτος· ἐκεῖ γὰρ κριθαὶ μὲν ἐν ἑξαμήνῳ πυροὶ δὲ ἐν τῷ ἑβδόμῳ θερίζονται· περὶ δὲ τὴν Ἑλλάδα κριθαὶ μὲν ἐν τῷ ἑβδόμῳ παρὰ δὲ τοῖς πλείστοις ὀγδόῳ, πυροὶ δὲ ἔτι προσεπιλαμβάνουσιν. οὐ μὴν οὐδὲ ἐκεῖ τό γε πᾶν πλῆθος οὕτως, ἀλλ' ὅσον εἰς ἀπαρχήν· κομίζεται γὰρ πρὸς ἱερῶν τινῶν χρείαν ἄλφιτα νέα τῷ ἕκτῳ μηνὶ καὶ ταῦτα ἐκ τῶν ἄνω τόπων ὑπὲρ Μέμφιν.

8 Λέγεται δὲ καὶ ἐν Σικελίᾳ τῆς Μεσσηνίας ἐν

---

¹ μέλινοι Ald.H.; ἔλυμοι Vo.Vin. *cf.* 8, 1, 1 n.

ENQUIRY INTO PLANTS, VIII. II. 5-8

Wherefore, at the time when some of the vetches are gathered, the lower seeds have already fallen, while the upper ones are still quite green.

After the flowering is over wheat and barley develop and mature in about forty days; one-seeded wheat and other such plants take about the same time. So too, they say, does the bean, which blooms and matures in a like number of days: but the others take fewer, and fewest of all the chick-pea, since, as some say, it takes only forty days from the time when it is sown to that when it is mature; and in any case it is clear that the plant as a whole develops very rapidly. Millet sesame Italian millet [1] and the summer crops in general, it is fairly well agreed, take the same number of days, that is, forty; though some say that they take less.

Of differences in development due to soil or climate.

[2] Again as to the development of the plant there are differences according to soil and climate. Some soils seem to produce the crop in fewer days; for instance, Egypt may be given as a specially conspicuous example; in that country barley is reaped in six months and wheat in seven: while in Hellas the barley [3] harvest is in the seventh month, or in most parts in the eighth, and wheat requires an even longer time. However even in Egypt the whole harvest is not gathered at such an early date, but only what is required for the first-fruits; for they gather new grain for the meal required in certain sacrifices in the sixth month, and that too in the regions high up the Nile, above Memphis.

It is said also that in the Messenian district in

[2] Plin. 18. 49.   [3] κριθαί conj. Sch.; πυροί UMAld.

ταῖς καλουμέναις Μύλαις ταχεῖάν τινα γίνεσθαι τὴν τελείωσιν τῶν ὀψίων· τὸν τῶν ὀσπρίων μὲν γὰρ σπορητὸν ἓξ μῆνας, τὸν δὲ τῷ ὑστάτῳ σπείραντα θερίζειν ἅμα τοῖς πρώτοις· ἀγαθὴν δὲ διαφερόντως εἶναι τὴν χώραν, ὥστε τριακοντάχοα ποιεῖν, ἔχειν δὲ καὶ νομὰς θαυμαστὰς καὶ ὕλην. ἐν Μήλῳ δέ τι θαυμασιώτερον λέγουσιν· ἐν γὰρ τριάκοντα ἢ τετταράκοντα ἡμέραις σπαρέντα θερίζουσι, δι' ὃ καὶ λέγειν αὐτοὺς ὅτι μέχρι τούτου δεῖ σπείρειν ἕως ἂν ἴδῃ τις δράγμα· γίνεσθαι δὲ οὔτε ὄσπρια τοιαῦτα οὔτε πολλὰ παρ' αὐτοῖς. δεινὴν δέ τινα διαδοῦναι τὴν χώραν τροφήν· καὶ γὰρ εἶναι σιτοφόρον μὲν καὶ ἐλαιοφόρον ἀγαθὴν ἀμπελοφόρον δὲ μετρίαν.

9 Ὑπερβάλλον δ' ἔτι τούτου καὶ πάντων θαυμασιώτερον τὸ περὶ Χαλκίαν τὴν νῆσον τὴν Ῥοδίων γινόμενον· ἐκεῖ γάρ φασιν εἶναί τινα τόπον πρώϊον οὕτω καὶ εὔφορον ὡς σπαρεισῶν κριθῶν ἅμα ταῖς ἄλλαις θερίσαντες ταύτας σπείρουσιν εἶτα πάλιν, εἶτα θερίζουσιν ἅμα τοῖς λοιποῖς· μεγίστη μὲν οὖν, εἴπερ ἀληθής, αὕτη διαφορά. τὸ γὰρ εἰς ἑτέραν χώραν μετενεχθέντα διαφέρειν, ὥσπερ ἐκ Κιλικίας φασὶν εἰς Καππαδοκίαν καὶ ὅλως τὴν ἐπέκεινα τοῦ Ταύρου, ἧττον ἄτοπον· φανερὰ γὰρ ἡ τῶν τόπων διάστασις.

---

[1] τριακοντάχοα conj. Sch.; τριάκοντα χοᾶς Ald.
[2] cf. C.P. 4. 11. 8.

ENQUIRY INTO PLANTS, VIII. 11. 8-9

Sicily at the place called Mylae the late sown crops mature rapidly; thus the sowing of pulses goes on for six months, but he that made the last sowing gathers his crop at the same time as the first: also that the soil is exceedingly good, so that it yields thirty-fold [1]; and there are also wonderful pastures and forest-land. They tell of an even more wonderful thing in Melos [2]; there they reap thirty or forty days after sowing; wherefore it is a saying of the islanders that "one should continue sowing till one sees a swathe." However it is said that pulses [3] in their country do not grow like this, nor are they abundant. Yet they say that the soil is wonderfully productive; for it is good both for corn and olives, and fairly good for vines.

However what occurs in Chalkia,[4] an island belonging to the Rhodians, goes even beyond this and is more extraordinary than all the instances given; there they say that there is a place which is so early and so fertile that, when the barley is sown after reaping the crop with the other crops, they then sow again, and then reap the crop thus sown at the same time as the remaining crops; this then, if it be true, marks a difference greater than we find anywhere else. For it is less surprising that there should be a difference in crops transferred [5] to another region, as they say occurs when they are transferred from Cilicia to Cappadocia or in general beyond the Taurus; for these regions are obviously very dissimilar.

[3] ὄσπρια τοιαῦτα I conj.; ὄψιμα ταῦτα UAld.; ἕψιμα ταῦτα M.G; P omits ταῦτα.
[4] cf. Thuc. 8. 41 foll.
[5] μετενεχθέντα διαφέρειν conj. Sch. and W. from G; μετεγκόντας σπείρειν Ald.

10 Τὸ δὲ τὴν αὐτὴν διφορεῖν, ἐν ᾧπερ γε ἅπαξ αἱ ἄλλαι, σύνορον οὖσαν καὶ μίαν θαυμασιώτατον· αὕτη μὲν οὖν ἐν μεγίστῃ διαφορᾷ.

Τὰ δὲ κατὰ τὰς ἄλλας χώρας οὐ πολλὴν ἢ οὐδεμίαν ὡς εἰπεῖν τοῦ γε χρόνου λαμβάνει διάστασιν· προτερεῖ γὰρ ταῖς ὥραις τὰ Ἀθήνησι τῶν περὶ Ἑλλήσποντον ἡμέραις τριάκοντα μάλιστα ἢ οὐ πολλῷ πλείοσιν· εἰ μὲν οὖν καὶ ὁ σπορητὸς πρότερον, μετάθεσις ἂν εἴη τῆς ὥρας· εἰ δ' ἅμα, δῆλον ὅτι πλείων ἂν ὁ χρόνος.

11 Οὐ μικρὰν δὲ ποιοῦσι διαφορὰν οὐδὲ οἱ τόποι, καίπερ ἔνιοι συνεγγὺς ὄντες· τὰ γὰρ ἐν Σαλαμῖνι προτερεῖ πολὺ τῶν ἄλλων τῶν ἐν τῇ Ἀττικῇ καὶ ὅλως τὰ ἐπιθαλάττια καὶ εἰς ταῦτα καὶ εἰς τοὺς ἄλλους καρπούς, ὡς τὰ περὶ τὴν Ἀκτὴν καλουμένην τῆς Πελοποννήσου καὶ τὰ ἐν Φαλύκῳ τῆς Μεγαρίδος· πλὴν ἐνταῦθά γε συμβάλλεται καὶ τὸ λεπτόγεων εἶναι καὶ ψαφαρὰν τὴν χώραν. καὶ τὰ μὲν περὶ τὴν γένεσιν καὶ τὴν τελείωσιν οὕτως ἔχει.

III. Διαφέρει δὲ καθ' ὅλα τὰ γένη τὰ διῃρημένα τῶν γενῶν, οἷον σῖτος χεδροπὰ τὰ θερινά, καὶ καθ' ἕκαστον γένος τὰ ὁμογενῆ. τὰ μὲν γὰρ σιτώδη τὸ φύλλον ἔχει καλάμου, τῶν δὲ χεδρο-

---

[1] i.e. and so in part account for the difference. εἴη τῆς ὥρας conj. Sch.; ᾖ τῆς χώρας MP; εἴη τῆς χώρας Ald.H.

[2] i.e. we cannot say how far the difference is due to climate without knowing whether the seed-time at either place is the same.

But that one particular land should produce two crops in the time that other lands to which it is close take to produce one, is very remarkable; wherefore Chalkia exhibits the greatest difference.

The crops grown in other regions show not much, if any, dissimilarity as to time; those grown at Athens are only about thirty days or not much more before those of the Hellespont region. Now, if the sowing should turn out to be also earlier, that would shift the season back [1]; if it is at the same time, it is plain that the difference of time would be greater.[2]

Again the particular district makes a considerable difference, even as between places which are not far apart; thus the crops of Salamis are far earlier than those of the rest of Attica, and so in general are those of places by the sea; and this applies to other fruits as well as these: for instance, those of the place called Akte in the Peloponnese and of Phalykos [3] in the Megarid are early; but here something is contributed by the fact that the soil is light and crumbling. Such are the facts in regard to growth and development.

Of differences between the parts of cereals, pulses, and summer crops respectively.

III. There are also differences between [4] the whole classes which we have mentioned, namely cereals leguminous plants [5] and summer crops, as well as between the several members [6] of the same class. Cereals have the leaf of a reed, while of

[3] ἐν Φαλύκῳ I conj.: cf. 2. 8. 1; ἐν Φαλήκῳ conj. W.; ἐφαλύκῳ U; ἐκ φαλήκῳ M; ἐκ φαλήκου Ald.
[4] καθ' conj. Sch.; καὶ Ald.H.   [5] cf. 8. 1. 1.
[6] ὁμογενῆ conj. Sch.; ὁμοιογενῆ Ald.

πῶν τὰ μὲν περιφερές, οἷον ὁ κύαμος καὶ σχεδὸν τὰ πλεῖστα, τὰ δὲ προμηκέστερον, οἷον ὁ πισὸς καὶ ὁ λάθυρος καὶ ὁ ὦχρος καὶ τὰ τοιαῦτα. καὶ τὰ μὲν ἰνώδη τὰ δ' ἄφλεβα καὶ ἄϊνα. τὸ δὲ σήσαμον καὶ τὸ ἐρύσιμον ἰδιώτερα παρὰ ταῦτα.

2 Πάλιν ὁ καυλὸς τῶν μὲν γονατώδης καὶ κοῖλος, δι' ὃ καὶ καλεῖται κάλαμος· ὁ δὲ τοῦ κυάμου κοῖλος, τῶν δ' ἄλλων χεδροπῶν ξυλωδέστερος, ξυλωδέστατος δὲ ὁ ἐρέβινθος· τῶν δὲ θερινῶν κέγχρου μὲν καὶ μελίνου καλαμώδης, σησάμου δὲ καὶ ἐρυσίμου ναρθηκώδης μᾶλλον. καὶ τὰ μέν ἐστιν ὀρθόκαυλα, καθάπερ πυρὸς καὶ κριθὴ καὶ ὅλως τὰ σιτώδη καὶ θερινά, τὰ δὲ πλαγιόκαυλα μᾶλλον, οἷον ἐρέβινθος ὄροβος φακός, τὰ δ' ἐπιγειόκαυλα, καθάπερ ὦχρος πισὸς λάθυρος· ὁ δὲ δόλιχος, ἐὰν παρακαταπήξῃ τις ξύλα μακρά, ἀναβαίνει καὶ γίνεται κάρπιμος, εἰ δὲ μή, φαῦλος καὶ ἐρυσιβώδης· μόνος δ' ἢ μάλιστα τῶν χεδροπῶν ὀρθόκαυλος ὁ κύαμος.

3 Ἔχει δὲ καὶ τὰ ἄνθη διαφορὰν καὶ τῇ φύσει καὶ τῇ θέσει, περὶ ὧν σχεδὸν ἐν τοῖς καθ' ὅλου διείλομεν, ὅτι τὰ μὲν χνοώδη, καθάπερ σίτου καὶ παντὸς τοῦ σταχνώδους· τὰ δὲ φυλλώδη, καθάπερ τῶν χεδροπῶν, καὶ τῶν πλείστων κολοβά·

---

[1] Plin. 18. 58. [2] *i.e.* 'summer crops'; *cf.* 8. 1. 1.
[3] Sc. but not jointed. W. suggests that the original text may have been τῶν δὲ διόλου κοῖλος οἷον ὁ τοῦ κυάμου.
[4] μελίνον Ald.H.; ἐλύμον V; ἐλύμου Vin. *cf.* 8. 1. 1; 8. 1. 6.
[5] ἢ add. St.; om. Ald.H.G. [6] 1. 13. 1 (?)
[7] χνοώδη. No rendering seems quite satisfactory: the

ENQUIRY INTO PLANTS, VIII. III. 1-3

leguminous plants some have a round leaf, as beans and most others, some a more oblong leaf, as pea *lathyros okhros* and the like. [1] Some again have fibrous leaves, others leaves without veins and fibres. Again sesame [2] and *erysimon* [2] have leaves quite distinct from these.

Again the stem of cereals is jointed and hollow; wherefore it is called the 'reed,' while that of the bean is hollow,[3] and that of the other leguminous plants is more woody, that of chick-pea woodiest of all; of the summer crops that of millet and Italian millet [4] is reed-like, that of sesame and *erysimon* is more like the stem of ferula. Some again have erect stems, as wheat barley and in general the cereals and summer crops; some have rather a crooked stem, as chick-pea vetch lentil; some a creeping stem as *okhros* pea *lathyros*; while calavance, if long stakes are set by it, climbs them and becomes fruitful, whereas otherwise the plant is unhealthy and liable to rust; the bean, most of all leguminous plants, if not [5] alone among them, has an erect stem.

The flowers also shew differences in character and in position (of which matters we have to some extent treated in our general account) [6]; thus some are 'downy,' [7] as those of corn [8] and of any plant that has an 'ear'; others are 'leafy,' [9] as those of leguminous plants, and in most cases they are irregular [10] flowers; for most of these have

meaning is that such flowers may be classed with those distinguished by this term in 1. 13. 1, as not being petaloid.

[8] σίτου καὶ παντὸς τοῦ σταχυώδους conj. Sch. from G, *ut omnium fere gerentium spicam*; που καὶ παντὸς τοῦ χυλώδους UMAld.   [9] Sc. petaloid.

[10] *cf.* 6. 5. 3. *i.e.* they depart from radial symmetry.

τὰ γὰρ πολλὰ κολοβανθῇ· χνοῶδες δὲ καὶ τὸ τοῦ κέγχρου καὶ μελίνου[1]· τοῦ δὲ σησάμου καὶ τοῦ ἐρυσίμου φυλλῶδες. καὶ ὅτι δὴ τὰ μὲν ἔχει περὶ αὐτὸν τὸν καρπόν, οἷον τὰ σιτώδη καὶ κεγχρώδη περὶ τὸν στάχυν· τὰ δὲ χεδροπὰ ἐξ αὐτοῦ πως τοῦ ἄνθους ἢ ἀπό γε τῆς αὐτῆς ἀρχῆς γίνεται. καὶ τὴν ἄνθησιν, ὅτι τὰ μὲν ἀθρόαν τὰ δὲ κατὰ μέρος ποιεῖται· καὶ τἆλλα δὲ τὰ παραπλήσια τούτοις.

4 Ὁμοίως δὲ καὶ τὰ κατὰ τοὺς καρπούς, ὅτι τὰ μὲν ἔχει στάχυν, τὰ δὲ χεδροπὰ λοβόν, τὰ δὲ κεγχρώδη φόβην· ἡ δὲ καλαμώδης ἀπόχυσις[4] φόβη. τὸ δ' ὅλον[5] ἐναγγειόσπερμα, τὰ δὲ ἐνυμενόσπερμα, τὰ δὲ γυμνόσπερμα· καὶ ἔτι τὰ μὲν ἀκρόκαρπα, τὰ δὲ πλαγιόκαρπα, καὶ ὅσα δὴ ἄλλα ταύτης ἔχεται τῆς θεωρίας.

Ὅλως δὲ πολυκαρπότερα καὶ πολυχούστερα τὰ χεδροπά, τούτων δ' ἔτι μᾶλλον τὰ θερινὰ κέγχρος καὶ σήσαμον, αὐτῶν δὲ τῶν χεδροπῶν 5 μάλιστα φακός. ἁπλῶς δὲ τὰ μικροσπερμότερα μᾶλλον ὡς εἰπεῖν, ὥσπερ καὶ τῶν λαχανωδῶν κύμινον ἁπάντων ὄντων πολυσπέρμων. ἰσχυρότερα δὲ πρὸς μὲν τὸν χειμῶνα καὶ ὅλως τὰ τοῦ ἀέρος τὰ σιτηρά, πρὸς δὲ τὴν τροφὴν τὰ

---

[1] μελίνου Ald.H.; ἐλύμου Vo.Vin. cf. 8. 3. 2 and reff.
[2] cf. 8. 3. 3 n.   [3] Plin. 18. 53.
[4] ἀπόχυσις conj. Sch. from G; ἀπόφυσις P₂Ald. cf. 4. 4. 10, ἀποχεῖται; 8. 10. 4; C.P. 3. 21. 5.
[5] τὸ δ' ὅλον: ? τὰ δ' οἷον.

ENQUIRY INTO PLANTS, VIII. III. 3-5

such flowers. Those of millet and Italian millet [1] are also 'downy,' [2] those of sesame and *erysimon* 'leafy.' Another difference is that in some cases the flowers are round the fruit; thus those of corn and millet are round the ear; while in leguminous plants the fruit comes as it were from the flower itself, or at least from the same starting-point. Another difference is that some produce all their flowers at once, others in succession. And there are other differences akin to these.

In like manner there are differences in the fruits; some have an 'ear,' leguminous plants a pod, and millet-like plants a 'plume' [3]—which is the name given to an inflorescence [4] such as reeds have. Again, generally speaking, [5] some have their seeds in a vessel, [6] some in pods, [7] some naked; and further some bear their fruit at the top, some at the sides; and there are other differences which bear on this enquiry.

In general the leguminous plants produce more fruit and are more prolific, and the summer crops millet and sesame are even more so than these, while among the leguminous plants themselves lentil is the most prolific. [8] Generally speaking, those that have small seeds are more prolific, as cummin among pot-herbs, which are all prolific of seed. The seeds of cereals are more robust as to standing winter and conditions of climate generally, while those of leguminous plants are stronger as to providing food. [9] However it may be that in this respect

[6] μὲν ἐναγγειόσπερμα conj. Sch.; μὲν ἐγγειόσπερμα P₂Ald. *cf.* C.P. 4. 7. 5.
[7] *cf.* 1. 11. 2.   [8] *cf.* C.P. 4. 15. 2.
[9] *i.e.* what has just been said perhaps applies only to *human* food. Sense fixed by 8. 9. 3 *ad fin.*: *cf.* Plin. 18. 50.

χεδροπά. τάχα δὲ τοῦτό γε ἡμῖν τοῖς ἄλλοις ἀνάπαλιν.

IV. Τὰ μὲν οὖν ὅλα γένη τοιαύτας ἔχει διαφοράς· τὰ δὲ ὁμογενῆ δῆλον ὅτι κατὰ τὴν τῶν μερῶν ἀνωμαλίαν, οἷον τῶν σιτωδῶν πυρὸς κριθῆς στενοφυλλότερον καὶ λειοκαυλότερον καὶ πυκνότερον καὶ γλισχρότερον ἔχει τὸν καυλὸν καὶ δυσθλαστότερον· ἅμα δὲ καὶ ὁ μὲν ἐν χιτῶσι πολλοῖς ἡ δὲ γυμνόν· μάλιστα γὰρ δὴ γυμνοσπέρματον ἡ κριθή. πολύλοπον δὲ καὶ ἡ τίφη καὶ ἡ ὄλυρα καὶ πάντα τὰ τοιαῦτα καὶ μάλιστα πάντων ὡς εἰπεῖν ὁ βρόμος. ἔστι δὲ καὶ ὑψηλότερος ὁ κάλαμος τοῦ πυροῦ ἢ τῆς κριθῆς, καὶ τὸν στάχυν ἀπηρτημένον ἔχει τοῦ φύλλου μᾶλλον ὁ πυρός.

Ἴδιον δὲ καὶ τὸ ἄχυρον τοῦ κριθίνου τὸ πύρινον· ἐγχυλότερον γὰρ καὶ μαλακώτερον. διαφέρει δὲ 2 ἡ κριθὴ καὶ τούτῳ τῶν πυρῶν· ἡ μὲν γὰρ στοιχειώδης, ὁ δὲ πυρὸς ἄστοιχος καὶ πανταχόθεν ὁμαλής τις.

Τῷ μὲν οὖν ὅλῳ γένει πρὸς γένος τοιαῦταί τινές εἰσι διαφοραί. καθ' ἑκάτερον δὲ τούτων πάλιν, οἷον πυρῶν καὶ κριθῶν, πολλὰ γένη καὶ τοῖς καρποῖς αὐτοῖς διαφέροντα καὶ τοῖς στάχυσι καὶ ταῖς ἄλλαις μορφαῖς καὶ ἔτι ταῖς δυνάμεσι

---

[1] cf. 7. 4. 9.
[2] After διαφορὰς UM add τὰ ὁμοιογενῆ, Ald. τὰ μὴ ὁμοιογενῆ; om. Sch. and W. after G.
[3] ὁμογενῆ conj. Sch.; ὁμοιογενῆ UMAld. cf. 8. 3. 1.
[4] δυσθλαστότερον conj. Scal. from G, ruptu difficiliorem; δυσαλθατώτερον UMAld.
[5] Plin. 18. 61. πολύλοπον conj. Salm.; πολύλοβον Ald.

the other animals are affected in the opposite[1] way to men.

Of the differences between cereals.

IV. There are then these differences[2] between the various classes; and as between plants of the same class[3] there are plainly differences due to the unlikeness in the various parts. Thus among cereals wheat as compared with barley has a narrower leaf, and a smoother stem of closer texture tougher and less brittle.[4] Again the seed of wheat has several coats,[5] that of barley is naked, that plant having its seeds specially naked. Also one-seeded wheat rice-wheat and all such plants have their seed in several coats, and above all, it may be said, is this true of oats.[6] Also the 'reed' of wheat is taller than[7] that of barley, and wheat has its ear less distant from the 'leaf.'

Further the husk of wheat is distinct[8] from that of barley, being less dry and softer. Barley also differs from wheat in this respect; it has grains in a regular row,[9] whereas those of wheat are not in a row, but the ear is as it were quite simple in form.[10]

Such then are the differences as between one whole kind and another. But in each of these kinds again, for instance in barley and wheat, there are many sub-divisions differing both in the actual fruits, in the ear, and in the other characteristic

---

[6] βρόμος conj. Scal. from Plin. *l.c.* and G; κρόμος PM; κρόκος Ald.; βρῶμος Vin.

[7] ἤ conj. Sch. from Plin. *l.c.* and G; καὶ Ald.H.

[8] ἴδιον Ald.; ἥδιον Vin.H. from G: so Sch. and W. *cf.* Col. 6. 3. 3.

[9] στοιχειώδης. ? στοιχώδης : *v.* LS.

[10] ὁμαλῆς conj. Sch.; ὁμαλὴ UMAld.

καὶ τοῖς πάθεσι. τῶν μὲν κριθῶν αἱ μέν εἰσι δίστοιχοι αἱ δὲ τρίστοιχοι αἱ δὲ τετράστοιχοι καὶ πεντάστοιχοι· πλεῖστον δ' ἑξάστοιχον, καὶ γὰρ τοιοῦτό τι γένος ἐστί. πυκνότεραι δὲ ἀεὶ κατὰ τὴν θέσιν ὡς ἐπὶ πᾶν αἱ πολυστοιχότεραι. διαφορὰ δὲ μεγάλη καὶ τὸ παραβλαστητικὴν εἶναι, καθάπερ εἴπομεν τὴν Ἰνδικήν. καὶ οἱ στάχυες δὲ τῶν μὲν μεγάλοι καὶ μανότεροι ταῖς κριθαῖς τῶν δὲ ἐλάττους καὶ πυκνότεροι, καὶ ἀπέχοντες δὲ τοῦ φύλλου τῶν μὲν πολὺ τῶν δὲ μικρόν, ὥσπερ τῶν Ἀχιλλείων καλουμένων. καὶ αὐτῶν δὲ τῶν κριθῶν αἱ μὲν στρογγυλότεραι καὶ ἐλάττους αἱ δὲ προμηκέστεραι καὶ μείζους καὶ μανότεραι κατὰ τὸν στάχυν. ἔτι δὲ αἱ μὲν λευκαί, αἱ δὲ μέλαιναι καὶ ἐπιπορφυρίζουσαι, αἵπερ καὶ πολυάλφιτοι δοκοῦσιν εἶναι καὶ πρὸς τοὺς χειμῶνας δὲ καὶ τὰ πνεύματα καὶ ὅλως τὸν ἀέρα τῶν λευκῶν ἰσχυρότεραι.

3 Πολλὰ δὲ γένη καὶ τῶν πυρῶν ἐστιν εὐθὺς ἀπὸ τῶν χωρῶν ἔχοντα τὰς ἐπωνυμίας, οἷον Λιβυκοὶ Ποντικοὶ Θρᾷκες Ἀσσύριοι Αἰγύπτιοι Σικελοί. διαφορὰς δὲ καὶ ταῖς χροιαῖς καὶ τοῖς μεγέθεσι καὶ τοῖς εἴδεσι καὶ ταῖς ἰδιότησιν ἔχουσι καὶ ἐν ταῖς δυνάμεσι ταῖς τε ἄλλαις καὶ μάλιστα ταῖς πρὸς τὴν σίτησιν. τινὲς καὶ ἀπ'

---

[1] Explained below, 8. 4. 4; cf. 8. 4. 3.
[2] πάθεσι: cf. l. l. 1 n.   [3] Plin. 18. 78.
[4] πλεῖστον δ' ἑξάστοιχον, καὶ γὰρ τοιοῦτο conj. W.; πλεῖστον ἑξάστοιχον τοιοῦτον UM; πλεῖσται καὶ ἑξάστοιχοι· καὶ γὰρ τοιοῦτον Ald. H.

ENQUIRY INTO PLANTS, VIII. iv. 2-3

features; and again in capacities[1] and properties.[2]
[3] Of barley different sorts have respectively two, three, four, and five rows of seeds; the largest number[4] known is six, for there is a kind which bears that number. And those which have more rows have generally the grains set closer together. Another great difference is that of having side-shoots, as we said of the Indian kind.[5] Again in barley[6] the ears are in some kinds large and of looser make, in some smaller and set closer; in some kinds the ear is some way from the 'leaf,' in some it is nearer to it, as in the kind called 'Achillean.'[7] Again of the grains themselves some are rounder and smaller, some more oblong and larger and set at wider intervals on the ear. Moreover some are white, some black or reddish, and the latter are thought to produce much meal and to be more robust than the white as to bearing winter wind or conditions of climate generally.

There are[8] also many kinds of wheat which take their names simply from the places where they grow,[9] as Libyan Pontic[10] Thracian Assyrian Egyptian Sicilian. They show differences[11] in colour size form and individual character, and also[12] as regards their capacities[13] in general and especially their value as food. Some again get

[5] Referred to 4. 4. 9, but without mention of this feature.
[6] ταῖς κριθαῖς conj. W.; τῆς κριθῆς Ald.
[7] cf. C.P. 3. 21. 3; 3. 22. 2.
[8] ἐστιν εὐθὺς conj. W.; εὐθύς ἐστιν Ald.
[9] cf. C.P. 3. 21. 2; Plin. 18. 2.
[10] Ποντικοὶ conj. Sch.; πόντιοι Ald.
[11] διαφορὰς conj. W.; διαφέροντες Ald.H.
[12] καὶ conj. W.; δὲ Ald.
[13] Explained below, 8. 4. 4: pace of growth.

ἄλλων τὰς προσηγορίας, οἷον καγχρυδίας στλεγγὺς¹ Ἀλεξάνδρειος· ὧν ἁπάντων ἐν τοῖς εἰρημένοις τὰς διαφορὰς ληπτέον. οὐχ ἥκιστα δ' οἰκεῖαι εἴ τις λαμβάνοι τὰς τοιαύτας²· οἷόν εἰσιν οἱ μὲν πρώιοι οἱ δὲ ὄψιοι, καὶ εὐαυξεῖς καὶ πολύχοι οἱ δὲ <μικροὶ>⁴ καὶ ὀλιγόχοι, καὶ μεγαλοστάχυες οἱ δὲ μικροστάχυες. καὶ οἱ μὲν ἐν κάλυκι πολὺν χρόνον οἱ δ' ὀλίγον ἔχοντες, ὥσπερ ὁ Λιβυκός. καὶ κάλαμον οἱ μὲν λεπτὸν οἱ δὲ παχύν· καὶ τοῦτο ὁ Λιβυκὸς ἔχει, παχὺν δὲ καὶ ὁ καγχρυδίας. ἔτι δὲ χιτῶνας οἱ μὲν ὀλίγους οἱ δὲ πολλούς, ὥσπερ ὁ Θρᾴκιος. καὶ ὁ μὲν μονοκάλαμος ὁ δὲ πολυκάλαμος, καὶ μᾶλλον δὲ καὶ ἧττον.

4   Ὁμοίως δὲ καὶ εἴ τι παραπλήσιον τούτοις ἢ τοῖς πρότερον εἰρημένοις κατὰ τὰς δυνάμεις. αἱ γὰρ τοιαῦται φυσικώταται δόξαιεν ἂν εἶναι τῶν διαφορῶν. ἐν αἷς καὶ τὸ τῶν τριμήνων καὶ τὸ τῶν διμήνων καὶ εἴ τι γένος ἐν ἐλάττοσιν ἡμέραις τελειοῦται, καθάπερ φασὶν εἶναι περὶ τὴν Αἰνείαν, οἳ τετταράκοντα ἡμέραις ἀπὸ τῆς σπορᾶς ἁδρύνονται καὶ τέλος ἴσχουσιν· εἶναι δ' ἰσχυρὸν τοῦτον καὶ βαρὺν οὐχ ὥσπερ τὸν τρίμηνον κοῦφον, δι' ὃ καὶ τοῖς οἰκέταις παρέχειν, καὶ γὰρ οὐδὲ πίτυρον ἔχειν πολύ. σπανιώτατος μὲν οὖν καὶ τάχιστος εἰς τελείωσιν

---

¹ στλεγγύς. Sir W. Thiselton-Dyer conjectures σίλιγνις: cf. Plin. 18. 184, LS. σιλίγνιον.
² cf. Geop. 3. 3. 11.   ³ i.e. colour, size, etc.
⁴ μικροὶ add. W. to correspond to εὐαυξεῖς (conj. Sch.; εὐαξεῖς Ald.).

their distinctive names for other reasons, as *kankhrydias stlengys*[1] 'Alexandrian'[2]; all of which must be distinguished by the above-mentioned[3] characters. Again, if one takes such differences as the following, they are quite characteristic—thus some are early, some late, some are vigorous and prolific, some are small[4] and produce little, some have a large, some a small ear. The ears of some remain[5] a long time in the sheath,[6] of some it remains but a short time, as that of the Libyan kind. [7] Again some have a slender, some a stout haulm; the Libyan kind has this characteristic also, and that of *kankhrydias* is also stout. Again the grain of some has few coats,[8] of some many, for instance the Thracian.[9] Some kinds have a single 'reed,' some more than one, and in the latter class the number varies.

[10] So too must we distinguish any differences like these or those mentioned above which are found in the several capacities; for these would seem to be the most essential differences. In this connexion we may distinguish kinds which mature in three or in two months, and those, if there be such, which take a less number of days; for instance, they say that in the region of Aineia there is a kind which ripens and attains perfection within forty days from the time of sowing; they say too that this grain is strong and heavy, not light like that which takes three months; wherefore they give it even to the servants, for it also does not contain much bran. Now this kind is the rarest and the quickest to

[5] ἔνοντες conj. W.; ἔχοντες Ald.
[6] *cf.* 8. 2. 4.   [7] Plin. 18. 69.   [8] *i.e.* glumes.
[9] *cf. C.P.* 4. 12. 5; Plin. *l.c.*   [10] Plin. 18. 70.

THEOPHRASTUS

οὗτος. εἰσὶ δὲ καὶ δίμηνοί τινες οἵπερ καὶ ἐκ Σικελίας ἐκομίσθησαν εἰς Ἀχαίαν· ὀλιγοχόοι δὲ καὶ ὀλιγόγονοι καὶ κοῦφοι κατὰ τὴν προσφορὰν καὶ ἡδεῖς. καὶ ἄλλοι δέ τινες οἱ περὶ Εὔβοιαν εἰσὶ καὶ μάλιστα ἐν τῇ Καρυστίᾳ. τρίμηνοι δὲ πολλοὶ καὶ πανταχοῦ κοῦφοι οὗτοι καὶ ὀλιγοχόοι καὶ μονοκάλαμοι κατὰ τὴν ἔκφυσιν καὶ τὸ ὅλον ἀσθενεῖς. κουφότατος μὲν οὖν ὡς ἁπλῶς εἰπεῖν πυρὸς ὁ Ποντικός· βαρύτερος δὲ τῶν εἰς τὴν Ἑλλάδα παραγινομένων ὁ Σικελός· τούτου δ' ἔτι βαρύτερος ὁ Βοιωτός· σημεῖον δὲ λέγουσιν ὅτι οἱ μὲν ἀθληταὶ ἐν τῇ Βοιωτίᾳ τρί¹ ἡμιχοίνικα μόλις ἀναλίσκουσιν, Ἀθήναζε δὲ ὅταν ἔλθωσι πένθ' ἡμιχοίνικα ῥᾳδίως. κοῦφος δὲ καὶ ὁ ἐν τῇ Λακωνικῇ. τούτων μὲν οὖν ἔν τε ταῖς χώραις καὶ τῷ ἀέρι τὸ αἴτιον· ἐπεὶ καὶ περὶ τὴν Ἀσίαν οὐ πόρρω Βάκτρων ἐν μέν τινι τόπῳ οὕτως ἁδρὸν εἶναί φασι τὸν σῖτον ὥστε πυρῆνος ἐλαίας μέγεθος λαμβάνειν, ἐν δὲ τοῖς Πισσάτοις καλουμένοις οὕτως ἰσχυρὸν ὥστ' εἴ τις πλεῖον προσενέγκοιτο διαρρήγνυσθαι, καὶ τῶν Μακεδόνων καὶ πολλοὺς τοῦτο παθεῖν. ἄτοπον δὲ καὶ ἀνομολογούμενον πρὸς τὴν τῶν τριμήνων κουφότητα τὸ περὶ τοὺς Ποντικοὺς συμβαῖνον· εἰσὶ γὰρ οἱ μὲν σκληροὶ ἠρινοὶ οἱ δὲ μαλακοὶ χειμερινοί· πολὺ γὰρ διαφέρει τῇ κουφότητι ὁ μαλακός. [ὁμοίως δὲ καὶ δύο ἀρότους ὡς ἔοικε

---
¹ τρί’ ἡμιχοίνικα conj. Sch.; τριημισχοίνικα ; τριημιχοίνικα P₂Ald.H.

ENQUIRY INTO PLANTS, VIII. iv. 4-6

mature. But there is also a kind which takes two months; this was brought to Achaia from Sicily; it is not however prolific nor fertile, though as food it is light and sweet. There is another such kind which grows in Euboea and especially in the region of Karystos. There are several kinds that take three months, and these, wherever they are found, are light and not prolific; their growth consists of a single 'reed,' and in general they are not robust. Lightest of all we may say is the Pontic wheat; the Sicilian is heavier than most of those imported into Hellas, but heavier still than this is the Boeotian; in proof of which it is said that the athletes in Boeotia consume scarcely three pints,[1] while, when they come to Athens, they easily manage five.[2] The Laconian kind is also light. The reason for these differences is to be found in the respective soils and in the climate; [3] for in Asia not far from Bactra they say that in a certain place the corn is so vigorous that the grains grow as large as an olive-stone, while in the country called that of the Pissatoi it is so strong that, if a man eats too much of it, he bursts, which was actually the fate of many of the Macedonians.[4] There is one curious thing about the corn of Pontus, which is an exception[5] to the rule as to the lightness of crops raised in three months; for there the hard crops are those of the spring, the soft ones those of the winter; for soft kinds are exceedingly light. Two sowings, as it appears, are made of all corn

---

[2] πενθ' ἡμιχοίνικα conj. Sch.; πενθημισχοίνικα M; πενθημιχοίνικα P₂Ald.H.   [3] Plin. 18. 70.

[4] *i.e.* in Alexander's army.

[5] ἀνομολογούμενον : *cf.* *C.P.* 4. 8. 2 ; Plat. *Gorg.* 495 A.

παντὸς τοῦ σίτου ποιοῦνται, τὸν μὲν χειμερινὸν τὸν δὲ ἠρινόν, ἐν ᾧ καὶ τὰ ὄσπρια καταβάλλουσιν].

Εἰσὶ δὲ καὶ οἱ μὲν καθαροὶ αἰρῶν, ὥσπερ ὁ Ποντικὸς καὶ ὁ Αἰγύπτιος· καθαρὸς δὲ ἐπιεικῶς καὶ ὁ Σικελὸς καὶ μάλιστα ὁ Ἀκραγαντῖνος οὐκ αἰρώδης.

Ὁ δὲ Σικελὸς ἴδιον ἔχει τὸ μελάμπυρον καλούμενον, ὅ ἐστιν ἀβλαβὲς καὶ οὐχ ὥσπερ ἡ αἶρα βαρὺ καὶ κεφαλαλγές. ἀλλὰ τὰ μὲν τοιαῦτα, καθάπερ ἐλέχθη, ταῖς χώραις ἀναθετέον καὶ ὅσον ἐπιβάλλει τοῖς γένεσιν.

V. Ἐν δὲ τοῖς ὀσπρίοις οὐχ ὁμοίως ἐστὶ λαβεῖν τὰς τοιαύτας διαφοράς, εἴτ᾽ οὖν διὰ τὸ μὴ ἐξετάζειν ὁμοίως εἴτε καὶ διὰ τὸ μονοειδέστερα τυγχάνειν· ἔξω γὰρ ἐρεβίνθου καὶ φακοῦ καὶ ἐπ᾽ ὀλίγου κυάμου καὶ ὀρόβου, καθ᾽ ὅσον ἡ τῶν χρωμάτων καὶ τῶν χυλῶν διαφορά, τῶν γ᾽ ἄλλων οὐ ποιοῦσιν ἰδέας. οἱ δὲ ἐρέβινθοι καὶ τοῖς μεγέθεσι καὶ τοῖς χυλοῖς καὶ τοῖς χρώμασι καὶ ταῖς μορφαῖς διαφέρουσιν, οἷον κριοὶ ὀροβιαῖοι οἱ ἀνὰ μέσον. ἐπὶ πᾶσι δὲ τὰ λευκὰ γλυκύτερα· καὶ γὰρ ὁ ὄροβος καὶ φακὸς καὶ ἐρέβινθος καὶ κύαμος καὶ σήσαμον· ἔστι γὰρ καὶ σήσαμον λευκόν.

2 Ἀλλὰ μᾶλλον ἐν τοῖς τοιοῖσδε ποιεῖν ἐστι τὰς διαφορὰς οἷον, ἐπεὶ πάντα ταῦτ᾽ ἔλλοβα, τὰ μὲν

---

[1] ὁμοίως . . . καταβάλλουσιν bracketed by Sch. as a gloss.
[2] But cf. 8. 8. 3.   [3] cf. Diosc. 4. 116.
[4] i.e. when it gets into the bread.
[5] cf. Plin. 18. 156 ; Diosc. 2. 100.
[6] ὅσον ἐπιβάλλει : cf. Arist. Pol. 1. 13. 8.

alike, one in winter and one in spring, at which time they also plant the seed of the pulses.[1]

Some kinds are free from darnel, as the Pontic and the Egyptian; the Sicilian is also fairly free from it, and that of Akragas is especially immune from darnel.

[2] Peculiar however to the Sicilian is the plant called *melampyron*,[3] which is harmless[4] and not, like the darnel, injurious and productive of headache.[5] However such peculiarities, as was said, must be ascribed to the soil, and to a certain extent[6] to the different characters of different kinds.

Of the differences between pulses.

V. In pulses we cannot find such differences to the same extent, whether for the want of equally careful enquiry or because there is actually less diversity in these plants. [7] For, apart from chick-pea lentil and to a certain extent bean and vetch (in so far as in these we find differences of colour and taste), among the rest[8] no distinct forms are recognised. Chick-peas however differ in size colour taste and shape; thus there are the varieties called 'rams,' 'vetch-like' chick-peas and the intermediate forms.[9] In all pulses the white are the sweeter, and this applies to vetch lentil chick-pea bean and sesame, of which also there is a white form.

[10] However it is more possible to recognise the differences in such points as these:—all these plants have pods,[11] but whereas the pods in some kinds have

[7] Plin. 18. 124. [8] γ' conj. Sch.; τ' Ald.H.
[9] οἱ after ὀριβιαῖοι add. Dalec. For ἀνὰ μέσον cf. 3. 18. 2.
[10] Plin 18. 125.
[11] ἔλλοβα conj. Scal. from G; ἐλλέβορα Ald.H.; ἐλλόβορα U; ἐπεὶ πάντα ταῦτ' conj. W.; ἐπὶ πάντα τὰ Ald.H.

αὐτῶν ἀδιάφρακτα καὶ ὥσπερ συμψαύοντα τυγχάνει, καθάπερ ὄροβος πισὸς καὶ σχεδὸν τὰ πλεῖστα, τὰ δὲ διαπεφραγμένα, καθάπερ θέρμος, ἔτι δὲ μᾶλλον καὶ ἰδίως τὸ σήσαμον. καὶ τὰ μὲν μακρόλοβα τὰ δὲ καὶ στρογγυλόλοβα, καθάπερ ὁ ἐρέβινθος. ἀνὰ λόγον δ᾽ ἀκολουθεῖ καὶ τὰ πλήθη τῶν σπερμάτων· ἐλάττω γὰρ ἐν τοῖς μικροῖς, ὥσπερ ἕν τε τῷ τῶν ἐρεβίνθων καὶ ἐν τῷ τῶν φακῶν.

3 Καὶ παραπλήσιαι δὲ ἴσως αἱ τοιαῦται καὶ ἃς ἐπὶ τῶν σιτηρῶν ἐλέγομεν περὶ τῶν σταχύων καὶ αὐτῶν τῶν καρπῶν· ἐπεὶ καὶ οἱ καλούμενοι λοβοὶ σχεδὸν ἀκόλουθοι τοῖς σπέρμασίν εἰσιν, οἱ μὲν ἐπιπλατεῖς, ὥσπερ οἱ τοῦ φακοῦ καὶ τῆς ἀφάκης, οἱ δὲ κυλινδρώδεις μᾶλλον, ὡς οἱ τοῦ ὀρόβου καὶ τοῦ πισοῦ· τὰ γὰρ σπέρματα ἑκατέρων τοιαῦτα τοῖς σχήμασι· ἀλλὰ τὰς μὲν τοιαύτας διαφορὰς πολλὰς ἄν τις εὕροι καθ᾽ ἕκαστον, ὧν αἱ μὲν κοιναὶ πᾶσιν αἱ δὲ ἴδιαι κατὰ γένος.

4 Ὅτι δὲ πάντα προσπέφυκε τοῖς λοβοῖς καὶ ἔχει καθάπερ ἀρχήν τινα, τὰ μὲν προέχουσαν, ὥσπερ ὁ κύαμος καὶ ὁ ἐρέβινθος, τὰ δὲ καὶ ἔγκοιλον, ὥσπερ θέρμος καὶ ἄλλ᾽ ἄττα, τὰ δὲ οὕτω μὲν οὐ φανερὰν ἐλάττω δὲ καὶ ὥσπερ ἀποσημαίνουσαν μόνον, δῆλον μὲν ἀπὸ τῆς ὄψεως· ἐξ ἧς καὶ ὅταν σπαρῇ βλαστάνει καὶ ῥιζοῦται, καθάπερ ἐλέχθη, κατ᾽ ἀρχὰς δὲ καὶ αὐτὰ τρέφεται προσηρτημένα τῷ λοβῷ, μέχρι οὗ ἂν

---

[1] ἀδιάφρακτα conj. Scal. from G, *non intersepta*; διάφρακτα Ald.H.   [2] *cf.* 1. 11. 5.
[3] διαπεφραγμένα conj. Sch.; λεῖα πεφραγμένα Ald.H.
[4] *cf.* 3. 18. 13.

no divisions,[1] but the seeds as it were touch one another,[2] as in vetch pea and most kinds, in some there are divisions,[3] as in lupin and still more in sesame, in which the divisions are of a peculiar kind.[4] Again some have long, some round pods, as chick-pea. And the number of seeds follows in proportion, since they are fewer in the small pods, as in those of chick-pea and lentil.

Possibly these differences correspond to those which we mentioned in the case of cereals as to the ears and the actual fruits; for what are called 'pods' also[5] fairly correspond to the shape of the seeds, some being flat, as those of lentil and tare, some more or less cylindrical, as those of vetch and pea[6]: for in the case of either pair of plants the seeds correspond in shape. However one might discover and distinguish many such differences, of which some are common to a whole kind,[7] others special to particular varieties.

In all cases the seeds are attached to the pods and have a sort of starting-point, which in some cases projects, as in bean and chick-pea, in some is hollow, as in lupin and some others, and in some is not thus conspicuous but smaller and, as it were, only indicated; this is plain from observation; it is from this point that the seeds germinate and take root when they are sown, as was said[8]: but to start with they are themselves nourished by being so attached to the pod until they are matured. This

[5] *i.e.* as does the form of the ear in cereals.
[6] καὶ τοῦ πισοῦ· τὰ γὰρ conj. Scal. from Plin. *l.c.* and G; τοῦ πισοῦ γὰρ τὰ UMAld.
[7] *i.e.* which either differentiate (*e.g.*) pea from lentil, or one variety of pea from another. *cf.* 8. 4. 2 n.
[8] 8. 2. 1.

τελειωθῇ· φανερὸν δέ ἐστι καὶ ἐκ τῶν νῦν καὶ ἐκ τῶν προειρημένων. περὶ μὲν οὖν τῶν κατὰ τὰς διαφορὰς ἅλις.

VI. Σπείρειν δὲ ξυμφέρει πάντα μάλιστα μὲν ἐν τοῖς ὡραίοις ἀρότοις· οὐ μὴν ἀλλὰ καὶ ἐν ξηρᾷ τινες καταβάλλουσι καὶ οὐχ ἥκιστα πυροὺς καὶ κριθὰς ὡς μάλιστα αὐταρκεῖν¹ δυνάμενα, ὅπου μὴ ὄρνισιν ἢ ἄλλοις θηρίοις ἐπισινὴς² ἡ χώρα. δοκεῖ γὰρ ὡς ἐπὶ πᾶν ὁ πρῶτος ἄροτος ἀμείνων εἶναι, χείριστος δὲ σπόρος ἐν ταῖς ἡμιβρόχοις· ἀπόλλυται γὰρ καὶ ἐκγαλακτοῦται τὰ σπέρματα, καὶ ἅμα ξυμβαίνει πόαν ἀναφύεσθαι πολλήν. μετὰ δὲ τὴν σπορὰν ὕδωρ³ ἐπιγίνεσθαι πᾶσι ξυμφέρει, πλὴν ὅσα δυσβλαστῆ γίνεται μᾶλλον, ὥσπερ ὅ τε κύαμος δοκεῖ καὶ τῶν θερινῶν σήσαμον καὶ κύμινον καὶ ἐρύσιμον.

2 Πυκνοσπορεῖν δὲ καὶ μανοσπορεῖν καὶ πρὸς τὰς χώρας βλέποντα χρή· πλεῖον γὰρ ἡ πίειρα καὶ ἀγαθὴ δύναται φέρειν τῆς ὑφάμμου τε καὶ λεπτῆς. καίτοι λέγεταί τις λόγος ὡς ὁτὲ μὲν πλέον ὁτὲ δὲ ἔλαττον ἡ αὐτὴ δέχεται χώρα· καὶ οἰωνίζονται τὸ πλέον ὡς οὐκ ἀγαθόν, πεινῆν γὰρ εὐθύς φασι τὴν γῆν· οὗτος μὲν οὖν ἴσως εὐηθέστερος λόγος. εἰ δέ τις πρὸς τὰ σπέρματα θεωροίη καὶ μάλιστά γε πρὸς αὐτοὺς τοὺς τόπους ἅμα τῷ ἐδάφει καὶ τὴν θέσιν ἀναθεωρῶν τήν γε πρὸς τὰ πνεύ-

---

¹ αὐταρκεῖν Ald.; ἀνταρκεῖν U.
² ἐπισινὴς conj. Dalec.; ἐπινὴς UMAld.; obnoxia G.
³ i.e. after the rains.

is clear both from what is said now and from what was said before. Enough then about the points of difference.

Of sowing, manuring, and watering.

VI. It is expedient to sow all these, if possible, at the early seed-time; however some plant the seed even in dry ground, and especially wheat and barley, on the theory that they are most likely to hold their own [1] at a time when the ground is not infested [2] with birds or other creatures. For it appears that in general the first sowing is better, and worst that which is made in half-soaked ground; [3] for then the seeds perish and become 'milky'; [4] moreover many weeds come up at that time. After the sowing however it is beneficial for all that rain should fall on them, except in the case of some which appear to germinate then with more difficulty, as seems [5] to be the case with beans, and among summer crops with sesame cummin and *erysimon*.

[6] As to sowing thickly or scantily one should have regard to the soil as well as to other considerations; for a fat good soil can bear more than one which is sandy and light. However there is a saying that the same soil can take at one time more, at another less seed; and in general the former condition is taken as an unfavourable omen, for then they say at once that the soil is hungry; however this is perhaps a rather foolish saying. If a man should have regard to the kind of the seed and especially to the actual situation, considering the aspect in respect of winds

[4] *cf.* γαλάκτωσις, *C.P.* 4. 4. 7 and 8.
[5] δοκεῖ conj. W.; ἐδόκει Ald.   [6] Plin. 18. 196.

ματα καὶ τὸν ἥλιον, οἰκειότερον ἂν λαμβάνοι τὰς διαφοράς.

3 Ἀνὰ λόγον δὲ ἔχει καὶ ἡ κόπρισις τοῖς σπόροις πρὸς τὰς χώρας· νειὸς δ᾽ ἀμείνων ἡ χειμέριος τῆς ἐαρινῆς. ἐνιαχοῦ δὲ οὐ ξυμφέρειν βαθεῖαν ἀροτριᾶν, ὥσπερ καὶ ἐν Συρίᾳ, δι᾽ ὃ μικροῖς ἀρότροις χρῶνται. παρ᾽ ἄλλοις δὲ τὸ λίαν ἐξεργάζεσθαι βλάπτει, καθάπερ ἐν Σικελίᾳ, δι᾽ ὃ καὶ τῶν ξένων ὡς ἔοικε πολλοὶ διαμαρτάνουσι. πάντα μὲν οὖν πρὸς τὰς χώρας.

4 Διαιροῦσι δὲ καὶ τὰ σπέρματα ποίᾳ ποῖον πρόσφορον· ἐν γὰρ ταῖς χειμεριναῖς πυρὸν μᾶλλον ἢ κριθήν, καὶ ὅλως σῖτον ἢ χεδροπὰ κελεύουσιν ἐν ταῖς χέρσοις καὶ διὰ χρόνου κινουμέναις· καὶ γὰρ αὗται πυρὸν φέρουσι μᾶλλον ἢ κριθήν. δέχεται δὲ καὶ ἐπομβρίαν μᾶλλον πυρὸς τῆς κριθῆς, καὶ ἐν τοῖς ἀκόπροις φέρει μᾶλλον. ὡσαύτως δὲ καὶ αὐτῶν τῶν πυρῶν ποῖος τῇ ποίᾳ πρόσφορος, οἷον ἀγαθῇ καὶ πιείρᾳ καὶ ψαφαρᾷ καὶ λεπτῇ <καὶ> ταῖς ἄλλαις ὁμοίως.

5 Ὕδωρ δὲ ὅταν μὲν χλοηφορήσῃ καὶ κυήσῃ πλεῖον ἅπασι ξυμφέρει· ἀνθοῦσι δὲ πυροῖς μὲν καὶ κριθαῖς καὶ τοῖς σιτώδεσι βλαβερόν· ἀπόλ-

---

[1] ἂν λαμβάνοι conj. Sch.; ἀναλαμβάνοι Ald. H.
[2] κόπρισις conj. Sch.; κόπρησις Ald.
[3] cf. C.P. 3. 20. 7.   [4] cf. C.P. 3. 20. 5.
[5] πάντα μὲν οὖν M; ταῦτα μὲν Ald. H.; ταῦτα μὲν οὖν conj. Sch. followed by W.
[6] κελεύουσιν conj. W.; καὶ ὅλως Ald. H.
[7] κινουμέναις conj. Sch. (cf. C.P. 3. 21. 4, ἡ διὰ χρόνου γεωργουμένη γῆ); κενουμέναις UAld; καινουμέναις Vin.

ENQUIRY INTO PLANTS, VIII. vi. 2-5

and sun, as well as the soil itself, he would more properly gauge[1] the differences.

Similarly manuring[2] for the sown crops should be done with regard to the soil; and it is better to turn up fallow[3] land in winter than in spring. And there are some[4] places in which deep ploughing is not expedient, as in Syria; wherefore they use small plough-shares. In other parts to work the ground too much is injurious, as in Sicily: wherefore many settlers in the country, it appears, make a mistake. From every point of view[5] therefore the soil must be considered.

The seeds are also classified according as each suits a particular soil; in wintry lands wheat is sown rather than barley, and in general they say[6] that corn rather than leguminous plants should be sown in barren soils which are only disturbed[7] at long intervals; and such soils bear wheat better than barley.[8] Moreover wheat welcomes abundant rain[9] more than barley, and bears better on land which is not manured.[10] In like manner they distinguish among wheats themselves which suits which kind of soil, namely which grows best in good[11] fat soil and which in crumbling light soil, and[12] so on with other kinds of soil.

[13] More abundant rain is beneficial to all crops when they have come into leaf and formed the flower; however it is harmful to wheats and barleys and other cereals when they are actually in flower; for

[8] τῆς κριθῆς conj. W.; καὶ κριθῆς UM; ἡ κριθὴ Ald.; ἢ κριθὴ H.
[9] cf. C.P. l.c.
[10] Explained C.P. l.c.
[11] ἀγαθῇ conj. Casaub: so Vin.; ἀγαθὴ Ald. (and so with the other datives). [12] καὶ add. St.
[13] Plin. 18. 151 and 152.

λυσι γάρ· ὀσπρίοις δ' ἀβλαβὲς πλὴν ἐρεβίνθων· οὗτοι γὰρ ἀποκλυσθείσης τῆς ἅλμης ἀπόλλυνται σφακελίζοντες καὶ ὑπὸ καμπῶν κατεσθιόμενοι· ἰσχυρότερος δὲ ὁ μέλας ἐρέβινθος καὶ ὁ πυρρὸς τοῦ λευκοῦ· συμφέρει δέ, φασίν, ἐν τοῖς ἐφύδροις τόποις ὀψὲ σπείρειν αὐτόν. κύαμος δὲ ἀνθῶν μάλιστα φιλεῖ βρέχεσθαι, δι' ὃ καὶ οὐκ ἐθέλουσιν ὀψισπορεῖν, ὥσπερ εἴπομεν, ὅτι πολὺν ἀνθεῖ· μετὰ δὲ τὴν ἀπάνθησιν ὀλίγου πάμπαν ὕδατος δεῖται· σύνεγγυς γὰρ ἡ τελείωσις. ἀλλ' ὅταν ἀδρυνθῇ καὶ βλάπτειν δοκεῖ τὰ σιτώδη καὶ κριθὴν δὲ πυροῦ μᾶλλον.

6 Ἐν Αἰγύπτῳ δὲ καὶ Βαβυλῶνι καὶ Βάκτροις, ὅπου μὴ ὕεται ἡ χώρα <ἢ> σπανίως, αἱ δρόσοι τὸ ὅλον ἐκτρέφουσιν. ἔτι καὶ οἱ περὶ Κυρήνην καὶ Εὐεσπερίδας τόποι. καιριώτατα δὲ πᾶσιν ὡς ἁπλῶς εἰπεῖν τὰ ἠρινά· δι' ὃ καὶ ἡ Σικελία πολύσιτος· πολλὰ γὰρ τοῦ ἦρος καὶ μαλακὰ γίνεται, τοῦ δὲ χειμῶνος ὀλίγα. ζητεῖ δὲ ἡ μὲν λεπτόγεως πολλὰ κατὰ μικρόν· ἡ δὲ πίειρα καὶ πλῆθος μὲν ἐνεγκεῖν δύναται καὶ αὐδρίαν—πρὸς δὲ τὴν χώρας αὐδρίαν πόντια πνεύματα καὶ αὖραι δοκοῦσι ξυμφέρειν, ἄλλα δὲ παρ' ἄλλοις τοιαῦτα, καθάπερ καὶ πρότερον εἴρηται,—ὡς ἐπὶ τὸ πᾶν δὲ μᾶλλον αὐχμὸς ἢ ἐπομβρία ξυμφέρει τῷ σίτῳ·

7 οἱ γὰρ ὄμβροι καὶ ἄλλως ἐναντίοι καὶ πολλάκις αὐτὰ τὰ σπέρματα διαφθείρουσιν, εἰ δὲ μὴ πλή-

---

[1] σφακελίζοντες: *cf.* 4. 14. 4.
[2] ὁ πυρρὸς τοῦ λευκοῦ conj. Scal. from G and Plin. 18. 124; ὁ λευκὸς τοῦ πυροῦ UAld; ὁ λ. τ. πυρροῦ H.; ὁ λ. τ. πυρὸς M.
[3] *cf.* C.P. 3. 22. 3.
[4] δεῖται conj. Sch.; δεῖσθαι Ald. H.

it destroys the flower. But to pulses it is harmless, except to chick-peas; for these, if the salt is washed off them, perish from rot[1] or from being eaten by caterpillars. However the black and the red[2] chick-pea are stronger than the white, and it is beneficial, they say, to sow this crop late in moist soil. The bean[3] likes especially to receive rain when it is in flower; wherefore men are unwilling, as we said, to sow it late, because it flowers for a long time; but after it has shed its flowers, it needs[4] very little water, since its time of maturity is now near. But, when cereals have matured, it appears that water actually injures them, and barley more than wheat.

In Egypt Babylon and Bactra, where the country receives no rain, or[5] but little, the dews are sufficient nourishment; and so is it also[6] in the regions about Cyrene and the Euesperides. However to all, generally speaking, it is the spring rains which are the most seasonable; and that is why Sicily is rich in corn; for there is abundance of soft rain in spring and little of it in winter. A light soil requires plenty of rain, but little at a time; while that which is fat can indeed bear both an abundance of rain and a drought; (for a droughty country sea-winds[7] and breezes seem to be helpful, and various breezes of this kind prevail in various countries, as has been said already). Yet in general drought suits corn better than excessive rain; for heavy showers, apart from the harm which they do in other ways, often actually destroy the seed, or at

---

[5] ἢ add. Scal. from G: so Vin.
[6] ἔτι conj. St. from G (?); ἐπεὶ Ald.
[7] πόντια conj. Sch.; πάντα Ald. *cf.* 8. 7. 6.

θός γε ποιοῦσι βοτάνης, ὥστε καταπνίγεσθαι καὶ ἀτροφεῖν.

VII. Τῶν μὲν οὖν ἄλλων σπερμάτων οὐδὲν εἰς ἄλλο πέφυκε μεταβάλλειν φθειρόμενον, πυρὸν δὲ καὶ κριθὴν εἰς αἶράν φασι καὶ μᾶλλον τὸν πυρόν, γίνεσθαι δὲ τοῦτ' ἐν ταῖς ἐπομβρίαις καὶ μάλιστα ἐν τοῖς εὐύδροις καὶ ὀμβρώδεσι χωρίοις. ὅτι δ' οὐκ ἔστιν ἠρινὸν ἡ αἶρα καθάπερ ἡ ἄλλη πόα, πειρῶνται γάρ τινες τοῦτο λέγειν, ἐκεῖθεν δῆλον· εὐθὺς γὰρ τοῦ χειμῶνος φανερὰ γίνεται πεφυκυῖα· καὶ διαφέρει πολλοῖς· ἔχει γὰρ τὸ φύλλον στενὸν καὶ δασὺ καὶ λιπαρόν, καὶ τούτων ἰδιώτατον τὸ λιπαρόν· ἡ γὰρ δασύτης καὶ τοῖς τοῦ αἰγίλωπος ὑπάρχει, ἀλλ' ἐκφανὴς γίνεται ἐπὶ τοῖς τοῦ αἰγίλωπος τοῦ ἦρος. τοῦτο μὲν οὖν ἴδιον τούτων, καὶ ἔτι τοῦ λίνου· καὶ γὰρ ἐκ τούτου φασὶ γίνεσθαι τὴν αἶραν.

2 Τοῦ δὲ ἐρεβίνθου πρὸς τὰ ἄλλα χεδροπὰ τό τε περὶ τὴν ἄνθησιν λεχθὲν καὶ τὸ τάχιστα τελειοκαρπεῖν ἰσχυρότατον ὂν καὶ ξυλωδέστατον, καὶ

---

[1] Plin 18. 149 and 150 ; cf. C.P. 4. 5. 2.
[2] πόα : ? grasses; cf. 8. 6. 1.
[3] πειρῶνται γάρ τινες H.; ἀπειρῶνται· αἰτιῶνται γάρ τινες U; ἃ πειρῶνται· αἰτιῶνται γάρ τινες PM : so also Ald.Bas.Cam. with mark of corruption.
[4] εὐθὺς γὰρ τοῦ conj. Sch.; εὐθὺς τὸ τοῦ Ald.

least cause a luxuriant growth of leafage, so that the grain is choked and becomes abortive.

Of the degeneration of cereals into darnel.

VII. [1] Now, while it is not the nature of any other of these seeds to degenerate and change into something else, they say that wheat and barley change into darnel, and especially wheat; and that this occurs with heavy rains and especially in well-watered and rainy districts. But that darnel is not a plant of the spring, like other weeds [2] (for some endeavour [3] to make this out) is clear from the following consideration: it springs up and becomes noticeable directly [4] winter comes; and it is distinguished in many ways; the foliage [5] is narrow abundant and glossy, and this gloss is the most marked of these differences; (the [6] leaves of *aigilops* [7] are indeed also abundant, [8] but this character does not shew itself in them till spring). This then is peculiar to the seeds of wheat and barley, and also to those of flax; for that too, they say, turns into darnel.

Of the peculiar character of chick-pea.

A peculiarity of chick-pea as compared with other leguminous plants is that which has been mentioned as to its flowering; and also the fact that it is the quickest to mature its fruit, being very strong and woody; and again there is the fact that in

[5] cf. *C.P.* 4. 4. 11. [6] τοῖς conj. Sch.; τῆς Ald.
[7] Plin. 18. 155.
[8] ἀλλ' . . . αἰγίλωπος: text a makeshift. Wanting in Ald. and all MSS. except U; ἀλλ' ἐκφανεῖς γίνονται καὶ τοῖς τοῦ αἰγ. U; ἐκφανὴς γίνεται conj. Sch.; ἐπὶ for καὶ conj. W.

τὸ ὅλον μὴ ποιεῖν νειὸν <ὡς> καρπιζόμενον· τὴν δὲ πόαν ἐξαπόλλυσι καὶ μάλιστα δὲ καὶ τάχιστα τὸν τρίβολον. ὅλως δὲ οὐδὲ ἡ τυχοῦσα δύναται φέρειν αὐτόν, ἀλλὰ μελάγγειόν τινα δεῖ καὶ πίειραν εἶναι. τῶν δὲ ἄλλων ἡ ἀρίστη νειὸς ἀπὸ τῶν κυάμων καίπερ πυκνοσπορουμένων καὶ πολὺν καρπὸν φερόντων.

3 Τὰ δὲ ἐν τοῖς θερινοῖς ἀρότοις ὀλίγου δεῖ πάντα, φασὶ δὲ καὶ τὰ ναματιαῖα συμφέρειν μᾶλλον αὐτοῖς τῶν ἐκ διός, μέλινοι δὲ καὶ κέγχροι ἐλάττους ὕδατος· ἐὰν γὰρ ἔχωσι πλεῖον φυλλοβολοῦσιν. ἰσχυρότερον δὲ ὁ κέγχρος· οἱ δὲ μέλινοι γλυκύτεροι καὶ ἀσθενέστεροι. σήσαμον δὲ οὐδὲν <ζῶον> ἐσθίει χλωρὸν οὐδὲ θερμόν. εἰ δὲ μηδ' ἐρύσιμον μηδὲ ὅρμινον σκεπτέον· καὶ ταῦτα πικρά. ἔστι δὲ τὸ μὲν ἐρύσιμον ὅμοιον σησάμῳ καὶ λίπος ἔχει· τὸ δὲ ὅρμινον κυμινῶδες μέλαν· σπείρεται δὲ ἅμα καὶ τὸ σήσαμον. περὶ μὲν οὖν τούτων σκεπτέον.

4 Ἐν δὲ ταῖς ἀγαθαῖς χώραις πρὸς τὸ μὴ φυλλομανεῖν ἐπινέμουσι καὶ ἐπικείρουσι τὸν σῖτον,

---

[1] Lit. 'does not make fallow land.' *cf. C.P.* 4. 8. 3.
[2] ὡς καρπιζόμενον I conj. after W. (καρπιζόμενον τὴν γῆν); καρπὸς U; καρπὸς M; καρπὸν Ald. *cf. C.P. l.c.* and 4. 8. 1 ; 4. 8. 3: μὴ καρπίζεσθαι τὴν γῆν ἀλλὰ νειὸν ποιεῖν (? <νέοις>) καρποῖς, 'for fresh crops.'
[3] δὲ conj. W.; γε Ald. [4] *cf. C.P.* 4. 8. 3.
[5] ἡ ἀρίστη νειὸς conj. W. (*cf.* 8. 9. 1 ; *C.P.* 4. 8. 1); χειρίστη νήπιος U; χειρίστην νήπιος MP; καλλίστη νειὸς Ald. *cf.* also *C.P.* 3. 20. 7.

general it does not reinvigorate the ground,[1] since it exhausts[2] it; but it destroys weeds,[3] and above all and soonest caltrop. And in general[4] it is not every kind of soil which suits it; the soil should be black and fat. Of the other leguminous plants the bean best[5] reinvigorates the ground, even if it is sown thick and produces much fruit.

Of special features of 'summer crops.'

All those crops[6] sown at the summer seed-time need little water,[7] and they say also that spring water is better for them than rain water; and Italian millet[8] and millet need less water, for, if they have too much, they shed their leaves. Millet is the robuster plant, Italian millet is sweeter and less robust. Sesame and lupin are not eaten green by any animal[9]; whether the same is true of *erysimon* and *horminon* is matter for enquiry; for these too are bitter. *Erysimon* is like sesame and is oily; *horminon* is like cummin and black, and is sown at the same time as sesame. These matters then require investigation.

Of treatment of cereals peculiar to special localities.

[10] In good soils to prevent the crop running wildly to leaf they graze and cut down the young corn,

[6] Plin. 18. 96 and 101.
[7] ὀλίγου, sc. ὕδατος, but the omission is strange; perhaps due to misunderstanding of ὀλίγου δεῖ by a scribe. Sch. joins the words τὰ δὲ ... πάντα to the last sentence, and supplies καρπίζεται τὴν γῆν (ὀλίγου δεῖ = almost).
[8] μέλινοι Ald. H.; ἔλυμοι Vin. *cf.* 8. 1. 1. n.
[9] ζῷον add. Sch. from G and Plin. 18. 96. *cf. C.P.* 6. 12. 12.
[10] Plin. 18. 157–162.

ὥσπερ καὶ ἐν Θετταλίᾳ. συμβαίνει δ' ἂν μὲν ἐπινέμωσιν ὁποσακισοῦν μηδὲν ἀλλοιοῦσθαι τὸν καρπόν, ἂν δὲ ἐπικείρωσιν ἅπαξ μόνον ἐξίστασθαι τὸν πυρὸν καὶ γίνεσθαι μακρὸν καὶ οὐχ ἁδρόν, ὃν καλοῦσι καμακίαν, καὶ οὐκ ἀποκαθίστασθαι πάλιν σπειρόμενον· τοῦτο μὲν οὖν ὡς παύροις συμβαῖνον Θετταλοὶ λέγουσιν. ἐν Βαβυλῶνι δὲ ἀεὶ καὶ ὥσπερ τεταγμένως ἐπικείρουσι μὲν δίς, τὸ δὲ τρίτον τὰ πρόβατα ἐπαφιᾶσιν· οὕτω γὰρ φύει τὸν καυλόν, εἰ δὲ μὴ φυλλομανεῖ· γίνεται δὲ μὴ καλῶς ἐργασαμένοις πεντηκονταχόα, τοῖς δὲ ἐπιμελῶς ἑκατονταχόα. ἡ δὲ ἐργασία τὸ ὡς πλεῖστον χρόνον ἐμμένειν τὸ ὕδωρ, ὅπως ἰλὺν ποιήσῃ πολλήν· πίειραν γὰρ οὖσαν καὶ πυκνὴν τὴν γῆν δεῖ ποιῆσαι μανήν. ὕλην δὲ οὐ φέρει καὶ πόαν ὥσπερ ἐν Αἰγύπτῳ. τὰ μὲν οὖν τοιαῦτα χώρας ἀρετῆς.

5 Φύεται δὲ καὶ ἀπὸ ῥιζῶν πυρὸς καὶ κριθὴ πολλαχοῦ τῷ ὑστέρῳ ἔτει· αὐτοετὴς δὲ καὶ ἀπὸ τῶν εἰς κράστιν κειρομένων ἑτέρου καλάμου παραβλαστάνοντος. ὡσαύτως δὲ κἂν ὑπὸ χειμῶνος ἐκπαγῇ· παραβλαστάνει γὰρ ὑδάτων ἐπιγινομένων· ὁ δὲ στάχυς ἀτελὴς καὶ μικρὸς ἀπὸ τῶν τοιούτων. βλαστάνουσι δὲ τῷ ὑστέρῳ ἔτει καὶ ἀπὸ τῶν καταπονουμένων καὶ συμπατουμένων, ὥστε μηδὲν εἶναι δῆλον ὡς εἰπεῖν, οἷον ὅταν

---

[1] ἰλὺν conj. Sch. from Plin. 18. 162 ; ὕλην Ald.H.
[2] Text perhaps defective : cf. Plin. l.c.

for instance in Thessaly. And the result is that, however often they graze it, the crop is not impaired; while if they cut it down not more than once, the wheat changes in character and becomes tall and weak—what they call 'long-shafted' corn, and, if seed of this is sown, it does not recover its character. This the Thessalians tell of as having occurred in a few cases. At Babylon however they cut it down twice always and as it were systematically, and after that they let the sheep on to it; for in that case it makes its straw, but otherwise it runs wildly to leaf; and, if the ground is ill cultivated, it produces fifty fold, if it is carefully cultivated, a hundred fold. And the 'cultivation' consists in letting the water lie on it as long as possible, so that it may make much silt[1]; for the soil being fat and close must be made open. And at Babylon[2] the ground does not produce weeds and grasses, as it does in Egypt. Such are the things which depend on the quality of the soil.

Of cereals which grow a second time from the same stock.

[3] Wheat and barley also in many places grow from the root in the next year, or in the same year from crops cut down for fodder, since a second haulm shoots up. The like happens also if the plant has been nipped by winter; for it shoots again when rain comes; but such plants produce an ear which is imperfect and under-sized. There is also new growth the next year from plants which are roughly treated or trodden down[4] so that hardly anything remains visible, as happens when an army has marched over

[3] *cf. C.P.* 4. 8. 5.   [4] *cf. C.P. l.c.*

διέλθῃ στρατόπεδον, καὶ οἱ στάχυες μικροὶ καὶ τούτων, οὓς καλοῦσιν ἄρνας· τῶν δὲ χεδροπῶν οὐδὲν δύναται τοιοῦτον ποιεῖν ἢ οὐχ ὁμοίως. καὶ αἱ βλαστήσεις τοσαυταχῶς.

6 Πρὸς αὔξησιν δὲ καὶ τροφὴν μέγιστα μὲν ἡ τοῦ ἀέρος κρᾶσις συμβάλλεται, καὶ ὅλως ἡ τοῦ ἔτους κατάστασις· εὐκαίρων γὰρ ὑδάτων καὶ εὐδιῶν καὶ χειμώνων γινομένων ἅπαντα εὔφορα καὶ πολύκαρπα, κἂν ἐν ἁλμώδεσι καὶ λεπτογείοις ᾖ· δι' ὃ καὶ παροιμιαζόμενοι λέγουσιν οὐ κακῶς ὅτι " ἔτος φέρει οὐχὶ ἄρουρα."

Μέγα δὲ καὶ αἱ χῶραι διαφέρουσιν οὐ μόνον τῷ πίειραι καὶ λεπταὶ καὶ ἔπομβροι καὶ αὐχμώδεις <εἶναι> ἀλλὰ καὶ τῷ ἀέρι τῷ περιέχοντι καὶ τοῖς πνεύμασιν· ἔνιαι γὰρ οὖσαι λεπταὶ καὶ φαῦλαι τελεσφοροῦσι διὰ τὸ πρὸς τὰ πνεύματα τὰ πόντια 7 κεῖσθαι καλῶς. ἄλλα δὲ ἄλλαις τοιαῦτα, καθάπερ πολλάκις εἴρηται· ταῖς μὲν γὰρ τὰ ζεφυρικὰ ταῖς δὲ τὰ βόρεια ταῖς δὲ τὰ νότια.

Συμβάλλεται δὲ καὶ οὐ μικρὰ ἡ ἐργασία καὶ μάλισθ' ἡ <πρὸ> τοῦ σπόρου· κατεργασθεῖσα γὰρ ῥᾳδίως ἐκφέρει. καὶ ἡ κόπρος δὲ μεγάλα βοηθεῖ τῷ διαθερμαίνειν καὶ συμπέττειν· προτρέχει γὰρ τὰ κοπριζόμενα τῶν ἀκόπρων καὶ εἴκοσιν ἡμέραις·

---

[1] τούτων conj. Sch.; τούτους Ald.H.
[2] cf. Lewis and Short s.v. agna.
[3] C.P. l.c. gives the reason.
[4] τοσαυταχῶς conj. Scal.; τοσαυταχεῖς Ald.H.

the field; the ears in such cases[1] too are undersized and are called 'lambs.'[2] But no kind of leguminous plant[3] can do anything of the kind, or at least not to the same extent. In these various ways[4] may new growth occur.

Of the effects of climate, soil, and manuring.

For growth and nourishment the climate is the most important factor, and in general the character of the season as a whole; for when rain, fair weather and storms occur opportunely, all crops bear well and are fruitful, even if they be in soil which is impregnated with salt or poor. Wherefore there is an apt proverbial saying[5] that "it is the year which bears and not the field."

But the soil also makes much difference, according as it is[6] fat or light, well watered or parched, and it also makes quite as much difference what sort of air and of winds prevails in that region; for some soils,[7] though light and poor, produce a good crop because the land has a fair aspect in regard to sea breezes. But, as has been repeatedly said already, the same breeze has not this effect in all places; some places are suited by a west, some by a north, some by a south wind.

Again the working of the soil and above all that which is done before[8] the sowing has an important effect; for when the soil is well worked it bears easily. Also dung is helpful by warming and ripening the soil, for manured land gets the start by as much as twenty days of that which has not been

[5] Quoted also *C.P.* 3. 23. 4.
[6] εἶναι add. Sch. [7] *cf. C.P.* 3. 23. 5.
[8] πρὸ add. W. *cf. C.P.* 3. 20. 6.

ἅπασι δὲ οὐ ξυμφέρει· καὶ χρήσιμος οὐ μόνον τοῖς περὶ τὸν σῖτον ἀλλὰ καὶ τοῖς ἄλλοις πλὴν πτερίδος, ταύτην δὲ φθείρειν φασὶν ἐπιβαλλομένην. ἀπόλλυται δὲ ἡ πτερὶς καὶ ἐπικοιμωμένων τῶν προβάτων, ὡς δέ τινες λέγουσι καὶ ἡ Μηδικὴ διὰ τὴν κόπρον καὶ τὸ οὖρον.

VIII. Τῶν δὲ σπερμάτων ἕκαστα καὶ πρὸς τὴν τῆς χώρας φύσιν ἁρμόττει, καὶ ὅλως γένη πρὸς γένος καὶ ἐν αὐτοῖς τοῖς ὁμογενέσιν, ἃ δὴ πειρῶνται διαιρεῖν. μεταβάλλει δὲ τὰ ξενικὰ τῶν σπερμάτων μάλιστα μὲν ἐν τρισὶν ἔτεσιν εἰς τὰ ἐπιχώρια. συμφέρει δὲ ἐκ τῶν ἀλεεινῶν εἰς τὰ μικρὸν ἧττον ἀλεεινὰ καὶ ἐκ τῶν ψυχεινῶν ἀνὰ λόγον ποιεῖσθαι τὴν μεταβολήν. τὰ δ' ἐκ τῶν δυσχειμερινῶν ἐν τοῖς πρωίοις ὀψὲ ἀποχεῖται, ὥστ' ἀπ' αὐχμοῦ φθείρεται, ἐὰν μὴ ὄψιον ὕδωρ σώσῃ. διὰ τοῦτο καὶ εὐλαβητέον φασὶ τὸ μίσγειν τὰ ξενικὰ τοῖς ἐπιχωρίοις ἐὰν μὴ ἐξ ὁμοίας, ὅτι ἀσύμφωνα τῇ χώρᾳ κατὰ τὸν σπόρον καὶ κατὰ τὴν γένεσιν, ὥστε καὶ ἐργασίας ἑτέρας δεῖται· τάς τε τῆς γῆς διαφορὰς καὶ τὰς τῶν σπερμάτων δυνάμεις καὶ ἔτι τὰς ἑκάστων ὥρας.

2 Ὅταν δὲ εὐετηρία γένηται, καὶ πολυνοστότερα τὰ σπέρματα γίνεται. Ἀθήνησι γοῦν αἱ κριθαὶ

---

[1] cf. Col. 2. 2. 13. The reference is perhaps to fern grown for litter, or possibly for medicinal use. cf. 9. 20. 5.
[2] χώρας conj. Sch.; ὥρας Ald.   [3] ἃ conj. Dalec.; ἂν Ald.
[4] ψυχεινῶν conj. W.; ψυχικῶν UM; ψυχρῶν Ald.
[5] ἀποχεῖται conj. Sch., cf. ἀπόχυσις 8. 3. 4; ἀποκεῖται Ald. cf. 4. 4. 10.

manured. However manure is not good for all crops; and further it is beneficial not only to corn and the like but to most other things, except fern,[1] which they say it destroys if it is put on. (Fern is also destroyed if sheep lie on it, and, as some say, lucerne is destroyed by their dung and urine.)

Of different qualities of seed.

VIII. There is a particular kind of soil[2] which best suits each kind of seed, whether we compare one class with another or those of the same class; and attempts are made to distinguish these.[3] Foreign seeds change into the native sorts in about three years. It is well that they should be imported from a warm climate to one that is rather less warm, or from a cold one[4] to one that is rather less cold. Those imported from a wintry climate, if they be those of early crops, are late in coming into ear,[5] so that they get destroyed by drought unless rain late in the season saves them. Wherefore they say that one should take good heed not to mix foreign with native seeds, unless they come from a similar place, since[6] they do not agree with the soil[7] as to the time of being sown and of germinating, and accordingly need different cultivation; and so that one should take good heed to the differences of soil, the properties of the seed, and further the seasons appropriate to each.

When however there is a good season, the grain also is fuller.[8] For instance at Athens the barley pro-

[6] ὅτι conj. Sch.; ἔτι UMAld.

[7] χώρᾳ conj. Sch.; ὥρᾳ UMAld.

[8] πολυνοστότερα: cf. νόστιμος, C.P. 4. 13. 2, Geop. 2. 16. 1, and other reff. in Sch.'s exhaustive note.

τὰ πλεῖστα ποιοῦσιν ἄλφιτα· κριθοφόρος γὰρ ἀρίστη· τοῦτο δ' οὐχ ὅταν πλεῖσται γένωνται ἀλλ' ὅταν λάβῃ τινὰ κρᾶσιν. ἐν δὲ τῇ Φωκίδι περὶ Ἐλάτειαν οἱ πυροὶ ποιοῦσιν ἡμιόλια τὰ ἄλευρα, καὶ ἐν Σόλοις τῆς Κιλικίας καὶ οἱ πυροὶ καὶ αἱ κριθαί· καὶ παρ' ἄλλοις ἄλλα πρὸς ἅπερ εὐφυὴς ἑκάστη. βελτίω μὲν οὖν καὶ χείρω τὰ σπέρματα καὶ διὰ τὴν ἐργασίαν καὶ διὰ τὴν γῆν γίνεται· καὶ γὰρ ἀπαγριοῦται καὶ ἡμεροῦται, καθάπερ τὰ δένδρα· καὶ ὅλως μεταβάλλει <κατὰ> τὴν χώραν, ὥσπερ τινὰ τῶν δένδρων εὐθὺς ἕστηκε πρὸς τὸ χεῖρον.

3 Γένος δ' ὅλον ἐξαλλάττειν εἰς ἕτερον οὐδὲν ἄλλο πέφυκε πλὴν τίφη καὶ ζειά, καθάπερ εἴπομεν ἐν τοῖς πρώτοις λόγοις, καὶ ἡ αἶρα δ' ἐκ τῶν πυρῶν καὶ κριθῶν διαφθειρομένων· ἢ εἰ μὴ τοῦτο ἀλλὰ φιλεῖ γε μάλιστα ἐν τοῖς πυροῖς γίνεσθαι, καθάπερ καὶ ὁ μελάμπυρος ὁ Ποντικὸς καὶ τὸ τῶν βολβῶν σπέρμα, καὶ ἄλλα δὲ ἐν ἄλλοις τῶν σπερμάτων· ἐπεὶ καὶ ὁ αἰγίλωψ δοκεῖ μᾶλλον ἐν ταῖς κριθαῖς, ἐν δὲ τοῖς φακοῖς ἄρακος τὸ τραχὺ καὶ σκληρόν, ἐν δὲ ταῖς ἀφάκαις ὁ πελεκῖνος ὅμοιον τῇ ὄψει τῷ πελέκει· σχεδὸν δὲ καθ' ἕκαστόν ἐστι τὸ συνεκτρεφόμενον καὶ συνανα-

---

[1] κατὰ add. W. cf. 2. 4. 1.
[2] τινα conj. W.; τε P; τῷ Ald.; τὸ H.Vin.Vo. cf. 2. 2. 6.
[3] ζειά conj. Scal.; ὕεα Ald. H.   [4] 2. 4. 1.

duces more meal than anywhere else, since it is an excellent land for that crop; and this is so, not merely when a very large crop is sown, but when the weather has been favourable for it. And in Phocis about Elateia the wheats produce half as much meal again as elsewhere; while at Soli in Cilicia this is true of both wheat and barley; and in other parts there are other crops for which the soil is severally well adapted. Wherefore grain turns out better or worse because of the soil as well as because of cultivation; for in some places it changes into the cultivated from the wild form, or the reverse, like trees; and in general it changes according[1] to the soil in which it is grown, just as some[2] trees, when transplanted, forthwith deteriorate.

Of degeneration of cereals, and of the weeds which infest particular crops.

But no kind can change altogether into another, except one-seeded wheat and rice-wheat,[3] as we said[4] in our previous discussions, and darnel which comes from degenerate wheat and barley: at least, if this is not the true account, darnel loves chiefly to appear among wheat, as does the Pontic[5] *melampyros* and the seed of purse-tassels,[6] even as other seeds appear in other crops; thus *aigilops* seems to grow for choice among barley, and among lentils the rough hard kind of *arakos*, while among tares occurs the axe-weed,[7] which resembles an axe-head in appearance. Indeed in the case of nearly every crop there is a plant which grows up with it and

[5] *cf.* 8. 4. 6, where μελάμπυρον was said to be peculiar to Sicily.   [6] *cf. C.P.* 4. 6. 1.

[7] Plin. 18. 155; 27. 121; Diosc. 3. 130; Hesych. *s.v.* βέλλεκυς.

μιγνύμενον εἴτε διὰ τὰς χώρας, ὅπερ οὐκ ἄλογον,
4 εἴτε δι' ἄλλην τινὰ αἰτίαν. ἔνια δὲ καὶ φανερῶς ἐστι κοινὰ πλειόνων, ἀλλὰ διὰ τὸ μάλιστα ἔν τισιν εὐθενεῖν ἴδια τούτων φαίνεται, καθάπερ ἡ ὀροβάγχη τῶν ὀρόβων καὶ ἡ ἀπαρίνη τῶν φακῶν· ἀλλὰ ἡ μὲν μάλιστα ἐπικρατεῖ τῶν ὀρόβων διὰ τὴν ἀσθένειαν· ἡ δὲ ἀπαρίνη μάλιστα ἐν τοῖς φακοῖς εὐτροφεῖ· τρόπον δέ τινα καὶ παραπλήσιόν ἐστι τῇ ὀροβάγχῃ <τῷ> ἐπιβάλλειν καὶ κατέχειν ὅλον ὥσπερ πλεκτάναις· ἀποπνίγει γὰρ οὕτως, ὅθεν καὶ τοὔνομα εἴληφε.

5 Τὸ δ' ὑποφυόμενον εὐθὺς ἐκ τῆς ῥίζης τῷ κυμίνῳ καὶ τῷ βουκέρῳ τὸ αἱμόδωρον καλούμενον μᾶλλον ἰδίᾳ. ἔστι δὲ τὸ αἱμόδωρον μονόκαυλον οὐκ ἀπεμφερές [τῷ καυλῷ], πλὴν βραχύτερόν τε πολύ, καὶ ἄνωθέν τι κεφαλῶδες ἔχει· ῥίζαν δὲ ὑποστρόγγυλον· οὐθὲν δὲ ἕτερον ἀφαναίνεται παρὰ τὸ βούκερας. γίνεται δὲ ταῦτα ἐν ταῖς λεπταῖς οὐκ ἐν ταῖς πιείραις, ὥσπερ καὶ τῆς Εὐβοίας ἐν τῷ Ληλάντῳ μὲν οὐ γίνεται περὶ δὲ τὸν Κάνηθον καὶ εἴ τις ἄλλος τοιοῦτος τόπος.

---

[1] ἄλλην τινα conj. Sch.; ἄλλης τινα U; ἄλλην Ald.
[2] τῷ add. Sch.
[3] πλεκτάναις conj. W.; πλεκτᾶνες U; πλεκτάνης M; πλεκτάνην Ald.; *veluti brachiis* G.
[4] Plin. 19. 176, who however calls this αἱμόδωρον. See Index App. (26).

ENQUIRY INTO PLANTS, VIII. VIII. 3-5

mingles with it, whether this is due to the soil, which is a reasonable explanation, or to some other[1] cause. Some plants of this character evidently attach themselves to more than one kind of crop, but, because they are specially vigorous in some one particular crop, they are thought to be peculiar to that one, as 'vetch-strangler' (dodder) to vetches and bedstraw to lentils. But the former gains the mastery over the vetches especially because of the weakness of that plant; and bedstraw is specially luxuriant among lentils; to some extent it resembles dodder, in that[2] it overspreads the whole plant and holds it fast as it were in coils,[3] for it is thus that dodder strangles the plant, and this is the origin of its name ('vetch-strangler').

[4]The plant which springs up straight from the roots of cummin and the plant called broom-rape which similarly attaches itself to 'ox-horn'[5] (fenugreek) are somewhat more peculiar in their habits.[6] Broom-rape has a single stem,[7] and is not unlike ... ,[8] but is much shorter and has on the top a sort of head, while its root is more or less round; and there is no other plant which it starves except fenugreek. These plants grow in light and not in fat soils; thus in Euboea they do not occur at Lelanton,[9] but only about Kanethos[10] and in districts of like character.

---

[5] Plin. 24. 184.
[6] ἰδίᾳ MSS.; ? ἴδια.
[7] cf. C.P. 5. 15. 5, where the same is said of λειμόδωρον (cf. Plin. 19. 176). But Ald.Bas.Cam. give αἱμόδωρον here; hemodorum G.
[8] τῷ καυλῷ probably conceals the name of a plant.
[9] cf. Strabo, 10. 1. 9. L. is the name of a Euboean *river* in Plin. 4. 64.
[10] cf. Strabo, 10. 1. 8, Ap. Rhod. 1. 77.

ταῦτα μὲν οὖν κοινὰ πλειόνων ὄντα κατισχύει μᾶλλον ἐν τοῖς εἰρημένοις διὰ τὴν ἀσθένειαν.

6 Τὸ δὲ τέραμον καὶ ἀτέραμον λέγεται μὲν ἐπὶ τῶν ὀσπρίων μόνον, οὐκ ἄλογον δὲ καὶ ἐπὶ τῶν σιτωδῶν παραπλήσιον ἢ καὶ ταὐτό τι συμβαίνειν, ἀλλὰ διὰ τὸ μὴ τὴν αὐτὴν εἶναι χρείαν οὐχ ὁμοίως ἐμφανές· ἐπεὶ οὐδ' ἐπὶ τούτων ἁπάντων ὁμοίως ἀλλὰ μάλιστα ἐπὶ τῶν κυάμων λέγεται καὶ φακῶν, εἴτ' οὖν καὶ μάλιστα πασχόντων εἴτε καὶ διὰ τὴν χρείαν φαινομένων. γίνεται γοῦν πλεοναχῶς· πολλαχοῦ γὰρ τόποι τινές εἰσιν οἳ ἀεὶ φέρουσι τεράμονα καὶ ἄλλοι πάλιν ἀτεράμονα· τὸ δὲ ὡς ἐπὶ πᾶν οἱ λεπτόγεω μᾶλλον 7 τεράμονα· καὶ ἀέρος κατάστασίς τις ποιεῖ τὴν τοιαύτην παραλλαγήν· σημεῖον δὲ ὅτι ταὐτὰ χωρία καὶ ὁμοίως ἐργασθέντα φέρει ποτὲ μὲν τεράμονα <ποτὲ δὲ ἀτεράμονα.> περὶ Φιλίππους δὲ ὁ κύαμος λικμώμενος, ἐὰν ὑπὸ πνεύματος ἐγχωρίου ληφθῇ, τεράμων ὢν ἀτεράμων γίνεται. ταῦτα μὲν οὖν μηνύει διότι πολλαχῶς τῶν αὐτῶν

---

[1] cf. 2. 4. 2; C.P. 4. 12; Plin. 18. 155, who makes *ateramum, teramum* plants.
[2] πλεοναχῶς· πολλαχοῦ I conj.; πλέον· πολλαχῶς MSS.
[3] ποτὲ δὲ ἀτεράμονα add. H. from G.
[4] cf. C.P. 4. 12. 8; Plut. *Quaest. Conv.* 7. 2. 3; Plin. *l.c.*

The reason then why these plants, which attach themselves to more than one kind, grow stronger when attached to the plants specified, is that the latter are not robust.

Of the conditions in the seeds of pulses known as 'cookable' and 'uncookable,' and their causes.

[1] The terms 'cookable' and 'uncookable' are only applied to pulses, but it is not unreasonable to suppose that conditions like those indicated, if not identical with them, occur also in cereals, though they are not so obvious, since these plants are not put to the same use. Indeed it is said that these terms are not applied even to all pulses alike, but chiefly to beans and lentils, either because these are specially subject to these conditions, or because the use to which they are put makes them more conspicuous. At all events the conditions occur for a variety of reasons; for in many parts [2] there are places which regularly produce seeds that are 'cookable,' while others again produce seeds that are 'uncookable'; in general however it is light soils which tend to produce the former. Now it is a certain condition of the climate which causes this variation; a proof of which is the fact that the same piece of land, tilled in the same manner, produces sometimes seeds that are 'cookable,' sometimes seeds that are 'uncookable.' [3] In the district of Philippi, if the beans, while being winnowed,[4] are caught by the prevailing wind of the country, they become 'uncookable,' having previously been 'cookable.' These facts prove that for various reasons, of districts [5] which are close together, have the same

[5] αὐτῶν conj. W.; δὲ τῶν Ald. *cf.* a similar expression 8. 2. 10.

χωρίων ἔνια σύνορα καὶ ὁμοίως καθήμενα καὶ οὐδεμίαν ἔχοντα κατὰ τὴν γῆν διαφορὰν τὸ μὲν τεράμονα τὸ δ' ἀτεράμονα φέρει, καὶ ἐνίοτε μόνον αὔλακος διοριζούσης.

IX. Καρπίζεται τὴν γῆν μάλιστα πυρὸς εἶτα κριθή, δι' ὃ καὶ ὁ μὲν ἀγαθὴν ζητεῖ χώραν ἡ δὲ κριθὴ δύναται καὶ ἐν ταῖς ψαφαρωτέραις ἐκφέρειν· τῶν δὲ χεδροπῶν μάλιστα ἐρέβινθος καίπερ ἐλάχιστον χρόνον ἐν τῇ γῇ μένων, ὁ δὲ κύαμος, ὥσπερ ἐλέχθη, καὶ ἄλλως οὐ βαρὺ καὶ ἔτι κοπρίζειν δοκεῖ τὴν γῆν διὰ μανότητα καὶ εὐσηψίαν· δι' ὃ καὶ οἱ περὶ Μακεδονίαν καὶ Θετταλίαν ὅταν ἀνθῶσιν ἀνατρέπουσι τὰς ἀρούρας.

2 Τῶν δὲ ὁμοιοπύρων καὶ ὁμοιοκρίθων, οἷον ζειᾶς τίφης ὀλύρας βρόμου αἰγίλωπος, ἰσχυρότατον καὶ μάλιστα καρπιζόμενον ἡ ζειά· καὶ γὰρ πολύρριζον καὶ βαθύρριζον καὶ πολυκάλαμον· ὁ δὲ καρπὸς κουφότατος καὶ προσφιλὴς πᾶσι τοῖς ζώοις. τῶν δὲ ἄλλων ὁ βρόμος· πολύρριζος γὰρ καὶ οὗτος καὶ πολυκάλαμος. ἡ δὲ ὀλύρα μαλακώτερον καὶ ἀσθενέστερον τούτων. ἡ δὲ τίφη πάντων κουφότατον· καὶ γὰρ καὶ μονοκάλαμον <καὶ λεπτοκάλαμον,> δι' ὃ καὶ χώραν ζητεῖ

---

[1] cf. C.P. 4. 12. 1.   [2] cf. C.P. 4. 8. 3.
[3] Plin. 18. 120; Varro 1. 23. 3; Col. 2. 10. 7.
[4] 8. 7. 2.
[5] i.e. dig in the bean-plants if the soil is poor, before the pods are formed enough to make it worth while to gather the beans. So Varro l.c.   [6] Cited by Galen.

aspect and shew no difference of soil, some bear 'cookable' some 'uncookable' seeds, and that sometimes when there is only [1] the breadth of a furrow between them.

Of the grains and pulses which most exhaust the soil, or which improve it.

IX. Wheat exhausts the land more than any other crop, and next to it barley; wherefore the former requires good soil, while barley will bear even on somewhat crumbling soils; [2] and of leguminous plants chick-pea is the most exhausting, although this crop is in the ground only a very short time. [3] Beans, as was said,[4] are in other ways not a burdensome crop to the ground, they even seem to manure it, because the plant is of loose growth and rots easily; wherefore the people of Macedonia and Thessaly turn over the ground when it is in flower.[5]

[6] Of the plants which resemble wheat or barley —such as *zeia* (rice-wheat) one-seeded wheat *olyra*[7] (rice-wheat) oats *aigilops*—*zeia* is the strongest[8] and most exhausts the ground; for it has many roots which run deep and many stems; but its fruit is the lightest and is welcome to all animals. Of the rest oats[9] is the most exhausting; for this too has many roots and many stems. *Olyra* is a more delicate plant and not so robust as these. But one-seeded wheat is the crop which is of all the least burdensome to the soil; for it has but a single slender stem[10]; wherefore also it requires a light soil and not, like

[7] See Index.
[8] ἰσχυρότατον conj. W. from Galen; ἰσχυρότερον Ald.
[9] βρόμος· πολύρριζος γὰρ conj. Sch.; β. πολ.· καὶ γὰρ Ald.
[10] καὶ λεπτοκάλαμον add. Bod. from Galen.

λεπτήν, οὐχ ὥσπερ ἡ ζειὰ πίειραν καὶ ἀγαθήν. ἔστι δὲ δύο ταῦτα καὶ ὁμοιότατα τοῖς πυροῖς ἥ τε <ζειὰ καὶ ἡ τίφη,> ὁ δ' αἰγίλωψ καὶ ὁ βρόμος ὥσπερ ἄγρι' ἄττα καὶ ἀνήμερα.

3 Ἐπικαρπίζεται δὲ σφόδρα καὶ ὁ αἰγίλωψ τὴν γῆν, καί ἐστι πολύρριζον καὶ πολυκάλαμον· ἡ δὲ αἶρα παντελῶς ἀπηγριωμένον. τῶν δὲ ἐν τοῖς θερινοῖς ἀρότοις τὸ σήσαμον δοκεῖ χαλεπώτατον εἶναι τῇ γῇ καὶ μάλιστα καρπίζεσθαι· καίτοι πολυκαλαμώτερον καὶ παχυκαλαμώτερον καὶ πολυρριζότερον κέγχρος. διαφέρει δὲ τά τε πρὸς τὴν γῆν κοῦφα καὶ τὰ πρὸς τὴν ἡμετέραν τροφήν. ἔνια γὰρ ἐναντίως, ὥσπερ τὰ χεδροπὰ καὶ οἱ κέγχροι· καὶ τὰ πρὸς ἡμᾶς δέ, ὥσπερ ἐλέχθη, καὶ τἆλλα ζῷα. καὶ περὶ μὲν τούτων ἅλις.

X. Νοσήματα δὲ τῶν σπερμάτων τὰ μὲν κοινὰ πάντων ἐστίν, οἷον ἡ ἐρυσίβη, τὰ δ' ἴδιά τινων, οἷον ὁ σφακελισμὸς τοῦ ἐρεβίνθου, καὶ τὸ ὑπὸ καμπῶν κατεσθίεσθαι καὶ ὑπὸ ψυλλῶν, τινὰ δὲ καὶ ὑπ' ἄλλων θηριδίων. ἔνια δὲ καὶ ψωριᾷ καὶ ἁλμᾷ, καθάπερ καὶ τὸ κύμινον. τὰ δ' ἐπιγινόμενα ζῷα μὴ ἐξ αὐτῶν ἀλλ' ἐκ τῶν ἔξωθεν οὐχ ὁμοίως βλάπτει. ἐπιγίνεται γὰρ ἡ μὲν κανθαρὶς

---

¹ ἥ τε ζειὰ καὶ ἡ τίφη add. W. from Galen.
² ὁ δ' conj. Scal.; ὅ τ' Ald.H.; ἤ τ' UMP.
³ καίτοι conj. W.; καὶ Ald. cf. C.P. 4. 15. 1.
⁴ τὰ add. St.   ⁵ 8. 3. 5 ad fin.
⁶ καὶ τὰ Ald.; καὶ οἱ τὰ UMP; ? καὶ αὖ τὰ W.

ENQUIRY INTO PLANTS, VIII. ix. 2–x. 1

zeia, one that is fat and good. These last two,[1] *zeia* and one-seeded wheat, are also those which are likest to wheat, while[2] *aigilops* and oats are as it were wild and uncultivated things.

Aigilops also greatly exhausts the land, having many roots and many stems; while darnel is a plant which has become altogether wild. Of the crops sown at the summer seed-time sesame seems to be most severe on the land and to exhaust it most; yet[3] millet has more numerous and stouter stems and more roots. Moreover there is a difference between crops which[4] are called 'light' in relation to the soil and those called 'light' in regard to human use. For some, such as leguminous plants and millet, are light in one sense but not in the other; and, as was said,[5] what[6] is light for men is not necessarily so for the other animals. Now enough of these matters.

Of the diseases of cereals and pulses, and of hurtful winds.

X. [7] As to diseases of seeds—some are common to all, as rust, some are peculiar to certain kinds; thus chick-pea is alone subject to rot[8] and to being eaten by caterpillars and by spiders[9]; and some seeds are eaten[10] by other small creatures. Some again are liable to canker and mildew,[11] as cummin. But creatures which do not come from the plant itself but from without do not do so much harm; thus the *kantharis*[12] is a visitor among wheat, the

---

[7] Plin. 18. 152 and 154. [8] *cf.* 4. 14. 2.
[9] ψυλλῶν: described by Arist. *H.A.* 9. 39. 1.
[10] δὲ add. Sch.; ? κατεσθίεσθαι· κατεσθίεται δὲ καὶ ὑπὸ ψ. W.
[11] ψωριᾷ καὶ ἀλμᾷ conj. W.; ψώραις καὶ ἄλμαις Ald. *cf.* 7.5.4 n.
[12] Plin. 18. 156.

ἐν τοῖς πυροῖς, τὸ δὲ φαλάγγιον ἐν ὀρόβοις, ἄλλα δ' ἐν ἄλλοις.

2 Ἐρυσιβᾷ δ' ὡς ἁπλῶς εἰπεῖν τὰ σιτώδη μᾶλλον τῶν ὀσπρίων· αὐτῶν δὲ τούτων κριθὴ μᾶλλον ἢ πυρός· καὶ τῶν κριθῶν ἕτεραι ἑτέρων, μάλιστα δ' ὡς εἰπεῖν ἡ Ἀχιλληΐς. διαφέρει δὲ καὶ ἡ τῶν χωρίων θέσις καὶ ἡ φύσις οὐ μικρόν· τὰ γὰρ προσήνεμα καὶ μετέωρα οὐκ ἐρυσιβᾷ ἢ ἧττον, ἀλλὰ τὰ ἔγκοιλα καὶ ἄπνοα· γίνεται δὲ ἡ ἐρυσίβη 3 πανσελήνοις μάλιστα. ἀπόλλυται δὲ καὶ ὑπὸ τῶν πνευμάτων καὶ πυρὸς καὶ κριθή, ὅταν ἢ ἀνθοῦντα ληφθῇ ἢ ἄρτι ἀπηνθηκότα καὶ ἀσθενῆ· μᾶλλον δὲ κριθή, πολλάκις δ' ἤδη ἐν τῷ ἀδρύνεσθαι οὖσα, ἐὰν μεγάλα καὶ πλείω χρόνον ἐπιγένηται· ξηραίνει γὰρ καὶ ἀφαναίνει, ὃ καλοῦσί τινες ἐξανεμοῦσθαι. διαπόλλυσι δὲ καὶ ἥλιος ὁ ἐκνέφελος ἄμφω καὶ μᾶλλον πυρὸν ἢ κριθήν, ὥστε μηδ' ἐπίδηλον εἶναι τὸν στάχυν τῇ ὄψει ὄντα κενόν.

4 Τὸν δὲ πυρὸν ἀπολλύουσι καὶ οἱ σκώληκες οἱ μὲν εὐθὺς κατεσθίοντες φυόμενοι τὰς ῥίζας, οἱ δὲ ὅταν αὐχμῶντες ἀποχυθῆναι μὴ δύνωνται· τότε γὰρ ἐγγινόμενος ὁ σκώληξ ἐσθίει τὸν ἀποπηνιζόμενον κάλαμον· ἐσθίει δὲ ἄχρι τοῦ στάχυος, εἶτ'

---

[1] Plin. 18. 154.
[2] ἐρυσιβᾷ conj. W.; ἐρυσίβαι Ald.; εἰς add. Sch.
[3] τὰ add. Sch.   [4] cf. C.P. 3. 22. 2.
[5] ἐρυσιβᾷ conj. Sch.; ἐρυσίβαι Ald.
[6] cf. C.P. 4. 13. 4 ; Plin. 18. 151.
[7] μέγαλα conj. Sch.; μεγάλη UMAld

ENQUIRY INTO PLANTS, VIII. x. 1-4

phalangion in vetches, and other pests in other crops.

[1] Generally speaking, cereals are more liable to rust [2] than pulses, and among these barley is more liable to it than wheat; while of barleys some kinds are more liable than others, and most of all, it may be said, the kind called 'Achillean.' Moreover the position and character of the land make no small difference in this respect; for lands which [3] are exposed to the wind [4] and elevated are not liable to rust,[5] or less so, while those that lie low and are not exposed to wind are more so. And rust occurs chiefly at the full moon. [6] Again wheat and barley are destroyed by winds, if they are caught by them either when in flower, or when the flower has just fallen and they are weak; and this applies specially to barley, indeed it occurs when the grain is already ripening, if the winds are violent [7] and last a long time; for they dry up and parch the grain, which some call being 'wind-bitten.' Also a hot sun after cloudy weather destroys both, and wheat more than barley, so that the ear is not even conspicuous, since it is empty.

Wheat is also destroyed by grubs; sometimes they eat the roots, as soon as they appear,[8] sometimes they do their work when by reason of drought the ear cannot be formed [9]; for at such times the grub is engendered, and eats the haulm as it is becoming unrolled [10]; it eats right up to the ear and then,

[8] φυόμενοι conj. Sch.; φυόμενον Ald. cf. C.P. 3. 22. 4.
[9] ἀποχυθῆναι conj. Sch. after Vin.Vo.G; ἀπολυθῆναι UM Ald. cf. C.P. 3. 22. 4; 4. 14. 1.
[10] ἀποπηνιζόμενον: lit. 'unwinding itself.' All edd. mark the word as corrupt.

ἐξαναλώσας ἀπόλλυται· καὶ ἐὰν μὲν ὅλον ἐκφάγῃ τελέως αὐτὸς ὁ πυρός, ἐὰν δὲ ἐπὶ θάτερον τοῦ καλάμου καὶ ἐκβιάσηται τὴν ἀπόχυσιν, τοῦτο μὲν αὖον τοῦ στάχυος θάτερον δὲ ὑγιές. γίνεται δὲ οὐ πανταχοῦ τὸ περὶ τοὺς πυρούς, οἷον ἐν Θετταλίᾳ, ἀλλὰ κατὰ χώρας τινάς, ὥσπερ ἐν τῇ Λιβύῃ καὶ τῆς Εὐβοίας ἐν τῷ Ληλάντῳ.

5 Σκώληκες δὲ γίνονται καὶ ἐν τοῖς ὤχροις καὶ τοῖς λαθύροις καὶ τοῖς πισοῖς, ὅταν ὑγρανθῶσι καὶ θερμημερίαι γένωνται, καθάπερ καὶ ἐν τοῖς ἐρεβίνθοις αἱ κάμπαι. πάντα δὲ ἐξαναλώσαντα τὰς τροφὰς ἀπόλλυται καὶ ἐν τοῖς χλωροῖς καὶ ἐν τοῖς ξηροῖς καρποῖς, οἷον οἵ τε ἶπες καὶ οἱ ἐν τοῖς κυάμοις ἐγγινόμενοι καὶ ἐν τοῖς ἄλλοις, ὥσπερ καὶ ἐν τοῖς δένδρεσι καὶ ἐν τοῖς ξύλοις ἐλέχθη, πλὴν τῶν κεραστῶν καλουμένων. πρὸς ἅπαντα δὴ ταῦτα μεγάλα διαφέρουσιν αἱ χῶραι οὐκ ἀλόγως· ὁ γὰρ ἀὴρ εὐθὺς διάφορος τῷ θερμὸς ἢ ψυχρὸς εἶναι ἢ ὑγρὸς ἢ ξηρός· οὗτος δ' ἦν ὁ γονεύων· δι' ὃ καὶ ἐν οἷς εἰώθασι γίνεσθαι οὐκ ἀεὶ γίνονται.

XI. Τῶν δὲ σπερμάτων οὐχ ἡ αὐτὴ δύναμίς ἐστιν εἴς τε τὴν βλάστησιν καὶ εἰς θησαυρισμόν.

---

[1] αὐτὸς: sc. the grain. ἀναίνεται conj. W.; ? αὐαίνεται αὐτὸς.
[2] θάτερον conj. Sch.; θατέρου Ald.
[3] cf. C P. 3. 22. 3. ὤχροις conj. St.; ὄχροις Ald. H.
[4] καθάπερ καὶ conj. Sch.; καὶ καθάπερ Ald.
[5] πάντα conj. W.; τὰ Ald.

having consumed it, perishes. And, if it has entirely eaten it, the wheat itself[1] perishes; if however it has only eaten one[2] side of the haulm and the plant has succeeded in forming the ear, half the ear withers away, but the other half remains sound. However it is not everywhere that the wheat is so affected; for instance this does not occur in Thessaly, but only in certain regions, as in Libya and at Lelanton in Euboea.

Grubs occur also in *okhros*[3] *lathyros* and peas, whenever these crops get too much rain and then hot weather supervenes; and caterpillars occur in chick-peas under the same conditions.[4] All[5] these pests perish, when they have exhausted their food, whether the fruit in which they occur be green or dry, just as wood-worms do and the grubs found in beans and other plants, as was said of the pests found in growing trees and in felled timber. But the creature called 'horned worm'[6] is an exception. Now in regard to all these pests the position makes a great difference, as might be expected. For the climate, it need hardly be said, makes a difference according as it is hot or cold, moist or dry; and it was the climate which gave rise to these pests[7]; wherefore they are not always found even in places in which they ordinarily occur.[8]

Of seeds which keep or do not keep well.

XI. The seeds have not all the same capacity for germination and for keeping well. Some germinate

[6] *cf.* 4. 14. 5; *C.P.* 5. 10. 5.
[7] δ' ἦν ὁ γονεύων I conj.; δ' ἦν ὁ νεύων UAld.; δ' ἠνονεύων M; δ' ἐστιν ὁ γονεύων conj. Sch.; δ' ὁ γονεύων conj. W.
[8] *i.e.* because the atmospheric conditions are not always favourable to the pest.

ἔνια μὲν γὰρ βλαστάνει καὶ τελειοῦται τάχιστα καὶ θησαυρίζεται κράτιστα, καθάπερ ἔλυμος καὶ κέγχρος· ἔνια δὲ βλαστάνει μὲν εὖ ταχέως δὲ σήπεται, καθάπερ ὁ κύαμος καὶ μᾶλλον ὁ τεράμων· ταχὺ δ' ἡ ἀφάκη καὶ ὁ δόλιχος· κριθὴ δὲ πυροῦ <θᾶττον·> θᾶττον δὲ καὶ ὁ κονιορτώδης σῖτος καὶ ὁ ἐν οἰκήμασι κονιατοῖς ἢ ἀκονιάτοις.

2 Γίνεται δὴ φθειρομένοις σπέρμασι ἴδια ζῷα, καθάπερ ἐλέχθη, πλὴν ἐρεβίνθου· μόνος γὰρ οὗτος οὐ ζωογονεῖ. καὶ σηπομένοις μὲν πᾶσι σκώληξ κοπτομένοις δὲ καθ' ἕκαστον ἴδιον. πάντων δὲ μάλιστα διαμένουσιν ἐρέβινθος καὶ ὄροβος, τούτων δ' ἔτι μᾶλλον ὁ θέρμος· ἀλλ' ἔοικέ γ' οὗτος ὥσπερ ἀγρίῳ.

3 Διαφέρει δὲ ὡς ἔοικε χώρα χώρας καὶ ἀὴρ ἀέρος εἰς τὸ κόπτεσθαι καὶ μὴ τὰ σπέρματα· ἐν Ἀπολλωνίᾳ γοῦν τῇ περὶ τὸν Ἰόνιον οὐκ ἐσθίεσθαί φασιν ὅλως κύαμον, δι' ὃ καὶ εἰς θησαυρισμὸν ἀποτίθεσθαι· διαμένει δὲ καὶ περὶ Κύζικον ἐπὶ πλείω. μέγα δὲ πρὸς διαμονὴν καὶ τὸ ξηρὰ θερίζειν· ἐλάττων γὰρ ἡ ὑγρότης· θερίζουσι δ' ἐγχυλότερα τὰ μὲν χεδροπὰ πρὸς τὸ μᾶλλον καὶ ῥᾷον συλλέξαι, ταχὺ γὰρ καταρρεῖ καὶ αὐανθέντα

---

[1] εὖ conj. W.; οὐ Ald.   [2] σήπεται add. W.
[3] cf. 8. 8. 6; C.P. 5. 18. 2.   [4] θᾶττον add. W.
[5] καὶ ὁ κον. . . . ἀκονιάτοις conj. W., cf. C.P. 4. 16. 1; ὁ κονιορτώδης καὶ ὁ κόνιορτος· καὶ ὁ ἐν τοῖς κονιορτοῖς ἐν ἅπασιν, οἷον κονιάτοις ἢ ἀκονιάτοις Ald.; so also UM, but omitting τοῖς; U gives κονιατοῖς· ἢ ὁ ἀκονίορθος for καὶ ὁ κόνιορτος mBas. cf. Plin. 18. 301, Varro 1. 57. 1, where the use of a cement of pounded marble is recommended.
[6] δὴ φθειρομένοις conj. Sch.; διαφθειρομένοις UMAld.

and mature very quickly, and keep excellently, as Italian millet and millet. Some germinate well,[1] but soon rot,[2] as beans, and especially those that are 'cookable[3]'; so do tare and calavance; also barley perishes sooner[4] than wheat; and dusty[5] grain and that which is kept in plastered store-rooms perishes sooner than that which is kept in unplastered rooms.

Again, as seeds decay,[6] they engender special creatures, except chick-pea, which alone engenders none. As they rot,[7] all produce a grub; but, as they get worm-eaten, each produces a special creature. Chick-pea and vetch keep best of all, and better still than these lupin; but this, as it were, is like a wild kind.[8]

[9] It appears that soil and climate make a difference as to whether the seed gets worm-eaten or not; at least they say that at Apollonia on the Ionian Sea beans do not get eaten in this way at all, and therefore they are put away and stored; and about Cyzicus they keep an even longer time. It also makes a great difference to keeping that the seed should be gathered dry, for then there is less moisture in it.[10] However the seeds of leguminous plants are gathered with a certain amount of moisture in them,[11] because then they can be collected in greater quantity and more easily; for otherwise they are soon shed and get shrivelled up and split[12];

[7] *i.e.* rot is produced in all cases by the same creature (σκώληξ), but the condition called being 'worm-eaten' is due in each plant to a different pest.

[8] *i.e.* and so the seed is hard and not liable to these attacks. *cf.* 8. 11. 8; *C.P.* 4. 16. 2.

[9] *cf. C.P.* 4. 16. 2.   [10] *i.e.* liability to rot.

[11] ἐγχυλότερα conj. Sch.; εὐχυλότερα Ald.H.Cam.; εὐχηλότερα Bas. *cf. C.P.* 4. 13. 3.   [12] Plin. 18. 125.

θρύπτεται, τοὺς δὲ πυροὺς καὶ γένος τι κριθῶν διὰ τὸ βελτίους εἰς τὰ ἄλφιτα γίνεσθαι μὴ ἀπεξηραμμένας.

4 Δι' ὃ καὶ εἰς θώμους συντιθέασι καὶ πυροὺς καὶ κριθάς, καὶ δοκοῦσιν ἁδρύνεσθαι ἐν θώμῳ μᾶλλον ἢ λιποσαρκεῖν. οὐκ ἐσθίεται δὲ σῖτος, ὅταν ὑσθεὶς θερισθῇ· ἀθέριστος δὲ μάλιστα διαμένει ὁ πυρός, ἔτι δὲ μᾶλλον ὁ θέρμος· οὐδὲ γὰρ θερίζουσι τοῦτον πρότερον ἢ ὕδωρ γενέσθαι, διὰ τὸ ἐκπηδᾶν θεριζόμενον καὶ ἀπόλλυσθαι τὸ σπέρμα.

5 Πρὸς ἔκφυσιν δὲ καὶ τὴν ὅλην σπορὰν ἄριστα δοκεῖ τὰ ἐνάενα· τὰ δὲ δίενα χείρω καὶ τὰ τρίενα, τὰ δ' ὑπερτείνοντα σχεδὸν ἄγονα, πρὸς δὲ τὴν σίτησιν ἀρκοῦντα. βίος γάρ ἐστιν ἑκάστοις ὡρισμένος εἰς γονήν. καίτοι καὶ ταύταις παραλλάττει ταῖς δυνάμεσι διὰ τοὺς τόπους ἐν οἷς ἂν θησαυρίζωνται. τῆς γοῦν Καππαδοκίας ἐν χωρίῳ τινὶ τῷ καλουμένῳ Πέτρᾳ καὶ τετταράκοντα ἔτη διαμένειν φασὶ γόνιμα καὶ χρήσιμα πρὸς σπόρον, εἰς δὲ τὴν σίτησιν ἑξήκοντα ἢ ἑβδομήκοντα· τὸ γὰρ ὅλον οὐ κόπτεσθαι· τὰ δὲ ἱμάτια καὶ τὴν ἄλλην

6 γάζαν κόπτεσθαι. τὸ γὰρ χωρίον ἄλλως τε ὑψηλὸν εἶναι καὶ εὔπνουν καὶ ἔναυρον αἰεὶ καὶ ἀπ' ἀνατολῆς ἔχουσι καὶ δύσεως καὶ μεσημβρίας.

---

[1] ἢ λιποσαρκεῖν conj. H.; ἡλίκα σωρῶν U; ἡλίκα σαρκῶν M. W. brackets as due to a gloss. *cf. C.P.* 4. 13. 6.

[2] ὅταν ὑσθεὶς conj. Scal.: so Vo.; ὁ τανυσθείς Bas.Cam. θερισθῇ conj. W.; περιφυῇ MSS.

ENQUIRY INTO PLANTS, VIII. xi. 3–6

and wheat and one kind of barley are gathered before they are dry, because then they are better for meal.

Wherefore the grain of wheat and barley is put into heaps, and it seems to ripen in a heap rather than to lose substance.[1] (However corn does not get worm-eaten when it is reaped after exposure to rain.)[2] Also corn lasts better than other things if it is left standing, and so does lupin to an even greater extent; indeed this crop is not even gathered till rain has fallen,[3] because, if it is gathered, the seed springs out and is lost.

Of the age at which seeds should be sown.

[4] For propagation and sowing generally seeds one year old seem to be the best; [5] those two or three years old are inferior, while those kept a still longer time are infertile, though they are still available as food. For each kind has a definite period of life in regard to reproduction. However these seeds too differ in their capacity according to the place in which they are stored. For instance, in Cappadocia at a place called Petra they say that seed remains even for forty years fertile and fit for sowing, while as food it is available for sixty or seventy years; for that it does not get worm-eaten at all like clothes and other stored-up articles: for that the region is, apart from this, elevated and always exposed to fair winds and breezes which prevail alike from [6] the east, the west, and the

---

[3] πρότερον ἢ conj. W.; τὸν τρόπον UAld. *cf. C.P.* 4. 13. 3; Plin. 18. 133.   [4] Plin. 18. 195.
[5] *cf.* 7. 5. 5; *Geop.* 2. 16.
[6] ἀπ' conj. Sch.; ἐπὶ P₂Ald.

φασὶ δὲ καὶ ἐν Μηδείᾳ καὶ ταῖς ἄλλαις ταῖς ὑψηλαῖς χώραις διαμένειν θησαυριζόμενα πολὺν χρόνον. ἐρέβινθον δὲ δὴ καὶ θέρμον καὶ ὄροβον καὶ κέγχρον καὶ τὰ τοιαῦτα δῆλον ὅτι πολλῷ πλείω τούτων, ὥσπερ καὶ ἐν τοῖς περὶ τὴν Ἑλλάδα τόποις. ἀλλὰ ταῦτα μέν, ὥσπερ εἴρηται, τῶν τόπων ἴδια.

7 Δοκεῖ δὲ καὶ γῆ τις εἶναι παρά τισιν ἣ διαπαττομένη συντηρεῖ τὸν πυρόν, ὥσπερ ἥ τε ἐν Ὀλύνθῳ καὶ ἐν Κηρίνθῳ τῆς Εὐβοίας· ποιεῖ δὲ χείρω μὲν εἰς τὴν σίτησιν ἁδρότερον δὲ τῇ προσόψει· παραπάττουσι δὲ χοίνικα εἰς τὸν μέδιμνον.

Πυρωθέντα πάντα τὰ σπέρματα ἀπόλλυται καὶ ἀβλαστῆ γίνεται· καίτοι περί γε Βαβυλωνά φασι τὰς κριθὰς καὶ τοὺς πυροὺς ἐπὶ τῆς ἅλω πηδᾶν, ὥσπερ τὰ φρυγόμενα· ἀλλὰ δῆλον ὅτι διαφορά τίς ἐστι τῆς θερμότητος, ἢ ἁπλῶς πως θερμασίᾳ γίνεται [καὶ] ἡ πήδησις. καὶ τὰ μὲν τοιαῦτα σχεδὸν ὡσπερεὶ κοινὰ δόξειεν ἂν εἶναι πάντων ἢ τῶν πλείστων.

---

[1] cf. C.P. 5. 18. 3; for millet-seed see J.H.S. vol. xxxv. part i. p. 22.
[2] διαπαττομένη conj. H.; διαπλαττομένη UMAld.; διακοπτομένη P₂. cf. Plin. 18. 305.
[3] παραπάττουσι conj. Sch., cf. Geop. 2. 21. 3; (ἐμπάσσειν); παρατάττουσι UMAld. cf. Varro 1. 57. 1.

ENQUIRY INTO PLANTS, VIII. xi. 6-7

south. They say that in Media [1] also and other elevated countries the seed when stored keeps for a long time. And it is plain that chick-pea lupin vetch millet and the like will keep a far longer time than these seeds, as they do even in districts of Hellas. However these peculiarities, as has been said, are due to the particular region.

Of artificial means of preserving seed.

There appears to be a kind of earth in some places, which when sprinkled [2] over the seed helps to make wheat keep, for instance, the earth found at Olynthos and at Kerinthos in Euboea; this makes the grain inferior for food, but fuller in appearance; the earth is sprinkled [3] in the proportion of one pint to twenty-four of grain.

Of the effect of heat on seeds.

All seeds if exposed to fire perish and become infertile. Yet they say that at Babylon [4] the grains of barley and wheat jump on the threshing-floor like corn which is being parched. However it is plain that it is some particular kind of warmth [5] which produces this effect: or else the jumping is simply another effect of heat.[6] Such behaviour would appear to be common to most,[7] if not to all kinds.

---

[4] *i.e.* the grain is there exposed to great *sun*-heat. *cf. de igne* 44.

[5] *i.e.* the sun's heat is different in kind, and therefore in effect, to that of a fire.

[6] θερμασίᾳ conj. Sch.; θερμασία Ald.H.

[7] ὡσπερεὶ κοινὰ conj. Sch. from G; ὥσπερ εἰκόνα UM; ὥσπερ εἰκόνες Ald.H.

THEOPHRASTUS

8 Ἔνια δὲ ἔχει τινὰ ἰδιότητα καὶ τῶν δοκούντων ὥσπερ ἀγρίων εἶναι καὶ κατὰ τὴν γένεσιν καὶ τὴν ἔκφυσιν, ὥσπερ ὁ θέρμος καὶ ὁ αἰγίλωψ· ὁ μὲν γὰρ θέρμος καίπερ ἰσχυρότατος ὢν ὅμως, ἐὰν μὴ εὐθὺς ἀπὸ τῆς ἅλω καταβληθῇ, κακοφυὴς γίνεται, καθάπερ ἐλέχθη, καὶ τὸ ὅλον δὲ οὐκ ἐθέλει κρύπτεσθαι τῇ γῇ, δι' ὃ καὶ οὐχ ὑπαροῦντες σπείρουσι· πολλάκις δὲ κἂν εἰς ὕλην ἢ βοτάνην τινὰ πέσῃ, διωσάμενος ταύτην συνάπτει τὴν ῥίζαν τῇ γῇ καὶ βλαστάνει. χώραν δὲ ὕφαμμον ζητεῖ καὶ φαύλην μᾶλλον, τὸ δὲ ὅλον οὐκ ἐθέλει φύεσθαι ἐν διειργασμένῃ.

9 Ὁ δὲ αἰγίλωψ ἀνάπαλιν· ἐν γὰρ τῇ γεωργουμένῃ κάλλιον· καὶ ἐνιαχοῦ δὲ πρότερον ἀβλαστὴς ὢν ἐὰν γεωργηθῇ βλαστάνει καὶ γίνεται πολύς, καὶ ὅλως δὲ φιλεῖ χώραν ἀγαθήν. ἴδιον δὲ αὐτοῦ λέγεται πρὸς τὰ ἄλλα τὰ σιτώδη σπέρματα καὶ ἡ παρ' ἐνιαυτὸν βλάστησις ἑκατέρου τῶν σπερμάτων. δι' ὃ καὶ οἱ βουλόμενοι τελέως φθεῖραι, δύσφθαρτον γὰρ δὴ φύσει τυγχάνει, τὰς ἀρούρας ἀνιᾶσιν ἀσπόρους ἐπὶ δύο ἔτη, καὶ ὅταν ἀναβλαστήσῃ τὰ πρόβατα ἐπαφιᾶσι πολλάκις, ἕως ἂν ἐκνεμηθῶσι, καὶ αὕτη γίνεται φθορὰ παντελής· ἅμα δὲ τοῦτο μαρτυρεῖ καὶ τὴν παρὰ μέρος βλάστησιν.

---

[1] καὶ conj. Sch.; ἢ Ald.H.
[2] cf. 8. 1. 3.   [3] cf. C.P. 4. 7. 3.
[4] ὑπαροῦντες conj. H.; ὑπαποροῦντες UMAld. cf. C.P. 3. 20. 8.

ENQUIRY INTO PLANTS, VIII. xi. 8-9

Of certain peculiarities of the seed of lupin and aigilops.

Some even[1] of those kinds which seem to be more or less wild have peculiarities as to their germination and growth, for instance, lupin and *aigilops*. For lupin, although it is very robust, unless it is planted immediately after leaving the threshing-floor,[2] turns out of poor growth, as was said, and refuses altogether to be buried in the ground;[3] wherefore they sow it without first ploughing[4] the land. And often if the seed has fallen amid thick undergrowth or herbage,[5] it thrusts this aside, fastens on to the earth with its root and grows vigorously. It seeks sandy and poor soil for choice, and will not grow at all in[6] cultivated[7] soil.

Aigilops has the opposite character; it grows better in tilled soil; and in some places where at first it would not grow, if the ground is tilled, it grows and yields a large crop, and in general it likes good soil. A peculiarity[8] mentioned in regard to it as compared with other cereal seeds is that one seed in two does not germinate for a year. Wherefore those who wish to destroy it entirely, (since it is naturally hard to destroy), leave the fields unsown for two years, and, when it springs up, send in[9] the sheep several times till they have grazed it down, and this is a way of completely destroying it. At the same time this testifies to the fact that the seed does not all germinate at once.

[5] *cf.* 1. 7. 3; Plin. 18. 134.   [6] ἐν conj. W.; τῇ Ald.H.
[7] *cf.* 8. 11. 2.   [8] *cf. C.P.* 4. 6. 1.
[9] ἐπαφιᾶσι conj. Sch., *cf.* 8. 7. 4; ἀφιῆσι M; ἀφίησι P; ἀφίασι Ald.

BOOK IX

I

1. Ἡ ὑγρότης οἰκεία τῶν φυτῶν, ἣν δὴ καλοῦσί τινες ὀπὸν ὀνόματι κοινῷ προσαγορεύοντες· δύναμιν δὲ ἔχει δῆλον ὅτι τὴν καθ' αὑτὴν ἑκάστη. χυμὸς δὲ ταῖς μὲν μᾶλλον ταῖς δ' ἧττον ἀκολουθεῖ, ταῖς δ' ὅλως οὐκ ἂν δόξειεν, οὕτως ἀσθενὴς καὶ ὑδαρής τίς ἐστι. πλείστη μὲν οὖν ὑπάρχει πᾶσι κατὰ τὴν βλάστησιν, ἰσχυροτάτη δὲ καὶ μάλιστα ἐκφαίνουσα τὴν ἑαυτῆς φύσιν ὅταν ἤδη παύσηται καὶ βλαστάνοντα καὶ καρπογονοῦντα. συμβαίνει δέ τισι τῶν φυτῶν καὶ χρόας ἰδίας ἔχειν· τοῖς μὲν λευκὰς οἷον τοῖς ὀπώδεσι, τοῖς δ' αἱματώδεις οἷον τῷ κενταυρίῳ καὶ τῇ ἀτρακτυλίδι καλουμένῃ ἀκάνθῃ, τοῖς δὲ χλωρόν, τοῖς δ' ἐν ἄλλῃ χρόᾳ. ἔνδηλα δὲ μᾶλλον ταῦτα ἐν τοῖς ἐπετείοις καὶ τοῖς ἐπετειοκαύλοις ἢ τοῖς δένδροις.

2. Ἡ δ' ὑγρότης τῶν μὲν πάχος ἔχει μόνον, ὥσπερ τῶν ὀπωδῶν· τῶν δὲ καὶ δακρυώδης γίνεται, καθάπερ ἐλάτης πεύκης τερεβίνθου πίτυος ἀμυγδαλῆς κεράσου προύμνης ἀρκεύθου κέδρου τῆς ἀκάνθης τῆς Αἰγυπτίας πτελέας, καὶ γὰρ αὕτη φέρει κόμμι

---

[1] cf. C.P. 6. 11. 16.
[2] I have omitted ἡ and restored δὲ before ἔχει (om. Scal.; found in UMAld.).
[3] τῷ κενταυρίῳ conj. Scal. cf. Plin. 25. 32; κενταυρίδι conj. St.; κευτηρίᾳ P₂Ald.G, cf. 9. 8. 7.

BOOK IX

OF THE JUICES OF PLANTS, AND OF THE MEDICINAL
PROPERTIES OF HERBS.

Of the various kinds of plant-juices and the methods of collecting them.

I. [1] Moisture belongs to plants as such and some call it the 'sap,' to give it a general name; and it plainly has [2] special qualities in each plant. This moisture is attended by a taste, in some cases more, in some less, while in some it would seem to have none, so weak and watery is it. Now all plants have most moisture at the time of making growth, but it is strongest and most shows its character when the plant has ceased to grow and to bear fruit. Again in some plants the juice has a special colour; in some it is white, as in those which have a milky juice; in some blood-red, as in centaury [3] and the spinous plant which is called distaff-thistle; in some green: and in some of other colours. And these qualities are more obvious in annual [4] plants and those with annual stems than in trees.

Again in some plants the juice is merely thick, as in those in which it is of milky character; but in some it is of gummy character, as in silver-fir fir terebinth Aleppo pine almond *kerasos* (bird-cherry) bullace Phoenician cedar prickly cedar acacia elm.[5] For

[4] ἐν inserted here by W. instead of before τοῖς ἐπετείοις.
[5] πτελέας after κέδρου P₂Ald.; transposed by Sch. after Tobias Aldinus. *cf.* Plin. 13. 67.

πλὴν οὐκ ἐκ τοῦ φλοιοῦ ἀλλ' ἐν τῷ κωρύκῳ, ἔτι δὲ ἀφ' ὧν ὁ λίβανος καὶ ἡ σμύρνα, δάκρυα γὰρ καὶ ταῦτα, καὶ τὸ βάλσαμον καὶ <ἡ> χαλβάνη καὶ εἴ τι τοιοῦτον ἕτερον, οἷόν φασι τὴν ἄκανθαν τὴν Ἰνδικήν, ἀφ' ἧς γίνεταί τι ὅμοιον τῇ σμύρνῃ· συνίσταται δὲ καὶ ἐπὶ τῆς σχίνου καὶ ἐπὶ τῆς ἀκάνθης τῆς ἰξίνης καλουμένης, ἐξ ὧν ἡ μαστίχη.

3 Ἅπαντα δὲ ταῦτα εὔοσμα καὶ σχεδὸν ὅσα πιότητά τινα ἔχει καὶ λῖπος· ὅσα δ' ἀλιπῆ ταῦτα δ' ἄοσμα, καθάπερ τὸ κόμμι καὶ τὸ τῆς ἀμυγδαλῆς. ἔχει δὲ δάκρυον καὶ ἡ ἰξία ἡ ἐν Κρήτῃ καὶ ἡ τραγάκανθα καλουμένη· ταύτην δὲ πρότερον ᾤοντο μόνον ἐν Κρήτῃ φύεσθαι, νῦν δὲ φανερὰ καὶ ἐν Ἀχαΐδι τῆς Πελοποννήσου καὶ ἄλλοθι καὶ τῆς Ἀσίας περὶ τὴν Μήδειαν. καὶ τούτων μὲν πάντων ἔν τε τοῖς καυλοῖς καὶ τοῖς στελέχεσι καὶ τοῖς ἀκρεμόσι τὸ δάκρυον· ἐνίων δ' ἐν ταῖς ῥίζαις, ὥσπερ τοῦ ἱπποσελίνου καὶ τῆς σκαμμωνίας καὶ ἄλλων πολλῶν φαρμακωδῶν. τῶν δὲ καὶ ἐν τῷ καυλῷ καὶ ἐν τῇ ῥίζῃ· καὶ γὰρ τὸν καυλὸν ὀπίζουσιν ἐνίων καὶ τὰς ῥίζας, ὥσπερ καὶ τοῦ σιλφίου.

4 Τὸ μὲν οὖν τοῦ ἱπποσελίνου παρόμοιον τῇ σμύρνῃ· καί τινες ἀκούσαντες ὡς ἐντεῦθεν ἡ σμύρνα ἡγοῦνται βλαστάνειν ἐξ αὐτῆς ἱπποσέ-

---

¹ κωρύκῳ conj. Sch.; ἀγγείῳ H.; ἀγείῳ P₂Ald. probably a gloss on κωρύκῳ, for which *cf.* 2. 8. 3 and reff. in note. Plin. *l.c.* has preserved the right word through an absurd blunder —*in Coryco monte Ciliciae.*

ENQUIRY INTO PLANTS, IX. I. 2–4

this last also produces a gum, though it does not exude from the bark, but is found in the 'bag'[1] of the leaves; there are also the juices from which come frankincense and myrrh; for these too are gums; so too are balsam of Mecca *khalbane*[2] and any others of the kind that there may be, such as, they say, the Indian *akantha*, from which comes something[3] resembling myrrh; and a similar substance forms on mastich and the spinous plant called *ixine* (pine-thistle), whence mastic-gum is made.

All these have a fragrant odour, as in general have those which contain a viscous substance and are fatty; while those that are not fatty have no scent, as gum and the juice which exudes from the almond. The pine-thistle[4] of Crete has also a gum, and so has the plant called tragacanth;[5] this was formerly supposed to grow only in Crete, but now it is well known to grow also in Achaia in the Peloponnese and elsewhere in Hellas and in Asia in the Median country. In all these plants the gum occurs in the stems the trunks and the branches, but in some plants it is found in the roots, as in alexanders scammony and many other medicinal plants. In some it is found in the stem and also in the root;[6] for of some[7] plants they tap the stem and the roots as well, as is done with silphium.

Now the juice of alexanders is like myrrh, and some, having heard that myrrh comes from it, have supposed that, if myrrh is sown, alexanders comes up

---

[2] *galbanum. cf.* Plin. 12. 121; 24. 21. Verg. *G.* 3. 415; 4. 264. See 9. 7. 2; 9. 9. 2 n.
[3] τι I conj.; τὸ MSS.   [4] ἰξία = ἰξίνη. See Index.
[5] Plin. 13. 115.   [6] *cf. C.P.* 6. 11. 15.
[7] ἐνίων καὶ conj. Sch.; καὶ ἐνίων Ald.

λινον· φυτεύεται γάρ, ὥσπερ ἐλέχθη, καὶ ἀπὸ δακρύου τὸ ἱπποσέλινον, καθάπερ ἡ κρινωνία καὶ ἄλλα. τὸ δὲ τοῦ σιλφίου δριμύ, καθάπερ αὐτὸ τὸ σίλφιον· ὁ γὰρ ὀπὸς καλούμενος τοῦ σιλφίου δάκρυόν ἐστιν. ἡ δὲ σκαμμωνία καὶ εἴ τι ἄλλο τοιοῦτον, ὥσπερ ἐλέχθη, φαρμακώδεις ἔχουσι τὰς δυνάμεις.

5 Πάντων δὲ τῶν εἰρημένων τὰ μὲν αὐτομάτως συνίσταται, τὰ δ' ἀπ' ἐντομῆς, τὰ δ' ἀμφοτέρωθεν· τέμνουσι δὲ δῆλον ὅτι τὰ χρήσιμα καὶ τὰ μᾶλλον ἐπιζητούμενα. τοῦ δ' ἀπὸ τῆς ἀμυγδαλῆς οὐδεμία χρεία δακρύου, δι' ὃ κοὺκ ἀφελκοῦσι. πλὴν ἐκεῖνό γε φανερὸν ὅτι ὧν αὐτόματος ἡ πῆξις 6 τούτων πλείων ἡ ἐπιρροὴ τῆς ὑγρότητος. οὐ τὴν αὐτὴν δ' ὥραν ἁπάντων αἱ ἐντομαὶ καὶ ἡ πῆξις, ἀλλὰ τὸ μὲν τῆς ἀμπέλου μάλιστα συνίστασθαί φασιν ἐὰν μικρὸν πρὸ τῆς βλαστήσεως τμηθῇ, τοῦ δὲ μετοπώρου καὶ ἀρχομένου τοῦ χειμῶνος ἧττον· καίτοι πρός γε καρποτοκίαν αἱ ὡραιόταται ταῖς γε πλείσταις αὗται. τῆς δὲ τερμίνθου καὶ τῆς πεύκης καὶ εἰ ἔκ τινων ἄλλων ῥητίνη γίνεται μετὰ τὴν βλάστησιν· τὸ δ' ὅλον οὐκ ἐπέτειος ἡ τούτων, ἀλλ' εἰς πλείω χρόνον ἡ ἐντομή. τὸν δὲ λιβανωτὸν καὶ τὴν σμύρναν ὑπὸ Κύνα φασὶ καὶ ταῖς θερμοτάταις ἡμέραις ἐντέμνειν· ὡσαύτως δὲ καὶ τὸ ἐν Συρίᾳ βάλσαμον.

7 Ἀκριβεστέρα δὲ καὶ ἐλάττων ἡ καὶ τούτων

---

[1] ἐξ αὐτῆς conj. Scal.: *cf.* Plin. 19. 162, where *smyrnium* is given as a synonym; ἐν αὐτοῖς Ald.
[2] *cf.* 2. 2. 1 ; 6. 6. 8 ; *C.P.* 1. 4. 6.
[3] 9. 1. 3.   [4] *cf. C.P.* 6. 11. 15.

ENQUIRY INTO PLANTS, IX. I. 4–7

from it;[1] for, as was said,[2] this plant can be grown from an exudation, like the *krinonia* (lily) and other plants. The juice of silphium is pungent like the plant itself; for what is called the 'juice' of silphium is a gum. Scammony and similar plants, as was said,[3] have medicinal properties.

In all the plants mentioned the juice either forms naturally, or when incisions are made, or in both ways,[4] but it is obvious that men only make incisions in plants whose juice is of use and is specially sought after.[5] Now there is no use in the gum which exudes from the almond, wherefore men do not tap it.[6] However it is plain that in plants whose gum forms naturally the flow of juice is greater. The incisions and the clotting of the juice do not take place at the same season in all cases;[7] but the juice of the vine clots best they say if the incision is made a little before budding begins, less well in the autumn or at the beginning of winter; (although in regard to production of fruit these[8] seasons are the best in the case of most[9] vines). However with terebinth fir or any other tree which produces resin the best time is after the period of budding; yet in general these trees are not cut every year, but at longer intervals. The frankincense and myrrh trees they say should be cut at the rising of the Dogstar and on the hottest days, and so also the 'Syrian balsam' (balsam of Mecca).

The cutting of these is also a more delicate matter

[5] μᾶλλον ἐπιζητούμενα· τοῦ δ' ἀπὸ τῆς ἀ. conj. W. supported by G; μᾶλλον· ἐπὶ γοῦν τὸ ἀπὸ τῆς ἀ. UMAld.
[6] κοὐκ ἀφελκοῦσιν conj. Scal., *cf.* 9. 2. 1; κἂν ἀφέλκουσιν U; κἂν ἀφέλκωσι MAld. *cf.* Plin. 24. 105.
[7] Plin. 24. 106. [8] αἱ conj. W.; καὶ UPAld.
[9] γε conj. Sch.; δὲ Ald. *cf. C.P.* 3. 13. 2.

ἐντομή· καὶ γὰρ ἡ συρροὴ τῆς ὑγρότητος ἐλάττων· ὧν δὲ καὶ ὁ καυλὸς ἐντέμνεται καὶ ἡ ῥίζα, τούτων ὁ καυλὸς πρότερον, ὥσπερ καὶ τοῦ σιλφίου, καὶ καλοῦσι δὲ τῶν ὀπῶν τούτων τὸν μὲν καυλίαν τὸν δὲ ῥιζίαν· καί ἐστι βελτίων ὁ ῥιζίας· καθαρὸς γὰρ καὶ διαφανὴς καὶ ξηρότερος. ὁ δὲ καυλίας ὑγρότερος· καὶ διὰ τοῦτο ἄλευρον αὐτῷ περιπάττουσι πρὸς τὴν πῆξιν. τὴν ὥραν τῆς ἐντομῆς ἴσασιν οἱ Λίβυες· οὗτοι γὰρ οἱ σίλφιον λέγοντες. ὡσαύτως δὲ καὶ οἱ ῥιζοτόμοι καὶ οἱ τοὺς φαρμακώδεις ὀποὺς συλλέγοντες· καὶ γὰρ οὗτοι τοὺς καυλοὺς ὀπίζουσι πρότερον. ἁπλῶς δὲ πάντες καὶ οἱ τὰς ῥίζας καὶ οἱ τοὺς ὀποὺς συλλέγοντες τὴν οἰκείαν ὥραν ἑκάστων τηροῦσι. καὶ τοῦτο μὲν δὴ κοινόν.

II. Ἡ δὲ ῥητίνη γίνεται τόνδε τὸν τρόπον· ἐν μὲν τῇ πεύκῃ ὅταν ἀφελκωθείσης ἡ δὰς ἐξαιρεθῇ, συρρεῖ γὰρ εἰς τὸ ἕλκωμα τοῦτο πλείων ἡ ὑγρότης, ἐν δὲ τῇ ἐλάτῃ καὶ τῇ πίτυϊ ὅταν γευσάμενοι τῶν ξύλων ἀφελκώσωσιν· οὐ γὰρ πᾶς ἀφορισμὸς ὁμοίως· ἀφελκοῦσι γὰρ καὶ τὰς τερμίνθους ἐν ἀμφοῖν καὶ ἐν τῷ στελέχει καὶ ἐν τοῖς ἀκρεμόσιν· αἰεὶ δὲ πλείων καὶ βελτίων ἡ εἰς τὸ στέλεχος συρρέουσα τῆς εἰς τοὺς ἀκρεμόνας.

2 Διαφέρουσι δὲ καὶ κατὰ τὰ δένδρα. βελτίστη μὲν γὰρ ἡ τερμινθίνη· καὶ γὰρ συνεστηκυῖα καὶ εὐωδεστάτη καὶ κουφοτάτη τῇ ὀσμῇ ἀλλ' ὀλίγη. δευτέρα δὲ ἡ ἐλατίνη καὶ πιτυΐνη, κουφότεραι γὰρ τῆς πευκίνης. πλείστη δὲ ἡ πευκίνη καὶ

---

[1] cf. 6. 3. 2; *C.P.* 6. 11. 16.
[2] σίλφιον conj. St.; σιλφίου UM; σιλφιολέγοντες PAld.

and is done on a smaller scale; for the flow of juice is less. In those plants whose stem and root are both cut the stem is cut first, as also with silphium; and the juices so obtained are called respectively stalk-juice and root-juice, of which the latter is the better, for it is clear transparent and less liquid. The stalk-juice is more liquid, and for this reason they sprinkle meal[1] over it to make it clot. The Libyans know the season for cutting, for it is they that gather the silphium.[2] So also do the root-diggers and those that collect medicinal juices, for these too tap the stems earlier. And in general all those who collect whether roots or juices observe the season which is appropriate in each case. And this remark applies generally.

Of resinous trees and the methods of collecting resin and pitch.

II. [3] Resin is made in the following manner:—in fir it is done by removing the resinous wood after the tree has been tapped; for then the juice flows into the hole so made in greater abundance; in silver-fir and Aleppo pine it is done by tapping the wood, after tasting it. For there is no fixed rule for all alike; thus with terebinth they tap both the stem and the branches; but the juice which runs into the stem is always more abundant and better than that which flows into the branches.

There are also differences in[4] the resin obtained from different trees. The best is that of terebinth; for it sets firm, is the most fragrant, and has the most delicate smell; but the yield is not abundant. Next comes that of silver-fir and Aleppo pine, for these are more delicate than that of the fir. But that of the fir

[3] Plin. 16. 57.   [4] κατὰ conj. W.; ταῦτα Ald.H.

βαρυτάτη καὶ πιττωδεστάτη διὰ τὸ μάλιστα
ἔνδαδον εἶναι τὴν πεύκην. ἄγεται δὲ ἐν ἀσκοῖς
ὑγρά, κἄπειτα οὕτω συνίσταται. καίτοι φασὶ καὶ
τὴν τέρμινθον πιττοκαυτεῖσθαι περὶ Συρίαν· ἔστι
γὰρ ὄρος, καθάπερ ἐν τοῖς ἔμπροσθεν εἴπομεν, μέγα
τερμίνθων μεστὸν ἅπαν μεγάλων.

3 Ἔνιοι δέ φασι καὶ τὴν πίτυν καὶ τὴν κέδρον δὲ
τὴν Φοινικικήν· ἀλλὰ ταῦτα μὲν ὡς ἐνδεχόμενα
ληπτέον διὰ τὸ σπάνιον· ἐπεὶ οἵ γε περὶ Μακεδονίαν οὐδὲ τὴν πεύκην πιττοκαυτοῦσιν ἀλλ' ἢ
τὴν ἄρρενα· καλοῦσι γὰρ ἄρρενα τὴν μὴ καρποφόρον. τῆς δὲ θηλείας ἐάν τινα τῶν ῥιζῶν
λάβωσιν· ἅπασα γὰρ ἔνδαδος πεύκη ταῖς ῥίζαις.
καλλίστη δὲ πίττα γίνεται καὶ καθαρωτάτη ἡ ἐκ
τῶν σφόδρα προσείλων καὶ προσβόρρων, ἐκ δὲ
τῶν παλισκίων βλοσυρωτέρα καὶ βορβορώδης· ἐν
γὰρ τοῖς σφόδρα παλισκίοις οὐδὲ φύεται πεύκη
τὸ παράπαν.

4 Ἔστι δὲ καὶ ἀφορία τις καὶ εὐφορία καὶ πλήθους καὶ καλλονῆς· ὅταν μὲν γὰρ χειμὼν μέτριος
γίνηται, πολλὴ γίνεται καὶ καλὴ καὶ τῷ χρώματι
λευκοτέρα, ὅταν δὲ ἰσχυρός, ὀλίγη καὶ μοχθηρο-
τέρα. καὶ ταῦτά γέ ἐστι τὰ ὁρίζοντα πλῆθος καὶ
καλλονὴν πίττης, οὐχ ἡ πολυκαρπία τῶν πευκῶν.

---

[1] πιττοκαυτεῖσθαι conj. Sch., *cf.* 9. 3. 4 ; πιττοκαυθῆσαι U; πιττωθεῖσαι Ald.   [2] 3. 2. 6.
[3] δὲ conj. W.; καὶ Ald.H. *cf.* 3. 12. 3.
[4] μὴ conj. W.; γε Ald.H.; γε μὴ Cod.Casaub.Vin.; γε μὴν Vo. (τὴν ἄκαρπον mBas.). *cf.* 3. 9. 2.
[5] Plin. 16. 59.

ENQUIRY INTO PLANTS, IX. II. 2-4

is the most abundant, the grossest and the most pitch-like, because this tree has the greatest amount of resinous wood. It is carried about in baskets in a liquid state, and so acquires the more solid form which we know. However they say that in Syria pitch is extracted even from the terebinth by burning[1]; for there is in that land a mountain which, as we said before,[2] is all covered with great terebinths.

Some[3] say the same of Aleppo pine and also of Phoenician cedar; but this must be taken as only indicating what *can* be done, the practice not being common; for the people of Macedonia do not extract pitch by burning even from fir, except from the 'male' kind (they call the kind which bears no fruit[4] the 'male'); the 'female' kind they only treat in this way when they have found roots containing pitch; for all firs have resinous wood extending to the roots. [5] The finest and purest pitch is that obtained from trees growing in a sunny position and facing north[6]; that obtained from trees growing in shade is coarser[7] and muddy; (in exceedingly shady places the fir does not even grow at all).

Again the yield may be either good or bad as to amount and as to quality; thus, when there is a moderate winter, it is abundant and good and whiter in colour, but, when there is a severe winter, it is scanty and of inferior quality. And it is these conditions, and not the tree's capacity for bearing fruit, which determine the amount and quality of pitch.

[6] Apparently because this is the dry quarter in the Balkan peninsula.

[7] βλοσωρωτέρα conj. Sch.; βροσηροτέρα M; βλοσηροτέρα Ald. *cf. C.P.* 6. 12. 5.

THEOPHRASTUS

5 Οἱ δὲ περὶ τὴν Ἴδην φασί, διαιροῦντες τὰς πεύκας καὶ τὴν μὲν καλοῦντες Ἰδαίαν τὴν δὲ παραλίαν, τὴν ἐκ τῆς Ἰδαίας πλείω καὶ μελαντέραν γίνεσθαι καὶ γλυκυτέραν καὶ τὸ ὅλον εὐωδεστέραν ὠμήν, ἑψηθεῖσαν δὲ ἐλάττω ἐκβαίνειν· πλείω γὰρ ἔχειν τὸν ὀρρόν, δι' ὃ καὶ λεπτοτέραν εἶναι. τὴν δὲ τῆς παραλίας ξανθοτέραν καὶ παχυτέραν ὠμήν, ὥστε καὶ τὴν ἄφεψιν ἐλάττω γίνεσθαι, δᾳδωδεστέραν δὲ τὴν Ἰδαίαν. ὡς δὲ ἁπλῶς εἰπεῖν ἐκ τῆς ἴσης δᾳδὸς πλείω καὶ ὑδαρεστέραν ἐν ταῖς ἐπομβρίαις γίνεσθαι ἢ ἐν τοῖς αὐχμοῖς, καὶ ἐκ τῶν χειμερινῶν καὶ παλισκίων τόπων ἢ ἐκ τῶν εὐείλων καὶ εὐδιεινῶν. ταῦτα μὲν οὖν οὕτως ἑκάτεροι λέγουσιν.

6 Ἀναπληροῦσθαι δὲ συμβαίνει τὰ κοιλώματα πρὸς τὸ πάλιν ἐξαιρεῖν τῶν μὲν ἀγαθῶν πευκῶν ἐνιαυτῷ, τῶν δὲ μετριωτέρων ἐν δυσὶν ἔτεσι, τῶν δὲ μοχθηρῶν ἐν τρισίν. ἡ δὲ ἀναπλήρωσις οὐ τοῦ ξύλου καὶ τῆς συμφύσεως ἀλλὰ τῆς πίττης ἐστίν· ἐπεὶ τὸ ξύλον ἀδύνατον συμφῦναι καὶ ἓν γενέσθαι πάλιν, ἀλλ' ἡ ἐργασία διὰ τοσούτου χρόνου γίνεται τῆς πίττης· ἀναγκαῖον δὲ δῆλον ὅτι καὶ τῷ ξύλῳ γίνεσθαί τινα πρόσφυσιν, εἴπερ

---

[1] Plin. 16. 60.
[2] εὐωδεστέραν H.; εὐκρινωδεστέραν UMAld.; ? εὐκρινεστέραν καὶ εὐωδεστέραν W. cf 3. 9. 2.
[3] Plin. l.c.

ENQUIRY INTO PLANTS, IX. II. 5–6

The people of Mount Ida distinguish different kinds of fir, calling one 'that of Mount Ida' (Corsican pine), another the 'fir of the seashore,' (Aleppo pine); [1] and they say that the pitch obtained from the former is more abundant blacker sweeter and generally more fragrant [2] in the raw state, but that boiling down reduces the amount; for that it contains a larger proportion of watery matter, wherefore it is less substantial; but that derived from the 'fir of the seashore' is browner and thicker in the raw state, so that the amount is less reduced by boiling down; that the 'fir of Mount Ida' however contains more resinous wood. And, speaking generally, they say that from an equal amount of resinous wood more pitch is obtained and in a more liquid state in wet weather than during a drought, and from a wintry and shady position than from one that is sunny and enjoys fair weather. Such is the account given by the peoples of Mount Ida and of Macedonia respectively.

[3] The holes for the pitch fill up, so that the pitch can be again removed,[4] in good firs in a year, in those of more moderate quality in two years, in poor trees in three. The filling-up is composed of the pitch; it is not caused by closing up of the wood; for the wood cannot close up and become one again, but the effect which takes the time mentioned is due to the formation of the pitch.[5] However it is clearly inevitable that there should be some new growth of the wood too, seeing that the resinous wood is

[4] ἐξαιρεῖν conj. Sch.; ἐξαίρειν Ald.H.
[5] So W. explains ἀλλὰ ... πιττῆς. Or perhaps (as Sch.), 'however this is the interval which must elapse before the pitch can be worked again.'

ἐξαιρουμένης τῆς δᾳδὸς καὶ καιομένης τῆς πίττης ἡ ἐκροή.   τοῦτο μὲν οὖν οὕτω ληπτέον.

7 Οἱ δὲ περὶ τὴν Ἴδην φασίν, ὅταν λεπίσωσι τὸ στέλεχος,—λεπίζουσι δὲ τὸ πρὸς ἥλιον μέρος ἐπὶ δύο ἢ τρεῖς πήχεις ἀπὸ τῆς γῆς—ἐνταῦθα τῆς ἐπιρροῆς γινομένης ἐνδᾳδοῦσθαι ἐνιαυτῷ μάλιστα, τοῦτο δ' ὅταν ἐκπελεκήσωσιν ἐν ἑτέρῳ πάλιν ἐνδᾳδοῦσθαι καὶ τὸ τρίτον ὡσαύτως, μετὰ δὲ ταῦτα διὰ τὴν ὑποτομὴν ἐκπίπτειν τὸ δένδρον ὑπὸ τῶν πνευμάτων σαπέν· τότε δ' ἐξαιρεῖν αὐτοῦ τὴν καρδίαν, τοῦτο γὰρ μάλιστα δᾳδῶδες, ἐξαιρεῖν δὲ ἐκ τῶν ῥιζῶν· καὶ γὰρ ταύτας, ὥσπερ εἴπομεν, ἐνδᾴδους πασῶν.

8 Εἰκὸς δὲ δῆλον ὅτι τὰς μὲν ἀγαθάς, ὥσπερ ἐλέχθη, συνεχῶς τοῦτο δρᾶν τὰς δὲ χείρονας διὰ πλείονος χρόνου· καὶ ταμιευομένων μὲν πλείω χρόνον ἀντέχειν, ἐὰν δὲ πᾶσαν ἐξαιρῶσιν ἐλάττω· δύναται δ' ὡς ἔοικε τρεῖς μάλιστα τοιαύτας ἐξαιρέσεις ὑπομένειν. οὐχ ἅμα δὲ καρποφοροῦσιν αἱ πεῦκαι καὶ δᾳδοφοροῦσι· καρποφοροῦσι μὲν γὰρ εὐθὺς νέαι, δᾳδοφοροῦσι δὲ ὕστερον πολλῷ πρεσβύτεραι γινόμεναι.

III. Τὴν δὲ πίτταν καίουσι τόνδε τὸν τρόπον· ὅταν κατασκευάσωσιν ὁμαλῇ τόπον ὥσπερ ἅλω

---

[1] i.e. and so this kind of wood at least is replaced by fresh growth. [2] Plin. 16. 57.
[3] τῆς ἐπιρροῆς γινομένης ἐνδᾳδοῦσθαι conj. W.; τὴν ἐπιρροὴν γινομένην ἔνδαδον Ald.; τῆς ἐπιρροῆς γινομένης ἔνδαδον γίνεσθαι conj. Sch.

removed[1] and burnt when the discharge of pitch takes place. So much for this account.

[2] The people of Mount Ida however say that, when they bark the stem,—and they bark the side towards the sun to a height of two or three cubits from the ground,—the flow of pitch takes place in that part,[3] and in about a year the wood becomes full of pitch; and that, when they have hewn this part out, pitch forms again in the next year, and in the third year in like manner; after which[4] that the tree, because it has been cut away underneath, is rotted by the winds and falls; and that then[5] they take out its heart, for that is especially full of pitch, and that they also extract pitch from[6] the roots; for that these too, as we said,[7] are full of pitch in all firs.

Now it is plainly to be expected that they should, as was said, repeatedly thus treat a good tree, but an inferior one at longer intervals, and that, if the tree is husbanded, the supply should hold out longer, while, if they remove all the pitch, it will not hold out so long; it appears as a matter of fact that the tree will stand about three such removals of its substance. [8] However firs do not produce both fruit and pitch at once; they begin to bear fruit when they are quite young, but they only produce pitch much later, when they are older.

Of the making of pitch in Macedonia and in Syria.

III. This is the manner in which they make pitch by fire :—having prepared a level piece of ground,

[4] μετὰ δὲ ταῦτα conj. Sch.; τὰ δὲ ταῦτα UM; τὰ δὲ τοιαῦτα Ald.
[5] τότε conj. Sch. from G; τοὺς Ald.
[6] ἐκ Ald.; καὶ conj. W.
[7] 9. 2. 3.   [8] *cf. C.P.* 5. 16. 2.

THEOPHRASTUS

ποιήσαντες ἔχουσαν εἰς τὸ μέσον συρροὴν καὶ ταύτην ἐδαφίσωσι, κατασχίσαντες τοὺς κορμοὺς συντιθέασι παραπλησίαν σύνθεσιν τῆς τῶν ἀνθρακευόντων, πλὴν οὐκ ἔμβοθρον· ἀλλὰ τὰς σχίζας ὀρθὰς πρὸς ἀλλήλας, ὥστε λαμβάνειν ὕψος αἰεὶ κατὰ πλῆθος· γίνεσθαι δέ φασιν, ὅταν ἡ σύνθεσις ᾖ κύκλῳ μὲν ὀγδοήκοντα καὶ ἑκατὸν πηχέων εἰς ὕψος δὲ ἑξήκοντα πλεῖστον ἢ πεντήκοντα ἢ ἑκατὸν ἀμφοτέροις, ἐάνπερ ἡ δᾳς τυγχάνῃ

2 πίειρα. συνθέντες οὖν αὐτὴν οὕτως καὶ κατασκεπάσαντες ὕλῃ γῆν ἐπιβαλόντες κατακρύπτουσιν ὅπως μηδαμῶς διαλάμψῃ τὸ πῦρ, ἀπόλλυται γὰρ ἡ πίττα τούτου συμβάντος. ὑφάπτουσι δὲ κατὰ τὴν ὑπολειπομένην δίοδον· εἶτα δὲ καὶ ταῦτα ἐπιφράξαντες τῇ ὕλῃ καὶ ἐπιχώσαντες τηροῦσιν ἀναβαίνοντες κατὰ κλίμακος, ᾗ ἂν ὁρῶσι τὸν καπνὸν ὠθούμενον, καὶ ἐπιβάλλουσιν αἰεὶ τῆς γῆς ὅπως μηδ' ἀναλάμψῃ. κατεσκεύασται δὲ ὀχετὸς τῇ πίττῃ διὰ τῆς συνθέσεως τῆς ἀπορροῆς εἰς βόθυνον ὅσον ἀπέχοντα πεντεκαίδεκα πήχεις· ἡ δ' ἀπορρέουσα τῆς πίττης

3 ψυχρὰ γίνεται κατὰ τὴν ἁφήν. καίεται δὲ μάλιστα δύο ἡμέρας καὶ νύκτας· τῇ γὰρ ὑστεραίᾳ πρὸ ἡλίου δύναντος ἐκκεκαυμένη γίνεται καὶ ἐνδέδωκεν ἡ πυρά· τοῦτο γὰρ συμβαίνει μηκέτι ῥεούσης. τοῦτον δὲ τὸν χρόνον ἅπαντα τηροῦσιν

---

[1] ἐδαφίσωσι: cf. 9. 4. 4.
[2] cf. 5 9, where however the 'pit' is not described.
[3] γίνεσθαι δέ. Something seems to have dropped out at the beginning of this clause. ? "and they say that the pile at largest is 180" . . .: so Sch. supplying μεγίστη after ᾖ. The omitted words might also throw light on the preceding sentence.

230

ENQUIRY INTO PLANTS, IX. III. 1–3

which they make like a threshing-floor with a slope for the pitch to run towards the middle, and having made it smooth,[1] they cleave the logs and place them in an arrangement like that used by charcoal-burners,[2] except that there is no pit; but the billets are set upright against one another, so that the pile goes on growing in height according to the number used. And they say that the erection is complete,[3] when the pile is 180 cubits in circumference, and fifty, or at most sixty, in height; or again when it is a hundred cubits in circumference and a hundred in height,[4] if the wood happens to be rich in pitch. Having then thus arranged the pile and having covered it in with timber they throw on earth and completely cover it, so that the fire may not by any means show through; for, if this happens, the pitch is ruined. Then they kindle the pile where the passage is left, and then, having[5] filled that part[6] up too with the timber and piled on earth, they mount a ladder and watch wherever they see the smoke pushing its way out, and keep on piling on the earth, so that the fire may not even shew itself. And a conduit is prepared for the pitch right though the pile, so that it may flow into a hole about fifteen cubits off, and the pitch as it flows out is now cold to the touch. The pile burns for nearly two days and nights; for on the second day before sunset it has burnt itself out and the pile has fallen in; for this occurs if the pitch is no longer flowing. All this time[7] they keep watch and do not

[4] ἀμφοτέροις : ἀμφοτέρωσε conj. Sch.
[5] ἐπιφράξαντες conj. Scal. from G, *stipant*; ἐπάξαντες UMAld.; ἐπισάξαντες H.
[6] ταῦτα Ald.; ταύτην W. after Sch.'s conj.
[7] τοῦτον δέ τὸν χρόνον conj. Sch. from G, *totum tempus*; τόνδε δὲ τὸν τρόπον UMAld.H.

ἀγρυπνοῦντες, ὅπως μὴ διαλάμπῃ, καὶ θύουσι δὲ καὶ ἑορτάζουσιν εὐχόμενοι πολλήν τε καὶ καλὴν γίνεσθαι τὴν πίτταν· οἱ μὲν δὴ περὶ Μακεδονίαν καίουσι τὸν τρόπον τοῦτον.

4 Ἐν δὲ τῇ Ἀσίᾳ φασὶ περὶ Συρίαν οὐκ ἐκπελεκῶντας τὴν δᾷδα ἀλλ' ἐπ' αὐτῷ τῷ δένδρῳ προσκαίειν φέροντας ὄργανόν τι περιπεποιημένον καὶ τούτῳ περιάπτοντας, εἶθ' ὅταν ἐκτήξωσι ταύτην πάλιν ἐπ' ἄλλο καὶ ἄλλο μεταφέρειν· ὅρος δέ ἐστιν αὐτοῖς τις καὶ σημεῖα τοῦ παύεσθαι καὶ μάλιστα δῆλον ὅτι τὸ μηκέτι ῥεῖν. πιττοκαυτοῦσι δέ, ὥσπερ καὶ πρότερον ἐλέχθη, τὰς τερμίνθους· πεύκην γὰρ οὐ φέρουσιν οἱ τόποι. τὰ μὲν οὖν περὶ τὴν ῥητίνην καὶ τὴν πίτταν οὕτως ἔχει.

IV. Περὶ δὲ λιβανωτοῦ καὶ σμύρνης καὶ βαλσάμου καὶ εἴ τι τοιοῦτον ἕτερον ὅτι μὲν καὶ ἀπ' ἐντομῆς γίνεται καὶ αὐτομάτως εἴρηται. ποία δέ τις ἡ τῶν δένδρων φύσις καὶ εἴ τι περὶ τὴν γένεσιν ἢ τὴν συλλογὴν ἢ τῶν ἄλλων ἴδιον αὐτοῖς ὑπάρχει πειρατέον εἰπεῖν, ὡσαύτως δὲ καὶ περὶ τῶν λοιπῶν εὐόσμων· σχεδὸν γὰρ τά γε πλεῖστα ἀπὸ τῶν τόπων ἐστὶ τῶν τε πρὸς μεσημβρίαν καὶ ἀνατολήν.

2 Γίνεται μὲν οὖν ὁ λίβανος καὶ ἡ σμύρνα καὶ ἡ κασία καὶ ἔτι τὸ κινάμωμον ἐν τῇ τῶν Ἀράβων Χερρονήσῳ περί τε Σαβὰ καὶ Ἀδραμύτα καὶ

---

[1] ἐκπελεκῶντας conj. W.; ἐκπλέκοντες U; ἐκπλέοντες M; ἐκ πλήττοντες Ald.
[2] The sense given to περιπεποιημένον is unexampled, and the word may be corrupt.

ENQUIRY INTO PLANTS, IX. III. 3–IV. 2

go to rest, in case the fire should come through; and they offer sacrifice and keep holiday, praying that the pitch may be abundant and good. Such is the manner in which the people of Macedonia make pitch by fire.

They say that in Asia in the Syrian region they do not extract the pitch by cutting[1] out of the tree the wood containing it, but use fire to the tree itself, applying an instrument fashioned on purpose,[2] with which they set fire to it.[3] And then, when they have melted out the pitch at once place, they shift the instrument to another. But they have a limit and indications when to stop, chiefly of course the fact that the pitch ceases to flow. They also, as was said before,[4] use fire to get pitch out of the terebinth; for the places where this tree grows do not produce the fir. Such are the facts about resin and pitch.

Of frankincense and myrrh: various accounts.

IV. As to frankincense myrrh balsam of Mecca and similar plants it has been said that the gum is produced both by incision and naturally. Now we must endeavour to say what[5] is the natural character of these trees and to mention any peculiarities as to the origin of the gum or its collection or anything[6] else. So too concerning the other fragrant plants; most of these come from places in the south and east.

Now frankincense myrrh cassia and also cinnamon are found in the Arabian peninsula[7] about

[3] τούτῳ περιάπτοντας seems to have been G's reading (Scal.); τοῦτο περιαλείφοντας MSS.   [4] 9. 2. 2.
[5] ποία conj. W.; πολλὴ Ald.   [6] τι before τῶν add. Sch.
[7] Χερρονήσῳ conj. Salm.; χώρᾳ νήσῳ Ald. *cf.* Plin. 6. 28.

233

THEOPHRASTUS

Κιτίβαινα καὶ Μαμάλι. φύεται δὲ τὰ τοῦ λιβανωτοῦ καὶ τῆς σμύρνης δένδρα τὰ μὲν ἐν τῷ ὄρει τὰ δ' ἐν ταῖς ἰδίαις γεωργίαις ὑπὸ τὴν ὑπώρειαν, δι' ὃ καὶ τὰ μὲν θεραπεύεται τὰ δ' οὔ· τὸ δ' ὄρος εἶναί φασιν ὑψηλὸν καὶ δασὺ καὶ νιφόμενον, ῥεῖν δ' ἐξ αὐτοῦ καὶ ποταμοὺς εἰς τὸ πεδίον. εἶναι δὲ τὸ μὲν τοῦ λιβανωτοῦ δένδρον οὐ μέγα, πεντάπηχυ δέ τι καὶ πολύκλαδον, φύλλον δ' ἔχειν ἐμφερὲς τῇ ἀπίῳ, πλὴν ἔλαττον πολὺ καὶ τῷ χρώματι ποῶδες σφόδρα, καθάπερ τὸ πήγανον· λειόφλοιον δὲ πᾶν, ὥσπερ τὴν δάφνην.

3 Τὴν δὲ σμύρναν ἔλαττον ἔτι τῷ μεγέθει καὶ θαμνωδέστερον δέ, τὸ δὲ στέλεχος ἔχειν σκληρὸν καὶ συνεστραμμένον ἐπὶ τῆς γῆς, παχύτερον δὲ ἢ κνημοπαχές· φλοιὸν δὲ ἔχειν λεῖον ὅμοιον τῇ ἀνδράχλῃ. ἕτεροι δὲ οἱ φάσκοντες τεθεωρηκέναι περὶ μὲν τοῦ μεγέθους σχεδὸν συμφωνοῦσιν· οὐδέτερον γὰρ εἶναι μέγα τῶν δένδρων, ἔλαττον δὲ τὸ τῆς σμύρνης καὶ ταπεινότερον· φύλλον δὲ ἔχειν τὸ τοῦ λιβανωτοῦ δαφνοειδὲς καὶ λειόφλοιον δ' εἶναι· τὸ δὲ τῆς σμύρνης ἀκανθῶδες καὶ οὐ λεῖον, φύλλον δὲ προσεμφερὲς ἔχειν τῇ πτελέᾳ, πλὴν οὖλον ἐξ ἄκρου δὲ ἐπακανθίζον, ὥσπερ τὸ τῆς πρίνου.

4 Ἔφασαν δὲ οὗτοι κατὰ τὸν παράπλουν ὃν ἐξ Ἡρώων ἐποιοῦντο κόλπου ζητεῖν ἐκβάντες ὕδωρ ἐν τῷ ὄρει καὶ οὕτω θεωρῆσαι τὰ δένδρα καὶ τὴν συλλογήν. εἶναι δ' ἀμφοτέρων ἐντετμημένα καὶ

---

[1] Plin. 12. 55 and 56. [2] Plin. 12. 67.
[3] ἔχειν conj. Sch.; ἔχει P₂Ald.
[4] λιβανωτοῦ δαφνοειδὲς καὶ λειόφλοιον δ' εἶναι conj. Sch.; λιβάνου τοῦ δαφνοειδὲς καὶ λειόφυλλον δ' εἶναι UM; λιβάνου δαφνοειδὲς

Saba Hadramyta Kitibaina and Mamali. The trees of frankincense and myrrh grow partly in the mountains, partly on private estates at the foot of the mountains; wherefore some are under cultivation, others not; the mountains, they say, are lofty, forest-covered and subject to snow, and rivers from them flow down to the plain. The frankincense-tree,[1] it is said, is not tall, about five cubits high, and it is much branched; it has a leaf like that of the pear, but much smaller and very grassy in colour, like rue; the bark is altogether smooth like that of bay.

The myrrh-tree[2] is said to be still smaller in stature and more bushy; it is said to have[3] a tough stem, which is contorted near the ground, and is stouter than a man's leg; and to have a smooth bark like that of andrachne. Others who say that they have seen it agree pretty closely about the size; neither of these trees, they say, is large, but that which bears myrrh is the smaller and of lower growth; however they say that, while the frankincense-tree[4] has a leaf like that of bay and smooth bark, that which bears myrrh is spinous and not smooth, and has a leaf like that of the elm, except that it is curly and spinous[5] at the tip like that of kermes-oak.

[6] These said that on the coasting voyage which they made from the bay of the Heroes they landed to look for water on the mountains and so saw these trees and the manner of collecting their gums. [7] They reported that with both trees incisions had

καὶ λειόφλοιον δ' εἶναι P₂; λιβάνου· ἀλλὰ τοῦ μὲν δαφνοειδὲς καὶ λειόφυλλον εἶναι Ald.H. *cf.* Plin. 12. 57.

[5] *cf.* 3. 10. 1; 3. 11. 3.
[6] *cf.* Arr. *Anab.* 3. 5. 4; 7. 20. 1 and 2.
[7] Plin. 12. 58–62.

τὰ στελέχη καὶ τοὺς κλάδους, ἀλλὰ τὰ μὲν ὥσπερ
ἀξίνῃ δοκεῖν τετμῆσθαι τοὺς δὲ λεπτοτέρας ἔχειν
τὰς ἐντομάς· τὸ δὲ δάκρυον τὸ μὲν καταπίπτειν
τὸ δὲ καὶ πρὸς τῷ δένδρῳ προσέχεσθαι. ἐνιαχοῦ
μὲν ὑποβεβλῆσθαι ψιάθους ἐκ φοινίκων πεπλεγ-
μένας, ἐνιαχοῦ δὲ τὸ ἔδαφος μόνον ἠδαφίσθαι καὶ
καθαρὸν εἶναι· καὶ τὸν μὲν ἐπὶ τῶν ψιάθων
λιβανωτὸν εἶναι καὶ καθαρὸν καὶ διαφανῆ, τὸν δ'
ἐπὶ τῆς γῆς ἧττον· τὸν δ' ἐπὶ τοῖς δένδροις προσ-
εχόμενον ἀποξύειν σιδήροις, δι' ὃ καὶ φλοιὸν ἐνίοις
5 προσεῖναι. τὸ δὲ ὄρος ἅπαν μεμερίσθαι τοῖς
Σαβαίοις, τούτους γὰρ εἶναι κυρίους, δικαίους δὲ
τὰ πρὸς ἀλλήλους, δι' ὃ καὶ οὐδένα τηρεῖν· ὅθεν
καὶ αὐτοὶ δαψιλῶς εἰς τὰ πλοῖα λαβόντες ἐνθέ-
σθαι τοῦ λιβανωτοῦ καὶ τῆς σμύρνης ἐρημίας
οὔσης καὶ ἀποπλεῖν. ἔλεγον δ' οὗτοι καὶ τόδε
καὶ ἔφασαν ἀκούειν, ὅτι συνάγεται πανταχόθεν ἡ
σμύρνα καὶ ὁ λιβανωτὸς εἰς τὸ ἱερὸν τὸ τοῦ ἡλίου·
τοῦτο δ' εἶναι μὲν τῶν Σαβαίων ἁγιώτατον δὲ
πολὺ τῶν περὶ τὸν τόπον, τηρεῖν δέ τινας Ἄρα-
6 βας ἐνόπλους· ὅταν δὲ κομίσωσιν, ἕκαστον σωρεύ-
σαντα τὸν αὑτοῦ καὶ τὴν σμύρναν ὁμοίως καταλιπεῖν τοῖς ἐπὶ τῆς φυλακῆς, τιθέναι δὲ ἐπὶ τοῦ
σωροῦ πινάκιον γραφὴν ἔχον τοῦ τε πλήθους τῶν
μέτρων καὶ τῆς τιμῆς ἧς δεῖ πραθῆναι¹ τὸ μέτρον
ἕκαστον· ὅταν δὲ οἱ ἔμποροι παραγένωνται, σκοπεῖν τὰς γραφάς, ὅστις δ' ἂν αὐτοῖς ἀρέσκῃ

---

¹ πραθῆναι conj. Sch. from G; πρασθῆναι U; προσθεῖναι P₂ Ald.

ENQUIRY INTO PLANTS, IX. iv. 4-6

been made both in the stems and in the branches, but that, while the stems looked as if they had been cut with an axe, in the branches the incisions were slighter; also that in some cases the gum was dropping, but that in others it remained sticking to the tree; and that in some places mats woven of palm-leaves were put underneath, while in some the ground underneath was merely made level and clean; and that the frankincense on the mats was clear and transparent, that collected on the ground less so; and that that which remained sticking to the trees they scraped off with iron tools, wherefore sometimes pieces of bark remained in it. The whole range, they said, belongs to the portion of the Sabaeans; for it is under their sway, and they are honest in their dealings with one another. Wherefore no one keeps watch; so that these sailors greedily took, they said, and put on board their ships some of the frankincense and myrrh, since there was no one about, and sailed away. They also reported another thing which they said they had been told, that the myrrh and frankincense are collected from all parts into the temple of the sun; and that this temple is the most sacred thing which the Sabaeans of that region possess, and it is guarded by certain Arabians in arms. And that when they have brought it, each man piles up his own contribution of frankincense and the myrrh in like manner, and leaves it with those on guard; and on the pile he puts a tablet on which is stated the number of measures which it contains, and the price for which each measure should be sold [1]; and that, when the merchants come, they look at the tablets, and whichsoever pile pleases them, they

μετρησαμένους τιθέναι τὴν τιμὴν εἰς τοῦτο τὸ χωρίον ἔνθεν ἂν ἕλωνται, καὶ τὸν ἱερέα παραγενόμενον τὸ τρίτον μέρος λαβόντα τῆς τιμῆς τῷ θεῷ τὸ λοιπὸν αὐτοῦ καταλιπεῖν καὶ τοῦτο σῶν εἶναι τοῖς κυρίοις, ἕως ἂν ἕλωνται παραγενόμενοι.

7 Ἄλλοι δέ τινες τὸ μὲν τοῦ λιβανωτοῦ δένδρον ὅμοιον εἶναί φασι σχίνῳ καὶ τὸν καρπὸν ταῖς σχινίσι φύλλον δὲ ὑπέρυθρον· εἶναι δὲ τὸν μὲν ἐκ τῶν νέων λιβανωτὸν λευκότερον καὶ ἀοδμότερον, τὸν δ' ἐκ τῶν παρηκμακότων ξανθότερον καὶ εὐοσμότερον· τὸ δὲ τῆς σμύρνης ὅμοιον τῇ τερμίνθῳ, τραχύτερον δὲ καὶ ἀκανθωδέστερον, φύλλον δὲ μικρῷ στρογγυλότερον, τῇ δὲ γεύσει διαμασωμένοις ὅμοιον τῷ τῆς τερμίνθου· εἶναι δὲ καὶ τούτων τὰ παρηκμακότα εὐοσμότερα.

8 Γίνεσθαι δὲ ἀμφότερα ἐν τῷ αὐτῷ τόπῳ· τὴν δὲ γῆν ὑπάργιλον καὶ πλακώδη, καὶ ὕδατα πηγαῖα σπάνια. ταῦτα μὲν οὖν ὑπεναντία τῷ νίφεσθαι καὶ ὕεσθαι καὶ ποταμοὺς ἐξιέναι· τὸ δὲ παρόμοιον εἶναι τὸ δένδρον τῇ τερμίνθῳ καὶ ἄλλοι τινὲς λέγουσιν, οἱ δὲ καὶ ὅλως τέρμινθον εἶναι· κομισθῆναι γὰρ ξύλα πρὸς Ἀντίγονον ὑπὸ τῶν Ἀράβων τῶν τὸν λιβανωτὸν καταγόντων, ἃ οὐδὲν διέφερε τῶν τῆς τερμίνθου· πλὴν οὗτοί γε μεῖζον

---

[1] Plin. 12. 66 and 67.   [2] Plin. 12. 53.
[3] πλακώδη: lit. 'with a crust'; so W., but the word does not seem to occur elsewhere in this sense.
[4] cf. 9. 4. 2.

measure, and put down the price on the spot whence they have taken the wares, and then the priest comes and, having taken the third part of the price for the god, leaves the rest of it where it was, and this remains safe for the owners until they come and claim it.

Others report that the tree which produces the frankincense is like mastich, and its fruit is like the fruit of that tree, but the leaf is reddish: also that the frankincense derived from young trees is whiter and less fragrant, while that derived from those which have passed their prime is yellower and more fragrant; also that the tree which produces myrrh is like the terebinth, but rougher and more thorny; that the leaf is somewhat rounder, and that, if one chews it, it resembles that of the terebinth in taste; also that of myrrh-trees too those that are past their prime give more fragrant myrrh.

[1] Both trees, it is said, grow in the same region; the soil is clayey[2] and caked,[3] and spring waters are scarce. Now these reports are contradictory to [4] that which says that the country is subject to snow and rain and sends forth rivers. However others make the statement[5] that the tree is like the terebinth; in fact some say that it is the same tree; for that logs of it were brought to Antigonus by the Arabs who brought the frankincense down to the sea, and that these did not differ at all from logs of terebinth.[6] However these informants were guilty

---

[5] *i.e.* the statement quoted of the myrrh-tree, § 7. The 'tree' is here the λιβανωτός, but these authorities did not distinguish it from the myrrh-tree. See below.

[6] τῶν conj. Sch.; τῷ UAld.

THEOPHRASTUS

ἕτερον ἀγνόημα προσηγνόουν· ᾤοντο γὰρ ἐκ τοῦ αὐτοῦ δένδρου τόν τε λιβανωτὸν γίνεσθαι καὶ τὴν

9 σμύρναν· διόπερ ἐκεῖνος ὁ λόγος πιθανώτερος ὁ παρὰ τῶν ἀναπλευσάντων ἐξ Ἡρώων πόλεως· ἐπεὶ καὶ τὸ ὑπὲρ Σάρδεων πεφυκὸς τοῦ λιβανωτοῦ δένδρον ἐν ἱερῷ τινι δαφνοειδὲς ἔχει τὸ φύλλον, εἴ τι δεῖ σταθμᾶσθαι τοῦτο· ὁ λιβανωτὸς δ' ἔχει, καὶ ὁ ἐκ τοῦ στελέχους καὶ ὁ ἐκ τῶν ἀκρεμόνων, ὁμοίως καὶ τῇ ὄψει καὶ τῇ ὀσμῇ θυμιώμενος τῷ ἄλλῳ λιβανωτῷ. πέφυκε δὲ τοῦτο μόνον τὸ δένδρον οὐδεμιᾶς τυγχάνειν ⟨θεραπείας⟩.

10 Ἔνιοι δὲ λέγουσιν ὡς πλείων μὲν ὁ λιβανωτὸς ἐν τῇ Ἀραβίᾳ γίνεται, καλλίων δὲ ἐν ταῖς ἐπικειμέναις νήσοις ὧν ἐπάρχουσιν· ἐνταῦθα γὰρ καὶ σχηματοποιεῖν ἐπὶ τῶν δένδρων οἷον ἂν θέλωσι· καὶ τάχα τοῦτό γε οὐκ ἀπίθανον· ἐνδέχεται ⟨γὰρ⟩ ὁποίαν ἂν βούλωνται ποιεῖν τὴν ἐντομήν. εἰσὶ δέ τινες καὶ μεγάλοι σφόδρα τῶν χόνδρων, ὥστ' εἶναι τῷ μὲν ὄγκῳ χειροπληθιαίους σταθμῷ δὲ πλέον ἢ τρίτον μέρος μνᾶς. ἀργὸς δὲ κομίζεται πᾶς ὁ λιβανωτός, ὅμοιος δὲ τῇ προσόψει φλοιῷ. τῆς σμύρνης δὲ ἡ μὲν στακτὴ ἡ δὲ πλαστή. δοκιμάζεται δ' ἡ μὲν ἀμείνων τῇ γεύσει, καὶ ἀπὸ ταύτης τὴν ὁμόχρων λαμβάνουσι. περὶ μὲν οὖν λιβανωτοῦ καὶ σμύρνης σχεδὸν τοσαῦτα ἀκηκόαμεν ἄχρι γε τοῦ νῦν.

---

[1] Plin. 15. 57 ; cf. 16. 135.
[2] δ' ἔχει conj. W.; δίεται P₂Ald.; om. H.
[3] θεραπείας add. Sch. from G ; οὐδεμιᾶς τυγχάνειν UMAld.; οὐ μιᾶς τυγχάνειν P₂. But there is no sign of a lacuna in the MSS., and οὐδεμίας is probably corrupt, as W. suggests. οὐδὲ θερ. is inconsistent with 9. 4. 2.
[4] πλείων conj. W.; ἡδίων P₂Ald.

of a further more important piece of ignorance; for they believed that the frankincense and the myrrh were produced by the same tree. Wherefore the account derived from those who sailed from the city of Heroes is more to be believed;[1] in fact the frankincense-tree which grows above Sardes in a certain sacred precinct has[2] a leaf like that of bay, if we may judge at all by this; and the frankincense derived both from its stem and its branches is like in appearance and in smell, when it is burnt as incense, to other frankincense. This is the only tree which can never be cultivated.[3]

Some say that the frankincense-tree is more abundant[4] in Arabia, but finer in the adjacent islands[5] over which the Arabians bear rule;[6] for there it is said that they mould the gum on the trees to any shape that they please. And perhaps this is not incredible, since it is possible to make any kind of incision that they like. Some of the lumps[7] of gum are very large, so that one is large enough in bulk to fill the hand and in weight is more than a third of a pound. All frankincense is gathered in the rough and is like bark in appearance. Myrrh[8] is either 'fluid'[9] (myrrh-oil) or 'solid' (agglutinated). That of better quality is tested by its taste, and of this they select that which is of uniform colour.[10] Now of frankincense and myrrh these are about all the facts that have come to our notice at present.

[5] Plin. 12. 60.
[6] ἐπάρχουσιν conj. Coraës; ἐνυπάρχουσι P₂Ald.
[7] τῶν χόνδρων conj. Sch.: τῷ χόνδρῳ P₂Ald.
[8] i.e. here the commodity so called.
[9] cf. Odor. 29.
[10] ὁμόχρων conj. Sch.; ὁμόχρῳ UM; ὁμόχρουν Ald.; τὴν ἑτέραν καταλαμβάνουσι P₂.

THEOPHRASTUS

V. Περὶ δὲ κιναμώμου καὶ κασίας τάδε λέγουσι· θάμνους μὲν ἀμφότερα ταῦτ' εἶναι οὐ μεγάλους ἀλλ' ἡλίκους ἄγνου, πολυκλάδους δὲ καὶ ξυλώδεις. ὅταν δὲ ἐκκόψωσιν, ὅλον τὸ κινάμωμον διαιρεῖν εἰς πέντε μέρη· τούτων δὲ τὸ πρῶτον πρὸς τοῖς βλαστοῖς βέλτιστον εἶναι, ὃ τέμνεται σπιθαμιαῖον ἢ μικρῷ μεῖζον· ἑπόμενον δὲ τὸ δεύτερον, ὃ καὶ τῇ τομῇ ἔλαττον· εἶτα τὸ τρίτον καὶ τέταρτον· ἔσχατον δὲ τὸ χείριστον τὸ πρὸς τῇ ῥίζῃ. φλοιὸν γὰρ ἐλάχιστον ἔχειν· χρήσιμος δὲ οὗτος, οὐ τὸ ξύλον· δι' ὃ καὶ τὸ ἀκροφυὲς κράτιστον, πλεῖστον γὰρ ἔχειν καὶ τὸν φλοιόν.   οἱ μὲν οὕτω λέγουσιν.

2   Ἄλλοι δὲ θαμνῶδες μὲν καὶ ἔτι μᾶλλον φρυγανῶδες εἶναί φασι· δύο δ' αὐτοῦ γένη, τὸ μὲν μέλαν τὸ δὲ λευκόν. λέγεται δέ τις καὶ μῦθος ὑπὲρ αὐτοῦ· φύεσθαι μὲν γάρ φασιν ἐν φάραγξιν, ἐν ταύταις δ' ὄφεις εἶναι πολλοὺς δῆγμα θανάσιμον ἔχοντας, πρὸς οὓς φραξάμενοι τὰς χεῖρας καὶ τοὺς πόδας καταβαίνουσι καὶ συλλέγουσιν, εἶθ' ὅταν ἐξενέγκωσι διελόντες τρία μέρη διακληροῦνται πρὸς τὸν ἥλιον, καὶ ἣν ἂν λάχῃ ὁ ἥλιος καταλείπουσιν· ἀπιόντες δ' εὐθὺς ὁρᾶν φασι καιομένην ταύτην· οὗτος μὲν οὖν τῷ ὄντι μῦθος.

3   Τὴν δὲ κασίαν φασὶ τὰς μὲν ῥάβδους παχυτέρας ἔχειν, ἰνώδεις δὲ σφόδρα καὶ οὐκ εἶναι περιφλεῦσαι· χρήσιμον δὲ καὶ ταύτης τὸν φλοιόν.

---

[1] Plin. 12. 85–94.
[2] A similar tale is told of frankincense by Herodotus (3. 107), who has an equally surprising tale about cinnamon (3. 111).

ENQUIRY INTO PLANTS, IX. v. 1–3

Of cinnamon and cassia: various accounts.

V. [1] Of cinnamon and cassia the following account is given: both are shrubs, it is said, and not of large size, but of the same size as bushes of chaste-tree, with many branches and woody. When they cut down the whole cinnamon-tree, they divide it into five parts; of these the first is that which grows next the branches and this is the best: this is cut in lengths a span long or a little longer; next comes the second kind, which is cut in shorter lengths; then come the third and the fourth, and last the least valuable wood, which grows next the root; for this has least bark, and it is the bark and not the wood which is serviceable; wherefore the part which grows high up the tree is the best, since it has the most bark. Such is the account given by some.

Others say that cinnamon is shrubby or rather like an under-shrub; and that there are two kinds, one black, the other white. [2] And there is also a tale told about it; they say that it grows in deep glens, and that in these there are numerous snakes which have a deadly bite; against these they protect their hands and feet before they go down into the glens, and then, when they have brought up the cinnamon, they divide it in three parts and draw lots for it with the sun; and whatever portion falls to the lot of the sun they leave behind; and they say that, as soon as they leave the spot, they see this take fire. Now this is sheer fable.

[3] Cassia, they say, has stouter branches, which are very fibrous and difficult to strip of the bark [4]; and it is the bark of this tree also which is serviceable.

---

[3] Plin. 12. 95–97.

[4] περιφλοῖσαι H.; περιφλεῦσαι UMP₂Ald. *cf. de igne* 72.

ὅταν οὖν τέμνωσι τὰς ῥάβδους, κατακόπτειν ὡς διδάκτυλα τὸ μῆκος ἢ μικρῷ μείζω, ταῦτα δ' εἰς νεόδορον¹ βύρσαν καταρράπτειν· εἶτ' ἐκ ταύτης καὶ τῶν ξύλων σηπομένων σκωλήκια γίνεσθαι, ἃ τὸ μὲν ξύλον κατεσθίει, τοῦ φλοιοῦ δ' οὐχ ἅπτεται διὰ τὴν πικρότητα καὶ δριμύτητα τῆς ὀσμῆς. καὶ περὶ μὲν κασίας καὶ κιναμώμου τοσαῦτα λέγεται.

VI. Τὸ δὲ βάλσαμον γίνεται μὲν ἐν τῷ αὐλῶνι τῷ περὶ Συρίαν.² παραδείσους δ' εἶναί φασι δύο μόνους, τὸν μὲν ὅσον εἴκοσι πλέθρων τὸν δ' ἕτερον πολλῷ ἐλάττονα. τὸ δὲ δένδρον μέγεθος μὲν ἡλίκον ῥόα μεγάλη πολύκλαδον δὲ σφόδρα· φύλλον δὲ ἔχειν ὅμοιον πηγάνῳ, πλὴν ἔκλευκον, ἀείφυλλον δὲ εἶναι· καρπὸν δὲ παρόμοιον τῇ τερμίνθῳ καὶ μεγέθει καὶ σχήματι καὶ χρώματι· εὐῶδες σφόδρα καὶ τοῦτο καὶ μᾶλλον τοῦ δακρύου.³

2 Τὸ δὲ δάκρυον ἀπὸ ἐντομῆς συλλέγειν, ἐντέμνειν δὲ ὄνυξι σιδηροῖς ὑπὸ τὸ ἄστρον, ὅταν μάλιστα πνίγη ὦσι, καὶ τὰ στελέχη καὶ τὰ ἄνω. τὴν δὲ συλλογὴν ὅλον τὸ θέρος ποιεῖσθαι· οὐκ εἶναι δὲ πολὺ τὸ ῥέον, ἀλλ' ἐν ἡμέρᾳ τὸν ἄνδρα συλλέγειν ὅσον κόγχην· τὴν δ' ὀσμὴν διαφέρουσαν καὶ πολλήν, ὥστε ἀπὸ μικροῦ πολὺν ἐφικνεῖσθαι τόπον. ἀλλ' οὐ φοιτᾶν ἐνταῦθα ἄκρατον ἀλλὰ τὸ συνηγμένον κεκραμένον· πολλὴν γὰρ δέχεσθαι

---

¹ νεόδορον conj. Sch.; νεόδερον P₂Ald.   ² Plin. 12 111–123.
³ εὐῶδες . . . τοῦτο P₂Ald.; εὐώδη . . . τοῦτον W. after Sch.'s conj. But the clause begins without a conjunction, and some

When then they cut off the branches, they chop them up into lengths of about two fingers' breadth or rather more, and these they sew up in raw[1] hide ; and then from the leather and the decaying wood little worms are engendered, which devour the wood but do not touch the bark, because it is bitter and has a pungent odour. This is all the information forthcoming about cinnamon and cassia.

Of balsam of Mecca.

VI. [2] Balsam of Mecca grows in the valley of Syria. They say that there are only two parks in which it grows, one of about four acres, the other much smaller. The tree is as tall as a good-sized pomegranate and is much branched ; it has a leaf like that of rue, but it is pale ; and it is evergreen ; the fruit is like that of the terebinth in size shape and colour, and this too is very fragrant,[3] indeed more so than the gum.

[4] The gum, they say, is collected by making incisions, which is done with bent pieces of iron at the time of the Dog-star, when there is scorching heat ; and the incisions are made both in the trunks and in the upper parts of the tree. The collecting goes on throughout the summer ; but the quantity which flows is not large ; in a day a single man can collect a shell-full[5] ; the fragrance is exceeding great and rich, so that that which comes from a small amount is perceived for a wide distance. However it does not reach us in a pure state ; what is collected is mixed with other things ; for it mixes freely with

words about the *flower* may have dropped out, to which this clause refers ; *cf.* however *Odor.* 32.
[4] Diosc. 1. 19.   [5] Plin. 12. 117.

κρᾶσιν· καὶ τὸ ἐν τῇ Ἑλλάδι πολλάκις εἶναι κεκραμένον· εὔοσμα δὲ σφόδρα καὶ τὰ ῥαβδία· καθαίρειν γὰρ καὶ τῶνδε ἕνεκα καί του¹ διαφόρου· πωλεῖσθαι <γὰρ>² τίμια. καὶ τὴν ἐργασίαν τὴν περὶ τὰ δένδρα σχεδὸν ἐν ταύτῃ αἰτίᾳ εἶναι καὶ τὴν βροχήν· βρέχεσθαι γὰρ συνεχῶς. συναιτίαν δὲ δοκεῖν εἶναι τοῦ μὴ μεγάλα γίνεσθαι τὰ δένδρα καὶ τὴν τῶν ῥαβδίων τομήν. διὰ γὰρ τὸ πολλάκις ἐπικείρεσθαι ῥάβδους ἀφιέναι καὶ οὐκ εἰς ἓν ἐκτείνειν τὴν ὁρμήν.

Ἄγριον δὲ οὐδὲν εἶναι βάλσαμον οὐδαμοῦ· γίνεσθαι δὲ ἐκ μὲν τοῦ μείζονος παραδείσου ἀγγεῖδια δώδεκα ὅσον ἡμιχοαῖα, ἐκ δὲ τοῦ ἑτέρου δύο μόνον· πωλεῖσθαι δὲ τὸ μὲν ἄκρατον δὶς πρὸς ἀργύριον τὸ δ' ἄλλο κατὰ λόγον τῆς μίξεως· καὶ τοῦτο μὲν διαφέρον τι φαίνεται κατὰ τὴν εὐοσμίαν.

VII. Ὁ δὲ κάλαμος γίνεται καὶ ὁ σχοῖνος ὑπερβάλλοντι τὸν Λίβανον μεταξὺ τοῦ τε Λιβάνου καὶ ἄλλου τινὸς ὄρους μικροῦ ἐν τῷ αὐλωνίσκῳ τούτῳ, καὶ οὐχ ὥς τινές φασι τοῦ Ἀντιλιβάνου· ὁ γὰρ Ἀντιλίβανος μακρὰν ἀπέχει τοῦ Λιβάνου καὶ μεταξὺ τούτων ἐστὶν ὃν αὐλῶνα καλοῦσι πεδίον πολὺ καὶ καλόν. ὅπου δὲ ὁ κάλαμος καὶ ὁ σχοῖνος φύεται λίμνη μεγάλη τυγχάνει, πρὸς ταύτην δὲ ἐν τῷ ἕλει τῷ ἀνεξηραμμένῳ πεφύκασι·

---

¹ του I conj.; τοῦ MSS. G's rendering shews that the explanation of the 'different reason' (*i.e.* to strengthen the tree) has dropped out of our texts.   ² γὰρ add. Sch.

other things; and what is known in Hellas is generally mixed with something else. The boughs are also very fragrant. In fact it is on account of these boughs, they say, that the tree is pruned (as well as for a different reason[1]), since[2] the boughs cut off can be sold for a good price. In fact the culture of the trees has the same motive[3] as the irrigation (for they are constantly irrigated). And the cutting of the boughs seems likewise to be partly the reason why the trees do not grow tall; for, since they are often cut about, they send out branches instead of putting out all their energy in one direction.

Balsam is said not to grow wild anywhere. From the larger park are obtained twelve vessels containing each about three pints, from the other only two such vessels; the pure gum sells for twice its weight in silver, the mixed sort at a price proportionate to its purity. Balsam then appears to be of exceptional fragrance.

Of other aromatic plants—all oriental, except the iris.

VII. [4] Sweet-flag and ginger-grass grow beyond the Libanus between that range and another small range, in the depression thus formed; and not, as some say, between Libanus and Anti-Libanus. For Anti-Libanus is a long way from Libanus, and between them is a wide fair plain called 'The Valley.' But, where the sweet-flag and ginger-grass grow, there is a large lake,[5] and they grow near it in the dried up marshes, covering an extent

---

[3] The 'motive' is the production of boughs. ἐν ταύτῃ αἰτίᾳ I conj.; ταύτην αἰτίαν UMP₂Ald.; ἐν ταύτῃ αἰτίᾳ conj. W.
[4] Plin. 12. 104 and 105.   [5] *cf.* C.P. 6. 18. 2.

τόπον δὲ ἔχουσι πλεῖον ἢ τριάκοντα σταδίων. οὐκ ὄζουσι δὲ χλωροὶ ἀλλὰ ξηρανθέντες, τῇ προσόψει δὲ οὐδὲν διαφέρουσι τῶν ἄλλων· εἰσβάλλοντι δ' εἰς τὸν τόπον εὐθὺς ὀσμὴ προσβάλλει· οὐ μὴν πορρωτέρω γε ἡ ἀποπνοὴ γίνεται, καθάπερ τινές φασι, ταῖς προσφερομέναις ναυσὶ πρὸς τὴν χώραν· καὶ γὰρ ὁ τόπος οὗτος ἀπὸ θαλάττης ἀπέχει πλείους ἢ ἑκατὸν πεντήκοντα σταδίους· ἀλλ' ἐν τῇ Ἀραβίᾳ τὴν ἀποπνοὴν εἶναί φασι τὴν ἀπὸ τῆς χώρας εὔοσμον.

Ἐν μὲν οὖν Συρίᾳ τὰ περιττὰ τῇ ὀσμῇ σχεδὸν ταῦτ' ἐστίν· ἡ γὰρ χαλβάνη βαρύτερον καὶ μᾶλλον φαρμακῶδες· ἐπεὶ καὶ αὕτη γίνεται περὶ Συρίαν ἐκ τοῦ πανάκους καλουμένου. τὰ δὲ ἄλλα πάντα τὰ εὔοσμα οἷς πρὸς τὰ ἀρώματα χρῶνται, τὰ μὲν ἐξ Ἰνδῶν κομίζεται κἀκεῖθεν ἐπὶ θάλατταν καταπέμπεται, τὰ δ' ἐξ Ἀραβίας, οἷον πρὸς τῷ κιναμώμῳ καὶ τῇ κασίᾳ καὶ κώμακον· ἕτερον δ' εἶναι τὸ κώμακον καρπόν· τὸ δ' ἕτερον παραμίσγουσιν εἰς τὰ σπουδαιότατα τῶν μύρων. τὸ δὲ καρδάμωμον καὶ ἄμωμον οἱ μὲν ἐκ Μηδείας, οἱ δ' ἐξ Ἰνδῶν καὶ ταῦτα καὶ τὴν νάρδον καὶ τὰ ἄλλα ἢ τὰ πλεῖστα.

Οἷς μὲν οὖν εἰς τὰ ἀρώματα χρῶνται σχεδὸν τάδε ἐστί· κασία κινάμωμον καρδάμωμον νάρδος ναῖρον βάλσαμον ἀσπάλαθος στύραξ ἶρις νάρτη

---

[1] οὐκ ὄζουσι conj. Guilandinus, cf. C.P. 6. 14. 8; οὐ δοκοῦσι P₂Ald.H.

[2] cf. C.P. 6. 18. 1. τῶν ἄλλων: sc. καλάμων καὶ σχοίνων.

[3] προσφερομέναις ναυσὶ πρὸς conj. Scal.; προσφ. εἶναι πρὸς P₂ Ald.

[4] cf. 9. 9. 2 n. 'The plant,' i.e. one of the plants so called.

[5] Plin. 12. 135; cf. 13. 18.

ENQUIRY INTO PLANTS, IX. vii. 1-3

of more than thirty furlongs. They have no fragrance[1] when they are green, but only when they are dried, and in appearance they do not differ[2] from ordinary reeds and rushes; but, as you approach the spot, immediately a sweet smell strikes you. However it is not true, as some say, that the fragrance is wafted to ships approaching[3] the country; for indeed this district is more than 150 furlongs from the sea. However it is said that in Arabia the breezes wafted from the land are fragrant.

Such then are the plants in Syria which have remarkable fragrance. For that of *khalbane* is more oppressive and somewhat medicinal;[4] for this perfume also is produced in Syria from the plant called allheal. As to all the other fragrant plants used for aromatic odours, they come partly from India whence they are sent over sea, and partly from Arabia, for instance, *komakon*[5]—as well as cinnamon and cassia. The *fruit* called *komakon* is said to be distinct[6] from this; the *komakon* of which we are speaking is a perfume which they mix with the choicest unguents. Cardamom and Nepaul cardamom some say come from Media; others say that these come from India, as well as spikenard and most, if not all, of the other species.

Now this is a general list of the plants used for perfumes:—cassia cinnamon cardamom spikenard *nairon* balsam of Mecca *aspalathos* storax iris *narte*

[6] εἶναι has no government, and W. considers the passage corrupt. Comparison of Plin. *l.c.* suggests that the original text may have been something like this: τὸ κώμακον καρπὸν ὄν· τὸν δὲ χυλὸν κ.τ.λ., *i.e.* ' *komakon* is of different character from these, being a fruit, whose juice—.' Plin. calls κ. a kind of cinnamon.

249

κόστος πάνακες κρόκος σμύρνα κύπειρον σχοῖνος κάλαμος ἀμάρακον λωτὸς ἄννητος. τούτων δὲ τὰ μὲν ῥίζαι τὰ δὲ φλοιοὶ τὰ δὲ κλῶνες τὰ δὲ ξύλα τὰ δὲ σπέρματα τὰ δὲ δάκρυα τὰ δὲ ἄνθη. καὶ τὰ μὲν πολλαχοῦ γίνεται, τὰ δὲ περιττότατα καὶ εὐοδμότατα πάντα ἐκ τῆς Ἀσίας καὶ ἐκ τῶν ἀλεεινῶν τόπων. ἐκ γὰρ αὐτῆς Εὐρώπης οὐδέν ἐστιν ἔξω τῆς ἴριδος.

4 Αὕτη δ' ἀρίστη ἐν Ἰλλυριοῖς, οὐκ ἐν τῇ πρὸς θάλατταν χώρᾳ, ἀλλ' ἐν τῇ ἀνακεχωρηκυίᾳ κειμένῃ δὲ μᾶλλον πρὸς ἄρκτον. τόποι δὲ τόπων διαφέρουσιν ἐν οἷς ἀμείνων· ἐργασία δὲ περὶ αὐτὴν οὐδεμία πλὴν τοῦ περικαθήραντα ἀναξηρᾶναι.

Τὰ γὰρ ἐν τῇ Θρᾴκῃ φυόμενα ῥιζία, καθάπερ τό τε τῇ νάρδῳ προσεμφερῆ τὴν ὀσμὴν ἔχον καὶ ἕτερ' ἄττα, μικράν τινα καὶ ἀσθενῆ τὴν εὐωδίαν ἔχει. καὶ περὶ μὲν τῶν εὐόσμων ἐπὶ τοσοῦτον εἰρήσθω.

VIII. Περὶ δὲ τῶν ὀπῶν ὅσα μὴ πρότερον εἴρηται, λέγω δ' οἷον εἴ τινες φαρμακώδεις ἢ καὶ ἄλλας ἔχουσι δυνάμεις, πειρατέον ὁμοίως εἰπεῖν· ἅμα δὲ καὶ περὶ ῥιζῶν, καὶ γὰρ τῶν ὀπῶν τινες ἐκ τούτων καὶ χωρὶς αὐταὶ καθ' αὑτὰς πολλὰς καὶ παντοίας ἔχουσι δυνάμεις, ὅλως δὲ περὶ πάντων φαρμακωδῶν, οἷον καρποῦ χυλισμοῦ φύλλων ῥιζῶν πόας· καλοῦσι γὰρ καὶ πόαν ἔνια τῶν φαρμακωδῶν οἱ ῥιζοτόμοι.

---

[1] cf. C.P. 6. 14. 8; 6. 18. 12; Plin. 21. 40.
[2] See Index App. (25).

kostos all-heal saffron-crocus myrrh *kypeiron* ginger-grass sweet-flag sweet marjoram *lotos* dill. Of these it is the roots, bark, branches, wood, seeds, gum or flowers which in different cases yield the perfume. Some of them grow in many places, but the most excellent and most fragrant all come from Asia and sunny regions. From Europe itself comes none of them except the iris.

[1] This is best in Illyria, not in the part near the sea, but in that which is further inland and lies more to the north. In different districts it varies in quality; no special attention is needed, except to scrape the roots clean and dry them.

As for the roots which grow in Thrace, such as one [2] which has a smell like spikenard and certain others, their fragrance is but slight and feeble. Let this suffice for an account of sweet-smelling plants.

Of the medicinal juices of plants and the collection of them: general account.

VIII. Now we must endeavour to speak in like manner of those juices which have not been mentioned already, I mean, such as are medicinal or have other properties; and at the same time we must speak of roots; for some of the juices are derived from roots, and apart from that roots have in themselves divers properties of all kinds; and in general we must discuss medicinal things of all kinds, as fruit, extracted juice,[3] leaves, roots, 'herbs'; for the herb-diggers call some medicinal things by this name.

[3] χυλισμοῦ P₂Ald.H; καυλοῦ conj. W. The list is of the aspects in which the herbalist would regard the plant, not of the parts of the plant.

251

THEOPHRASTUS

Τῶν δὲ ῥιζῶν πλείους μέν εἰσιν αἱ δυνάμεις καὶ πρὸς πλείω· ζητοῦνται δὲ μάλιστα αἱ φαρμακώδεις ὡς χρησιμώταται, διαφέρουσαι τῷ τε μὴ πρὸς ταὐτὰ καὶ τῷ μὴ ἐν τοῖς αὐτοῖς ἔχειν τὴν δύναμιν. ὡς δ' οὖν ἐπὶ πᾶν αἱ πλεῖσται μὲν ἐν αὐταῖς ἔχουσι καὶ τοῖς καρποῖς καὶ τοῖς ὀποῖς, ἔνιαι δὲ καὶ ἐν τοῖς φύλλοις· τὰς δὲ φυλλώδεις δυνάμεις τὰς πολλὰς σχεδὸν πόας καλοῦσιν, ὥσπερ εἴρηται μικρῷ πρότερον, οἱ ῥιζοτόμοι.

2   Ὁ μὲν οὖν ὀπισμὸς γίνεται τῶν ὀπιζομένων ὡς ἐπὶ τὸ πολὺ τοῦ θέρους, τῶν μὲν ἐνισταμένου τῶν δὲ προεληλυθότος. ἡ δὲ ῥιζοτομία γίνεταί τινων καὶ ὑπὸ πυροτομίαν καὶ μικρῷ πρότερον, οὐ μὴν ἀλλ' ἥ γε πλείων τοῦ μετοπώρου μετ' Ἀρκτοῦρον, ὅταν φυλλορροήσωσιν, ὅσων δὲ καὶ ὁ καρπὸς χρήσιμος, ὅταν ἀμερθῶσι τὸν καρπόν. ἔστι δὲ ὁ ὀπισμὸς ἢ ἀπὸ τῶν καυλῶν, ὥσπερ τοῦ τιθυμάλλου καὶ τῆς θριδακίνης καὶ σχεδὸν τῶν πλείστων, ἢ ἀπὸ τῶν ῥιζῶν, ἢ τρίτον ἀπὸ τῆς κεφαλῆς, ὥσπερ τῆς μήκωνος· ταύτης γὰρ μόνης οὕτω καὶ τοῦτ' ἴδιον αὐτῆς. τῶν μὲν οὖν καὶ αὐτόματος ὁ ὀπὸς

---

[1] From this point to 9. 19. 4, γίνεσθαι, the text is repeated in U, with considerable variations, as a tenth book. Ald. also repeats the first few lines of this passage (down to end of §1) as a fragment of a tenth book. The two Medicean MSS also repeat 9. 8. 1, τῶν δὲ ῥιζῶν, down to 9. 10. 3, βέλτιστοι δὲ καὶ οἷς, as part of a tenth book. The 'tenth book' readings in each case are distinguished by a *.

[2] ῥιζῶν: ῥίζα signifies a medicinal plant in general (cf. ῥιζότομοι) as well as 'root'; the double sense makes translation of this section awkward: I print it 'root' where it has the wider sense.

ENQUIRY INTO PLANTS, IX. VIII. 1-2

[1] The properties of 'roots'[2] are numerous and they have numerous uses; but those which have medicinal virtues are especially sought after, as being the most useful; and they differ in not all being applied to the same[3] purposes and in not all having their virtue in the same parts of them.[4] To speak generally, most 'roots' have it in themselves[5]; or else it is found in the fruits or the juices of the plant; and in some cases in the leaves as well, and it is to the virtues of the leaves in most cases that the herb-diggers refer, when they speak, as has just been said,[6] of 'herbs.'

The collection of the juice from plants from which it is collected is mostly done in summer, in some cases at the beginning of that season, in others when it is well advanced. The digging of roots is done in some cases at the time of wheat-harvest or a little earlier, but the greater part of it in autumn after the rising of Arcturus when the plants have shed their leaves, and, in the case of those whose fruit is serviceable, when they have lost their fruit. The collection of juice is made either from the stalks,[7] as with *tithymallos* (spurge) wild lettuce and the majority of plants, or from the roots, or thirdly from the head, as in the case of the poppy; for this is the only plant which is so treated[8] and this is its peculiarity. In some plants the juice collects of its own accord in

---

[3] ταὐτὰ conj. Scal. from G; ταῦτα Ald.
[4] After δύναμιν U*Ald.* add ὥσπερ εἴρηται μικρῷ πρότερον; omitted here by Sch.: see below. [5] Sc. in the roots.
[6] ὥσπερ ... πρότερον inserted here by Sch.: see above; ὥσπερ εἴρηται Ald.
[7] καυλῶν Vin.Vo.Cod.Cas.: so also G; καρπῶν Ald.HM*.
[8] μόνης οὕτω καὶ conj. W.; μόνης καὶ Ald.H; μόνον οὕτω καὶ M*.

253

συνίσταται δακρυώδης τις, ὥσπερ καὶ τῆς τραγακάνθης· ταύτην γὰρ οὐδὲ τέμνειν ἐστί· τῶν δὲ πλείστων ἀπὸ τῆς ἐντομῆς. ὧν ἐνίους μὲν εὐθὺς εἰς ἀγγεῖα συνάγουσιν, ὥσπερ καὶ τὸν τοῦ τιθυμάλλου ἢ μηκωνίου, καλοῦσι γὰρ ἀμφοτέρως, καὶ ἁπλῶς ὅσα πολύοπα τυγχάνει· τῶν δὲ μὴ πολυόπων ἐρίῳ λαμβάνουσιν ὥσπερ καὶ τῆς θριδακίνης.

3 Ἐνίων δ' οὐδ' ὀπισμὸς ἀλλ' οἷον χυλισμός ἐστιν, ὥσπερ ὅσα κόψαντες ἢ τρίψαντες καὶ ὕδωρ ἐπιχέαντες ἀπηθοῦσι καὶ λαμβάνουσι τὴν ὑπόστασιν· ξηρὸς δὲ δῆλον ὅτι καὶ ἐλάττων ὁ χυλὸς τούτων. ἔστι δὲ τῶν μὲν ἄλλων ῥιζῶν τὸ χύλισμα ἀσθενέστερον τοῦ καρποῦ, τοῦ κωνείου δὲ ἰσχυρότερον, καὶ τὴν ἀπαλλαγὴν ῥᾷω ποιεῖ καὶ θᾶττω μικρὸν πάνυ καταπότιον δοθέν· ἐνεργότερον δὲ καὶ εἰς τὰς ἄλλας χρείας. ἰσχυρὸν δὲ καὶ τὸ τῆς θαψίας. τὰ δὲ ἄλλα πάντα ἀσθενέστερα. οἱ μὲν οὖν ὀπισμοὶ σχεδὸν τοσαυταχῶς γίνονται.

4 Τῆς δὲ ῥιζοτομίας οὐκ ἔστι τοιαύτη διαφορὰ πλὴν ἐν ταῖς ὥραις οἷον θέρους ἢ μετοπώρου, καὶ τῷ τάσδε ἢ τάσδε τῶν ῥιζῶν· οἷον τοῦ ἐλλεβόρου τὰς κάτω τὰς λεπτάς· τὴν γὰρ ἄνω τὴν παχεῖαν τὴν κεφαλώδη φασὶν ἀχρεῖον εἶναι καὶ διδόναι

---

[1] *cf.* Diosc. 2. 136 ; Plin. 20. 58.
[2] *cf.* Diosc. 3. 7.
[3] ῥᾴω conj. Sch.; ῥαιω U; ῥᾳδίως M*Ald.

the form of a sort of gum, as with tragacanth; for incision of this plant cannot be made; but in most it is obtained by incision. In some cases the juice is collected straight into vessels, for instance that of *tithymallos* (spurge) or *mekonion* (for the plant has both names) and in general the juice of specially juicy plants is so collected. But that of those which do not yield abundant juice is taken with a piece of wool, as also that of wild lettuce.[1]

[2] In some cases there can be no collection of juice, but there is a sort of extraction of it, for instance in the case of plants which are cut down or bruised; they then pour water over them and strain off the fluid, keeping the sediment; but it is plain that in these cases the juice obtained is dry and less copious. In most 'roots' the juice thus extracted is less powerful than that of the fruit, but in hemlock it is stronger and it causes an easier [3] and speedier [4] death even when administered in a quite small pill; and it is also more effective for other uses. [5] That of *thapsia* is also powerful, while all the rest are less so. Such then is a general account of the various ways of obtaining the juices of plants.

Of the cutting of roots for medicinal purposes, and of certain superstitions connected therewith.

As to cutting of the roots there is no such diversity of practice, except as to the season, which may be summer or autumn, and as to the particular roots selected. [6] Thus in hellebore the slender lower roots are taken, for they say that the thick upper part [7] which forms a sort of head is useless, and that

---

[4] θάττω conj. Sch.; ἐλάττω UM; θᾶττον U*M*Ald.
[5] *cf.* Plin. 13. 125.   [6] Plin. 25. 53.   [7] *i.e.* rhizome.

ταῖς κυσὶν ὅταν βούλωνται καθαίρειν. καὶ ἐφ ἑτέρων δέ τινων τοιαύτας λέγουσι διαφοράς.

5 Ἔτι δὲ ὅσα¹ οἱ φαρμακοπῶλαι καὶ οἱ ῥιζοτόμοι τὰ μὲν ἴσως οἰκείως τὰ δὲ καὶ ἐπιτραγῳδοῦντες λέγουσι. κελεύουσι γὰρ τὰς μὲν κατ' ἄνεμον ἱσταμένους τέμνειν, ὥσπερ ἑτέρας τέ τινας καὶ τὴν θαψίαν, ἀλειψάμενον λίπα· τὸ γὰρ σῶμα ἀνοιδεῖν ἐὰν ἐξ ἐναντίας. κατ' ἄνεμον δὲ καὶ τοῦ κυνοσβάτου τὸν καρπὸν συλλέγειν, εἰ δὲ μὴ κίνδυνον εἶναι τῶν ὀφθαλμῶν. τὰς δὲ νύκτωρ τὰς δὲ μεθ' ἡμέραν, ἐνίας δὲ πρὶν τὸν ἥλιον ἐπιβάλλειν, οἷον καὶ τὸ καλούμενον κλύμενον.

6 Καὶ ταῦτα μὲν καὶ τὰ παραπλήσια τούτοις τάχ' ἂν οὐκ ἀλλοτρίως δόξειεν λέγειν· ἐπισινεῖς γάρ τινων αἱ δυνάμεις· ἐξάπτειν γάρ φασιν ὥσπερ πῦρ καὶ κατακαίειν· ἐπεὶ καὶ ὁ ἑλλέβορος ταχὺ καρηβαρεῖν ποιεῖ, καὶ οὐ δύνανται πολὺν χρόνον ὀρύττειν, δι' ὃ καὶ προεσθίουσι σκόροδα καὶ ἄκρατον ἐπιπίνουσιν. ἀλλὰ τὰ τοιαῦτα ὥσπερ ἐπίθετα καὶ πόρρωθεν, οἷον τὴν παιωνίαν, οἱ δὲ γλυκυσίδην καλοῦσι, νύκτωρ κελεύουσιν ὀρύττειν· ἐὰν γὰρ ἡμέρας καὶ ὀφθῇ τις ὑπὸ δρυοκολάπτου τὸν μὲν καρπὸν ἀπολέγων κινδυνεύειν τοῖς ὀφθαλμοῖς, τὴν δὲ ῥίζαν τέμνων ἐκπίπτειν τὴν ἕδραν.

---

¹ ἔτι δὲ ὅσα conj. Sch. from G ; ἔτι δ' ὡς U*; ἔτι δὲ ὡς Ald. H ; ἐστι δὲ ὡς M*.

ENQUIRY INTO PLANTS, IX. viii. 4-6

it is only given to dogs when it is desired to purge them. And in certain other plants also such differences are mentioned.

Further we may [1] add statements made by druggists and herb-diggers, which in some cases may be to the point, but in others contain exaggeration. Thus they enjoin that in cutting some roots one should stand to windward,—for instance, in cutting *thapsia* among others, and that one should first anoint oneself with oil,[2] for that one's body will swell up if one stands the other way. Also that the fruit of the wild rose must be gathered standing to windward, since otherwise there is danger to the eyes. Also that some roots should be gathered at night, others by day, and some before the sun strikes on them, for instance those of the plant called honeysuckle.[3]

These and similar remarks may well seem to be not off the point, for the properties of these plants are hurtful; they take hold, it is said, like fire and burn; [4] for hellebore too soon makes the head heavy, and men cannot go on digging it up for long; wherefore they first eat garlic and take a draught of neat wine therewith. On the other hand the following ideas may be considered far-fetched and irrelevant; [5] for instance they say that the peony, which some call *glykyside,* should be dug up at night, for, if a man does it in the day-time and is observed by a woodpecker while he is gathering the fruit, he risks the loss of his eyesight; and, if he is cutting the root at the time, he gets *prolapsus ani.*

[2] Plin. 13. 124; Diosc. 4. 153.
[3] *cf.* 9. 18. 6.   [4] Plin. 25. 50.
[5] Plin. 27. 85; 25. 29.

THEOPHRASTUS

7   Φυλάττεσθαι δὲ καὶ τὴν κενταυρίδα τέμνοντα τριόρχην, ὅπως ἂν ἄτρωτος ἀπέλθῃ. καὶ ἄλλας δέ τινας αἰτίας. τὸ δ' ἐπευχόμενον τέμνειν οὐθὲν ἴσως ἄτοπον· ἀλλ' εἴ τι καὶ ἄλλο προστιθέασιν, οἷον ὅταν τὸ πάνακες τὸ Ἀσκληπίειον καλούμενον· ἀντεμβάλλειν γὰρ τῇ γῇ παγκαρπίαν <καὶ> μελιττοῦταν· ὅταν δὲ τὴν ξίριν, τριμήνου μελιττούτας ἀντεμβάλλειν μισθόν· τέμνειν δὲ ἀμφήκει ξίφει περιγράψαντα εἰς τρίς· καὶ ὅτι ἂν πρῶτον τμηθῇ μετέωρον ἔχειν εἶθ' οὕτω τὸ ἕτερον τέμνειν.

8   Καὶ ἄλλα δὲ τοιαῦτα πλείω. περιγράφειν δὲ καὶ τὸν μανδραγόραν εἰς τρὶς ξίφει, τέμνειν δὲ πρὸς ἑσπέραν βλέποντα. τὸν δ' ἕτερον κύκλῳ περιορχεῖσθαι καὶ λέγειν ὡς πλεῖστα περὶ ἀφροδισίων. τοῦτο δ' ὅμοιον ἔοικε τῷ περὶ τοῦ κυμίνου λεγομένῳ κατὰ τὴν βλασφημίαν ὅταν σπείρωσι. περιγράφειν δὲ καὶ τὸν ἐλλέβορον τὸν μέλανα καὶ τέμνειν ἱστάμενον πρὸς ἕω καὶ κατευχόμενον· ἀετὸν δὲ φυλάττεσθαι καὶ ἐκ δεξιᾶς καὶ ἐξ ἀριστερᾶς· κίνδυνον γὰρ εἶναι τοῖς τέμνουσιν, ἐάνπερ ἐγγὺς ἐπιγένηται ὁ ἀετός, ἀπο-

---

[1] Plin. 25. 69 adds that this plant was therefore also called τρίορχις. cf. Diosc. 4. 162.
[2] καὶ ... αἰτίας U*M*; ? καὶ ἄλλα δὲ τοιαῦτα W.
[3] Plin. 25. 30 and 31.

ENQUIRY INTO PLANTS, IX. viii. 7–8

It is also said that, while cutting feverwort[1] one must beware of the buzzard-hawk, if one wishes to come off unhurt; and other reasons for caution[2] are also given. That one should be bidden to pray while cutting is not perhaps unreasonable, but the additions made to this injunction are absurd; for instance as to cutting the kind of all-heal which is called that of Asklepios;[3] for then it is said that one should put in the ground in its place an offering made of all kinds of fruits and a cake; and that, when one is cutting gladwyn,[4] one should put in its place to pay for it cakes of meal from spring-sown wheat,[5] and that one should cut it with a two-edged sword, first making a circle round it three times,[6] and that the piece first cut must be held up in the air while the rest is being cut.

And many similar notions are mentioned. Thus it is said that one should draw three circles round mandrake with a sword, and cut it with one's face towards the west; and at the cutting of the second piece one should dance round the plant and say as many things as possible about the mysteries of love. (This seems to be like the direction given about cummin,[7] that one should utter curses at the time of sowing.) One should also, it is said, draw a circle round the black hellebore and cut it standing towards the east and saying prayers, and one should look out for an eagle both on the right and on the left; for that there is danger to those that cut, if your eagle should come near, that

[4] *cf.* Plin. 21. 42, who read ἴριν. *cf.* Diosc. 4. 22, where ξυρίς is called a kind of ἴρις; so also Plin. 21. 142.
[5] τριμήνου conj. Salm.; τριμήνους M*Ald.H.
[6] τρὶς conj. Sch.; τρεῖς U*M*P₂Ald. So also in next section.
[7] *cf.* 7. 3. 3.

259

θνήσκειν ἐνιαυτῷ.   ταῦτα μὲν οὖν ἐπιθέτοις
ἔοικεν, ὥσπερ εἴρηται.   τρόποι δ' οὐκ εἰσὶ τῶν
ῥιζοτομιῶν πλὴν οὓς εἴπομεν.

IX. Ἔστι δέ, ὥσπερ ἐλέχθη, τῶν μὲν πάντα χρήσιμα καὶ ἡ ῥίζα καὶ ὁ καρπὸς καὶ ὁ ὀπός, ὥσπερ ἄλλων τε καὶ τοῦ πανάκους· τῶν δὲ ἡ ῥίζα καὶ ὁ ὀπός, οἷον τῆς σκαμμωνίας καὶ τοῦ κυκλαμίνου καὶ τῆς θαψίας καὶ ἑτέρων, καθάπερ καὶ τοῦ μανδραγόρου· τοῦ γὰρ μανδραγόρου τὸ φύλλον χρήσιμον εἶναί φασι πρὸς τὰ ἕλκη μετ' ἀλφίτου, τὴν δὲ ῥίζαν πρὸς ἐρυσίπελας ξυσθεῖσάν τε καὶ ὄξει δευθεῖσαν καὶ πρὸς τὰ ποδαγρικὰ καὶ πρὸς ὕπνον καὶ πρὸς φίλτρα· διδόασι δ' ἐν οἴνῳ ἢ ὄξει· τέμνουσι δὲ τροχίσκους ὥσπερ ῥαφανῖδος καὶ ἐνείραντες ὑπὲρ γλεύκους ἐκρέμασαν ἐπὶ καπνῷ.

2   Ὁ δὲ ἐλλέβορος ἐπὶ ταὐτὰ τῇ τε ῥίζῃ καὶ τῷ καρπῷ χρήσιμος, εἴπερ οἱ ἐν Ἀντικύρᾳ, καθάπερ φασί, τῷ καρπῷ καθαίρουσιν· ἔχει δὲ <τὸν> σησαμώδη τοῦτον.

Πλείω δὲ καὶ τοῦ πανάκους τὰ χρήσιμα καὶ οὐ πάντα πρὸς τὰ αὐτά· ἀλλ' ὁ μὲν καρπὸς πρὸς τὰς ἐξαμβλώσεις καὶ τὰς δυσουρίας, ὁ δὲ ὀπὸς ἡ χαλβάνη καλουμένη πρός τε τὰς ἐξαμβλώσεις καὶ πρὸς τὰ σπάσματα καὶ τοὺς τοιούτους πόνους, ἔτι δὲ πρὸς τὰ ὦτα καὶ τὰς φωνασκίας·

---

[1] ἐάνπερ ἐγγὺς ἐπιγένηται conj. W.; ἐὰν δὲ ὁ ἐγγὺς μὴ ἀποτέμνῃ ὁ ἀετὸς ἀποθνήσκει ἐνιαυτῷ UMAld. Similar confusion with variations in U*M*PP₂: restoration a makeshift. cf. Plin. 25. 50.   [2] 9. 8. 1.
[3] Diosc. 4. 75; Plin. 26. 104 and 121.
[4] cf. 9. 12. 1.

they may die[1] within the year.   These notions then seem to be irrelevant, as has been said.   There are however no methods of root-cutting besides those which we have mentioned.

Of the medicinal uses of divers parts of plants.

IX. As was said,[2] of some plants the root, fruit and juice are all serviceable, as of all-heal among others; of some the root and the juice, as of scammony[3] cyclamen thapsia and others, such as mandrake; for the leaf of this, they say, used with meal, is useful for wounds, and the root for erysipelas, when scraped and steeped in vinegar, and also for gout, for sleeplessness, and for love potions. It is administered in wine or vinegar; they cut little balls of it, as of[4] radishes, and making a string of them hang them up in the smoke over must.

[5] Of hellebore both root and fruit are useful for the same purposes,[6]—if it is true, as is said, that the people of Anticyra use the fruit as a purge; this fruit contains the well-known[7] drug called *sesamodes*.

Various parts of all-heal are also useful, and not all for the same purposes; the fruit is used in cases of miscarriage and for disorders of the bladder, while the juice,[8] which is called *khalbane*, is used in cases of miscarriage and also for sprains and such-like troubles; also for the ears, and to strengthen the

---

[5] Diosc. 4. 108, 109 and 162.

[6] ταὐτὰ conj. Sch. from G; ταῦτα U*M*Ald.

[7] I have inserted τὸν. *cf.* 9. 14. 4; Plin. 22. 133; 25. 52 and 64; Diosc. *l.c.* The drug was actually called σησαμοειδές or σησαμοειδής. For the sense of τοῦτον *cf.* 3. 7. 3; 3. 8. 3 and reff.

[8] This seems to be a mistake. *cf.* 9. 7. 2; Diosc. 3. 83; Plin. 12. 126.

ἡ δὲ ῥίζα πρός τε τοὺς τόκους καὶ τὰ γυναικεῖα καὶ πρὸς ὑποζυγίων φύσας· χρησίμη δὲ καὶ πρὸς τὸ ἴρινον μύρον διὰ τὴν εὐωδίαν· ἰσχυρότερον δὲ τὸ σπέρμα τῆς ῥίζης. γίνεται δὲ περὶ Συρίαν καὶ τέμνεται περὶ πυραμητόν.

3 Τοῦ δὲ κυκλαμίνου ἡ μὲν ῥίζα πρός τε τὰς ἐκπυήσεις¹ τῶν φλεγμονῶν καὶ πρόσθετον γυναιξὶ καὶ πρὸς τὰ ἕλκη ἐν μέλιτι· ὁ δὲ ὀπὸς πρὸς τὰς ἀπὸ κεφαλῆς καθάρσεις ἐν μέλιτι ἐγχεόμενος, καὶ πρὸς τὸ μεθύσκειν, ἐὰν ἐν οἴνῳ διαβρέχων διδῷ τις πίνειν. ἀγαθὴν δὲ τὴν ῥίζαν καὶ ὠκυτόκιοι περίαπτον καὶ εἰς φίλτρα·² ὅταν δὲ ὀρύξωσι, κατακαίουσιν εἶτ' οἴνῳ δεύσαντες τροχίσκους ποιοῦσιν, ὥσπερ τῆς τρυγὸς ᾗ ῥυπτόμεθα.

4 Καὶ τοῦ σικύου δὲ τοῦ ἀγρίου τὴν μὲν ῥίζαν <πρὸς> ἀλφοὺς καὶ ψώρας βοσκημάτων· τὸ δὲ σπέρμα χυλισθὲν ποιεῖ τὸ ἐλατήριον. συλλέγεται δὲ τοῦ φθινοπώρου· τότε γὰρ βέλτιστον.

5 Τῆς δὲ χαμαίδρυος³ τὰ μὲν φύλλα πρὸς τὰ ῥήγματα καὶ πρὸς τὰ τραύματα ἐν ἐλαίῳ τριβόμενα καὶ πρὸς τὰ νεμόμενα ἕλκη· τὸν δὲ καρπὸν καθαίρειν χολήν· ἀγαθὸν δὲ καὶ ὀφθαλμοῖς· πρὸς δὲ τὰ ἄργεμα προσάγειν τὸ φύλλον τρίψαντα ἐν ἐλαίῳ. ἔχει δὲ φύλλα μὲν οἷάπερ δρῦς, τὸ δὲ ἀνάστημα τῆς ὅλης ὅσον σπιθαμιαῖον· εὔοσμον δὲ καὶ ἡδύ.

Τὸ μὲν οὖν μὴ πρὸς ταὐτὸ πάντα τὰ μέρη χρήσιμα τυγχάνειν οὐκ ἴσως ἄτοπον· τὸ δὲ τῆς

---

¹ ἐκπυήσεις conj. Coraës from Plin. 26. 120, *eruptiones*; ἐκπνεύσεις M*Ald.
² Diosc. 2. 164 ; Plin. 25. 133 ; 26. 149.
³ *cf.* Plin. 23. 63.

ENQUIRY INTO PLANTS, IX. ix. 2–5

voice. The root is used in childbirth, for diseases of women, and for flatulence in beasts of burden. It is also useful in making the iris-perfume because of its fragrance; but the seed is stronger than the root. It grows in Syria and is cut at the time of wheat-harvest.

Of cyclamen the root is used for suppurating[1] boils; also as a pessary for women and, mixed with honey, for dressing wounds; the juice for purgings of the head,[2] for which purpose it is mixed with honey and poured in; it also conduces to drunkenness, if one is given a draught of wine in which it has been steeped. They say also that the root is a good charm for inducing rapid delivery and as a love potion;[3] when they have dug it up, they burn it, and then, having steeped the ashes in wine, make little balls like those made of wine-lees which we use as soap.

[4] Of 'wild cucumber' (squirting cucumber) the root is used for[5] white leprosy and for mange in sheep, while the extracted juice makes the drug called 'the driver.'[6] It is collected in autumn, for then it is best.

Of germander the leaves pounded up in olive-oil are used for fractures and wounds and for spreading sores; the fruit purges bile, and is good also for the eyes; for ulcers[7] in the eye they pound up the leaf in olive-oil before applying it. It has leaves like the oak, but its entire growth is only about a palm high; and it is sweet both to smell and taste.

Now that all parts are not serviceable for the same purpose is perhaps not strange; it is more

[4] Diosc. 4. 150; Plin. 20. 3.   [5] πρὸς add. St.
[6] cf. 9. 14. 1 and 2.   [7] cf. 7. 6. 2; Diosc. 3. 98.

αὐτῆς ῥίζης τὸ μὲν ἄνω τὸ δὲ κάτω καθαίρειν θαυμασιώτερον, οἷον καὶ τῆς θαψίας καὶ τῆς ἰσχάδος, οἱ δ' ἄπιον καλοῦσι, καὶ τῆς λιβανωτίδος· ὅτι γὰρ αὖ καὶ κάτω καὶ ἄνω ταὐτὰ δύναται [καθαίρειν], καθάπερ τὸ ἐλατήριον, οὐθὲν ἄτοπον.

6 Ἔχει δὲ ἡ θαψία φύλλον μὲν ὅμοιον τῷ μαράθῳ πλὴν πλατύτερον καυλὸν δὲ ναρθηκώδη ῥίζαν δὲ λευκήν.

Ἡ δ' ἰσχὰς ἢ ἄπιος φύλλον μὲν ἔχει πηγανῶδες βραχὺ καυλοὺς δ' ἐπιγείους τρεῖς ἢ τέτταρας ῥίζαν δὲ οἵανπερ ὁ ἀσφόδελος πλὴν λεπυριώδη· φιλεῖ δὲ ὀρεινὰ χωρία καὶ κοχλακώδη. συλλέγεται δὲ τοῦ ἦρος. τοῦτο μὲν οὖν ἴδιον τῶν εἰρημένων.

X. Ὁ δὲ ἐλλέβορος ὅ τε μέλας καὶ ὁ λευκὸς ὥσπερ ὁμώνυμοι φαίνονται· περὶ δὲ τῆς ὄψεως διαφωνοῦσιν· οἱ μὲν γὰρ ὁμοίους εἶναι, πλὴν τῷ χρώματι μόνον διαφέρειν τὴν ῥίζαν τοῦ μὲν λευκὴν τοῦ δὲ μέλαιναν· οἱ δὲ τοῦ μὲν μέλανος τὸ φύλλον δαφνῶδες τοῦ δὲ λευκοῦ πρασῶδες, τὰς δὲ ῥίζας ὁμοίας πλὴν τῶν χρωμάτων. οἱ δ' οὖν ὁμοίους λέγοντες τοιάνδε φασὶν εἶναι τὴν μορφήν· καυλὸν μὲν ἀνθερικώδη βραχὺν σφόδρα· φύλλον δὲ πλατύσχιστον, παρόμοιον σφόδρα τῷ τοῦ νάρθηκος, μῆκος δ' ἔχον· εὐθὺ δ' ἐκ τῆς ῥίζης

---

[1] ὅτι γὰρ conj. W.; ὅσα γὰρ UU*M*; τὰ γὰρ M; τὸ γὰρ Ald.
[2] Diosc. 4. 153; Plin. 13. 124.   [3] Diosc. 4. 175.

surprising that part of the same 'root' should purge upwards and another part downwards, as is the case with *thapsia* and *iskhas* – which some call *apios* (spurge) —and with *libanotis*; for [1] it is not strange that on the other hand the same parts should purge both upwards and downwards, as is the case with ' the driver.'

[2] *Thapsia* has a leaf like fennel, but broader, a stalk like that of ferula, and a white root.

[3] *Iskhas* (or *apios*) has a leaf like rue and short, three or four prostrate stems, and a root like that of asphodel, except that it is composed of scales; [4] it loves mountain districts with a gravelly soil. It is collected in spring. Now this account applies only to the above-mentioned plants.

Of hellebores, the white and the black: their uses and distribution.

X. [5] The white and the black hellebore appear to have nothing in common except the name. But accounts differ as to the appearance of the plants; some say that the two are alike and differ only in colour, the root of the one being white, of the other black; some however say that the leaf of the 'black' is like that of bay, that of the white like that of the leek, but that the roots are alike except for their respective colours. Now those who say that the two plants are alike describe the appearance [6] as follows :—the stem is like that of asphodel and very short; the leaf has broad divisions, and is extremely like that of ferula, but is long; it is closely attached to the root and

[4] *cf.* Diosc. 3. 134.
[5] Plin. 25. 47–61. See Index. *cf.* 9. 11. 5 n.
[6] *i.e.* of the two plants regarded as one; but the text of the following description seems to be hopelessly confused.

ἠρτημένον καὶ ἐπιγειόφυλλον· πολύρριζον δ' εὖ μάλα ταῖς λεπταῖς καὶ χρησίμοις.

2 Ἀναιρεῖν δὲ τὸν μὲν μέλανα καὶ ἵππους καὶ βοῦς καὶ ὗς, δι' ὃ καὶ οὐδὲν νέμεσθαι τούτων· τὸν δὲ λευκὸν νέμεσθαι τὰ πρόβατα καὶ ἐκ τούτου πρῶτον συνοφθῆναι τὴν δύναμιν καθαιρομένων ἐκείνων· ὡραῖος δὲ μετοπώρου, τοῦ δ' ἦρος ἄωρος· ἀλλὰ πρὸς τὴν πυλαίαν οἱ ἐκ τῆς Οἴτης συλλέγουσι· πλεῖστος γὰρ ἐνταῦθα φύεται καὶ ἄριστος· μοναχοῦ δὲ φύεται τῆς Οἴτης περὶ τὴν Πυράν.

Μίσγεται δὲ πρὸς τὴν πόσιν, ὅπως εὐεμὲς ᾖ, τὸ τῆς ἐλλεβορίνης σπέρμα· τοῦτο δ' ἐστὶ ποάριον.

3 Φύεται δὲ ὁ μὲν μέλας πανταχοῦ· καὶ γὰρ ἐν τῇ Βοιωτίᾳ καὶ ἐν Εὐβοίᾳ καὶ παρ' ἄλλοις πολλοῖς· ἄριστος δὲ ὁ ἐκ τοῦ Ἑλικῶνος, καὶ ὅλως τὸ ὄρος εὐφάρμακον. ὁ δὲ λευκὸς ὀλιγαχοῦ· βέλτιστοι δὲ καὶ οἷς χρῶνται μάλιστα τέτταρες ὁ Οἰταῖος ὁ Ποντικὸς ὁ Ἐλεάτης ὁ Μαλιώτης. φασὶ δὲ τὸν Ἐλεάτην ἐν τοῖς ἀμπελῶσι φύεσθαι καὶ ποιεῖν τὸν οἶνον οὕτω διουρητικὸν ὥστε λαγαροὺς εἶναι πάνυ τοὺς πίνοντας.

4 Ἄριστος δὲ πάντων καὶ τούτων καὶ τῶν ἄλλων ὁ Οἰταῖος. ὁ δὲ Παρνάσιος καὶ ὁ Αἰτωλικός, γίνεται γὰρ καὶ ἐνταῦθα καὶ πολλοὶ καὶ ὠνοῦνται καὶ πωλοῦσιν οὐκ εἰδότες, [οὐχ ὅτε] σκληροὶ καὶ

---

[1] Which were held apparently at Thermopylae regularly in autumn and sometimes in spring: the meeting would give opportunities for sale. ἀλλὰ implies a spring meeting.

ENQUIRY INTO PLANTS, IX. x. 1-4

creeps on the ground; the plant has numerous roots, to wit, the slender roots which are serviceable.

Also they say that the black is fatal to horses oxen and pigs, wherefore none of these animals eat it; while the white is eaten by sheep, and from this circumstance the virtue of the plant was first observed, since it purges them; it is at its prime in autumn, and past its prime when spring comes. However the people of Mount Oeta gather it for the meetings[1] of the Amphictyons; for it grows there in greatest abundance and best, though at only one place in the district of Oeta, namely about Pyra.

(The seed of rupture-wort is mixed with the potion given to promote easy vomiting; this plant is a small herb).

The black kind of hellebore grows everywhere; it is found in Boeotia, in Euboea and in many other places; but best is that from Mount Helicon, which mountain is in general rich in medicinal herbs. The white occurs in few places; the best and that which is most used comes from one of four places, Oeta,[2] Pontus, Elea, and Malea.[3] They say that that of Elea grows in the vineyards and makes the wine so diuretic that those who drink it become quite emaciated.

But best of all these and better than that found anywhere else is that of Mount Oeta, while that of Parnassus and that of Aetolia (for the plant is common in these parts too and men buy and sell it, not knowing[4] the difference) are tough and ex-

[2] ὁ before Οἰταῖος add. Sch.

[3] Μαλιώτης conj. Hahnemann, cf. Strabo 9. 3. 3; Μασσαλιώτης Ald. Plin. l.c. gives Parnassus as the fourth locality: cf. § 4.

[4] The words οὐχ ὅτε may have arisen from οὐκ εἰδότες.

ἄγαν περισκελεῖς. ταῦτα μὲν οὖν ὅμοια ταῖς μορφαῖς ὄντα ταῖς δυνάμεσι διαφέροντα.

Καλοῦσι δὲ τὸν μέλανά τινες ἔκτομον Μελαμπόδιον, ὡς ἐκείνου πρῶτον τεμόντος καὶ ἀνευρόντος. καθαίρουσι δὲ καὶ οἰκίας αὐτῷ καὶ πρόβατα συνεπᾴδοντές τινα ἐπῳδὴν καὶ εἰς ἄλλα δὲ πλείω χρῶνται.

XI. Πολλὰ δέ ἐστι καὶ τὰ πανάκη καὶ οἱ τιθύμαλλοι καὶ ἕτερ' ἄττα. πάνακες γὰρ καλοῦσι πρῶτον μὲν τὸ ἐν Συρίᾳ, περὶ οὗ μικρῷ πρότερον εἴρηται. ἄλλα δὲ τὰ τρία, τὸ μὲν Χειρώνειον καλούμενον τὸ δ' Ἀσκληπίειον τὸ δ' Ἡράκλειον. ἔχει δὲ τὸ μὲν Χειρώνειον φύλλον μὲν ὅμοιον λαπάθῳ μεῖζον δὲ καὶ δασύτερον, ἄνθος δὲ χρυσοειδές, ῥίζαν δὲ μικράν· φιλεῖ δὲ μάλιστα τὰ χωρία τὰ πίονα· χρῶνται δὲ πρός τε τοὺς ἔχεις καὶ τὰ φαλάγγια καὶ τοὺς σῆπας καὶ τὰ ἄλλα ἑρπετὰ διδόντες ἐν οἴνῳ καὶ ἀλείφοντες μετ' ἐλαίου· τοῦ δ' ἔχεως τὸ δῆγμα καὶ καταπλάττοντες καὶ ἐν ὀξίνῃ πιεῖν διδόντες· ἀγαθὴν δέ φασι καὶ ἑλκῶν ἐν οἴνῳ καὶ ἐλαίῳ καὶ φυμάτων ἐν μέλιτι.

2 Τὸ δ' Ἀσκληπίειον τὴν ῥίζαν μῆκος μὲν ὡς σπιθαμὴν λευκὴν δὲ καὶ παχεῖαν σφόδρα, καὶ φλοιὸν παχὺν καὶ ἁλυκώδη· καυλὸν δὲ ἔχει γονατώδη πανταχόθεν, φύλλον δὲ οἷόνπερ ἡ θαψία πλὴν παχύτερον· ἀγαθὸν δὲ εἶναί φασι ἑρπετῶν

---

[1] From this phrase ἔκτομον came to be used as a synonym for 'black hellebore.' cf. Plin. 25. 47; Diosc. 4. 149; Hesych. and Galen, *Lex. Hipp. s.v.*
[2] 9. 9. 2.   [3] Plin. 25. 32; 26. 139.
[4] μικρὰν conj. H. from Plin. 25. 32. *radix parva*; μακρὰν U* Ald.

ceeding harsh. These plants then, while resembling the best form in appearance, differ in their virtues.

Some call the black the 'hellebore of Melampus,'[1] saying that he first cut and discovered it. Men also purify horses and sheep with it, at the same time chanting an incantation; and they put it to several other uses.

Of the various kinds of all-heal.

XI. There are also several kinds of all-heal *tithymallos* (spurge) and other herbs. To begin with, one plant called all-heal is the one found in Syria, of which we have recently spoken.[2] [3] Then come the three other kinds, known as that of Chaeronea, that of Asclepios, and that of Heracles. That of Chaeronea has a leaf like monk's rhubarb, but larger and rougher, a golden flower, and a small [4] root; and it specially loves rich ground; they use it for the bites of snakes, spiders, vipers [5] and other reptiles, administering it in wine or anointing the place with it mixed with olive-oil. In treating a snake-bite they use a plaster of it, and also give a draught of it mixed with vinegar [6]; and they also say that it is good for sores [7] when mixed with wine and olive-oil, and for tumours when mixed with honey.

[8] The kind called after Asklepios has a white and very stout root about a span long and a thick bark which is crusted with salt [9]; its stem is jointed all the way up, its leaf like that of *thapsia*, but thicker; it is said that it is good to scrape and drink

---

[5] σήπας conj. Scal., *cf.* Arist. *Mir. Ausc.* 164; σήτας Ald.
[6] καὶ ἐν ὀξίνῃ conj. Sch., *cf.* 9. 13. 3; ἐν ὀξίνῃ καὶ PAld.
[7] For the genitive *cf.* §§ 2, 3; Xen. *Mem.* 3. 8. 3.
[8] Plin. 25. 30; Diosc. 3. 49.
[9] ἁλυκώδη: ? 'has a briny taste.'

τε ξύοντα πίνειν, καὶ σπληνὸς ὅταν αἷμα περὶ αὐτὸν ἐν μελικράτῳ, καὶ κεφαλαίας¹ τρίβοντα ἐν ἐλαίῳ ἀλείφειν καὶ ἄλλο τι ἐὰν πονῇ τις ἀφανές, καὶ γαστρὸς ὀδύνης ἐν οἴνῳ ξύοντα. δύνασθαι δὲ καὶ τὰς μακρὰς ἀρρωστίας ἐκκλίνειν. ἔπειτα τῶν ἑλκῶν² τῶν μὲν ὑγρῶν ξηρὸν ἐπιπάττοντα προκατακλύζοντα ἐν οἴνῳ θερμῷ, τῶν δὲ ξηρῶν ἐν οἴνῳ δεῦσαι καὶ καταπλάττειν.

3 Τὸ δ' Ἡράκλειον³ φύλλον μὲν ἔχει μέγα καὶ πλατὺ καὶ τρισπίθαμον πανταχῇ, ῥίζαν δὲ ὡς δακτύλου τὸ πάχος δίκραν ἢ τρίκραν⁴, τῇ γεύσει μὲν ὑπόπικρον τῇ δ' ὀσμῇ καθάπερ λιβανωτοῦ καθαροῦ· ἀγαθὴν δὲ τῆς ἱερᾶς νόσου μιγνυμένην φώκης πιτύᾳ ὅσον τεταρτημόριον πίνειν, καὶ ὀδύνης κατὰ γαστέρα ἐν οἴνῳ γλυκεῖ, καὶ ἑλκῶν τῶν μὲν ὑγρῶν ξηρὰν τῶν δὲ ξηρῶν ἐν μέλιτι. αὗται μὲν οὖν ταύτας ἔχουσι διαφοράς τε καὶ δυνάμεις.

4 Ἄλλα δὲ πανάκη τὸ μὲν λεπτόφυλλον τὸ δὲ οὔ· αἱ δὲ δυνάμεις ἀμφοῖν αἱ αὐταί, πρόσθετόν τε γυναιξὶ καὶ κατάπλασμα μετ' ἀλφίτου καὶ πρὸς τὰ ἕλκη τὰ ἄλλα καὶ πρὸς τὰ νεμόμενα.

5 Συνώνυμοι δὲ καὶ οἱ στρύχνοι καὶ οἱ τιθύμαλ-

---

¹ κεφαλαίας conj. Sch.; κεφαλῆς Ald.
² τῶν ἑλκῶν conj. Sch.; τῶν δὲ ἑλκώδων (sic) U*; τῶν ἑλκωδῶν Ald.H. *cf.* § 3.
³ Plin. 25. 32; Diosc. 3. 48.
⁴ δίκραν ἢ τρίκραν conj. Sch.; δίκραν ἢ τρίκαν UM; δικρανῇ πικρὰν U*; δικρανῇ ἢ τρικρανῇ Ald.

it against bites of reptiles, to take it in a posset of honey for disorders of the spleen, when the blood collects about it, and against headache [1] to pound it up in olive-oil and anoint the head; that it is of use also in other obscure troubles, and against stomach-ache, if scraped and taken in wine. It is said also to be able to prevent long periods of sickness. Again for running sores [2] one may sprinkle it on in hot wine, first washing the place, while for dry sores one may soak it in wine and apply a plaster.

[3] The kind named after Herakles has a large broad leaf, three spans each way, a root as thick as a man's finger, forking in two or three [4]; in taste it is somewhat bitter, in smell like pure frankincense [5]; [6] it is good to drink it against epilepsy, mixed with the rennet of a seal in the proportion of one to four, or in sweet wine against pain [7] in the stomach; it may be used dry [8] for running sores, and mixed with honey for dry ones.   Such are the special features about these plants and their respective virtues.

[9] There are also other kinds of all-heal, of which one has a fine leaf, the other not; the properties of both kinds are the same; namely they are used as a pessary for women, and a plaster may be made of them mixed with meal for spreading sores as well as for ordinary sores.

Of the various plants called strykhnos.

As to *strykhnos* again and *tithymallos* (spurge) there is in either case more than one form of the plant

[5] λιβανωτοῦ καθαροῦ conj. Sch.; λιβανωτὸν καθαρὸν UM; λιβανωτοῦ U*; λιβανωτοῦ καθαρῶν Ald.H   καθαροῦ perhaps due to καθάπερ.   [6] *cf.* Fr. 175; Diosc. 2. 75.

[7] ὀδύνης conj. W.; ὀδύναι UMU*Ald.

[8] ξηρὰν conj. Sch.; ξηρὰ U*Ald.; ξηραὶ M.   [9] Plin. 35. 33.

λοι. τῶν γὰρ στρύχνων ὁ μὲν ὑπνώδης ὁ δὲ μανικός. καὶ ὁ μὲν ὑπνώδης ἐρυθρὰν ἔχων τὴν ῥίζαν ὥσπερ αἷμα ξηραινομένην, ὀρυττομένην δὲ λευκήν, καὶ καρπὸν ἐρυθρότερον κρόκου, φύλλον δὲ τιθυμάλλῳ ὅμοιον ἢ μηλέᾳ τῇ γλυκείᾳ καὶ αὐτὸ δασὺ καὶ πυγμὴν μέγα. τούτου τῆς ῥίζης τὸν φλοιὸν κόπτοντες λίαν καὶ βρέχοντες ἐν οἴνῳ ἀκράτῳ διδόασι πιεῖν καὶ ποιεῖ καθεύδειν. φύεται δὲ ἐν χαράδραις καὶ τοῖς μνήμασιν.

6 Ὁ δὲ μανικός, οἱ δὲ θρύορον καλοῦσιν αὐτὸν οἱ δὲ περιττόν, λευκὴν ἔχει τὴν ῥίζαν καὶ μακρὰν ὡς πήχεως καὶ κοίλην. δίδοται δ' αὐτῆς, ἐὰν μὲν ὥστε παίζειν καὶ δοκεῖν ἑαυτῷ κάλλιστον εἶναι, δραχμὴ σταθμῷ· ἐὰν δὲ μᾶλλον μαίνεσθαι καὶ φαντασίας τινὰς φαίνεσθαι, δύο δραχμαί· ἐὰν δ' ὥστε μὴ παύεσθαι μαινόμενον τρεῖς, καὶ συμπαραμιγνύναι φασὶν ὀπὸν κενταυρίου· ἐὰν δὲ ὥστε ἀποκτεῖναι, τέτταρες. ἔχει δὲ τὸ μὲν φύλλον ὅμοιον εὐζώμῳ πλὴν μεῖζον, τὸν δὲ καυλὸν ὥσπερ ὀργυίας, κεφαλὴν δὲ ὥσπερ γηθύου μείζω δὲ καὶ δασυτέραν· ἔοικε δὲ καὶ πλατάνου καρπῷ.

---

[1] cf. 7. 15. 4, where a third στρύχνος is mentioned, which is ὁμώνυμος, not συνώνυμος, i.e. which has nothing in common with these two στρύχνοι except the name. cf. also 9. 15. 5.

[2] κρόκου conj. Dalec. from Diosc. 4. 72, καρπὸν ... κροκίζοντα; κόκκου MSS.

[3] πυγμὴν μέγα U; πυθμὴν μέγας U*Ald.H.; W. adopts Bod.'s conjecture σπιθαμὴν μέγα.

[4] Plin. 21. 177–179; Diosc. 4. 73.

[5] θρύορον Ald.H.; θρυόρον U*; βρυόρον U; βρύορον MmBas.;

ENQUIRY INTO PLANTS, IX. xi. 5–6

denoted by the name. [1] Of the plants called *strykhnos* one induces sleep, the other (thorn-apple) causes madness. The first-mentioned has a root which becomes red like blood as it dries, but when first dug up it is white; its fruit is a deeper orange than saffron,[2] its leaf like that of *tithymallos* or the sweet apple; and it is itself rough, and about a foot high.[3] The 'bark' of the root of this they bruise severely, and soaking it in neat wine give it as a draught, and it induces sleep. It grows in water-courses and on tombs.

[4] The kind which produces madness (which some call *thryoron*[5] and some *peritton*[6]) has a white hollow root about a cubit long. Of this three twentieths of an ounce in weight is given, if the patient is to become merely sportive and to think himself a fine fellow; twice this dose[7] if he is to go mad outright and have delusions[8]; thrice the dose, if he is to be permanently insane; (and then they say that the juice of centaury is mixed with it); four[9] times the dose is given, if the man is to be killed. The leaf is like that of rocket, but larger, the stem about a fathom long; the 'head'[10] is like that of a long onion, but larger and rougher. And it also resembles the fruit of the plane-tree.

---

briorem G. Plin. *l.c.* seems to have read ἐρυθρὸν; Diosc. *l.c.* βρύον.

[6] περιττὸν Ald.H., *i.e.* 'violent'; *pissum* G; Plin *l.c. perisson*; Diosc. *l.c.* πέρσιον.

[7] δραχμαὶ conj. Sch.; δραχμὰς Ald.

[8] καὶ ... φαίνεσθαι om. UM: ungrammatical, and possibly a gloss; but *cf.* Diosc. and Plin. *l.c.*

[9] τέτταρες conj. Sch.; τέτταρας Ald; τέσσαρας U*.

[10] 7. 4. 10 it was said that γήθυον has no 'head,' *i.e.* bulb; here the 'head' seems to be the inflorescence. *cf.* Diosc. and Plin. *l.c.*

THEOPHRASTUS

7 Τῶν δὲ τιθυμάλλων ὁ μὲν παράλιος καλούμενος κόκκινον φύλλον ἔχει περιφερές, καυλὸν δὲ καὶ τὸ ὅλον μέγεθος ὡς σπιθαμῆς· τὸν δὲ καρπὸν λευκόν. ἀμᾶται δὲ ὅταν ἄρτι περκάζῃ σταφυλή, καὶ ξηρανθεὶς ὁ καρπὸς δίδοται πίνειν τριφθεὶς ὅσον τρίτον μέρος ὀξυβάφου.

8 Ὁ δ' ἄρρην καλούμενος τὸ μὲν φύλλον ἐλαιῶδες ἔχει, τὸ δὲ ὅλον μέγεθος πηχυαῖον. τοῦτον ὀπίζουσιν ἅμα τρυγητῷ καὶ θεραπεύσαντες οὕτως ὡς δεῖ χρῶνται· καθαίρει δὲ κάτω μᾶλλον.

9 Ὁ δὲ μυρτίτης καλούμενος τιθύμαλλος λευκός· τὸ μὲν φύλλον ἔχει καθάπερ ὁ μύρρινος, πλὴν ἀκανθῶδες ἀπ' ἄκρου· κλήματα δ' ἀφίησιν ἐπὶ τὴν γῆν ὡς σπιθαμιαῖα, ταῦτα δ' οὐχ ἅμα φέρει τὸν καρπὸν ἀλλὰ παρ' ἔτος, τὰ μὲν νῦν τὰ δ' εἰς νέωτα, πεφυκότα ἀπὸ τῆς αὐτῆς ῥίζης. φιλεῖ δὲ ὀρεινὰ χωρία. ὁ δὲ καρπὸς αὐτοῦ καλεῖται κάρυον. ἀμῶσι δ' ὅταν ἁδρύνωνται αἱ κριθαὶ καὶ ξηραίνοντες καὶ ἀποκαθαίροντες· αὐτὸν τὸν καρπὸν πλύναντες ἐν ὕδατι καὶ πάλιν ξηράναντες διδόασι πιεῖν συμμιγνύντες δύο μέρη τῆς μελαίνης μήκωνος, τὸ δὲ συναμφότερον ὅσον ὀξύβαφον· καθαίρει δὲ φλέγμα κάτω· ἐὰν δὲ τὸ κάρυον αὐτὸ διδῶσι, τρίψαντες ἐν οἴνῳ γλυκεῖ διδόασιν ἢ ἐν σησάμῳ πεφρυγμένῳ κατατραγεῖν. ταῦτα μὲν οὖν τοῖς τε φύλλοις καὶ τοῖς ὀποῖς καὶ τοῖς καρποῖς χρήσιμα.

---

[1] Plin. 26. 68.
[2] κόκκινον conj. W.; κόκκος MSS. cf. Plin. l.c. ramis rubentibus. [3] Diosc. 4. 164; Plin. 26. 62-65.
[4] οὕτως ὡς δεῖ χρ. conj. Sch.; οὕτως ὡς δὴ χρ. U*; οὕτως χρ. Ald.

ENQUIRY INTO PLANTS, IX. xi. 7–9

Of the various kinds of tithymallos.

[1] Of the various plants called *tithymallos* (spurge) that which is called sea-spurge has a round scarlet [2] leaf; the stem (and the size of the plant generally) is about a span long, and the fruit is white. It is gathered when the grape is just turning, and the dried fruit is given in a draught, the dose being the twenty-fourth part of a pint.

[3] That which is called the 'male' has a leaf like the olive, and the height of the whole plant is a cubit. Of this they collect the juice at the time of vintage, and, after preparing it, use it as occasion demands [4]; and it purges chiefly downwards.

[5] The kind of *tithymallos* called 'myrtle-like' is white; it has a leaf like the myrtle, but spinous at the tip; it puts out earthward twigs about a span long, and these bear the fruit [6] not all at the same time but in alternate years, so that from the same root grow fruits partly this and partly next year. It loves hill-country. The fruit of it is called a 'nut.' They gather it when the barley is ripening and dry and clean it; (it is the actual fruit [7] which they clean); they wash it in water and, after drying it again, give it in a draught, mixing with it two parts of 'black [8] poppy'; and the whole dose amounts to about an eighth of a pint. It purges phlegm downwards. If they administer the 'nut' itself, they first pound it up in sweet wine, or give it in parched sesame to bite up. These plants then have leaves juices or fruits which are serviceable.

---

[5] Diosc. 4. 164; Plin. 26. 66.   [6] *cf. C.P.* 4. 6. 9.
[7] W. adds δὲ after αὐτὸν. The treatment of the leaves has perhaps dropped out. *cf.* Plin. *l.c.* G's version is even shorter.
[8] μέλαινα must here mean 'dark,' *i.e.* red. See Index.

10 Τῶν δὲ λιβανωτίδων, δύο γάρ εἰσιν, ἡ μὲν ἄκαρπος ἡ δὲ κάρπιμος, ἡ μὲν καὶ τῷ καρπῷ καὶ τῷ φύλλῳ χρησίμη ἡ δὲ μόνον τῇ ῥίζῃ. καλεῖται δὲ ὁ καρπὸς κάχρυ. ἔχει δὲ αὕτη τὸ μὲν φύλλον ἐοικὸς σελίνῳ ἐλείῳ μεῖζον δὲ πολύ, καυλὸν δὲ μέγεθος πήχεως ἢ μείζω, ῥίζαν δὲ μεγάλην καὶ παχεῖαν λευκὴν ὄζουσαν ὥσπερ λιβανωτοῦ, καρπὸν δὲ λευκὸν τραχὺν προμήκη· φύεται δὲ μάλιστα ὅπου ἂν αὐχμηρὰ χωρία ᾖ καὶ πετρώδη· χρησίμη δὲ ἡ μὲν ῥίζα πρός τε τὰ ἕλκη καὶ πρὸς τὰ γυναικεῖα πινομένη ἐν οἴνῳ αὐστηρῷ μέλανι· ὁ δὲ καρπὸς πρός τε τὰς στραγγουρίας καὶ πρὸς τὰ ὦτα καὶ ἄργεμα καὶ πρὸς ὀφθαλμίας καὶ ὥστε γάλα γυναιξὶν ἐμποιεῖν.

11 Ἡ δὲ ἄκαρπος ἔχει τὸ φύλλον ὅμοιον θριδακίνης τῆς πικρᾶς τραχύτερον δὲ καὶ λευκότερον, ῥίζαν δὲ βραχεῖαν. φύεται δὲ ὅπουπερ ἐρείκη πλείστη. δύναται δὲ ἡ ῥίζα καθαίρειν καὶ ἄνω καὶ κάτω· τὸ μὲν γὰρ πρὸς τὴν βλάστην ἄνω, τὸ δὲ πρὸς τὴν γῆν κάτω· κωλύει δὲ καὶ εἰς ἱμάτια τιθεμένη τοὺς σῆτας. συλλέγεται δὲ περὶ πυροτομίας.

XII. Χαμαιλέων δὲ ὁ μὲν λευκὸς ὁ δὲ μέλας· αἱ δὲ δυνάμεις τῶν ῥιζῶν καὶ αὐταὶ δὲ αἱ ῥίζαι τοῖς εἴδεσι διάφοροι. τοῦ μὲν γὰρ λευκὴ καὶ παχεῖα καὶ γλυκεῖα καὶ ὀσμὴν ἔχουσα βαρεῖαν· χρήσιμον δέ φασι πρός τε τοὺς ῥοῦς, ὅταν ἑψηθῇ

---

[1] Diosc. 3. 74; Plin 19. 187.
[2] cf. Plin. 24. 99 and 101.
[3] cf. 7. 6. 2; 9. 9. 5.
[4] ὅπουπερ ἐρείκη conj. Dalec. etc. from Diosc. l.c.; ὅπουπερ εἴρηται Ald.H.; ὁπότε ἐρείκη U*.

ENQUIRY INTO PLANTS, IX. xi. 10-xii. 1

Of the two herbs called libanotis.

[1] Of the plants called *libanotis*, (for there are two) one is barren, the other fruitful, the latter having both fruit and leaves that are serviceable, the former only a serviceable root. The fruit is called *kakhry*.[2] This plant has a leaf like marsh celery, but much larger, a stem a cubit long or more, a large stout white root, which smells like frankincense, and a white rough elongated fruit. It grows chiefly wherever there is parched and rocky soil; the root is serviceable for sores, and for diseases of women when given in a draught of dry black wine. The fruit is good for strangury, for the ears, for ulcers[3] on the eye, for ophthalmia and for producing milk in women.

The barren kind has a leaf like that of the bitter lettuce, but rougher and paler; the root is short. It grows where there is abundance of heather.[4] The root can purge both upwards and downwards, the upper part being used for the former, that nearer the ground for the latter purpose. Also, if it is put among clothes, it prevents moth. It is gathered at the time of wheat-harvest.

Of the two kinds of chamaeleon.

XII. [5] Of chamaeleon there is the white kind and the dark; the properties of the roots are different, and the roots also differ[6] in appearance. In the one case the root is white stout and sweet, and it has a heavy smell; they say that when cooked it is serviceable against flux; it is chopped up like

[5] Diosc. 3. 8; Plin. 22. 45 and 46, who explains the name (*mutat cum terra colores*). See Index.

[6] διαφέρουσαι conj. W.; διαφέρει U; διαφέρουσι δὲ M; διάφορον Ald.

κατατμηθεῖσα καθάπερ ῥαφανὶς ἐνειρομένη ἐφ ὁλοσχοίνου, καὶ πρὸς τὴν ἕλμινθα τὴν πλατεῖαν, ὅταν ἀσταφίδα προφάγῃ πίνειν ἐπιξύοντα ταύτην ὅσον ὀξύβαφον ἐν οἴνῳ αὐστηρῷ. ἀναιρεῖ δὲ καὶ κύνα καὶ σῦν· κύνα μὲν ἐν ἀλφίτοις ἀναφυραθεῖσα μετὰ ἐλαίου καὶ ὕδατος, σῦν δὲ μετὰ ῥαφάνων μεμιγμένη τῶν ὀρείων. γυναικὶ δὲ δίδοται ἐν τρυγὶ γλυκείᾳ ἢ ἐν οἴνῳ γλυκεῖ. καὶ ἐὰν βούληταί τις ἀσθενοῦντος ἀνθρώπου διαπειρᾶσθαι εἰ βιώσιμος, λούειν κελεύουσι τρεῖς ἡμέρας, κἂν περιενέγκῃ βιώσιμος. φύεται δὲ ὁμοίως πανταχοῦ, καὶ ἔχει τὸ φύλλον ὅμοιον σκολύμῳ μεῖζον δέ· αὐτὸ δὲ πρὸς τῇ γῇ τινα κεφαλὴν ἔχει ἀκανοειδῆ μεγάλην, οἱ δὲ καὶ ἄκανον καλοῦσιν.

2   Ὁ δὲ μέλας τῷ μὲν φύλλῳ παρόμοιος, σκολυμῶδες γὰρ ἔχει πλὴν ἔλαττον καὶ λειότερον, αὐτὸς δ' ὅλος ἐστὶν ὥσπερ σκιάδιον, ἡ δὲ ῥίζα παχεῖα καὶ μέλαινα διαρραγεῖσα δὲ ὑπόξανθος. χωρία δὲ φιλεῖ ψυχρὰ καὶ ἀργά· δύναται δὲ λέπραν τε ἐξελαύνειν ἐν ὄξει τριβόμενος καὶ ξυσθεὶς ἐπαλειφόμενος καὶ ἀλφὸν ὡσαύτως· ἀναιρεῖ δὲ καὶ τοὺς κύνας.

3   Μήκωνες δ' εἰσὶν ἄγριαι πλείους· ἡ μὲν κερατῖτις καλουμένη μέλαινα· ταύτης τὸ φύλλον ὥσ-

---

[1] cf. 9. 9. 1.
[2] cf. Pseudo-Diosc. 4. 175 and Index.
[3] ἀκανοειδῆ conj. Sch.; κονοειδῆ U*; κωνοειδῆ mP; ὁμοίαν ἀκάνῳ PAld.
[4] δὲ after τὴν om. Sch.; ? τινα κεφαλὴν W.

radishes[1] and the pieces strung on a rush; it is also good against the broad maw-worm; the patient first eats a bunch of raisins and then drinks about an eighth of a pint of this scraped up in a draught of dry wine. It is fatal to dogs and pigs; to kill a dog it is well mixed up in a meal paste with oil and water, to kill a pig it is mixed with 'mountain cabbage' (spurge).[2] It is given to a woman in sweet wine-lees or sweet wine. And if one wishes to discover whether a man that is sick will recover, they say that he should be washed with this for three days, and, if he survives the experience, he will recover. It grows everywhere alike and has a leaf like the golden thistle, but larger; the plant itself has a large thistle-like[3] head[4] close to the ground; some actually[5] call it the thistle.

[6] The dark kind resembles the other in leaf, which is like that of the golden thistle but smaller and smoother; the plant itself is in general appearance like a sunshade; the root is stout and black, and when broken is yellowish. It likes cold uncultivated soil: it has the property of expelling leprosy; for this it is given pounded up in vinegar, or else scrapings of it are made into a plaster; and it is also used for the white leprosy. This plant is also fatal to dogs.[7]

Of the various plants called 'poppy.'

[8] There are several kinds of wild poppy: the one called the horned poppy is black: the leaf of this is

[5] δὲ καὶ ἄκανον I conj.; δ' ἄκανθαν U*mPar., so also Diosc. *l.c.*; δ' ἄκανον PAld.G.

[6] Diosc. 3. 9; Plin. *l.c.*

[7] κύνας: κυνορραΐστας, dog-ticks, conj. Reinesius from Plin. *ricinos canum.*

[8] Diosc. 4. 64; Plin. 20. 205 and 206.

περ φλόμου τῆς μελαίνης ἧττον δὲ μέλαν, τοῦ δὲ καυλοῦ τὸ ὕψος ὡς πηχυαῖον, ῥίζα δὲ παχεῖα καὶ ἐπιπόλαιος, ὁ δὲ καρπὸς καμπύλος ὥσπερ κεράτιον· συλλέγεται δὲ περὶ πυροτομίας. δύναται δὲ καθαίρειν κοιλίαν, τὸ δὲ φύλλον ἄργεμα προβάτοις ἀφαιρεῖν. φύεται δὲ παρὰ θάλατταν, οὗ ἂν ᾖ πετρώδη χωρία.

4 Ἑτέρα δὲ μήκων ῥοιὰς καλουμένη παρομοία κιχορίῳ τῷ ἀγρίῳ, δι᾽ ὃ καὶ ἐσθίεται· ἐν τοῖς ἀρουραίοις δὲ φύεται, μάλιστα ἐν ταῖς κριθαῖς· ἄνθος δ᾽ ἔχει ἐρυθρὸν κωδύαν δ᾽ ὅσην ὄνυχα τοῦ δακτύλου. συλλέγεται δὲ πρὸ τοῦ θερισμοῦ τῶν κριθῶν, ἐγχλωροτέρα δὲ μᾶλλον. καθαίρει δὲ κάτω.

5 Ἑτέρα δὲ μήκων Ἡράκλεια καλεῖται τὸ μὲν φύλλον ἔχουσα οἷον στρουθός, ᾧ τὰ ὀθόνια λευκαίνουσι, ῥίζαν δὲ λεπτὴν ἐπιπόλαιον, τὸν δὲ καρπὸν λευκόν. ταύτης ἡ ῥίζα καθαίρει ἄνω· χρῶνται δέ τινες πρὸς τοὺς ἐπιλήπτους ἐν μελικράτῳ. ταῦτα μὲν οὖν ὥσπερ ὁμωνυμίᾳ τινὶ συνείληπται.

XIII. Τῶν δὲ ῥιζῶν καὶ ἐν τοῖς χυμοῖς αἱ διαφοραὶ καὶ ἐν ταῖς ὀσμαῖς· αἱ μὲν γάρ εἰσι δριμεῖαι αἱ δὲ πικραὶ αἱ δὲ γλυκεῖαι, καὶ αἱ μὲν εὔοδμοι αἱ δὲ βαρεῖαι. γλυκεῖα μὲν ἥ τε νυμφαία καλουμένη· φύεται δ᾽ ἐν ταῖς λίμναις καὶ ἐν τοῖς ἑλώδεσιν, οἷον ἔν τε τῇ Ὀρχομενίᾳ καὶ Μαραθῶνι

---

[1] ὥσπερ κεράτιον conj. W.; ὥσπερ κέρας UM; ὥσπερ τῶν κερατίων U*Ald.

[2] Diosc. l.c.; Plin. 19. 167–169.

[3] Diosc. 4. 66; Plin. 20. 207, cf. 19. 21.

like that of the black mullein, but it is not so black; the stem grows about a cubit high, the root is stout and shallow, the fruit is twisted like a little horn[1]: it is gathered at the time of wheat harvest. It has the property of purging the belly, and the leaf is used for removing ulcers on sheep's eyes. It grows by the sea, wherever there is rocky ground.

[2] Another kind of poppy is that called *rhoias*, which is like wild chicory, wherefore it is even eaten: it grows in cultivated fields and especially among barley. It has a red flower, and a head as large as a man's finger-nail. It is gathered before the barley-harvest, when it is still somewhat green. It purges downwards.

[3] Another kind of poppy is called *Herakleia*: it has a leaf like soap-wort, with which [4] they bleach linen: the root is slender and does not run deep, and the fruit is white. The root of this plant purges upwards: and some use it in a posset of mead for epileptics.

[5] These kinds then are distinct plants, though they come under one name.

Of roots possessing remarkable taste or smell.

XIII. The differences between roots are shown in their tastes [6] and in their smells: some are pungent, some bitter, some sweet: some again have a pleasant, others a disagreeable smell. The plant called yellow water-lily [7] is sweet: it grows in lakes and marshy places, as in the district of Orchomenus, at

[4] This appears to refer to στρουθός, not to Ἡρακλεία, as Plin. takes it. *cf.* 6. 4. 3 and Index, στρούθιον (2).

[5] *i.e.* ῥοιάς and Ἡρακλεία are popularly called 'poppies.' ὁμωνυμίᾳ τινι conj. W.; ὁμώνυμα τινα Ald.; ὁμώνυμα τινι U*. *cf.* 7. 15. 4.

[6] χυμοῖς conj. Sch.; χυλοῖς Ald.   [7] Plin. 25. 75.

καὶ περὶ Κρήτην· καλοῦσι δ' αὐτὴν οἱ Βοιωτοὶ μαδωνάϊν καὶ τὸν καρπὸν ἐσθίουσιν. ἔχει δὲ τὸ φύλλον μέγα ἐπὶ τοῦ ὕδατος· εἶναι δέ φασιν ἴσχαιμον, ἐάν τρίψας τις ἐπὶ τὴν πληγὴν ἐπιθῇ· χρησίμη δὲ καὶ πρὸς δυσεντερίαν πινομένη.

2 Γλυκεῖα δὲ καὶ ἡ Σκυθική· καὶ ἔνιοι δὲ καλοῦσιν εὐθὺς γλυκεῖαν αὐτήν· γίνεται δὲ περὶ τὴν Μαιῶτιν· χρησίμη δὲ πρός τε τὰ ἄσθματα καὶ πρὸς τὴν βῆχα ξηρὰν καὶ ὅλως τοὺς περὶ τὸν θώρακα πόνους· ἔτι δὲ πρὸς τὰ ἕλκη ἐν μέλιτι· δύναται δὲ καὶ τὴν δίψαν παύειν, ἐάν τις ἐν τῷ στόματι ἔχῃ· δι' ὃ ταύτῃ τε καὶ τῇ ἱππακῇ διάγειν φασὶ τοὺς Σκύθας ἡμέρας καὶ ἔνδεκα καὶ δώδεκα.

3 [Ἡ δὲ ἀριστολοχία τῇ ὀσφρήσει μὲν εὔοδμος τῇ δὲ γεύσει πικρὰ σφόδρα τῇ χροιᾷ δὲ μέλαινα. φύεται δὲ ἐν τοῖς ὄρεσιν ἡ βελτίστη· φύλλον δὲ ἔχει προσεμφερὲς τῇ ἀλσίνῃ πλὴν στρογγυλότερον· χρησίμη δὲ πρὸς πολλά, καὶ ἀρίστη πρὸς κεφαλῆς· ἀγαθὴ δὲ καὶ πρὸς τὰ ἄλλα ἕλκη, καὶ πρὸς τὰ ἑρπετὰ καὶ πρὸς ὕπνον καὶ πρὸς ὑστέραν. τὰ μὲν οὖν προσάγειν κελεύουσιν ἐν ὕδατι ἀναδεύσαντα καὶ καταπλάττοντα, τὰ δὲ ἄλλα εἰς μέλι ἐνξύσαντα καὶ ἔλαιον· πρὸς δὲ τὰ τῶν ἑρπετῶν ἐν οἴνῳ ὀξίνῃ πίνειν καὶ ἐπὶ τὸ δῆγμα ἐπιπλάττειν· εἰς ὕπνον δὲ ἐν οἴνῳ μέλανι αὐστηρῷ κνίσαι· ἐὰν δὲ αἱ μῆτραι προπέσωσι, τῷ ὕδατι ἀποκλύζειν.]

---

[1] Diosc. 3. 5; Plin. 25. 82.

[2] γλυκεῖαν : γλυκύρριζαν conj. Dalec., whence 'liquorice.' cf. Plin. 11. 284.

[3] cf. Plin. l.c., who took ἱππακή to be a plant.

Marathon and in parts of Crete: the Boeotians, who eat the fruit, call it *madonaïs*. It has a large leaf which lies on the water: and it is said that it acts as a styptic if it is pounded up and put on the wound: it is also serviceable in the form of a draught for dysentery.

[1] 'Scythian root' (liquorice) is also sweet; some indeed call it simply 'sweet-root.' [2] It is found about Lake Maeotis: it is useful against asthma or a dry cough and in general for troubles in the chest: also, administered in honey, for wounds: also it has the property of quenching thirst, if one holds it in the mouth: wherefore they say that the Scythians, with the help of this and mares' milk cheese [3] can go eleven or twelve days without drinking.

[4] [Birthwort is fragrant to the smell but in taste is very bitter: in colour it is black. The best grows on the mountains: it has a leaf like *alsine*, but rounder: it is useful for many purposes, and is best for sores on the head [5] and other sores, also for bites of reptiles, for inducing sleep and for disorders of the womb.[6] It is directed that it should be applied as a plaster, steeped in water, and for the other purposes should be given shredded into honey and olive-oil: for snake-bites it should be taken in sour wine and also used as a plaster on the bite: to induce sleep it should be scraped up [7] and administered in black dry wine; in cases of *prolapsus uteri* a lotion of it mixed with water should be applied.]

[4] Diosc. 3. 4; Plin. 25. 95. This section is repeated 9. 20. 4. with considerable variations: that seems to be its proper place.

[5] κεφαλῆς conj. W.; κεφαλὴν Ald. *cf.* § 20, κεφαλόθλαστα.

[6] ὑστέραν conj. W., *cf.* below, ἐὰν δὲ αἱ μῆτραι κ.τ.λ. and the duplicate passage § 20; ἕτερα MSS.

κνίσαι conj. W.; κνίσας U*Ald.

4 Αὗται μὲν οὖν γλυκεῖαι. ἄλλαι δὲ πικραί, αἱ δὲ βαρεῖαι τῇ γεύσει. γίνονται δέ τινες τῶν γλυκειῶν αἱ μὲν ἐκστατικαί, καθάπερ ἡ ὁμοία τῷ σκολύμῳ περὶ Τέγεαν, ἣν καὶ Πάνδειος ὁ ἀνδριαντοποιὸς φαγὼν ἐργαζόμενος ἐν τῷ ἱερῷ ἐξέστη. αἱ δὲ θανατηφόροι, καθάπερ ἡ περὶ τὰ μέταλλα ἐν τοῖς ἔργοις τοῖς ἐν Θρᾴκῃ· κούφη δὲ καὶ ἡδεῖα πάνυ τῇ γεύσει καὶ τὸν θάνατον ὑπνώδη τινὰ ποιοῦσα καὶ ἐλαφρόν. ἔχουσι δὲ καὶ τοῖς χρώμασι διαφοράς οὐ τῷ λευκῷ καὶ μέλανι καὶ ξανθῷ μόνον, ἀλλ᾽ ἔνιαι καὶ οἰνοχρῶτες, αἱ δ᾽ ἐρυθραί, καθάπερ ἡ τοῦ ἐρευθεδανοῦ.

5 Ἡ δὲ τοῦ πενταφύλλου ἢ πενταπετοῦς, καλοῦσι γὰρ ἀμφοτέρως, ὀρυττομένη ἐρυθρὰ ξηραινομένη δὲ μέλαινα γίνεται καὶ τετράγωνος· ἔχει δὲ τὸ φύλλον ὥσπερ οἴναρον μικρὸν δὲ καὶ τὴν χροιὰν ὅμοιον· καὶ αὐξάνεται καὶ φθίνει ἅμα τῇ ἀμπέλῳ· πάντα δὲ πέντε τὰ φύλλα, δι᾽ ὃ καὶ ἡ προσηγορία· καυλοὺς δὲ ἐπὶ γῆν ἵησι λεπτοὺς καὶ κνήμας ἔχει.

6 Τὸ δὲ ἐρευθεδανὸν φύλλον ὅμοιον κιττῷ πλὴν στρογγυλότερον· φύεται δ᾽ ἐπὶ γῆς ὥσπερ ἄγρωστις, φιλεῖ δὲ παλίσκια χωρία. οὐρητικὴ δέ, δι᾽ ὃ καὶ χρῶνται πρὸς τὰ τῆς ὀσφύος ἀλγήματα καὶ πρὸς τὰς ἰσχιάδας.

Ἔνιαι δὲ ἰδιόμορφοί τινες, ὥσπερ ἥ τε τοῦ σκορπίου καλουμένου καὶ ἡ τοῦ πολυποδίου. ἡ

---

[1] These words shew that § 3 is out of place.
[2] cf. C.P. 6. 4. 5.
[3] ἡ conj. Sch.; αἱ U*Ald.
[4] ἐν τοῖς ἔργοις τοῖς W. from U*. ? a gloss on μέταλλα. τὰ μετ. τὰ ἐν Θ. Ald. H.
[5] Plin. 25. 139.

ENQUIRY INTO PLANTS, IX. xiii. 4-6

[1] These then are sweet: other roots are bitter, and some unpleasant to the taste. Of those that are sweet [2] there are some that cause mental derangement, as the plant like the golden thistle which grows near Tegea: of this Pandeios the sculptor ate, and went mad while he was working in the temple. Others have fatal effects, as that [3] which grows near the mines in the fields of [4] Thrace: this however is inoffensive and quite sweet to the taste, and the death which it causes is easy and like falling asleep. There are also differences in colour, not merely as to being black or white or yellow, but some are quite wine-coloured and some are red, as the root of madder.

[5] The root of *pentaphyllon* or *pentapetes* (cinquefoil) [6] (for the plant bears both names) is red when it is dug up, but as it dries it becomes black and square: its leaf is like a vine-leaf, and it is small and like it in colour: it grows and fades along with the vine. It only has five leaves in all, whence its name: it sends out long slender stems on the ground, and it has joints.[7]

[8] Madder has a leaf like ivy, but it is rounder: it grows along the ground like dog's-tooth grass and loves shady spots. It has diuretic properties, wherefore it is used for pains in the loins or hip-disease.

Some roots are of peculiar shape, as that of the plant called 'scorpion-plant' (leopard's bane) [9] and that of polypody. For the former is like a scorpion

---

[6] πενταπετοῦς conj. Sch.; πενταπέτου UAld.; πεντεπέτου M U*. cf. Diosc. 4. 42.

[7] καὶ κνήμας ἔχει U*; και κν. ἐ. πυκνάς Ald.; καὶ κνίκας ἔχει πυκνάς UM. cf. πολύκνημος, Diosc. 3. 94. Text probably defective, as nothing is said of the plant's medicinal use.

[8] Diosc. 3. 143; Plin. 19. 47.

[9] cf. 9. 18. 2.

μὲν γὰρ ὁμοία σκορπίῳ καὶ χρησίμη δὲ πρὸς τὴν πληγὴν αὐτοῦ καὶ πρὸς ἄλλ' ἄττα. ἡ δὲ τοῦ πολυποδίου δασεῖα καὶ ἔχουσα κοτυληδόνας, ὥσπερ αἱ τοῦ πολύποδος πλεκτάναι. καθαίρει δὲ κάτω· κἂν περιάψηταί τις οὔ φασιν ἐμφύεσθαι πολύπουν. ἔχει δὲ φύλλον ὅμοιον τῇ πτερίδι τῇ μεγάλῃ καὶ φύεται ἐν ταῖς πέτραις.

XIV. Πασῶν δὲ τῶν ῥιζῶν αἱ μὲν πλείω χρόνον αἱ δὲ ἐλάττω διαμένουσιν. ὁ μὲν γὰρ ἐλλέβορος καὶ τριάκοντα ἔτη χρήσιμος, ἡ δὲ ἀριστολοχία πέντε ἢ ἕξ, χαμαιλέων δὲ ὁ μέλας τετταράκοντα, κενταυρὶς δὲ δέκα ἢ δώδεκα· πίειρα δὲ ἡ ῥίζα καὶ πυκνή· πευκέδανον δὲ πέντε ἢ ἕξ, ἀμπέλου δὲ ἀγρίας ἐνιαυτόν, ἐὰν ἐν σκιᾷ ᾖ καὶ ἄπληκτος, εἰ δὲ μή, σαπρὰ καὶ σομφώδης· ἄλλαι δὲ ἄλλους ἔχουσαι χρόνους. πάντων δὲ ὅλως τῶν φαρμάκων πλεῖστον διαμένει χρόνον τὸ ἐλατήριον, καὶ τὸ παλαιότατον ἄριστον. ἰατρὸς δ' οὖν τις ἔλεγεν οὐκ ἀλαζὼν οὐδὲ ψεύστης ὡς εἴη παρ' αὐτῷ καὶ διακοσίων ἐτῶν θαυμαστὸν δὲ τῇ 2 ἀρετῇ, δοῦναι δὲ αὐτῷ τινα δῶρον. αἰτία δὲ τῆς χρονιότητος ἡ ὑγρότης· διὰ γὰρ ταύτην καὶ ὅταν κόψωσι τιθέασι εἰς τέφραν ὑγρόν, καὶ οὐδ' ὣς γίνεται ξηρόν, ἀλλ' ἄχρι πεντήκοντα ἐτῶν σβέννυσι προσαγόμενον τοὺς λύχνους. φασὶ δὲ μόνον

---

[1] *cf.* the mediaeval doctrine of 'signatures.'
[2] Diosc. 4. 186.
[3] τις οὐ conj. Sch.; τις ὡς Ald.; τίς UM; τίς ὣς U*.
[4] Plin. 27. 14?. [5] *cf.* 9. 8. 7. [6] *cf.* 9. 20. 3.
[7] ἄπληκτος : ? by worms. *cf.* ἄκοπος.

and is also useful[1] against the sting of that creature
and for certain other purposes. [2] The root of polypody
is rough and has suckers like the tentacles of the
polyp. It purges downwards: and, if one wears it
as an amulet, they say that one[3] does not get a
polypus. It has a leaf like the great fern, and it
grows on rocks.

Of the time for which roots can be kept without losing their virtue.

XIV. [4] Some roots keep a longer, some a shorter
time. Hellebore retains its usefulness for as much
as thirty years, birthwort five or six, the black
chamaeleon for forty, feverwort[5] (whose root is thick
and compact) for ten or twelve. Sulphur-wort keeps
five or six years, the root of the 'wild vine'[6] (bryony)
for a year, if it be kept in the shade and not
damaged:[7] otherwise it rots and becomes spongy.[8]
Others keep for various periods. But, to speak
generally, of all plants used as drugs the 'driver'[9]
keeps longest, and, the older it is, the better it is.
At least a certain physician, who was no boaster nor
liar, said that he had some which was 200 years old
and of marvellous virtue, and that it was a present
to him from some one. The cause of its keeping so
long is its moisture: [10] for to secure this, as soon as
they have cut it, they put it among ashes without
drying it, and not even so does it become dry, but
up to fifty years it will put the lamp[11] out if it is
brought near it. And they say that alone of all

[8] σομφώδης conj. Sch.; σογκώδης Ald.H.
[9] A manufactured drug. *cf.* 9. 9. 4.
[10] Diosc. 4. 150; Plin. 20. 5.
[11] λύχνους conj. Sch.: so Vin.Cod.Cas.GPlin. *l.c.*; αὐχμοὺς U*Ald.; χρόνους UM.

ἢ μάλιστα ὑπέρινον ἄνω ποιεῖν τῶν φαρμάκων· αὕτη μὲν οὖν ἰδιότης τις δυνάμεως.

3 Τῶν δὲ ῥιζῶν ὅσαι μὲν γλυκύτητά τινα ἔχουσι ξυμβαίνει θριπηδέστους γίνεσθαι χρονιζομένας, ὅσαι δὲ δριμεῖαι, τοῦτο μὲν μὴ πάσχειν ἀμαυροῦσθαι δ' αὐτῶν τὰς δυνάμεις μανουμένων καὶ κενουμένων. τῶν δ' ἔξω θηρίων ἄλλο μὲν οὐδὲν ἅπτεται ῥίζης δριμείας, ἡ δὲ σφονδύλη πασῶν· τοῦτο μὲν οὖν ἴδιον τῆς τοῦ ζῴου φύσεως.

4 Πᾶσαν δὲ χείρω γίνεσθαι ῥίζαν, ἐὰν ἐάσῃ τις τελειωθῆναι καὶ ἁδρυνθῆναι τὸν καρπόν· ὡσαύτως δὲ καὶ τὸν καρπόν, ἐὰν ὀπίσῃς τὴν ῥίζαν· ὡς ἐπὶ τὸ πολὺ δὲ αἱ φαρμακώδεις οὐκ ὀπίζονται, ὧν δ' ἂν τὰ σπέρματα φαρμακώδη, αὗται δὴ ὀπίζονται· χρῆσθαι δέ τινές φασι μᾶλλον ταῖς ῥίζαις, ὅτι ἰσχυρότερος ὁ καρπὸς ὥσθ' ὑπενεγκεῖν τὸ σῶμα. φαίνεται δὲ οὐ καθ' ὅλου τοῦτο ἀληθές· ἐπεὶ καὶ οἱ ἐν Ἀντικύρᾳ τοῦ σησαμώδους [ἐλλεβόρου] διδόασιν, [ὅτι ὁ καρπὸς ὅμοιος σησάμῳ].

XV. Φαρμακώδεις δὲ δοκοῦσιν εἶναι τόποι μάλιστα τῶν μὲν ἔξω τῆς Ἑλλάδος οἱ περὶ τὴν Τυρρηνίαν καὶ τὴν Λατίνην, ἐν ᾗ καὶ τὴν Κίρκην εἶναι λέγουσιν· καὶ ἔτι μᾶλλόν γε, ὡς Ὅμηρός

---

[1] Plin. 27. 143.
[2] *i.e.* not engendered in the root.
[3] A beetle? *cf.* Arist. *H.A.* 5. 8.
[4] This section is omitted in U*. Plin. 27. 144.

drugs, or to a greater degree than any, it effects a thorough purge upwards: this then is a virtue peculiar to it.

Those roots which contain any sweetness become worm-eaten in course of time, but those that are pungent are not so affected, though their virtues diminish as they become flabby and waste away. [1] No creature coming from without [2] touches a pungent root, but the *sphondyle* [3] attacks them all; this then is a peculiarity of this creature.

[4] Any root, they say, deteriorates if one lets the fruit grow to maturity and ripen: and so in like manner does the fruit, if you drain the root of its juice: and in general roots with medicinal properties do not have the juice of their roots taken, and only those whose seeds are medicinal are thus treated. But some say that they use the roots for choice, because the fruit is too powerful for the human body to be able to bear it. However this does not appear to be true as a universal rule, seeing that the people of Anticyra administer [5] doses of the drug [6] *sesamodes* made from hellebore, which is so called because its fruit is like sesame.

Of the localities which specially produce medicinal herbs.

XV. The places outside Hellas which specially produce medicinal herbs seem to be the parts of Tyrrhenia and Latium (where they say that Circe dwelt), and still more parts of Egypt, as Homer says:

[5] *i.e.* and it is in this case the *fruit* which is used. The drug in question, as well as the plant, was called σησαμοειδές or σησαμοειδής. *cf.* 9. 9. 2 n.; Diosc. 4. 149.

[6] Or (if ἐλλεβόρου is sound) ' of the sesame-like hellebore,' *i.e.* he 'black.' ὅτι ... σησάμῳ I have bracketed, as a gloss on σησαμώδους: ἐλλεβόρου is probably also a gloss.

THEOPHRASTUS

φησι, τὰ περὶ Αἴγυπτον· ἐκεῖθεν γὰρ τὴν Ἑλένην φησὶ λαβεῖν "ἐσθλὰ τά οἱ Πολύδαμνα πόρεν Θῶνος παράκοιτις Αἰγυπτίη· τόθι πλεῖστα φύει ζείδωρος ἄρουρα φάρμακα, πολλὰ μὲν ἐσθλὰ τετυγμένα πολλὰ δὲ λυγρά." ὧν δὴ καὶ τὸ νηπενθὲς ἐκεῖνό φησιν εἶναι καὶ ἄχολον, ὥστε λήθην ποιεῖν καὶ ἀπάθειαν τῶν κακῶν. καὶ σχεδὸν αὗται μὲν ἐοίκασιν ὥσπερ ὑπὸ τῶν ποιητῶν ὑποδεδεῖχθαι. καὶ γὰρ Αἰσχύλος ἐν ταῖς ἐλεγείαις ὡς πολυφάρμακον λέγει τὴν Τυρρηνίαν· "Τυρρηνὸν γενεάν, φαρμακοποιὸν ἔθνος."

2 Οἱ δὲ τόποι πάντες πως φαίνονται μετέχειν τῶν φαρμάκων, ἀλλὰ τῷ μᾶλλον καὶ ἧττον διαφέρειν· καὶ γὰρ οἱ πρὸς ἄρκτον καὶ μεσημβρίαν καὶ οἱ πρὸς ἀνατολὰς ἔχουσι θαυμαστὰς δυνάμεις. ἐν Αἰθιοπίᾳ γὰρ ᾗ τοὺς ὀϊστοὺς χρίουσι ῥίζα τίς ἐστι θανατηφόρος. ἐν δὲ Σκύθαις αὕτη τε καὶ ἕτεραι πλείους, αἱ μὲν παραχρῆμα ἀπαλλάττουσαι τοὺς προσενεγκαμένους, αἱ δ' ἐν χρόνοις αἱ μὲν ἐλάττοσιν αἱ δ' ἐν πλείοσιν, ὥστ' ἐνίους καταφθίνειν. ἐν Ἰνδοῖς δὲ καὶ ἕτερα γένη πλείω, περιττότατα δέ, εἴπερ ἀληθῆ λέγουσιν, ἥ τε δυναμένη τὸ αἷμα διαχεῖν καὶ οἷον ὑποφεύγειν, καὶ πάλιν ἡ συνάγουσα καὶ πρὸς ἑαυτὴν ἐπισπωμένη, ἃ δή φασιν εὑρῆσθαι πρὸς τὰ τῶν ὀφιδίων τῶν θανατηφόρων δήγματα.

3 Περὶ δὲ τὴν Θρᾴκην εἶναι μὲν καὶ ἑτέρας οὐκ ὀλίγας, ἰσχυροτάτην δὲ ὡς εἰπεῖν τὴν ἴσχαιμον, ἣν δὴ λέγουσιν οἱ μὲν κεντηθείσης τῆς φλεβὸς

---

[1] *Od.* 4. 221 foll.
[2] ὧν δὴ conj. Sch.; ωἱ δὴ U*; ἐν οἷς δὴ PAld.

ENQUIRY INTO PLANTS, IX. xv. 1-3

for thence he says[1] that Helen brought "things of virtue which Polydamna, the Egyptian wife of Thon, gave her; there the grain-bearing earth produces most drugs, many that are good, and many baneful." Among these[2] he says was *nepenthes*, the famous drug which cures sorrow and passion, so that it causes forgetfulness and indifference to ills. So these lands seem to have been pointed out, as it were, by the poets. For Aeschylus too in his elegies speaks of Tyrrhenia as rich in drugs, for he tells of the "Tyrrhenian stock, a nation that makes drugs."

It seems that almost all places take their share in producing drugs, but that they differ in the extent to which they do so; for the regions of the North, South, and East have herbs of marvellous virtue. Thus in Ethiopia there is a certain deadly root[3] with which they smear their arrows. And in Scythia there is this and there are also others, some of which kill at once those who eat them, some after an interval, shorter or longer, so that in the latter case men have a lingering death. In India there are many other kinds,[4] but the most extraordinary,[5] if they tell the truth, are these: there is one which has the power to make the blood disperse and as it were to put it to flight,[6] and another which collects it and draws it to itself; these they say were discovered as remedies for the bites of deadly serpents.

In Thrace it is said there are fairly numerous other kinds, but that about the most powerful is 'blood-stancher,'[7] which stops and prevents the flow

---

[3] Somali arrow-poison. Index App. (27).
[4] γένη conj. Dalec.; μέρη Ald.
[5] περιττότατα conj. W.; περιττοτάτη Ald.
[6] ? add ποιεῖν after ὑποφεύγειν.   [7] Plin. 25. 83.

οἱ δὲ καὶ σφοδροτέρως διατμηθείσης ἴσχειν καὶ κωλύειν τὴν χύσιν. [ταῦτα μὲν οὖν, ὥσπερ εἴπομεν, ἔοικε δηλοῦν τὸ κοινόν.] τῶν μὲν οὖν ἔξω τόπων οἱ φαρμακωδέστατοι οὗτοι.

4 Τῶν δὲ περὶ τὴν Ἑλλάδα τόπων φαρμακωδέστατον τό τε Πήλιον τὸ ἐν Θετταλίᾳ καὶ τὸ Τελέθριον τὸ ἐν Εὐβοίᾳ καὶ ὁ Παρνασός, ἔτι δὲ καὶ ἡ Ἀρκαδία καὶ ἡ Λακωνική· καὶ γὰρ αὗται φαρμακώδεις ἀμφότεραι· δι' ὃ καὶ οἵ γε Ἀρκάδες εἰώθασιν ἀντὶ τοῦ φαρμακοποτεῖν γαλακτοποτεῖν περὶ τὸ ἔαρ, ὅταν οἱ ὀποὶ μάλιστα τῶν τοιούτων φύλλων ἀκμάζωσι· τότε γὰρ φαρμακωδέστατον τὸ γάλα· πίνουσι δὲ βόειον· δοκεῖ γὰρ πολυνομώτατον καὶ παμφαγώτατον εἶναι πάντων ὁ βοῦς.

5 Φύεται δὲ παρ' αὐτοῖς ὅ τε ἐλλέβορος ἀμφότερος καὶ ὁ λευκὸς καὶ ὁ μέλας· ἔτι δὲ δαῦκον δαφνοειδὲς κροκόεν, καὶ ἣν ἐκεῖνοι μὲν ῥάφανον ἀγρίαν καλοῦσι τῶν δ' ἰατρῶν τινες κεράϊν, καὶ ἣν οἱ μὲν ἀλθαίαν ἐκεῖνοι δὲ μαλάχην ἀγρίαν, καὶ ἡ ἀριστολοχία καὶ τὸ σέσελι καὶ τὸ ἱπποσέλινον καὶ τὸ πευκέδανον καὶ ἡ Ἡράκλεια καὶ ὁ στρύχνος ἀμφότερος ὅ τε φοινικοῦν ἔχων τὸν καρπὸν καὶ ὁ μέλανα.

6 Φύεται δὲ καὶ ὁ σίκυος ὁ ἄγριος, ἐξ οὗ τὸ ἐλατήριον συντίθεται· καὶ ὁ τιθύμαλλος, ἐξ οὗ τὸ ἱπποφαές· ἄριστον δὲ τοῦτο περὶ Τεγέαν κἀκεῖνο μάλιστα σπουδάζεται· φύεται δ' ἐκεῖ ἐπὶ πλέον·

---

[1] I omit ταῦτα . . . κοινόν as apparently out of place and a duplicate of the last sentence of § 8.
[2] Plin. 25. 94; cf. 4. 5. 2.   [3] Plin. 25. 110.

ENQUIRY INTO PLANTS, IX. xv. 3-6

of blood, some say if the vein is merely pricked, others even if it is deeply cut into.[1] These then of the places outside Hellas are those that are most productive of drugs.

[2] Of places in Hellas those most productive of drugs are Pelion in Thessaly, Telethrion in Euboea, Parnassus, and also Arcadia and Laconia, for both these states produce medicinal herbs; wherefore the Arcadians are accustomed, instead of drinking medicine, to drink milk in spring when the juices of such plants are at their best, for then the milk has most medicinal virtue. It is cows' milk that they drink, since it appears that the cow eats more than any other animal and is more impartial as to what she eats.

[3] Both kinds of hellebore, the white and the black, grow in their country, and also carrot,[4] a saffron-coloured plant like bay, and a plant which the Arcadians call 'wild cabbage'[5] (spurge) but some physicians *keraïs*; also a plant called by some marsh mallow,[6] also birthwort hartwort alexanders sulphurwort *Herakleia*, and both kinds of *strykhnos*,[7] that which has a scarlet and that which has a black fruit.

There also grow there the 'wild cucumber' (squirting cucumber), of which the drug 'driver'[8] is compounded, and the *tithymallos* (spurge) of which *hippophaës*[9] is made; this is best about Tegea, and that kind is much sought after; it grows there in

[4] δαῦκον. This name recurs §8 and 9. 20. 2. Text must be defective here: the epithets are unintelligible, and perhaps belong to another plant whose name has dropped out. See Index.

[5] *cf.* 9. 12. 1. and Index.

[6] ἀλθαίαν conj. Sch., *cf.* 9. 18. 1; ἀλθέαν Ald. *cf.* Plin. 20. 222.

[7] *cf.* 9. 11. 5.   [8] *cf.* 9. 9. 4; 9. 14. 1.

[9] ἱπποφάες is elsewhere the name of a plant: *cf.* Diosc. 4. 159. ἐξ οὗ may be corrupt, or the text defective.

293

πλεῖστον δὲ καὶ κάλλιστον φύεται περὶ τὴν Κλειτορίαν.

7 Ἡ δὲ πανάκεια γίνεται κατὰ τὸ πετραῖον περὶ Ψωφίδα καὶ πλείστη καὶ ἀρίστη. τὸ δὲ μῶλυ περὶ Φενεὸν καὶ ἐν τῇ Κυλλήνῃ. φασὶ δ' εἶναι καὶ ὅμοιον ᾧ ὁ Ὅμηρος εἴρηκε, τὴν μὲν ῥίζαν ἔχον στρογγύλην προσεμφερῆ κρομύῳ τὸ δὲ φύλλον ὅμοιον σκίλλῃ· χρῆσθαι δὲ αὐτῷ πρός τε τὰ ἀλεξιφάρμακα καὶ τὰς μαγείας· οὐ μὴν ὀρύττειν γ' εἶναι χαλεπόν, ὡς Ὅμηρός φησι.

8 Τὸ δὲ κώνειον ἄριστον περὶ Σοῦσα καὶ ἐν τοῖς ψυχροτάτοις τόποις. γίνεται δὲ καὶ ἐν τῇ Λακωνικῇ τὰ πολλὰ τούτων· καὶ γὰρ αὕτη πολυφάρμακος. ἐν Ἀχαΐᾳ δὲ ἥ τε τραγάκανθα πολλὴ καὶ οὐδὲν χείρων ὡς οἴονται τῆς Κρητικῆς ἀλλὰ καὶ τῇ ὄψει καλλίων· καὶ δαῦκον περὶ τὴν Πατραϊκὴν διαφέρον· τοῦτο δὲ θερμαντικὸν φύσει, ῥίζαν δὲ ἔχει μέλαιναν. φύεται δὲ τὰ πολλὰ τούτων καὶ ἐν τῷ Παρνασῷ καὶ περὶ τὸ Τελέθριον. καὶ ταῦτα μὲν κοινὰ πλειόνων χωρῶν.

XVI. Τὸ δὲ δίκταμνον ἴδιον τῆς Κρήτης, θαυμαστὸν δὲ τῇ δυνάμει καὶ πρὸς πλείω χρήσιμον μάλιστα δὲ πρὸς τοὺς τόκους τῶν γυναικῶν. ἔστι δὲ τὸ μὲν φύλλον παρόμοιον τῇ βληχοῖ, ἔχει δέ τι καὶ κατὰ τὸν χυλὸν ἐμφερὲς τὰ δὲ κλωνία λεπτότερα. χρῶνται δὲ τοῖς φύλλοις, οὐ τοῖς κλωσὶν οὐδὲ τῷ καρπῷ· χρήσιμον δὲ πρὸς πολλὰ μὲν καὶ ἄλλα, μάλιστα δέ, ὥσπερ ἐλέχθη, πρὸς

---

[1] Plin. 25. 30–33.   [2] κατὰ conj. St.; καὶ Ald.H.
[3] Plin. 25. 151.
[4] Σοῦσα: cf. 9. 16. 8; Λοῦσα (a town in Arcadia) conj. Sch. (usually Λοῦσοι), the other places mentioned being all in

considerable abundance, but in greatest abundance and best about Kleitoria.

[1] All-heal grows in great abundance and best in [2] the rocky ground about Psophis, moly about Pheneos and on Mount Kyllene. They say that this plant is like the moly mentioned by Homer, that it has a round root like an onion and a leaf like squill, and that it is used against spells and magic arts, but that it is not, as Homer says, difficult to dig up.

[3] Hemlock is best about Susa[4] and in the coldest spots. Most of these plants occur also in Laconia, for this too is a land rich in medicinal herbs. In Achaia tragacanth[5] is abundant and is as good as that of Crete, it is believed, and even fairer in appearance. *Daukon*[6] again is excellent in the country about Patrai[7]; this is by nature healing, and it has a black root. Most of these grow also on Mount Parnassus and about Telethrion. So these plants are common to several lands.

Of the medicinal herbs peculiar to Crete.

XVI. [8] But dittany is peculiar to Crete. This plant is marvellous in virtue and is useful for many purposes, but especially for women in child-birth. Its leaf is like pennyroyal, to which it also bears some resemblance in taste; but the twigs are slenderer. They use the leaves, not the twigs nor the fruit: and the leaf is useful for many other purposes, but above all,

Hellas. But Plin. 25. 154 has Susa: it can hardly be the Persian town.

[5] Plin. 13. 115.

[6] Repeated 9. 20. 2; *cf.* 9. 15. 5 and Index.

[7] Πατραϊκὴν conj. Sch., *cf.* 9. 20. 2; πατρικὴν Ald.; σπαρτιακὴν U*; σπαρτικὴν MP; *Patrensi agro* G.

[8] Plin. 25. 92.

τὰς δυστοκίας τῶν γυναικῶν· ἢ γὰρ εὐτοκεῖν φασι ποιεῖν ἢ παύειν γε τοὺς πόνους ὁμολογουμένως· δίδοται δὲ πίνειν ἐν ὕδατι. σπάνιον δέ ἐστι· καὶ γὰρ ὀλίγος ὁ τόπος ὁ φέρων, καὶ τοῦτον αἱ αἶγες ἐκνέμονται διὰ τὸ φιληδεῖν. ἀληθὲς δέ φασιν εἶναι καὶ τὸ περὶ τῶν βελῶν, ὅτι φαγούσαις ὅταν τοξευθῶσι ἐκβάλλει.[1] τὸ μὲν οὖν δίκταμνον τοιοῦτόν τε καὶ τοιαύτας ἔχει τὰς δυνάμεις.

2 Τὸ δὲ ψευδοδίκταμνον τῷ μὲν φύλλῳ ὅμοιον τοῖς κλωνίοις δ' ἔλαττον τῇ δυνάμει δὲ πολὺ λειπόμενον. βοηθεῖ μὲν γὰρ καὶ ταῦτά, χεῖρον δὲ πολλῷ καὶ ἀσθενέστερον. ἔστι δὲ εὐθὺς ἐν τῷ στόματι φανερὰ τοῦ δικτάμνου ἡ δύναμις· διαθερμαίνει γὰρ ἀπὸ μικροῦ σφόδρα. τιθέασι δὲ τὰς δεσμίδας ἐν νάρθηκι ἢ[3] καλάμῳ πρὸς τὸ μὴ ἀποπνεῖν· ἀσθενέστερον γὰρ ἀποπνεῦσαν. λέγουσι δέ τινες ὡς ἡ μὲν φύσις μία ἡ τοῦ δικτάμνου καὶ ἡ τοῦ ψευδοδικτάμνου, διὰ δὲ τὸ ἐν εὐγειοτέροις φύεσθαι τόποις χεῖρον γίνεσθαι, καθάπερ καὶ ἄλλα πολλὰ πλείω τούτων κατὰ τὰς δυνάμεις. τὸ γὰρ δίκταμνον φιλεῖ χώραν τραχεῖαν.

3 Ἔστι δὲ καὶ ἕτερον δίκταμνον ὥσπερ ὁμώνυμον, οὔτε τὴν ὄψιν οὔτε τὴν δύναμιν ἔχον τὴν αὐτήν· φύλλον γὰρ ἔχει ὅμοιον σισυμβρίῳ τοὺς δὲ κλώνας μείζους· ἔτι δὲ τὴν χρείαν καὶ τὴν δύναμιν οὐκ ἐν τοῖς αὐτοῖς. τοῦτο μὲν οὖν, ὥσπερ ἐλέχθη, θαυμαστὸν ἅμα καὶ ἴδιον τῆς νήσου. φασὶ δέ

---

[1] ἐκβάλλει conj. Sch.; ἐκβάλλειν Ald.
[2] Plin. 25. 93.
[3] νάρθηκι ἢ conj. Sch.; ναρθηκίδη ἢ U; ναρθηκίδι ἢ M; νάρθηκι καὶ Ald.

as was said, against difficult labour in women; for it is said that either it makes labour quite easy or at least it confessedly makes the pains to cease: it is given as a draught in water. It is a scarce plant: for the region which bears it is not extensive, and the goats graze it down because they are fond of it. The story of the arrows is also said to be true,—that, if goats eat it when they have been shot, it rids them [1] of the arrow. Such then is dittany and such its properties.

[2] 'False dittany' is like it in leaf, but has smaller twigs, and in virtue is far inferior. For it is of service in the same ways, but is feebler and not nearly so powerful. The virtue of dittany is perceived directly it is taken into the mouth: for a small piece of it has a very warming effect. The bunches of it are put in the hollow stem of ferula [3] or a reed, so that it may not exhale its virtue: for, if it does so, it is less effective. Some say that dittany and 'false dittany' are essentially the same plant, but that the latter is an inferior form produced by growing in places with richer soil; just as many other things [4] become inferior in their properties for the same cause. For dittany loves rough ground.

[5] There is also another plant called 'dittany,' though it has nothing in common with these except the name. This has neither the same appearance nor the same virtue; for its leaf is like bergamot-mint and its twigs are larger, and further its use and virtue are differently shewn. The true plant is, as was said, marvellous, and is also peculiar to the island of Crete. Indeed some say that the plants

---

[4] πλείω τούτων Ald., probably a duplicate of ἄλλα πολλὰ; not represented in G; ἀλλοιοῦται conj. W. [5] Plin.25. 94.

τινες ὅλως τῶν φύλλων καὶ τῶν ὀροδάμνων καὶ ἁπλῶς τῶν ὑπὲρ γῆς τὰ ἐν Κρήτῃ διαφέρειν, τῶν δὲ ἄλλων τῶν γε πλείστων τὰ ἐν τῷ Παρνασῷ.

4 Τὸ δ' ἀκόνιτον γίνεται μὲν καὶ ἐν Κρήτῃ καὶ ἐν Ζακύνθῳ, πλεῖστον δὲ καὶ ἄριστον ἐν Ἡρακλείᾳ τῇ ἐν Πόντῳ. ἔχει δὲ φύλλον μὲν κιχοριῶδες, ῥίζαν δὲ ὁμοίαν τῷ σχήματι καὶ τῷ χρώματι καρίδι, τὴν δὲ δύναμιν τὴν θανατηφόρον ἐν ταύτῃ· τὸ δὲ φύλλον καὶ τὸν καρπὸν οὐθέν φασι ποιεῖν· καρπὸς δέ ἐστι πόας οὐχ ὑλήματος. βραχεῖα δὲ ἡ πόα καὶ οὐδὲν ἔχουσα περιττόν, ἀλλὰ παρομοία τῷ σίτῳ· τὸ δὲ σπέρμα οὐ σταχυηρόν. φύεται δὲ πανταχοῦ καὶ οὐκ ἐν ταῖς Ἀκόναις μόνον, ἀφ' ὧν ἔχει τὴν προσηγορίαν· αὕτη δέ ἐστι κώμη τις τῶν Μαριανδυνῶν· φιλεῖ δὲ μάλιστα τοὺς πετρώδεις τόπους· οὐ νέμεται δὲ οὔτε πρόβατον οὔτ' ἄλλο
5 ζῶον οὐδέν. συντίθεσθαι δὲ τρόπον τινὰ πρὸς τὸ ἐργάζεσθαι καὶ οὐ παντὸς εἶναι· δι' ὃ καὶ τοὺς ἰατροὺς οὐκ ἐπισταμένους συντιθέναι σηπτικῷ τε χρῆσθαι καὶ πρὸς ἄλλα ἄττα· πινόμενον δ' οὐδεμίαν αἴσθησιν ποιεῖν οὔτ' ἐν οἴνῳ οὔτ' ἐν μελικράτῳ· συντίθεσθαι δὲ ὥστε κατὰ χρόνους τακτοὺς ἀναιρεῖν, οἷον δίμηνον τρίμηνον ἑξάμηνον ἐνιαυτόν, τοὺς δὲ καὶ δύο ἔτη· χείριστα δὲ ἀπαλ-

---

[1] ὀροδάμνων: this word seems to occur only here in T.
[2] Diosc. 4. 76 and 77; Plin. 27. 9 and 10.
[3] καρίδι conj. W.; καρίαι U; καρύα Ald. cf. Diosc. l.c.
[4] cf. 9. 8. 1.
[5] Plin. 6. 4, portus Acone veneno aconito dirus. But in 27. 10. he apparently did not recognise Ἀκόναις as a proper name,

of Crete are superior in leaves boughs[1] and in general all the parts above ground to those of other places; while those of Parnassus are superior to most of those found elsewhere.

Of wolf's-bane and its habitat, and of meadow-saffron.

[2] Wolf's-bane grows in Crete and in Zakynthos, but is most abundant and best at Herakleia in Pontus. It has a leaf like chicory, a root like in shape and colour to a prawn,[3] and in this root resides its deadly property, whereas they say that the leaf and the fruit produce no effects. The fruit is that of a herb,[4] not that of a shrub or tree. It is a low-growing herb and shows no special feature, but is like corn, except that the seed is not in an ear. It grows everywhere and not only at Akonai,[5] from whence it gets its name (this is a village of the Mariandynoi)[6]: and it specially likes rocky ground. Neither sheep nor any other animals eat it.[7] In order to be effective it is said that it must be compounded in a certain manner, and that not everyone can do this: and so that physicians, not knowing how to compound it, use it as a septic and for other purposes: and [8] that, if drunk mixed in wine or a honey-posset, it produces no sensation: but that it can be so compounded as to prove fatal at a certain moment which may be in two three or six months, or in a year, or even in two years: and that the

and translates it *in nudis cautibus*, misled perhaps by τοὺς πετρώδεις τόπους below.

[6] Μαριανδυνῶν conj. Meurs.; περιανδύνων U*Ald.H.

[7] U* adds here βοηθεῖαι δὲ τοῖς ἐνεγκαμένοις εἰσί and omits §§ 5, 6 ... εἰδέναι, continuing πολλάκις γὰρ φασὶ τὰ ἀνδράποδα.

[8] δὲ add. Sch.

λάττειν τοὺς ἐν πλείστῳ χρόνῳ καταφθίνοντος τοῦ σώματος, ῥᾷστα δὲ τοὺς παραχρῆμα. λυτικὸν δὲ φάρμακον οὐχ εὑρῆσθαι, καθάπερ ἀκούομεν ἑτέρων τι φύεσθαι. ἀλλὰ τοὺς ἐγχωρίους ἀνασώζειν τινὰς μέλιτι καὶ οἴνῳ καὶ τοιούτοις τισί, σπανίως δὲ καὶ τούτους καὶ ἐργωδῶς.

6 Ἀλλὰ τοῦ ἐφημέρου τὸ φάρμακον εὑρῆσθαι· ἕτερον γάρ τι ῥίζιον εἶναι ὃ ἐφήμερον ἀπαλλάττει· τοῦτο δὲ φύλλον ὅμοιον ἔχειν τῷ ἐλλεβόρῳ ἢ τῷ λειρίῳ· καὶ τοῦτο πάντας εἰδέναι· δι' ὃ καὶ τὰ ἀνδράποδά φασι πολλάκις παροργισθέντα χρῆσθαι, κἄπειτα ἰατρεύειν αὐτὰ πρὸς τοῦτο ὁρμῶντα, καὶ γὰρ οὐδὲ ταχεῖαν ποιεῖσθαι τὴν ἀπαλλαγὴν οὐδὲ ἐλαφρὰν ἀλλὰ δυσχερῆ καὶ χρόνιον· εἰ μὴ ἄρα διὰ τὸ εὐθεράπευτον εἶναι καὶ ἀκατασκεύαστον ὡς δεῖ. φασὶ γοῦν καὶ παραχρῆμα ἀπαλλάττεσθαι καὶ ὕστερον χρόνῳ τοὺς δὲ καὶ εἰς ἐνιαυτὸν ἄγειν, καὶ τὰς δόσεις ἀβοηθήτους εἶναι.

---

[1] *i.e.* no *herb* having that effect.
[2] ἑτέρων conj. Sch.; ἕτερόν τι φύεσθαι UAld.H.; ἕτερόν τι φυόμενον conj. W. G seems to have had a fuller text.
[3] ἀλλὰ τοὺς ἐγχ. UM; ἀλλά τινες τοῖς ἐνεγκαμένοις βοήθειαι εὕρηνται· τοὺς γὰρ ἐγχ. Ald.H., which the indicative εὕρηνται shews to be a gloss.
[4] τοῦ ἐφημέρου U; τὰ ἐφ' ἡμέρον M; καὶ τὸ ἐφήμερον Ald. The passage about ἐφήμερον, which interrupts the account of ἀκόνιτον, is confused, and the text probably defective; translation a makeshift. The sense of εἰ μὴ ... ὡς δεῖ being perhaps irrecoverable, the connexion of what follows is obscured. W. gives up the passage.

longer the time the more painful the death, since the body then wastes away, while, if it acts at once, death is quite painless. And it is said that no antidote [1] which can counteract it has been discovered, like the natural antidotes to other [2] poisonous herbs of which we are told: though the country-folk [3] can sometimes save a man with honey and wine and such like things, only however occasionally and with difficulty.

(On the other hand they say that for meadow-saffron [4] the antidote has been found: for that there is another root which counteracts that herb: [5] and that it [6] has a leaf like hellebore [7] or the madonna lily: [8] and that this [9] is generally known. Wherefore they say that slaves often take meadow-saffron when greatly provoked, and then themselves have recourse [10] to the antidote and effect a cure,—seeing that the poison does not cause a speedy and easy death, but [11] one that is lingering and slow,—unless indeed, merely because the cure is so easy,[12] the antidote has not been properly prepared.[13] At least they say that though death may ensue at once, sometimes it only occurs after a considerable interval, which in some cases extends to a year, and that in these latter cases the dose given has incurable effects: and that

[5] ὁ ἐφήμερον PH.; ὁ ἐφημεραῖον U; ὁ ἐφ' ἡμεραῖον M; ὁ οὐκ ἐφήμερον Ald.

[6] τοῦτο δὲ Ald.; τόνδε δὲ καὶ U; τῶνδε δὲ καὶ M.

[7] *i.e.* the 'black': see Index.

[8] λειρίῳ conj. Guilandinus from Diosc. 4. 84 (κρίνῳ); αἰρίῳ Ald.H. [9] τοῦτο Ald.; τοῦτο μὲν UM.

[10] After ὁρμῶντα UM add καὶ τοὺς οἰκέτας ἐπὶ τοῦτο ὁρμᾶν and omit καὶ γὰρ ... θανατηφόρων.   [11] ἀλλὰ Ald.; οὐδὲ U*.

[12] εὐθεράπευτον Ald.; ἀθεράπευτον U*P.

[13] In which case apparently the slave outwits himself as well as his master by 'dying on him.'

ταῦτα δὲ ἐξακριβωθῆναι μάλιστα παρὰ τοῖς Τυρ-
7 ρηνοῖς τοῖς ἐν Ἡρακλείᾳ. τοῦτο μὲν <οὖν> οὐδὲν
ἄτοπον, εἰ τρόπον μέν τινα ἀβοήθητον ἄλλως δὲ
βοηθήσιμον, ὥσπερ καὶ ἕτερα τῶν θανατηφόρων.

Τὸ δὲ ἀκόνιτον ἄχρηστον, ὥσπερ εἴρηται, τοῖς
μὴ ἐπισταμένοις· οὐδὲ κεκτῆσθαι δὲ ἐξεῖναι, ἀλλὰ
θάνατον τὴν ζημίαν· τὴν δὲ τῶν χρόνων διαφορὰν
ἀκολουθεῖν κατὰ τὰς συλλογάς· ἰσοχρόνους γὰρ
τοὺς θανάτους γίνεσθαι τοῖς ἀπὸ τῆς συλλογῆς
χρόνοις.

8 Θρασύας δ' ὁ Μαντινεὺς εὑρήκει τι τοιοῦτον,
ὥσπερ ἔλεγεν, ὥστε ῥᾳδίαν ποιεῖν καὶ ἄπονον τὴν
ἀπόλυσιν τοῖς ὁποῖς χρώμενος κωνείου τε καὶ
μήκωνος καὶ ἑτέρων τοιούτων, ὥστε εὔογκον εἶναι
σφόδρα καὶ μικρὸν ὅσον εἰς δραχμῆς ὁλκήν.
ἀβοήθητον δὲ πάντῃ καὶ δυνάμενον διαμένειν
ὁποσονοῦν χρόνον καὶ οὐδὲν ἀλλοιούμενον. ἐλάμ-
βανε δὲ τὸ κώνειον οὐχ ὅθεν ἐτύγχανεν ἀλλ' ἐκ
Σούσων καὶ εἴ τις ἄλλος τόπος ψυχρὸς καὶ πα-
λίσκιος· ὡσαύτως δὲ καὶ τἆλλα. συνετίθει δὲ
καὶ ἕτερα φάρμακα πολλὰ καὶ ἐκ πολλῶν. δεινὸς
δὲ καὶ Ἀλεξίας ὁ μαθητὴς αὐτοῦ καὶ οὐχ ἧττον
ἔντεχνος ἐκείνου· καὶ γὰρ τῆς ἄλλης ἰατρικῆς
ἔμπειρος.

---

[1] οὖν add. W.
[2] ἀκολουθεῖν κατὰ conj. W.; ἀκούειν εἶναι κατὰ Ald. H.; ἀκουεῖν εἶναι καὶ M.

these facts have been most carefully ascertained among the Tyrrhenians of Herakleia. Now [1] it is not surprising that in some circumstances the effects of the poison should be incurable, and yet in others that a cure should be possible, this being also the case with other deadly poisons.)

To return—wolf's bane, as has been said, is useless to those who do not understand it; in fact it is said that it is not lawful even to have it in one's possession, under pain of death; also that the length of time which it takes to produce its effects depends on [2] the time when it is gathered; for that the time which it takes to kill is equal to that which has elapsed since it was gathered.

Of two famous druggists and of the virtues of hemlock.

Thrasyas of Mantineia had discovered, as he said, a poison which produces an easy and painless end; he used the juices of hemlock poppy and other such herbs, so compounded as to make a dose of conveniently small size, weighing only somewhat less than a quarter of an ounce. For the effects of this compound there is absolutely no cure, and it will keep any length of time without losing its virtue at all. He used to gather his hemlock, not just anywhere, but at Susa [3] or some other cold and shady spot; and so too with the other ingredients; he also used to compound many other poisons, using many ingredients. His pupil Alexias was also clever and no less skilful than his master, being also versed in the science of medicine generally.

[3] Σούσων MSS.; Λούσων conj. Sch. *cf.* 9. 15. 8 n. The mention of Mantineia makes it likely that a place in Arcadia is intended.

9 Ταῦτα μὲν οὖν εὑρῆσθαι δοκεῖ πολλῷ μᾶλλον νῦν ἢ πρότερον. ὅτι δὲ διαφέρει τὸ χρῆσθαί πως ἑκάστῳ φανερὸν ἐκ πολλῶν· ἐπεὶ καὶ Κεῖοι τῷ κωνείῳ πρότερον οὐχ οὕτω ἀλλὰ τρίβοντες ἐχρῶντο, καθάπερ οἱ ἄλλοι· νῦν δ' οὐδ' ἂν εἷς τρίψειεν, ἀλλὰ περιπτίσαντες καὶ ἀφελόντες τὸ κέλυφος, τοῦτο γὰρ τὸ τὴν δυσχέρειαν παρέχον δυσκατέργαστον¹ ὄν, μετὰ ταῦτα κόπτουσιν ἐν τῷ ὅλμῳ καὶ διαττήσαντες² λεπτὰ ἐπιπάττοντες ἐφ' ὕδωρ πίνουσιν, ὥστε ταχεῖαν καὶ ἐλαφρὰν γίνεσθαι τὴν ἀπαλλαγήν.

XVII. Ἁπάντων δὲ τῶν φαρμάκων αἱ δυνάμεις ἀσθενέστεραι τοῖς συνειθισμένοις τοῖς δὲ καὶ ἀνενεργεῖς τὸ ὅλον. ἔνιοι γὰρ ἐλλέβορον ἐσθίοντες πολὺν ὥστε ἀναλίσκειν δέσμας ὅλας οὐδὲν πάσχουσιν· ὅπερ ἐποίει καὶ Θρασύας δεινότατος ὢν ὡς ἐδόκει περὶ τὰς ῥίζας. ποιοῦσι δὲ τοῦθ' ὡς ἔοικε καὶ τῶν νομέων τινές· δι' ὃ καὶ πρὸς τὸν φαρμακοπώλην τὸν θαυμαζόμενον ὡς κατῆσθιε ῥίζαν μίαν ἢ δύο παραγενόμενος ὁ ποιμὴν καὶ ἀναλώσας ὅλην τὴν δέσμην ἐποίησεν ἀδόκιμον· ἐλέγετο δ' ὅτι καθ' ἑκάστην ἡμέραν τοῦτο ποιεῖ καὶ αὐτὸς καὶ ἕτεροι.

2 Κινδυνεύει γὰρ ἔνια τῶν φαρμάκων τῇ ἀσυνηθείᾳ φάρμακα γίνεσθαι, τάχα δὲ ἀληθέστερον

---

¹ δυσκατέργαστον: cf. C.P. 1 14. 4.
² διαττήσαντες conj. Hoffmann from G; διαπτήσαντες Ald.H.; διηθήσαντες U*mP.

ENQUIRY INTO PLANTS, IX. XVI. 9–XVII. 2

Now these things seem to have been ascertained far better in recent than in former times. And many things go to shew that the method of using the various drugs makes a difference; thus the people of Ceos formerly did not use hemlock in the way described, but just shredded it up for use, as did other people; but now not one of them would think of shredding it, but they first strip off the outside and take off the husk, since this is what causes the difficulty, as it is not easily assimilated [1]; then they bruise it in the mortar, and, after putting it through a fine sieve,[2] sprinkle it on water and so drink it; and then death is made swift and easy.

How use diminishes the efficacy of drugs, and how they have not the same effect on all constitutions.

XVII. [3] The virtues of all drugs become weaker to those who are accustomed to them, and in some cases become entirely ineffective. Thus some eat enough hellebore to consume whole bundles and yet suffer no hurt; this is what Thrasyas did, who, as it appeared, was very cunning in the use of herbs. And it appears that shepherds sometimes do the like; wherefore the shepherd who came before the vendor of drugs (at whom men marvelled because he ate one or two roots) and himself consumed the whole bundle, destroyed the vendor's reputation: it was said that both this man and others did this every day.

For it seems that some poisons become poisonous because they are unfamiliar, or perhaps it is a more accurate way of putting it to say that familiarity makes

[3] Plin. 27. 144.

εἰπεῖν ὡς τῇ συνηθείᾳ οὐ φάρμακα· προσδεξαμένης γὰρ τῆς φύσεως καὶ κατακρατούσης οὐκέτι φάρμακα, καθάπερ καὶ Θρασύας ἔλεγεν· ἐκεῖνος γὰρ ἔφη τὸ αὐτὸ τοῖς μὲν φάρμακον εἶναι τοῖς δ' οὐ φάρμακον, διαιρῶν τὰς φύσεις ἑκάστων· ᾤετο γὰρ δεῖν καὶ ἦν δεινὸς διαγνῶναι. ποιεῖ δέ τι δῆλον ὅτι πρὸς τῇ φύσει καὶ τὸ ἔθος. Εὔδημος γοῦν ὁ φαρμακοπώλης εὐδοκιμῶν σφόδρα κατὰ τὴν τέχνην συνθέμενος μηδὲν πείσεσθαι πρὸ ἡλίου δύναντος κατέφαγε μέτριον πάνυ καὶ οὐ
3 κατέσχεν οὐδ' ἐκράτησεν. ὁ δὲ Χῖος Εὔδημος πίνων ἐλλέβορον οὐκ ἐκαθαίρετο. καί ποτε ἔφη πιεῖν ἐν μιᾷ ἡμέρᾳ δύο καὶ εἴκοσι πόσεις ἐν τῇ ἀγορᾷ καθήμενος ἐπὶ τῶν σκευῶν καὶ οὐκ ἐξαναστῆναι πρὸ τοῦ δείλην γενέσθαι· τότε δ' ἐλθὼν καὶ λούσασθαι καὶ δειπνεῖν ὥσπερ εἰώθει καὶ οὐκ ἐξεμέσαι· πλὴν οὗτός γε βοήθειάν τινα παρασκευασάμενος κατέσχε· κίσσηριν γὰρ ἐπιπάττων ἐπ' ὄξος δριμὺ πιεῖν ἔφη μετὰ τὴν ἑβδόμην πόσιν, καὶ πάλιν ὕστερον ἐν οἴνῳ τὸν αὐτὸν τρόπον· τὴν δὲ τῆς κισσήριδος οὕτως ἰσχυρὰν εἶναι δύναμιν ὥστ' ἐάν τις εἰς πίθον ζέοντα <οἴνου> ἐμβάλῃ παύειν τὴν ζέσιν οὐ παραχρῆμα μόνον ἀλλὰ καὶ ὅλως καταξηραίνουσάν τε δῆλον ὅτι καὶ ἀναδεχομένην τὸ πνεῦμα καὶ τοῦτο διϊεῖσαν. οὗτος μὲν οὖν τό γε πλῆθος ταύτῃ τῇ βοηθείᾳ κατέσχεν.
4 Ὅτι δὲ καὶ τὸ ἔθος ἰσχυρὸν φανερὸν ἐκ πολλῶν·

---

[1] This story is quoted by Apollonius, *Hist. Mirab.* 50.

poisons non-poisonous; for, when the constitution has accepted them and prevails over them, they cease to be poisons, as Thrasyas also remarked; for he said that the same thing was a poison to one and not to another; thus he distinguished between different constitutions, as he thought was right; and he was clever at observing the differences. Also, besides the constitution, it is plain that use has something to do with it. At least Eudemus, the vendor of drugs, who had a high reputation in his business, after making a wager that he would experience no effect before sunset, drank a quite moderate dose, and it proved too strong for his power of resistance: [1] while the Chian Eudemus took a draught of hellebore and was not purged. And on one occasion he said that in a single day he took two and twenty draughts in the market-place as he sat at his stall, and did not leave the place till it was evening, and then he went home and had a bath and dined, and was not sick. However this man was able to hold out because he had provided himself with an antidote; for he said that after the seventh dose he took a draught of tart vinegar with pumice-stone dust in it, and later on took a draught of the same in wine in like manner; and that the virtue of the pumice-stone dust is so great that, if one puts it into a boiling pot of wine,[2] it causes it to cease to boil, not merely for the moment, but altogether, clearly because it has a drying effect and it catches the vapour and passes it off. It was then by this antidote that Eudemus was able to contain himself in spite of the large quantity of hellebore which he took.

However many things go to show that use makes

[2] οἴνου add. Sch., *cf.* Plin. 36. 42; 14. 138.

THEOPHRASTUS

ἐπεὶ καὶ τὸ ἀψίνθιον τὰ μὲν ἐνταῦθα πρόβατα οὔ φασί τινες νέμεσθαι, τὰ δ' ἐν τῷ Πόντῳ νέμεται καὶ γίνεται πιότερα καὶ καλλίω καί, ὥς δή τινες λέγουσιν, οὐκ ἔχοντα χολήν. ἀλλὰ γὰρ ταῦτα μὲν ἑτέρας ἄν τινος εἴη θεωρίας.

XVIII. Αἱ δὲ ῥίζαι καὶ τὰ ὑλήματα, καθάπερ εἴρηται, πολλὰς ἔχουσι δυνάμεις οὐ πρὸς τὰ ἔμψυχα σώματα μόνον ἀλλὰ καὶ πρὸς τὰ ἄψυχα. λέγουσι γὰρ ἄκανθάν τινα εἶναι ἢ πήγνυσι τὸ ὕδωρ ἐμβαλλομένη· πηγνύναι δὲ καὶ τὴν τῆς ἀλθαίας ῥίζαν, ἐάν τις τρίψας ἐμβάλῃ καὶ θῇ ὑπαίθριον· ἔχει δὲ ἡ ἀλθαία φύλλον μὲν ὅμοιον τῇ μαλάχῃ πλὴν μεῖζον καὶ δασύτερον, τοὺς δὲ καυλοὺς μαλακούς, ἄνθος δὲ μήλινον, καρπὸν δ' ὅμοιον τῇ μαλάχῃ, ῥίζαν δὲ ἰνώδη λευκὴν τῇ γεύσει δὲ ὥσπερ τῆς μαλάχης ὁ καυλός· χρῶνται δὲ αὐτῇ πρός τε τὰ ῥήγματα καὶ τὰς βῆχας ἐν οἴνῳ γλυκεῖ καὶ ἐπὶ τὰ ἕλκη ἐν ἐλαίῳ.

2 Ἑτέραν δέ τινα συνεψομένην τοῖς κρέασι συνάπτειν εἰς ταὐτὸ καὶ οἷον πηγνύναι· τὰς δὲ καὶ ἕλκειν, ὥσπερ ἡ λίθος καὶ τὸ ἤλεκτρον. καὶ ταῦτα μὲν ἐν τοῖς ἀψύχοις.

Τὸ δὲ θηλύφονον, οἱ δὲ σκορπίον καλοῦσι διὰ τὸ τὴν ῥίζαν ὁμοίαν ἔχειν τῷ σκορπίῳ, ἐπιξυνό-

---

[1] cf. Plin. 27. 45.
[2] ὑλήματα: here a general term for shrubs and under-shrubs. cf. 9. 20. 6.
[3] Diosc. 3. 146; Plin. 20. 84.

much difference; [1] thus some say that the sheep of some places do not eat wormwood; yet those of Pontus not only eat it but become fatter and fairer and, as some say, have no bile. But these things may be said to belong to a different enquiry.

Of plants that possess properties affecting lifeless objects.

XVIII. Herbs and shrubs,[2] as has been said, have many virtues which are shown in their effects not only on living bodies but on lifeless ones. Thus they say that there is a kind of *akantha* (gum arabic) which thickens water, when it is put in it; [3] and that so also does the root of marsh-mallow if one shreds it and puts it in and stands the water in the open air. Marsh-mallow has a leaf like mallow, but larger and rougher; the stems are soft, the flower yellow, the fruit like that of mallow, the root fibrous and white, with a taste like that of the stem of mallow. They use it for fractures and for coughs in sweet wine, and for sores in olive-oil.

[4] They say that there is another kind which, if cooked with meat, combines with it and as it were sets it hard; and there are others that attract things to them, like the magnet or amber. So much for effects produced on lifeless things.

Of plants whose properties affect animals other than man.

[5] Wolf's bane, which some call 'scorpion-plant because it has a root like a scorpion, kills that animal

[4] Referred to by Apollon. *Hist. Mirab.* 41. *cf.* Diosc. 3. 147; Plin. 27. 42; 25. 67.
[5] Referred to by Ael. *H.A.* 9. 27; Apollon. *Hist. Mirab.* 41. *cf.* Plin. 25. 122 (*cf.* 27. 6); Diosc. 4. 76. This is evidently a different plant to the σκορπίος mentioned 9. 13. 6. See Index.

μενον ἀποκτείνει τὸν σκορπίον· ἐὰν δέ τις ἐλλέβορον λευκὸν καταπάσῃ, πάλιν ἀνίστασθαί
φασιν· ἀπόλλυσι δὲ καὶ βοῦς καὶ πρόβατα καὶ
ὑποζύγια καὶ ἁπλῶς πᾶν τετράπουν ἐὰν εἰς τὰ
αἰδοῖα τεθῇ ἡ ῥίζα ἢ τὰ φύλλα αὐθήμερον·
χρήσιμον δὲ πρὸς σκορπίου πληγὴν πινόμενον.
ἔχει δὲ τὸ μὲν φύλλον ὅμοιον κυκλαμίνῳ τὴν δὲ
ῥίζαν, ὥσπερ ἐλέχθη, σκορπίῳ. φύεται δὲ ὥσπερ
ἡ ἄγρωστις καὶ γόνατα ἔχει· φιλεῖ δὲ χωρία
σκιώδη. εἰ δὲ ἀληθῆ τὰ περὶ τὸν σκορπίον ἤδη
καὶ τἆλλα, οὐκ ἀπίθανα τὰ τοιαῦτα. καὶ τὰ
3 μυθώδη δὲ οὐκ ἀλόγως συγκεῖται. ἐν δὲ τοῖς
ἡμετέροις σώμασι χωρὶς τῶν πρὸς ὑγείαν καὶ
νόσον καὶ θάνατον καὶ πρὸς ἄλλα δυνάμεις ἔχειν
φασὶν οὐ μόνον τῶν σωματικῶν ἀλλὰ καὶ τῶν
τῆς ψυχῆς. . . . .

XIX. Πρὸς δὲ τὴν ψυχὴν τὸν μὲν στρύχνον
ὥστε παρακινεῖν καὶ ἐξιστάναι, καθάπερ ἐλέχθη
πρότερον, ἡ δὲ τοῦ ὀνοθήρα ῥίζα δοθεῖσα ἐν οἴνῳ
πραότερον καὶ ἱλαρώτερον ποιεῖ τὸ ἦθος. ἔχει
δὲ ὁ μὲν ὀνοθήρας τὸ μὲν φύλλον ὅμοιον ἀμυγδαλῇ
μικρότερον δέ, τὸ δὲ ἄνθος ἐρυθρὸν ὥσπερ ῥόδον·
αὐτὸς δὲ μέγας θάμνος· ῥίζα δὲ ἐρυθρὰ καὶ
μεγάλη, ὄζει δὲ αὐανθείσης ὥσπερ οἴνου· φιλεῖ
δὲ ὀρεινὰ χωρία. φαίνεται δὲ οὐ τοῦτο ἄτοπον·
οἷον γὰρ προσφορά τις γίνεται δύναμιν ἔχοντος
οἰνώδη.

---

[1] σκορπίῳ conj. W.; σκορπίου Ald.
[2] 18. 3, λέγω δὲ σωματικῶν . . . 18. 11 (the account of the physical effects) is here omitted.
[3] 9. 11. 6.

if it is shredded over him; while if one then sprinkles him with white hellebore, they say that he comes to life again. It is also fatal to oxen sheep beasts of burden and in general to any fourfooted animal, and kills them the same day if the root or leaf is put on the genitals; and it is also useful as a draught against a scorpion's sting. It has a leaf like cyclamen, and a root, as was said, like a scorpion.[1] It grows like dog's-tooth grass, and is jointed, and it loves shady places. Now if what has been told already about the scorpion be true, then other similar tales are not incredible. (Indeed fabulous tales are not composed without some reason).[2] And in relation to our own persons, apart from their effects in regard to health disease and death, it is said that herbs have also other properties affecting not only the bodily but also the mental powers. . . . .

Of plants possessing properties which affect the mental powers.

XIX. As to those which affect the mind, *strykhnos*, as was said before,[3] is said to upset the mental powers and make one mad;[4] while the root of *onotheras* (oleander) administered in wine makes the temper gentler and more cheerful. This plant has a leaf like the almond, but smaller,[5] and the flower is red like a rose. The plant itself (which loves hilly country) forms a large bush; the root is red and large, and, if this is dried, it gives off a fragrance like wine. And this does not seem surprising, since there is a sort of 'bouquet' given off by a thing which has the peculiar quality of wine.

[4] Diosc. 4. 117; Plin. 26. 111.
[5] μικρότερον conj. W.; πικρότερον UM; πλατύτερον Ald. (so also Diosc. *l.c.*). G seems to have read μακρότερον.

2 Ἀλλὰ τάδε εὐηθέστερα καὶ ἀπιθανώτερα¹ τά τε τῶν περιάπτων καὶ ὅλως τῶν ἀλεξιφαρμάκων λεγομένων τοῖς τε σώμασι καὶ ταῖς οἰκίαις. καὶ ὡς δή φασι τὸ τριπόλιον² καθ' Ἡσίοδον καὶ Μουσαῖον εἰς πᾶν πρᾶγμα σπουδαῖον χρήσιμον εἶναι, δι' ὃ καὶ ὀρύττουσιν αὐτὸ νύκτωρ σκηνὴν πηξάμενοι. καὶ τὰ περὶ τῆς εὐκλείας δὲ καὶ εὐδοξίας ὁμοίως ἢ καὶ μᾶλλον· εὔκλειαν γάρ φασι ποιεῖν τὸ ἀντίρρινον καλούμενον· τοῦτο δ' ὅμοιόν ἐστι τῇ ἀπαρίνῃ· ῥίζα δὲ οὐχ ὕπεστιν· ὁ δὲ καρπὸς ὥσπερ μόσχου ῥῖνας ἔχει. τὸν δ' ἀπὸ 3 τούτου ἀλειφόμενον εὐδοξεῖν. εὐδοξεῖν δὲ καὶ ἐάν τις τοῦ ἑλειοχρύσου τῷ ἄνθει στεφανῶται μύρῳ ῥαίνων ἐκ χρυσίου ἀπύρου. ἔχει δὲ ὁ ἑλειόχρυσος τὸ μὲν ἄνθος χρυσοειδές, φύλλον δὲ λευκὸν καὶ τὸν καυλὸν δὲ λεπτὸν καὶ σκληρὸν ῥίζαν δὲ ἐπιπόλαιον καὶ λεπτήν. χρῶνται δὲ αὐτῷ πρὸς τὰ δακετὰ ἐν οἴνῳ καὶ πρὸς τὰ πυρίκαυστα κατακαύσαντες καὶ μίξαντες μέλιτι. τὰ μὲν οὖν τοιαῦτα, καθάπερ καὶ πρότερον ἐλέχθη, συναύξειν βουλομένων ἐστὶ τὰς ἑαυτῶν τέχνας.

4 Αἱ δὲ τῶν ῥιζῶν καὶ τῶν καρπῶν καὶ τῶν ὀπῶν φύσεις ἐπεὶ πολλὰς ἔχουσι καὶ παντοίας δυνάμεις, ὅσαι ταὐτὸ δύνανται καὶ τῶν αὐτῶν αἰτίαι,

---

¹ ἀπιθανώτερα conj. Scal. after G: so also Cod. Cas. Vin. Vo.; πιθανώτερα U*; πιθανότερα Ald.
² τριπόλιον UMU*Ald.; G from Plin. 21. 44 has *polium*. It may be observed that τριπόλιον can hardly have occurred in a hexameter. Hesych., however, gives τρισπόλιον as the name

ENQUIRY INTO PLANTS, IX. xix. 2–4

Of plants said to have magical properties.

On the other hand what is said of amulets and charms in general for the body or the house is somewhat foolish and incredible.[1] Thus they say that *tripolion*[2] according to Hesiod and Musaeus is useful for every good purpose, wherefore they dig it up by night, camping on the spot. So too what is said of good or fair fame as affected by plants is quite as foolish or more so: for they say that the plant called snapdragon[3] produces fair fame. This plant is like bedstraw but it has no root: and the fruit has what resembles a calf's nostrils. The man who anoints himself with this they say wins fair fame. [4] And they say that the same result follows, if he crowns himself with the flower of gold-flower, sprinkling it with unguent from a vessel of unfired gold. The flower of gold-flower is like gold, the leaf is white. The stem also is white[5] and hard, the root is slender and does not run deep. [6] Men use it in wine against the bites of serpents, and to make a plaster for burns after burning it and mixing the ashes with honey. Such tales then, as was said before, proceed from men who desire to glorify their own crafts.

A problem as to cause and effect.

Now since the natural qualities of roots fruits and juices have many virtues of all sorts, some having the same virtue and causing the same result, while

of a plant. Plin. *l.c.* seems to combine Diosc.'s account of πόλιον (3. 110) with his account of τριπόλιον (4. 132).

[3] τὸ ἀντίρρινον conj. St. from Diosc. 4. 130; Plin. 25. 129; τὸ ἀντίρριζον Ald.H.; τὸν ἀντ. UM; τὸ ἀντίριζον U*.

[4] Diosc. 4. 57; Plin. 21. 66. Cited also by Athen. 15. 27.

[5] λευκὸν conj. Sch.; λεπτὸν UMU*Ald.G.

[6] Diosc. *l.c.*; Plin. 21. 168 and 169.

καὶ πάλιν ὅσαι τὰ ἐναντία, διαπορήσειεν ἄν τις κοινὸν ἴσως ἀπόρημα καὶ ἐφ' ἑτέρων ἀπόρων, πότερον ὅσα τῶν αὐτῶν αἴτια κατὰ μίαν τινὰ δύναμίν ἐστιν, ἢ καὶ ἀφ' ἑτέρων ἐνδέχεται ταὐτὸ γίνεσθαι. τοῦτο μὲν οὖν ταύτῃ ἠπορήσθω· εἰ δέ τινων καὶ ἄλλων τὰς φύσεις ἢ τὰς δυνάμεις ἔχομεν εἰπεῖν, ταῦτα ῥητέον.

XX. Τὸ δὴ πέπερι καρπὸς μέν ἐστι διττὸν δὲ αὐτοῦ τὸ γένος· τὸ μὲν γὰρ στρογγύλον ὥσπερ ὄροβος, κέλυφος ἔχον καὶ σάρκα καθάπερ αἱ δαφνίδες, ὑπέρυθρον· τὸ δὲ πρόμηκες μέλαν σπερμάτια μηκωνικὰ ἔχον· ἰσχυρότερον δὲ πολὺ τοῦτο θατέρου· θερμαντικὰ δὲ ἄμφω· δι' ὃ καὶ πρὸς τὸ κώνειον βοηθεῖ ταῦτά τε καὶ ὁ λιβανωτός.

2 Ὁ δὲ Κνίδιος κόκκος στρογγύλον ἐρυθρὸν τῇ χροιᾷ μεῖζον δὲ τοῦ πεπέριος ἰσχυρότερον δὲ πολὺ τῇ θερμότητι, δι' ὃ καὶ ὅταν δίδωσι κατάποτον, διδόασι γὰρ πρὸς κοιλίας λύσιν, ἐν ἄρτῳ ἢ στέατι περιπλάττοντες· κάει γὰρ ἄλλως τὸν φάρυγγα.

Θερμαντικὸν δὲ καὶ ἡ τοῦ πευκεδάνου <ῥίζα,> δι' ὃ καὶ ἄλειμμά τι ποιοῦσιν ἐξ αὐτῆς ἱδρωτικὸν ὥσπερ καὶ ἐξ ἄλλων. δίδοται δὲ ἡ τοῦ πευκε-

---

[1] ἀφ' conj. Sch.; ἐφ' U*P; Ald. omits the preposition.
[2] Cited by Athen. 2. 73; cf. Diosc. 2. 159.
[3] Plin. 27. 70.

others have opposite virtues, one might raise a question which is perhaps equally perplexing in regard to other matters, to wit, whether those that produce the same effect do so in virtue of some single virtue which is common to them all, or whether the same result may not come about also from [1] different causes.—Let us be content to put the question thus : but now we must proceed to speak of the natural qualities or virtues of any other plants that we can mention.

Of certain plants, not yet mentioned, which possess special properties.

XX. [2] Pepper is a fruit, and there are two kinds : one is round like bitter vetch, having a case and flesh like the berries of bay, and it is reddish : the other is elongated and black and has seeds like those of poppy : and this kind is much stronger than the other. Both however are heating : wherefore these, as well as frankincense, are used as antidotes for poisoning by hemlock.

[3] The 'Cnidian berry' is round, red in colour, larger than that of pepper, and far stronger in its heating power ; wherefore, when it is given as a pill [4] (for it is given to open the bowels) they knead it up in a piece of bread or dough : otherwise it burns the throat.

[5] The root [6] of sulphur-wort is also heating, wherefore they make of it an ointment to produce a sweat, as with other things so used. This root [6] is also

---

[4] καταπότον conj. Sch.; κατὰ πότον Ald. *cf.* καταπότιον 9. 8. 3.
[5] *cf.* 9. 14. 1 ; Plin. 25. 117.
[6] ῥίζα add. W.

δάνου ῥίζα καὶ πρὸς τοὺς σπλῆνας· τὸ δὲ σπέρμα οὐ χρήσιμον οὐδὲ ὁ ὀπὸς αὐτῆς· γίνεται δὲ ἐν Ἀρκαδίᾳ.

Δαῦκον δὲ περὶ Πατραϊκὴν τῆς Ἀχαΐας διαφέρον, θερμαντικὸν φύσει· ῥίζαν δὲ ἔχει μέλαιναν.

3 Θερμαντικὸν δὲ καὶ δριμὺ καὶ τῆς ἀμπέλου τῆς ἀγρίας ῥίζα· δι' ὃ καὶ εἰς ψίλωθρον χρήσιμον καὶ ἐφηλίδας ἀπάγειν· τῷ δὲ καρπῷ ψιλοῦσι τὰ δέρματα. τέμνεται δὲ πᾶσαν ὥραν ὀπώρας δὲ μάλιστα.

Δρακοντίου δὲ ῥίζα βῆχας ἐν μέλιτι διδομένη παύειν χρησίμη. καυλὸν δὲ ἔχει ποικίλον ὀφιώδη· σπέρματι δ' οὐ χρῶνται.

Ἡ δὲ τῆς θαψίας ἐμετική· ἐὰν δέ τις κατάσχῃ, καθαίρει καὶ ἄνω καὶ κάτω· δύναται δὲ καὶ τὰ πελιώματα ἐξαιρεῖν· ὑπώπια δὲ ποιεῖ ἄλλα ἔκλευκα. ὁ δὲ ὀπὸς ἰσχυρότερος αὐτῆς καθαίρει καὶ ἄνω καὶ κάτω· σπέρματι δ' οὐ χρῶνται· γίνεται δὲ καὶ ἄλλοθι μὲν ἀτὰρ καὶ ἐν τῇ Ἀττικῇ· καὶ τὰ βοσκήματα ταύτης οὐχ ἅπτεται τὰ ἐγχώρια, τὰ δὲ ξενικὰ βόσκεται καὶ διαρροίᾳ διαφθείρεται.

4 Τὸ δὲ πολυπόδιον μετὰ τὰ ὕδατα ἀναβλαστεῖ σπέρμα δὲ οὐ φύει.

Τὸ δὲ τῆς ἐβένου ξύλον κατὰ μὲν τὴν πρόσοψιν ὅμοιον πύξῳ φλοϊσθὲν δὲ μέλαν γίνεται· χρήσιμον δὲ πρὸς ὀφθαλμίας ἀκόνῃ τριβόμενον.

---

[1] cf. 9. 15. 5. [2] cf. 9. 15. 8. n.
[3] cf. 9. 14. 1 ; Diosc. 4. 181–183 ; Plin. 23. 19 and 21.
[4] cf. 7. 12. 2 ; Diosc. 2. 167. cf. Plin. 24. 89.
[5] Diosc. 4. 153 ; Plin. 13. 125 and 126.

ENQUIRY INTO PLANTS, IX. xx. 2–4

given for the spleen : but neither its seed nor its juice is of use : it grows in Arcadia.[1]

[2] *Daukon* of excellent quality grows in the district of Patrai in Achaia, and is heating by nature : it has a black root.

[3] The root of the 'wild vine' (bryony) is also heating and pungent : wherefore it is useful as a depilatory and to remove freckles : and the fruit is used for smoothing hides. It is cut at any season, but especially in autumn.

[4] The root of edderwort given in milk is useful for stopping a cough. It has a variegated snake-like stem : the seed is not used.

[5] The root of *thapsia* has emetic properties : and, if one retains it, it purges both upwards and downwards. It is also able to remove bruises : and it restores other contusions to a pale colour.[6] Its juice is stronger and purges both upwards and downwards : the seed is not used. It grows especially in Attica, but also in other places : the cattle of the country do not touch it, but imported cattle feed on it and perish of diarrhoea.[7]

[8] Polypody springs up [9] after rain, and produces no seed.

[10] The wood of ebony is in appearance like box, but when barked it becomes black : it is useful against ophthalmia, and is rubbed on a whetstone for that use.

---

[6] ὑπώπια . . . ἔκλευκα : text perhaps defective.

[7] διαρροίᾳ conj. Sch.; διάρροια ἢ UM : διάρροια αὐτοῖς γίνεται ἢ Ald.G.

[8] *cf.* 9. 13. 6 ; *C.P.* 2. 17. 4. The account of the virtues of this plant is evidently missing.

[9] ἀναβλαστεῖ conj. W.; αἰεὶ βάλλει Ald.

[10] Diosc. 1. 98 ; Plin. 24. 89.

Ἡ δὲ ἀριστολοχία παχεῖα καὶ ἐσθιομένη πικρὰ τῷ χρώματι μέλαινα καὶ εὔοσμος, τὸ δὲ φύλλον στρογγύλον, οὐ πολὺ δὲ τὸ ὑπὲρ τῆς γῆς. φύεται δὲ καὶ μάλιστα ἐν τοῖς ὄρεσι· καὶ αὕτη βελτίστη. τὴν δὲ χρείαν αὐτῆς εἰς πολλὰ καταριθμοῦσιν· ἀρίστη μὲν πρὸς τὰ κεφαλόθλαστα, ἀγαθὴ δὲ καὶ πρὸς τὰ ἄλλα ἕλκη καὶ πρὸς τὰ ἑρπετὰ καὶ πρὸς ὕπνον καὶ πρὸς ὑστέραν ὡς πεσσός, τὰ μὲν σὺν ὕδατι ἀναδευομένη καὶ καταπλαττομένη, τὰ δ' ἄλλα εἰς μέλι ξυομένη καὶ ἔλαιον· τῶν δὲ ἑρπετῶν ἐν οἴνῳ ὀξίνῃ πινομένη καὶ ἐπὶ τὸ δῆγμα ἐπιπαττομένῃ· εἰς ὕπνον δὲ ἐν οἴνῳ μέλανι αὐστηρῷ κνισθεῖσα· ἐὰν δὲ αἱ μῆτραι προπέσωσι, τῷ ὕδατι ἀποκλύζειν. αὕτη μὲν οὖν ἔοικε διαφέρειν τῇ πολυχρηστίᾳ.

5 Τῆς δὲ σκαμμωνίας ὥσπερ ἐξ ἐναντίας ὁ ὀπὸς μόνον χρήσιμος ἄλλο δ' οὐδέν.

Ἡ δὲ τῆς πτερίδος ῥίζα μόνον τῷ χυλῷ γλυκύστρυφνος· ἕλμινθα δὲ πλατεῖαν ἐκβάλλει· σπέρμα δὲ οὐκ ἔχει οὐδὲ ὀπόν· τέμνεσθαι δὲ ὡραίαν μετοπώρου φασίν.

Ἡ δ' ἕλμις σύμφυτον ἐνίοις ἔθνεσιν· ἔχουσι γὰρ ὡς ἐπὶ πᾶν Αἰγύπτιοι Ἄραβες Ἀρμένιοι Ματαδίδες Σύροι Κίλικες· Θρᾷκες δ' οὐκ ἔχουσιν οὐδὲ Φρύγες· τῶν δὲ Ἑλλήνων Θηβαῖοί τε οἱ περὶ τὰ γυμνάσια καὶ ὅλως Βοιωτοί· Ἀθηναῖοι δ' οὔ.

Πάντων δὲ τῶν φαρμάκων ὡς ἁπλῶς εἰπεῖν βελτίω τὰ ἐκ τῶν χειμερινῶν καὶ προσβόρρων

---

[1] cf. 9. 13. 3.   [2] καὶ μάλιστα conj. W.; μάλιστα καὶ Ald.
[3] αὕτη conj. Scal.; αὐτὴ Ald.

ENQUIRY INTO PLANTS, IX. xx. 4–5

[1] Birthwort is a stout plant and is bitter to the taste : it is black in colour and fragrant ; the leaf is round. However there is not much of the plant above ground. It grows especially [2] on mountains, and then [3] it is best. Many uses of it for various purposes are enumerated ; it is best for bruises on the head, good also for other wounds, against snake-bites, to produce sleep, for the womb as a pessary : for some purposes it is soaked with water and applied as a plaster, for others it is scraped into honey and olive-oil : against snake-bites it is drunk in sour wine and also sprinkled over the bite ; to induce sleep it is given pounded up in black dry wine : [4] in cases of *prolapsus uteri* it is used in water as a lotion. This plant then seems to have a surpassing variety of usefulness.

[5] Of scammony, as though by contrast, only the juice is useful and no other part.

Of male-fern no part but the root is useful and it has a sweet astringent taste. It expels the flat worm. It has no seed nor juice : and they say it is ripe for cutting in autumn.

[6] (This worm naturally infests certain races : speaking generally the following are liable to it— the Egyptians, the Arabians, the Armenians, the Matadides, the Syrians, the Cilicians : the Thracians have it not, nor the Phrygians. Among the Hellenes those Thebans who frequent wrestling-schools and the Boeotians generally are liable to it : but not the Athenians.)

Of all drugs, to speak generally, those are better which come from places that are wintry, face the

[4] Cited by Apollon. *Hist. Mirab.* 29.
[5] Diosc. 4. 170 ; Plin. 27. 78–80.   [6] Plin. 27. 145.

καὶ ξηρῶν· δι' ὃ καὶ τῶν ἐν Εὐβοίᾳ τὰ ἐν ταῖς Αἰγαῖς ἢ τὰ ἐν τῷ Τελεθρίῳ φασί· ξηρότερα γάρ· τὸ δὲ Τελέθριον σύσκιον.

6 Περὶ μὲν οὖν τῶν ῥιζῶν ὅσαι φαρμακώδεις καὶ ὁποιασοῦν ἔχουσι δυνάμεις εἴτε ἐν αὐταῖς εἴτε ἐν τοῖς ὀποῖς ἢ καὶ ἄλλῳ τινὶ τῶν μορίων, καὶ τὸ ὅλον εἴ τι φρυγανικὸν ἢ ποῶδες ἔχει τοιαύτας δυνάμεις, καὶ περὶ τῶν χυλῶν τῶν τε εὐόσμων καὶ τῶν ἀόσμων καὶ ὅσας ἔχουσι διαφοράς, αἵπερ οὐθὲν ἧττον φυσικαί εἰσιν, εἴρηται.

north and are dry: wherefore of those which grow in Euboea best, they say, are the drugs of Aigai or Telethrion, these places being dry, while Telethrion is also shady.

[1] Thus we have spoken of drugs, those that are medicinal and those that have virtues of whatsoever kind, whether in the root itself, or in the juice, or in any other of their parts, and in general of all the shrubby or herbaceous plants which have such virtues, as well as their tastes, whether they be fragrant or without fragrance, with the differences between them, which are equally part of their essential character.

[1] This section begins a tenth book in UMAld.H.G; *cf.* 9. 8. 1 n. The concluding words can hardly represent the original text.

MINOR WORKS

INTRODUCTION TO THE TREATISES CONCERNING ODOURS AND CONCERNING WEATHER SIGNS

The text of the two *opuscula* given here is reprinted from that of Wimmer in the Teubner series, 1862, and in the Didot edition, 1866; the latter is very carelessly printed: a few slight alterations are mentioned in the notes. Both works are included in the Aldine edition (1497), and in that of Camotius (see p. x). For the *de odoribus* two MSS., Cod. Vaticanus (A) and Cod. Parisiensis (Q) were collated by Brandis. The text of the *de signis* is considered by Wimmer to be very corrupt and defective: he has admitted some emendations made by Schneider from an old Latin translation published at Bologna in 1516. Schneider's commentary makes frequent reference to an edition of the *opuscula* of Theophrastus by Turnebus and Daniel Furlanus, printed at Hanau in Prussia in 1605, and reprinted there in 1615.

The *de signis* was one of Aratus' authorities for his *Diosemeia*: I have only however made reference to that work where it appears to throw light on the text of Theophrastus. These and most other references for the two fragments I owe chiefly to Schneider.

CONCERNING ODOURS

CONTENTS

SECTIONS.
- 1–3. Introductory: Of odours in general and the classification of them.
- 4. Of natural odours: Of those of animals and of the effect of odours on animals.
- 5. Of smell and taste.
- 6. Of odours in plants.
- 7–13. Of artificial odours in general and their manufacture: especially of the use of perfumes in wine.
- 14–20. Of the oils used as the vehicle of perfumes.
- 21–26. Of the spices used in making perfumes and their treatment.
- 27–31. Of the various parts of plants used for perfumes, and of the composition of various notable perfumes.
- 32–35. Of the properties of various spices.
- 35–36. Of the medicinal properties of certain perfumes.
- 37–41. Of rules for the mixture of spices, and of the storing of various perfumes.
- 42–50. Of the properties of certain perfumes.
- 51–56. Of other properties and peculiarities of perfumes.
- 57–59. Of the making of perfume-powders and compound perfumes.
- 61–63. Of the characteristic smells of animals, and of certain curious facts as to the smell of animal and vegetable products.
- 64–69. Of odours as compared with other sense-impressions.

ΠΕΡΙ ΟΣΜΩΝ

I. Αἱ ὀσμαὶ τὸ μὲν ὅλον ἐκ μίξεώς εἰσι καθάπερ οἱ χυμοί· τὸ γὰρ ἄμικτον ἅπαν ἄοδμον ὥσπερ ἄχυμον, διὸ καὶ τὰ ἁπλᾶ ἄοδμα, οἷον ὕδωρ ἀὴρ πῦρ· ἡ δὲ γῆ μάλιστ᾽ ἢ μόνη ὀδμὴν ἔχει διὸ μάλιστα μικτή.

Τῶν δ᾽ ὀδμῶν αἱ μὲν ὥσπερ ἀειδεῖς καὶ ὑδαρεῖς καθάπερ ἐπὶ τῶν χυμῶν, αἱ δ᾽ ἔχουσαί τινας ἰδέας. αἱ δ᾽ ἰδέαι δοκοῦσι μὲν ἀκολουθεῖν ταῖς τῶν χυμῶν, οὐ μὴν ἔχουσί γε πᾶσαι τὰς αὐτὰς προσηγορίας, ὥσπερ ἐν τοῖς πρότερον εἴπομεν, οὐδ᾽ ὅλως οὕτω διωρισμέναι τοῖς εἴδεσιν ὥσπερ οἱ χυμοὶ ἀλλ᾽ ὡς ἂν τοῖς γένεσιν, ὅτι τὰ μὲν 2 εὔοσμα τὰ δὲ κάκοσμα. τῆς δ᾽ εὐωδίας καὶ κακωδίας οὐκέτι τὰ εἴδη κατωνόμασται καίπερ ἔχοντα διαφορὰς μεγάλας ἐπί γ᾽ αὐτῶν τῶν γλυκέων καὶ πικρῶν, ἀλλὰ δριμεῖα λέγεται καὶ ἰσχυρὰ καὶ μαλακὴ καὶ γλυκεῖα καὶ βαρεῖα ὀδμή· κοιναὶ δ᾽ ἔνιαι τούτων καὶ τῶν κακωδῶν.

---

[1] *i.e.* there is not one set of terms applied to the varieties of 'good' and another distinct set applied to the varieties of 'evil' odours, but we get a cross-division, some terms (such as 'strong') being applied to varieties of *both* classes. *cf.* 64–66.

CONCERNING ODOURS

Introductory: Of odours in general and the classification of them.

I. ODOURS in general, like tastes, are due to mixture: for anything which is uncompounded has no smell, just as it has no taste: wherefore simple substances have no smell, such as water air and fire: on the other hand earth is the only elementary substance which has a smell, or at least it has one to a greater extent than the others, because it is of a more composite character than they.

Of odours some are, as it were, indistinct and insipid, as is the case with tastes, while some have a distinct character. And these characters appear to correspond to those of tastes, yet they have not in all cases the same names, as we said in a former treatise; nor in general are they marked off from one another by such specific differences as are tastes: rather the differences are, one may say, in generic character, some things having a good, some an evil odour.[1] But the various kinds of good or evil odour, although they exhibit considerable differences, have not received further distinguishing names, marking off one particular kind of sweetness or of bitterness from another: we speak of an odour as pungent, powerful, faint, sweet, or heavy, though some of these descriptions apply to evil-smelling things as well as to those which have a good odour.

THEOPHRASTUS

Ἡ δὲ καθόλου καὶ ὥσπερ ἐπὶ πᾶσι τοῖς διαφθειρομένοις σαπρότης. ἅπαν γὰρ τὸ σηπόμενον κακῶδες, εἰ μή τις τὴν ὀξύτητα λέγει τοῦ οἴνου
3 σαπρότητα τῇ ὁμοιότητι τῆς φθορᾶς. ἐν ἅπασι δ' ἐστὶν ἡ τοῦ σαπροῦ κακωδία καὶ ἐν φυτοῖς καὶ ἐν ζώοις καὶ ἐν τοῖς ἀψύχοις· ἐν ἅπασι δὲ διαφθειρομένοις ὧν μὴ ἡ σύστασις εὐθὺς ἐκ τοιαύτης ὕλης· ἔχει γὰρ ἔνια καὶ τὴν τῆς ὕλης ὀσμήν, οὐ μὴν ἐπὶ πάντων γε τοῦτ' ἀκολουθεῖ. πολλὰ γὰρ οὐ κακώδη τὰ ἐκ τῶν σαπρῶν, ὡς οὐδ' οἱ μύκητες οἱ ἐκ τῆς κόπρου φυόμενοι· τὰ δ' ἐκ σήψεως φυόμενα καὶ συνιστάμενα κακώδη. εὔοσμα μὲν οὖν ὡς ἁπλῶς εἰπεῖν τὰ πεπεμμένα καὶ λεπτὰ καὶ ἥκιστα γεώδη· τὸ γὰρ τῆς ὀσμῆς ἐν ἀναπνοῇ· κακώδη δὲ δηλονότι τἀναντία. πολλὰ δὲ ὥσπερ τῶν γλυκέων ἐμφαίνει τινὰ πικρότητα, καὶ τῶν εὐωδῶν βαρύτητα ταῖς ὀσμαῖς.

4 II. Ἔχει δὲ ἕκαστον ὀσμὴν ἰδίαν καὶ ζώων καὶ φυτῶν καὶ τῶν ἀψύχων ὅσα ὀσμώδη· πολλὰ

---

[1] And so here we have a term which possibly is applied only to the one class of 'evil' odours.

[2] Which is not an 'evil' odour.

[3] *i.e.* putridity is a quality which things acquire as they decay, and does not necessarily imply that they are themselves formed out of decaying matter. In fact things so produced are not always 'putrid.'

[4] The sense is apparently that 'lighter' (or less solid) things exhale a lighter and pleasanter odour because in their

CONCERNING ODOURS, 2-4

Putridity however is a general term, applied, one may say, to anything which is subject to decay[1]: for anything which is decomposing has an evil odour,—unless indeed the name putridity be extended to sourness[2] in wine because the change in the wine is analogous to decomposition. The evil odour of putridity is found in all things, alike in plants in animals and in inanimate things: it attends the decay of things which are not formed directly out of a substance which is decaying: for some things have also the odour of that substance, though it is not found in every case.[3] Thus in many instances things which are produced by decaying matter have no evil odour: for instance, mushrooms which grow from dung have none: but things which grow from decay and are actually formed out of it have such an odour. To speak generally then, things that have been cooked, delicate things, and things which are least of an earthy nature have a good odour,[4] (odour being a matter of exhalation), and it is obvious that those of an opposite character have an evil odour. But, even as many things pleasant to the taste present a certain bitterness, so many things that have a good odour have a kind of heavy scent.

Of natural odours; of those of animals and of the effect of odours on animals.

II. Every plant animal or inanimate thing that has an ódour has one peculiar to itself: but in many

case exhalation is easier. The sense given to ἀναπνοή requires illustration (the passages cited by LS. are not in point). Sch., construing apparently as W. does, 'since smell depends on breathing' (? inhalation), admits that he does not see the point of this clause.

THEOPHRASTUS

δ' ἡμῖν οὐ φαίνεται διὰ τὸ χειρίστην ἔχειν τὴν αἴσθησιν ταύτην ὡς εἰπεῖν. ἐπεὶ τοῖς γε ἄλλοις καὶ τὰ παντελῶς ἄοδμα φαινόμενα δίδωσί τινα ὀσμήν, ὥσπερ αἱ κριθαὶ τοῖς ὑποζυγίοις αἱ ἐκ τῆς Κεδροπόλιος, ἃς οὐκ ἐσθίουσιν διὰ τὴν κακωδίαν. ἡμᾶς δὲ καὶ αἱ τῶν ζῴων λανθάνουσιν τῶν ὀσμωδῶν δοκούντων. εὐωδίᾳ μὲν οὖν οὐθὲν φαίνεται καθ' αὑτὸ χαίρειν ὡς εἰπεῖν, ἀλλ' ὅσα πρὸς τὴν τροφὴν καὶ τὴν ἀπόλαυσιν. πονεῖν δ' ἔνια φαίνεται ταῖς ὀσμαῖς καὶ ταῖς εὐωδίαις,[2] εἴπερ ἀληθὲς τὸ ἐπὶ τῶν γυπῶν καὶ τῶν κανθάρων. τοῦτο δὲ δῆλον ὡς δι' ἐναντίωσιν τῆς ἐν αὑτοῖς φύσεως. ὡς δὲ καθ' ἕκαστον ἅμα δεῖ τὴν γε κρᾶσιν τὴν ἑκάστου καὶ τὴν τῆς ὀσμῆς λαμβάνειν δύναμιν.

5 Εἰσὶ μὲν οὖν ἔνιαι τῶν εὐόσμων καὶ ἐν ταῖς τροφαῖς, οἷον αἱ τῶν ἀκροδρύων καὶ ἀπίων καὶ μήλων· αὗται γὰρ ἄνευ τῆς προσφορᾶς ἡδεῖαι, καὶ μᾶλλον ὡς εἰπεῖν. οὐ μὴν ἀλλ' ὥς γ' ἁπλῶς διελεῖν αἱ μέν εἰσι καθ' αὑτὰς αἱ δὲ κατὰ συμβεβηκός· αἱ μὲν τῶν χυλῶν καὶ τῆς τροφῆς κατὰ συμβεβηκός, αἱ δ' ὥσπερ τῶν ἀνθῶν καθ' αὑτάς. ὡς δ' ἐπίπαν τὰ εὔοσμα, καθάπερ καὶ πρότερον ἐλέχθη, δύσχυμα καὶ στρυφνὰ καὶ ὑπόπικρα.

---

[1] In Thrace. cf. Arist. H.A. 9. 36. Turn. quotes an illustration from Scriptor θαυμασίων ἀκουσμάτων 126.

[2] εὐωδίαις. ? εὐώδεσι.

cases it is not obvious to us because, one might almost say, our sense of smell is inferior to that of all other animals. Thus things which appear to us to have no odour give forth an odour of which other animals are conscious : for instance beasts of burden can smell the barley of Kedropolis,[1] and refuse to eat it because of its evil odour. Also we are unaware of the odour of animals which appear to possess one. Now no animal appears to take pleasure in a good odour for its own sake, so to speak, but only in the odour of things which conduce to its nurture and enjoyment. Indeed some animals seem to be annoyed by odours, even good[2] ones, if what is said of vultures and beetles be true ; the explanation is that their natural character is antipathetic to odours. To appreciate this in particular cases one should take into consideration the temperament of the animal in question and also its power of smell.

Of smell and taste.

Now the odour of some things which have a good odour resides in things which are used for food, for instance that of stone-fruits[3] pears and apples, the smell of which is sweet even if one does not eat them ; indeed it may be said to be sweeter in that case. However, to make a general distinction, some odours exist independently, while others are incidental;[4] those of juices and things used for food are incidental, those of flowers exist independently. And, as was said above,[5] things which have a good odour are generally of unpleasant, astringent or

[3] ἀκροδρύων here apparently plums, peaches, etc.
[4] *i.e.* the smell is a kind of 'accident,' or by-product of the taste.   [5] l. 3.

ἔνια δὲ τῶν εὐχύμων καὶ κακώδη, καθάπερ καὶ τὸ Αἰγύπτιον καλούμενον σῦκον, γλυκὺ ὄν, καὶ εἰ μὴ πανταχοῦ ἀλλ' ἐνιαχοῦ. καὶ ἡ ἄρκευθος ἐμφαίνει τινὰ τῇ μασήσει κακωδίαν γλυκεῖα οὖσα· τὸ δ' οὖρον ποιεῖ εὐῶδες.

6 Ἐπεὶ δὲ τῶν ὀσμῶν αἱ μὲν ἐν φυτοῖς καὶ τοῖς τούτων μορίοις, οἷον κλωσὶ φύλλοις φλοιοῖς καρποῖς δακρύοις, αἱ δὲ ὥσπερ διείλομεν ἐν ζώοις [καὶ φυτοῖς] καὶ τοῖς ἀψύχοις, αὗται μὲν φανερὸν ὅτι πέψιν ἔκασται λαμβάνουσιν ἐν τοῖς οἰκείοις [αἷς]· καὶ τὸ εὐῶδες καὶ κακῶδες ἀκολουθεῖ κατὰ τὰς οἰκείας φύσεις, ἡ δὲ πέψις τῷ οἰκείῳ θερμῷ. ἐν δὲ τοῖς ἀψύχοις ταῖς τῶν ἁπλῶν δυνάμεσι καὶ γίνονται καὶ μεθίστανται καθάπερ οἱ χυμοί.

7 III. Ὅσαι δὲ δὴ κατὰ τέχνην καὶ ἐπίνοιαν γίνονται περὶ τούτων πειρατέον εἰπεῖν ὥσπερ καὶ περὶ τῶν χυλῶν. ἐν ἀμφοῖν δὲ δῆλον ὡς ἀεὶ πρὸς τὸ βέλτιον [ἦν] ἡμῖν ἡ ἀναφορά· πᾶσα γὰρ τέχνη στοχάζεται τούτου. εἰσὶ μὲν οὖν καὶ τοῖς ἀμίκτοις ὀσμαί τινες πρὸς ἃς συνεργεῖν πειρῶνται καὶ ταῖς παρα<σκευαῖς, ὡς καὶ> πρὸς τὰς τῶν χυμῶν εὐστομίας. οὐ μὴν ἀλλ' ὥς γ' ἁπλῶς

---

[1] cf. *H.P.* 1. 11. 2.
[2] *i.e.* the berry: Sch. would read ἀρκευθίς. cf. *H.P.* 3. 12. 4, with which this statement is inconsistent. Sch. suggests punctuating—γλυκὺ ὄν. καὶ εἰ μὴ πανταχοῦ ἀλλ' ἐνιαχοῦ καὶ ἡ ἄρκευθος κ.τ.λ. [3] καὶ φυτοῖς om. Turn.
[4] αἷς I omit; ἢ καὶ τὸ εὐ. conj. Turn.

somewhat bitter taste. Again some things which have a good taste have also an evil odour, such as the carob,[1] which is sweet (this is true of some regions, if not of all). Again the Phoenician cedar,[2] though it is sweet to the taste, when chewed produces a sort of evil odour, though it makes the water fragrant.

Of odours in plants.

Some odours being found in plants or in their parts—as twig, leaf, bark, fruit, gum—and others, as we distinguished, in animals [3] and in inanimate things, it is plain that the former are matured each of them in the part to which it belongs; and [4] a good or evil odour follows according to the natural character of that part, the maturing being due to the warmth which is found in it. On the other hand in inanimate things the odour, like the taste, is formed and modified by the properties of the simple substances of which the thing is made.

Of artificial odours in general and their manufacture: especially of the use of perfumes in wine.

III. Next we must endeavour to speak of those odours, and also those tastes, which are artificially [5] and deliberately produced. In either case it is clear that improvement is always what we have in view; for that is the aim of every artificial process. Now even uncompounded substances have certain odours, which men endeavour to assist by artificial means,[6] even as they try to assist nature in producing palatable tastes. However, to speak generally, the

[5] The same phrase occurs in similar connexion *C. P.* 6. 11. 2.
[6] Text defective. Ald.Bas.Vo. have marks of omission. W. after Turn. gives καὶ ταῖς παρα<σκευαῖς, ὡς καὶ> ταῖς τῶν χυμῶν εὔστο<μίαις>, which I have slightly altered.

εἰπεῖν ἐν μίξει τὸ πλέον, καὶ οὕτως αἱ <μίξεις> δυοῖν μὲν ὡς τῷ γένει λαβεῖν, ὑγροῦ καὶ ξηροῦ· τριχῶς δὲ γί<νονται>, ὅταν ἢ ὁμογενὲς ὁμογενεῖ, ἢ παράλλαττον τῷ παραλλάττοντι, ἢ ὑγρῷ ὑγρὸν ἢ ξηρῷ ξηρόν, <ἢ ὑγρῷ ξηρόν>.

8 Ἐκ δυοῖν γὰρ τούτων καὶ ἡ τῶν χυλῶν καὶ τῶν ὀσμῶν γένεσις· ὡς μὲν οἱ τὰ ἀρώματα καὶ τὰ διαπάσματα συντιθέντες ξηροῖς πρὸς ξηρά· ὡς δ' οἱ τὰ μύρα κεραννύντες ἢ τῷ οἴνῳ ἐπιχέοντες ὑγροῖς πρὸς ὑγρά. τὸ δὲ τρίτον, ὃ καὶ πλεῖστόν ἐστιν, ὡς οἱ μυρεψοὶ ξηροῖς πρὸς ὑγρά· παντὸς γὰρ μύρου καὶ χρίσματος ἡ σύνθεσις αὕτη. δεῖ δ' εἰδέναι ποῖαι ποίοις εὔμικτοι καὶ ποῖαι ποίοις συνεργοῦσιν εἰς τὸ ποιεῖν μίαν ὥσπερ ἐπὶ τῶν χυμῶν. καὶ γὰρ ἐκεῖ ταὐτὸ τοῦτο ζητοῦσιν οἱ μιγνύντες καὶ οἷον ἀρτύοντες. ταῦτα μὲν οὖν ἐν οἷς καὶ δι' ὧν αἱ τέχναι ποιοῦνται τὰ τέλη.

9 Μίγνυνται δὲ τὰ μὲν αὐτῆς τῆς ὀσμῆς ἕνεκα καὶ πρὸς ταύτην τὴν αἴσθησιν, τὰ δ' ὥσπερ ἡδύνειν βουλόμενα τὴν γεῦσιν, οἷον ὡς οἱ τὰ μύρα τοῖς οἴνοις ἐπιχέοντες ἢ τὰ ἀρώματα ἐμβάλλοντες.

---

[1] I have supplied μίξεις to fill the lacuna marked by W. after οὕτως αἱ : the text to the end of the section is defective, but a makeshift restoration and rendering seem possible : the sense of οὕτως is obscure.

CONCERNING ODOURS, 7-9

result is usually obtained by a mixture, and accordingly[1] such mixtures are of two things (or classes of things), a liquid and a solid: but there are three ways in which the result may be reached (the combination[2] being one either of like with like, or of unlike substances), according as a liquid is compounded with another liquid, a solid with another solid, or a solid with a liquid.

For tastes and odours alike are derived from these two things: the method of the makers of spices and perfume-powders[3] is to mix solid with solid, that of those who compound unguents or flavour wines is to mix liquid with liquid: but the third method, which is the commonest, is that of the perfumer, who mixes solid with liquid, that being the way in which all perfumes[4] and ointments are compounded. Further one must know which odours will combine well with which, and what combination makes a good blend, just as in the case of tastes: for there too those who make combinations and, as it were, season their dishes, are aiming at this same object. So much for the ingredients and the methods whereby these arts attain their ends.

The object of the mixture is in the one case simply the production of a particular odour and the gratification of the corresponding sense, in the other there is a desire to produce, as it were, a pleasanter taste: this for instance is the object of flavouring wine with perfumes or of putting spices into it.

[2] *i.e.* given two components we have three possible combinations, A with A, B with B, or A with B.
[3] διαπάσματα. *cf.* Plin. 13. 19; 21. 125.
[4] The difference between μύρον and χρίσματος does not appear; μύρον seems to be loosely used, as just above it was used of an entirely liquid mixture.

αἱ γὰρ αἰσθήσεις σύνεγγυς οὖσαι ποιοῦσί τινα ἀπόλαυσιν ἀλλήλων, ὅθεν καὶ αὐτοῖς τοῖς γευστοῖς ζητοῦσι τὰς εὐοσμίας.

10 Ἀπορήσειε δ' ἄν τις ἴσως διὰ τί ποτε μύρον καὶ τἆλλα εὔοσμα τοὺς μὲν οἴνους ἡδύνει τῶν δὲ βρωμάτων οὐδέν, ἀλλὰ πάντα λυμαίνεται καὶ ἀπύρωτα καὶ πεπυρωμένα. τὸ δ' αἴτιον ὑποληπτέον ὅτι συμβαίνει τῶν μὲν ξηρῶν ἀφαιρεῖσθαί τε τὸν οἰκεῖον χυλὸν διὰ τὴν ἰσχὺν καὶ ἅμα συνεπιφαίνειν τὸν αὐτοῦ ὄντα στρυφνὸν καὶ ὑπόπικρον· ἅπαν γὰρ τὸ εὔοσμον τοιοῦτον, διαμασωμένοις δὲ καὶ μᾶλλον ἐμφανὲς διά τε τὴν 11 θλῖψιν καὶ τομὴν καὶ ἔτι τῷ χρονίζεσθαι. τὸν δ' οἶνον οὐδέτερον ποιεῖ· καὶ γὰρ ὁ χυλὸς ἰσχυρότατος καὶ πλείων εἰς τὸ μὴ κρατεῖσθαι καὶ οὐδένα τῇ γεύσει χρόνον ἐπιδιατρίβων ἀλλ' ὅσον ἐπιθιγγάνων, ὥστε τὸ μὲν ἡδὺ ἐνδιδόναι τῇ αἰσθήσει τὸ δὲ πικρὸν καὶ δύσχυμον τῇ γεύσει μὴ ἐμφαίνειν, ἀλλὰ συμβαίνειν τῷ ὄντι καθάπερ ἥδυσμα γίνεσθαι τῷ πόματι τὴν ὀσμήν· τῷ μὲν γὰρ γλυκεῖ καὶ μάλιστα δεομένῳ διὰ τὸ μηδὲν ἔχειν, τοῖς δ' ἄλλοις ὥσπερ μιᾶς ἐξ ἀμφοῖν γενομένης

---

[1] cf. 67; Arist. de Sens. 5.   [2] cf. Arist. l.c.
[3] As opposed to wine. Sch., misunderstanding this, thinks ξηρῶν corrupt.
[4] I have restored καὶ, which Sch. and W. omit, missing the point of the antithesis μὲν . . . δέ.

CONCERNING ODOURS, 9-11

[1] For the two senses of taste and smell being akin to one another, each provides in a way for the enjoyment of the other: wherefore it is through things which appeal to the taste, as well as those which appeal to the sense of smell, that men try to discover fragrant odours.

The question may perhaps be raised why perfume and other fragrant things, while they give a pleasant taste to wine,[2] yet have not this effect on any other article of food, but in all cases spoil food, whether it be cooked or not. The explanation we must take to be that this is what happens—the perfume if mixed with solid[3] things is in any case powerful enough to deprive them of their proper taste, and at the same time it makes obtrusive its own taste, which is astringent and somewhat bitter,—all perfumeries having that character,—while, if one bites up the food, this effect is even[4] more apparent because the food is crushed and broken up, and also because it remains longer in the mouth. But on wine neither effect is produced, since in this the taste is very strong and too generally diffused to be overpowered: also wine does not linger on the palate for any length of time, but merely touches it, so that, while it makes one conscious of its own pleasant taste, it does not make the palate feel the bitter unpalatable taste of the perfume: in fact the odour of this acts as a sort of relish[5] to the draught. This effect indeed[6] it has on wine which is sweet and specially needs the addition of perfume, because it has no 'relish' of its own; while with other wines the reason is that, as the effect of the mixture, the

[5] Sc. 'bouquet.'
[6] I have restored γὰρ, omitted by Sch. and W.

διὰ τὴν μῖξιν. ὁ γὰρ οἶνος, ὥσπερ καὶ πρότερον ἐλέχθη, δεινὸς δέξασθαι τὰς ὀσμάς.

12 Ἔχει δ' ἀπόρησιν καὶ τόδε, διὰ τί τὰ μὲν ἄνθη καὶ τὰ στεφανώματα ἀσθενέστερα ὄντα ταῖς ὀσμαῖς καὶ πόρρωθεν ὄζει, ἡ δ' ἶρις καὶ τὸ νάρδον καὶ τἆλλα τὰ εὔοσμα τῶν ξηρῶν ἰσχυρότερα ἐγγύθεν· καὶ ἔνιά γε προσενεγκαμένοις, ἔνια δὲ καὶ τρίψεως προσδεῖται καὶ διαιρέσεως, τὰ δὲ καὶ πυρώσεως, ὥσπερ ἡ σμύρνα καὶ ὁ λιβανωτὸς

13 καὶ πᾶν τὸ θυμιατόν. αἴτιον δ' ὅτι τῶν μὲν ἀνθῶν ἐπιπολῆς τὸ ποιοῦν τὴν ὀσμὴν ἅτε μανῶν ὄντων καὶ οὐκ ἐχόντων βάθος, τῶν δὲ ῥιζῶν καὶ πάντων τῶν στερεῶν ἐν βάθει, τὰ δ' ἔξωθεν ἀπεξηραμμένα καὶ πεπυκνωμένα· διὸ καὶ ἀφιᾶσι πόρρω τὰς ἀποπνοίας, τὰ δ' οἷον ἀνοίξεως δέονται τῶν πόρων, ὅθεν διαιρούμενα καὶ κοπτόμενα πάντ' εὐωδέστερα, τὰ δ' ἄνθη κακωδέστερα τριβόμενα· τὰ μὲν γὰρ ἐκφαίνει τὸ οἰκεῖον τὰ δὲ προσλαμβάνει τὸ ἀλλότριον. ὁ δὲ λιβανωτὸς καὶ ἡ σμύρνα πυκνοτέραν ἔτι τὴν φύσιν ἔχοντα προσδέονται πυρώσεως μαλακῆς, ἣ κατὰ μικρὸν ἐκθερμαίνουσα ποιήσει τὴν ἀναθυμίασιν. ἐὰν γὰρ κόπτῃ τις ἢ τρίβῃ ταῦτα, προσοίσονται μὲν

---

[1] *i.e.* of the unadulterated wine and of the perfume.
[2] *C.P.* 6. 19. 2. Sch.'s reasons for bracketing this sentence seem inadequate.
[3] *i.e.* fragrant leaves, etc. *cf. H.P.* 1. 12. 4.
[4] Made from the rhizomes: *cf. H.P.* 1. 7. 2, and Index.

two [1] odours combine, as it were, to form one. Wine indeed, as was said before,[2] has a special property of assimilating odours.

Another question also suggests itself,—why it is that, while the smell of flowers and other [3] things used for garlands, though it is not so strong, can be perceived even at a great distance, the iris-perfume,[4] spikenard and other fragrant solids smell stronger at a short distance: and of some of these the smell is only perceived when they are eaten, while some need even to be bruised and broken up, and others to be subjected to fire, as myrrh frankincense and anything that is burnt as incense. The explanation is that, whereas in flowers that which causes the smell is on the surface, seeing that the texture of flowers is open and they are not substantial, in all such solid substances as roots the power of producing smell is diffused through a substantial mass, while the exterior parts are dried up and of close texture: and this is why flowers emit the scent which exhales from them to a long distance, while things like roots need an opening of their passages. Hence, when these are broken up or bruised, they are in all cases more fragrant, while, [5] if flowers are crushed, they have a comparatively evil smell: for under such treatment roots give forth the property which belongs to them, but flowers acquire a property which is not their own. Again frankincense and myrrh, since they are by nature of even closer texture than roots, need a gentle application of fire, which, by gradually warming them, will cause the scent to be exhaled. For, if these substances are bruised or crushed, they will indeed present an odour, but it will not be so

[5] *cf.* Arist. *Probl.* 12. 9; 13. 3 and 11.

THEOPHRASTUS

ὀσμὴν οὐχ ὁμοίως δὲ ἡδεῖαν οὐδ' εὐταμίευτον. τούτων μὲν οὖν τοιαῦταί τινες αἱ αἰτίαι.

14 IV. Τῶν δὲ μύρων ἡ σύνθεσις καὶ ἡ κατασκευὴ τὸ ὅλον οἷον εἰς θησαυρισμόν ἐστι τῶν ὀσμῶν· διόπερ εἰς τοὔλαιον τίθενται· τοῦτο γὰρ χρονιώτατον καὶ ἅμα πρὸς τὴν χρείαν μάλισθ' ἁρμόττον. ἐπεὶ φύσει ἥκιστα δεκτικὸν ὀσμῆς διὰ τὴν πυκνότητα καὶ τὸ λίπος, αὐτῶν δὲ τούτων τὸ λιπαρώτατον, οἷον τὸ ἀμυγδάλινον· τὸ δὲ σησάμινον καὶ τὸ ἐκ τῶν ἐλαιῶν μάλιστα.

15 Χρῶνται δὲ μάλιστα τῷ ἐκ τῆς βαλάνου τῆς Αἰγυπτίας καὶ Συρίας, ἥκιστα γὰρ λιπαρόν· ἐπεὶ καὶ τῷ ἐκ τῶν ἐλαιῶν μάλιστα χρῶνται τῷ ὠμοτριβεῖ τῆς φαυλίας· δοκεῖ γὰρ ἀλιπέστατον ἔχειν καὶ λεπτότατον· καὶ τούτῳ νέῳ καὶ μὴ παλαιῷ· τὸ γὰρ ὑπὲρ ἐνιαυτὸν ἀχρεῖον παχύτερον καὶ λιπαρώτερον γενόμενον. ἔλαιον μὲν οὖν τὸ τοιοῦτον οἰκειότατον, ἀλιπέστατον γάρ. φασὶ δέ τινες καὶ <ἐν> τῷ χρίσματι τὸ ἐκ τῶν πικρῶν ἀμυγδάλων· πολλὰ δὲ γίνεται περὶ Κιλι-
16 κίαν καὶ ποιοῦσιν ἐξ αὐτῶν χρίσμα. φασὶ δὲ καὶ εἰς τὰ σπουδαῖα τῶν μύρων ἁρμόττειν, ὥσπερ καὶ τὸ ἐκ τῆς βαλάνου καὶ αὐτό· ποιεῖ δὲ <τὰ> κελύφη αὐτῶν εὔοσμον εἰς τὸ ἔλαιον ἐμβαλλό-

---

[1] This passage was misunderstood by Plin. 13. 19. The sense seems to be that the viscous character of oil, though preservative of perfume, is not easily receptive of it.

[2] cf. H.P. 4. 2. 1; 4. 2. 6. βάλανος, balanites aegyptiaca. See Index.

CONCERNING ODOURS, 13-16

sweet nor so lasting as when they have been subjected to fire. Such are the explanations of these difficulties.

Of the oils used as the vehicle of perfumes.

IV. Now the composition and preparation of perfumes aim entirely, one may say, at making the odours last. That is why men make oil the vehicle of them, since it keeps a very long time and also is most convenient for use. [1] By nature indeed oil is not at all well suited to take in an odour, because of its close and greasy character: and of particular oils this is specially true of the most viscous, such as almond-oil, while sesame-oil and olive-oil are the least receptive of all.

The oil most used is that derived from the Egyptian [2] or Syrian *balanos*, since this is the least viscous; the olive-oil which is most used is that which is pressed from 'coarse olives' [3] in the raw state, since this is thought to be the least greasy and the least coarse: this is used while it is new, not when it is old, for that which is kept above a year is useless, having become thick and viscous. This then is the kind of olive-oil which is most suitable, since it is the least greasy. Some say that for unguent the oil derived from bitter almonds is best: these are abundant in Cilicia, where an unguent is made from them. It is said that this is suitable for choice perfumes, like the oil of the Egyptian *balanos*: this is suitable in itself,[4] however the shells of the fruit are thrown into the oil to give it a good odour: indeed they are also thrown into

[3] cf. *H.P.* 2. 2. 12; *C.P.* 6. 8. 3 and 5.
[4] αὐτὸ conj. Sch.; τοῦτο Vulg.W.

μενα· ἐπεὶ καὶ τὸ τῶν πικρῶν. ἤδη δὲ πῶς οὐκ ἐναντίον ἅμα μὲν τὸ ἀοσμότατον ζητεῖν, ὥσπερ καὶ τὸ ὠμοτριβὲς ἐκ τῶν φαυλιῶν, ἅμα δ' ἐν τούτοις ποιεῖν; δριμύτητα γὰρ ἔχει τὸ τῶν ἀμυγδάλων· εἰ μὴ ἄρ' ὅτι τὸ ἔλαιον ἑψόμενον κακῶδες. ταῦτα μὲν οὖν ἐπισκεπτέον.

17 Χρῶνται δὲ πρὸς πάντα τοῖς ἀρώμασι, τοῖς μὲν ἐπιστύφοντες τὸ ἔλαιον τοῖς δὲ καὶ τὴν ὀσμὴν ἐκ τούτων ἐμποιοῦντες. ὑποστύφουσι γὰρ πᾶν εἰς τὸ δέξασθαι μᾶλλον τὴν ὀσμήν, ὥσπερ τὰ ἔρια εἰς τὴν βαφήν. ὑποστύφεται δὲ τοῖς ἀσθενεστέροις τῶν ἀρωμάτων, εἶθ' ὕστερον ἐμβάλλουσιν ἀφ' οὗ ἂν βούλωνται τὴν ὀσμὴν λαβεῖν· ἐπικρατεῖ γὰρ ἀεὶ τὸ ἔσχατον ἐμβαλλόμενον καὶ ἂν ἔλαττον ᾖ· οἷον ἐὰν εἰς κοτύλην σμύρνης ἐμβληθῇ μνᾶ καὶ ὕστερον ἐμβληθῶσι κιναμώμου δραχμαὶ δύο, κρατοῦσιν αἱ τοῦ κιναμώμου δύο δραχμαί.

18 Θαυμάσειε δ' ἄν τις ἴσως τοῦτό τε καὶ διὰ τί ποτε τὰ ἀρώματα προεμβαλλόμενα δεκτικώτερον ποιεῖ τοὔλαιον ὀσμὴν ἔχοντα· δεῖ γὰρ ἀῶδες εἶναι τὸ δεξόμενον, τὸ δὲ κατειλημμένον ὑφ' ἑτέρου οὐκ ἀῶδες, ὥσθ' ἧττον ἐχρῆν εἶναι δεκτικόν. αἴτιον δ' ἀμφοτέρων ἢ πάντων τὸ αὐτό. ξηρὰ γὰρ ὄντα τὸ λίπος ἕλκει πρὸς ἑαυτὰ

---

[1] τὸ conj. Sch.; τὰ Vulg. W. Sch. also adds ἀμυγδάλων after πικρῶν.
[2] *i.e.* those derived from the Egyptian *balanos* and bitter almonds.

CONCERNING ODOURS, 16-18

that[1] which is made from bitter almonds. Once more, is it not inconsistent to seek the vehicle which has the least odour of its own, such as the oil which is pressed raw from 'coarse olives,' and yet at the same time to use the above-mentioned[2] oils as vehicles? (for oil of almonds has a pungent smell). Possibly the explanation is that it is only by being cooked that oil acquires an evil smell.[3] These matters then are subject for enquiry.

They use spices in the making of all perfumes; some to thicken[4] the oil, some in order to impart their odour. For in all cases they thicken the oil to some extent to make it take the odour better, just as they treat wool for dyeing. The less powerful spices are used for the thickening, and then at a later stage they put in the one whose odour they wish to secure. [5] For that which is put in last always dominates, even if it is in small quantity; thus, if a pound of myrrh is put into a half-pint of oil, and at a later stage a third of an ounce of cinnamon is added, this small amount dominates.

At this one may well wonder; and also why it is that the previous addition of spices, which have an odour of their own, renders the oil more receptive: for the vehicle should be scentless, but a substance over which another substance has thus prevailed, cannot be scentless, so that it ought, one would think, to have become *less* receptive. However both facts, or rather all of them, may be accounted for in the same way:—the spices, being solid, attract to

---

[3] Sc. 'and these oils are used in the raw state' (?). I do not see how Furlanus' explanation, quoted by Sch., is to be found in the text. The following sentence shews that T. does not claim to have settled the question.

[4] *i.e.* to make it less volatile.   [5] *cf.* Plin. 13. 19.

καὶ ἀναδέχεται, διὸ καὶ τὴν συνέχειαν ἐξαιρεῖ· μανὸν δὲ γενόμενον [καὶ][1] τοῦ λίπους ἀφαιρεθέντος ἐν ᾧ καὶ ἡ οἰκεία μάλιστα ὀσμή, δεκτικώτερον ἐγένετο τοῦ ἐπιβαλλομένου διὰ τὸ μὴ ἀντιστατεῖν.

19 Ἡ δὲ ἀπὸ τῶν ἀρωμάτων ὀδμὴ καὶ ἀσθενὴς ἅτε εἰς τὸ λιπαρὸν ἀνηλωμένη, καὶ ἔτι κατέχεται τούτῳ διὰ τὸ πληρῶσαι τοὺς πόρους. ὥστε κατὰ λόγον κἂν ἔλαττον ᾖ τὸ ἐπιβαλλόμενον ἐπικρατεῖν τὴν τούτου ὀσμήν· εἰς ἀσθενέστατον γὰρ ἐμπίπτει καὶ δεκτικώτερον. ἀνὰ λόγον δ' ἔχει καὶ ἡ πολυχρονιότης ἡ ἐν ἑκάστῳ καὶ ἡ πρὸς τὴν πύρωσιν εὐσθένεια καὶ τἆλλα τὰ τοιαῦτα. τὸ γὰρ δεκτικώτατον, οἷον τῆς βαλάνου, καὶ χρονιώτατον, καὶ διὰ τὴν αὐτὴν αἰτίαν· μάλιστα γὰρ ὥσπερ ἓν γίνεται καὶ συμφυὲς τὸ μάλιστα δεχόμενον· ἀεὶ γὰρ τὸ τοιοῦτον διαμονώτατον, διὸ καὶ πυρούμενον μάλιστα ἀπαθές.

20 Ὡσαύτως δὲ καὶ τῶν ἄλλων τὸ σησάμινον, τοῦτο γὰρ δεκτικώτατον· τὸ δὲ ἀμυγδάλινον παρακμάζει ταχὺ καὶ ὀλιγοχρονιώτατον διὰ τὴν ἐναντίαν αἰτίαν· τὸ γὰρ ἥκιστα δεξάμενον τάχιστα μεθίησι. τοῦ ῥοδίνου δὲ μάλιστα δεκτικὸν τὸ

---

[1] I have bracketed καί.
[2-2] This passage is omitted, apparently by accident, in both W.'s texts, though represented in his Latin version. I

themselves the viscid part of the oil, and so it attaches itself to them; thus the density of the oil is destroyed: the oil, thus becoming thinner by the removal[1] of its viscid part which chiefly contains the characteristic odour, becomes more receptive of the spice which is added to it, because it does not now offer resistance.

Again that odour which is due to the spices becomes less powerful as it is spent on the viscid part of the oil, while at the same time it is preserved by this because it has entirely filled up its passages. Wherefore it naturally follows that, even if the added spice is in small quantity, its odour predominates, since it passes into a vehicle which is in itself not at all powerful and which is more receptive than itself. A corresponding account may be given of the keeping quality of the several oils, of their power of resisting fire, and other such qualities. Thus that oil which is most receptive, for instance, that of the Egyptian *balanos*, will also keep longest, and for the same reason; namely that that oil which is most receptive unites, more than others, into one single substance, as it were, with the spices. Such a substance will always last longer than others; which also explains why, if exposed to fire, it is less affected than others.

Of the other oils the same applies to that of sesame, this being specially receptive; [2] but, for the contrary reason, almond-oil soon loses its virtue and keeps for a shorter time than any other, for that oil which has been least receptive parts soonest with the property received. Sesame-oil however receives rose-perfume better than other oils [2] because of its

have printed it from Sch.'s text. The omission is evidently due to the double occurrence of τὸ σήσαμινον.

345

σησάμινον διὰ τὴν λιπαρότητα· πυρούμενον δὲ ἐξόζει σησάμου καθάπερ ἀναλυόμενον. αἱ μὲν οὖν τῶν ἐλαίων φύσεις καὶ δυνάμεις τοιαῦται.

21 V. Τὰ δ' ἀρώματα πάντα σχεδὸν καὶ εὔοσμα πλὴν τῶν ἀνθῶν ξηρὰ καὶ θερμὰ καὶ στυπτικὰ καὶ δηκτικά. τὰ δὲ καὶ ἔχοντά τινα πικρότητα, καθάπερ καὶ ἐν τοῖς πρότερον εἴπομεν, ὥσπερ ἶρις σμύρνα λιβανωτός, ὡς δ' ἁπλῶς εἰπεῖν καὶ τὰ μύρα. κοινόταται δὲ τῶν δυνάμεων τό τε στυπτικὸν καὶ τὸ θερμαντικόν, ἃ δὴ καὶ ἐργάζονται.

22 Ὑποστύφονται μὲν οὖν πάντα πυρούμενα, τὰς δ' ὀσμὰς τὰς κυρίας ἔνια λαμβάνει ψυχρὰ καὶ ἀπύρωτα. καὶ ἔοικεν ὥσπερ τῶν ἀνθῶν τὰ μὲν ψυχροβαφῆ τὰ δὲ θερμοβαφῆ παραπλησίως ἔχειν καὶ ἐπὶ τῶν ὀσμῶν. πάντων δὲ ἡ ἕψησις εἴς τε τὴν ὑπόστυψιν καὶ τὰς κυρίας ὀσμὰς ἐνισταμένων τῶν ἀγγείων ὕδατι γίνεται καὶ οὐκ αὐτῷ τῷ πυρὶ χρωμένων· τοῦτο δέ, ὅτι μαλακὴν εἶναι δεῖ τὴν θερμότητα, καὶ ἀπουσία πολλὴ γένοιτ' ἂν τῇ φλογὶ χρωμένων, καὶ ἔτι καῦσιν ἂν ὄζοι.

23 Ποιεῖ δ' ἐλάττω τὴν ἀπουσίαν ὅσα πυρούμενα λαμβάνει τὰς κυρίας ὀσμὰς μᾶλλον ἢ ὅσα ψυχρὰ διὰ τὸ προφυρᾶσθαι τὰ πυρούμενα, τὰ μὲν οἴνῳ εὐώδει, τὰ δὲ ὕδατι· ἧττον γὰρ ἀναπίνει· τὰ δὲ ψυχρὰ ξηρὰ ὄντα μᾶλλον, καθάπερ ἶρις κοπεῖσα.

viscid quality ; and, when subjected to fire, it gives out a smell of sesame, as though it were being disintegrated. Such are the special characters and properties of the various oils.

Of the spices used in making perfumes and their treatment.

V. Almost all spices and sweet scents except flowers are dry hot astringent and mordant. Some also possess a certain bitterness, as we said above, as iris, myrrh, frankincense, and perfumes in general. However the most universal qualities are astringency and the production of heat ; they actually produce these effects.

All spices are given their astringent quality by exposure to fire, but some of them assume their special odours even when cold and not exposed to fire ; and it also appears that, just as with vegetable dyes some are applied hot and some cold, so is it with odours. But in all cases the cooking, whether to produce the astringent quality or to impart the proper odour, is done in vessels standing in water and not in actual contact with the fire ; the reason being that the heating must be gentle, and there would be considerable waste if these were in actual contact with the flames ; and further the perfume would smell of burning.

However there is less waste when the perfume obtains its proper odour by exposure to fire than when it does so in a cold state, since those perfumes which are subjected to fire are first steeped either in fragrant wine or in water : for then they absorb[1] less : while those which are treated in a cold state, being dry, absorb[1] more, for instance bruised iris-root. Thus, if

[1] ἀναπίνει. So Sch. explains. *cf.* ἐκπίνωσιν, 24.

λαμβάνοντος γὰρ τοῦ ἀμφορέως ξηρᾶς ἴριδος κεκομμένης μέδιμνον καὶ δύο ἡμίεκτα πολλὴν ποιεῖν φασὶν ἀπουσίαν, ἐὰν δὲ μετρίως φυράσῃ λείπειν ὅσον δύο χόας, τοῖς δὲ πολλοῖς ἔλαττον.

24 Γίνεται δὲ τὸ βέλτιον ἴρινον ἐὰν ᾖ ξηρὰ καὶ ἀπύρωτος ἡ ἶρις· ἀκρατεστέρα γὰρ ἡ δύναμις ἢ ἐὰν φυραθεῖσα καὶ πυρουμένη. συμβαίνει δὲ ὥσπερ καὶ ἐκθλίβεσθαι μᾶλλον ἐκ τῶν προπεφυραμένων διὰ τὸ ἧττον ἀναδέχεσθαι καὶ ἕλκειν εἰς αὑτό· προστύφοντες δὲ οὐ πολὺν χρόνον ἐῶσι τὰ ἀρώματα ἀλλ' ἐξαιροῦσιν, ὅπως μὴ πολὺ ἐκπίνωσιν.

25 Πρὸς ἕκαστον δὲ τῶν μύρων ἐμβάλλουσι τὰ πρόσφορα τῶν ἀρωμάτων, οἷον εἰς μὲν τὴν κύπρον καρδάμωμον ἀσπάλαθον ἀναφυράσαντες τῷ εὐώδει. εἰς δὲ τὸ ῥόδινον σχοῖνον ἀσπάλαθον κάλαμον. ἡ δ' ἀναφύρασις ὁμοίως. καὶ τοῖς ἄλλοις ἀεὶ τὰ ἁρμόττοντα. τῷ ῥοδίνῳ δ' ἐμβάλλονται καὶ ἅλες πολλοὶ καὶ τοῦτ' ἴδιον παρὰ τἆλλα, διὸ καὶ πλείστη ἀπουσία γίνεται· μίγνυται γὰρ εἰς τὸν ἀμφορέα δύο μέδιμνοι.

26 Τῆς δὲ κύπρου ἡ μὲν ἐργασία παραπλησία τῇ

---

[1] Dry measure: the equivalents given are, of course, only approximate.
[2] τὸ βέλτιον ἴρινον W. after Sch.; τὸ βέλτιον τὸ ἴρινον vulg. The article must be omitted in one place or the other.

κύπρος, called from a tree of that name: not mentioned in *H.P.* cf. Plin. 12. 119.

[4] cf. *H.P.* 9. 7. 2 and 3. [5] cf. *H.P.* 9. 7. 3.

into eight and a half gallons of oil we put thirteen gallons[1] of dry and bruised iris-root, they say that much loss is caused, while if one does not steep it too much, only about eleven pints and a half are wasted: and in the case of most perfumes the waste is less.

However the superior[2] iris-perfume is made by using the root dry and not subjecting it to fire: for then its virtue asserts itself more completely than when it is steeped in a liquid or subjected to fire. It also comes to pass that, if the perfumes have been first steeped, their virtues are, as it were, squeezed out of them to a greater extent, because they take in and absorb less: and so, when they are making them astringent, they do not leave the spices in the oil for long, but take them out, so that they should not absorb an excessive amount.

For making each perfume they put in the suitable spices. Thus to make *kypros*[3] they put in cardamom[4] and *aspalathos*,[5] having first steeped them in sweet wine.[6] To make rose-perfume they put in ginger-grass *aspalathos* and sweet-flag: and these are steeped as in the case of *kypros*. So too into each of the others are put the spices which suit them. Into rose-perfume moreover is put a quantity of salt[7]: this treatment is peculiar to that perfume, and involves a great deal of waste, twenty-three gallons[8] of salt being put to eight gallons and a half of the perfume.

The manufacture of *kypros* resembles that of

[6] τῷ εὐώδει here evidently means the same as τῷ γλυκεῖ, 44, where τῷ οἴνῳ τῷ εὐώδει occurs just above: *cf.* μελικράτῳ ἢ γλυκεῖ, *C.P.* 6. 17. 2.

[7] To prevent decay, as Diosc. 2. 53 explains.

[8] Turn. suggests that μέδιμνοι should be μναῖ, the initial M having been misunderstood by a copyist.

τοῦ ῥοδίνου· πλὴν ἀλλ' ἐάν τις μὴ ταχέως ἐξαίρῃ καὶ ἀποθλίβῃ σῆψις ἐγγινομένη φθείρει τὰ μύρα διὰ τὴν δυσωδίαν· ποιεῖ γὰρ σῆψιν ἀνυγραινομένη. παραπλησία δ' ἐργασία καὶ τοῦ μηλίνου· προστυφέντος γὰρ ἐλαίου καὶ τὰ μῆλα ἐμβάλλουσιν εἰς ψυχρόν, εἶτ' ἐξαιροῦσι πάλιν πρὸ τοῦ μελαίνεσθαι κατὰ πάσας τὰς ἐμβολάς· μελαινομένων γὰρ σῆψις διὰ τὸ ἀνυγραίνεσθαι, καθάπερ καὶ ἐπὶ τῆς κύπρου.

27 VI. Ἅπαντα δὲ συντίθενται τὰ μύρα τὰ μὲν ἀπ' ἀνθῶν τὰ δὲ ἀπὸ φύλλων τὰ δὲ ἀπὸ κλωνὸς τὰ δ' ἀπὸ ῥίζης τὰ δ' ἀπὸ ξύλων τὰ δ' ἀπὸ καρποῦ τὰ δ' ἀπὸ δακρύων. μικτὰ δὲ πάνθ' ὡς εἰπεῖν. ἀπ' ἀνθῶν μὲν οἷον τὸ ῥόδινον καὶ τὸ λευκόϊνον. καὶ τὸ σούσινον· καὶ γὰρ τοῦτο ἐκ τῶν κρίνων· ἔτι δὲ τὸ σισύμβρινον καὶ τὸ ἑρπύλλινον καὶ ἡ κύπρος καὶ πρὸς τούτοις τὸ κρόκινον· βέλτιστος δ' ἐν Αἰγίνῃ καὶ Κιλικίᾳ. ἀπὸ δὲ τῶν φύλλων οἷον τό τε μύρρινον καὶ τὸ οἰνάνθινον· αὕτη δ' ἐν Κύπρῳ φύεται ὀρεινὴ καὶ πολυόδμος· ἀπὸ δὲ τῆς ἐν τῇ Ἑλλάδι οὐ γίνεται διὰ τὸ ἄοδμον.

28 Ἀπὸ ῥιζῶν δὲ τό τε ἴρινον καὶ τὸ νάρδινον καὶ τὸ ἀμαράκινον ἐκ τοῦ κόστου· τοῦτο γὰρ ὀνο-

---

[1] cf. Diosc. 1. 58.
[2] I have bracketed καὶ as suggested by Sch.
[3] This passage, with some variations, is quoted by Athen. 15. 39. [4] cf. Plin. 13. 11.
[5] cf. H.P. 6. 6. 11. for the plant, and for the perfume Athen. 15. 38.

rose-perfume, except that, unless one soon takes out the flowers and squeezes them out, decay sets in and ruins the perfume by giving it a disagreeable smell, since they cause decay as they get soaked. Similar also is the manufacture of quince-perfume[1]: the oil is first made astringent, and is cold when the quinces[2] are put into it: then they take them out before they turn black, removing each batch before the next is put in: for, as they turn black, decay ensues because they get soaked through—just as in the case of *kypros*.

Of the various parts of plants used for perfumes, and of the composition of various notable perfumes.

VI. [3] Perfumes are compounded from various parts of the plant, flowers leaves twigs root wood fruit and gum: and in most cases the perfume is made from a mixture of several parts. Rose and gilliflower perfumes are made from the flowers: so also is the perfume called *susinon*,[4] this too being made from flowers, namely, lilies: also the perfumes named from bergamot-mint and tufted thyme, *kypros*, and also the saffron-perfume; the crocus which produces this is best in Aegina and Cilicia. Instances of those made from the leaves are the perfumes called from myrtle and drop-wort[5]: this grows in Cyprus on the hills and is very fragrant: that which grows in Hellas yields no perfume, being scentless.

[6] From roots are made the perfumes named from iris spikenard and sweet marjoram,[7] an ingredient in which is *koston*; for it is the root to which this name

---

[6] Instances of perfumes made from twigs seem to be missing.   [7] *cf.* 30. Text perhaps defective.

μάζουσι τὴν ῥίζαν. τὸ δὲ χρίσμα τὸ Ἐρετρ...
ἐκ τοῦ κυπείρου. κομίζεται δὲ ἀπὸ τῶν Κυκλά-
δων τὸ κύπειρον. ἀπὸ ξύλου δὲ ὁ φοῖνιξ καλού-
μενος· ἐμβάλλουσι γὰρ τὴν ὀνομαζομένην σπάθην
ξηράναντες. ἀπὸ καρπῶν δὲ τό τε μήλινον καὶ
τὸ μύρτινον καὶ τὸ δάφνινον· τὸ δ' Αἰγύπτιον ἐκ
πλειόνων, ἔκ τε τοῦ κιναμώμου καὶ ἐκ σμύρνης καὶ
ἐξ ἄλλων.

29 Ἔτι δ' ἐκ πλειόνων τούτου τὸ μεγαλεῖον· καὶ
γὰρ ἐκ κιναμώμου ... καὶ ἐκ τῆς σμύρνης κοπτο-
μένης ἔλαιον ῥεῖ· στακτὴ γὰρ καλεῖται διὰ τὸ
<κατὰ> μικρὸν στάζειν. ὃ δὴ μόνον τινές φασιν
ἁπλοῦν εἶναι καὶ ἀσύνθετον τῶν μύρων τὰ δ'
ἄλλα πάντα σύνθετα, πλὴν τὰ μὲν ἐκ πλειόνων
τὰ δ' ἐξ ἐλαττόνων, ἐξ ἐλαχίστων δὲ τὸ ἴρινον. οἱ
μὲν οὖν οὕτω λέγουσιν, οἱ δὲ τὴν ἐργασίαν τῆς
στακτῆς εἶναι τοιάνδε· τὴν σμύρναν ὅταν κόψωσι
καὶ διατήξωσι ἐν ἐλαίῳ βαλανίνῳ πυρὶ μαλακῷ
ὕδωρ ἐπιχεῖν θερμόν· συνιζάνειν δ' εἰς βυθὸν τὴν
σμύρναν καὶ τοὔλαιον καθάπερ ἰλύν· ὅταν δὲ
τοῦτο συμβῇ, τὸ μὲν ὕδωρ ἀπηθεῖν τὴν δ' ὑπό-
στασιν ἀποθλίβειν ὀργάνοις.

30 Τὸ δὲ μεγαλεῖον ἐκ ῥητίνης κεκαυμένης συντίθεσθαι καὶ ἐλαίου βαλανίνου· μίγνυσθαι δὲ
κασίαν κινάμωμον σμύρναν. πλείστην δὲ πραγ-
ματείαν περὶ τὸ μεγαλεῖον καὶ τὸ Αἰγύπτιον εἶναι,

---

[1] cf. H.P. 9. 7. 3 ; C.P. 6. 11. 13.
[2] cf. H.P. 2. 8. 4. σπάθην appears to be a conj. of W. for vulg. πλάτην : ἐλάτην Turn. cf. LS. s.v.
[3] Said to be called after the inventor, one Megallos : cf. Plin. 13. 13.

CONCERNING ODOURS, 28-30

is applied. The Eretrian unguent is made from the root of *kypeiron*,[1] which is obtained from the Cyclades as well as from Euboea. From wood is made what is called 'palm-perfume': for they put in what is called the 'spathe,'[2] having first dried it. From fruits are made the quince-perfume, the myrtle, and the bay. The 'Egyptian' is made from several ingredients, including cinnamon and myrrh.

Again from several parts of the plant is made the perfume called *megaleion*,[3] which is made from cinnamon and . . . .[4] and from the myrrh when it is bruised flows an oil: it is in fact called *stakte*[5] (in drops) because it comes in drops slowly. Some indeed say that this is the only simple uncompounded perfume, and that all the others are compound, though made from a larger or smaller number of ingredients, and that iris-perfume is made from the smallest number of all. Some assert this, but others declare that the manufacture of *stakte* (myrrh-oil) is as follows: having bruised the myrrh and dissolved it in oil of *balanos* over a gentle fire, they pour hot water on it: and the myrrh and oil sink to the bottom like a deposit; and, as soon as this has occurred, they strain off the water and squeeze the sediment in a press.

Megaleion, these authorities say, is compounded of burnt resin[6] and oil of *balanos*, with which are mixed cassia cinnamon and myrrh. They add that this perfume and the Egyptian are the most troublesome

---

[4] The end of the account of *megaleion* and the beginning of that of myrrh-perfume seem to be missing. ? Supply καὶ κασίας καὶ σμύρνης.

[5] *cf. H.P.* 9. 4. 10.   [6] *cf.* Plin. 13. 7.

πλείστων γὰρ μίξιν καὶ πολυτελεστάτων. τῷ δὲ μεγαλείῳ καὶ τὸ ἔλαιον ἕψεσθαι δέχ' ἡμέρας καὶ δέκα νύκτας, εἶτα οὕτως τὴν ῥητίνην ἐμβάλλεσθαι καὶ τἆλλα· δεκτικώτερον γὰρ ἀφεψηθέν. τὸ δ' ἀμαράκινον τὸ χρηστὸν ἐκ τῶν βελτίστων ἀρωμάτων συντίθεσθαι χωρὶς ἀμαράκου· τούτῳ δ' οὐ χρῆσθαι μόνῳ τῶν ἀρωμάτων τοὺς μυρεψοὺς οὐδ' εἰς ἓν μύρον, ἀλλὰ ψευδώνυμός τις ἡ ἐπίκλησις.

31 Ποιοῦσι δὲ καὶ τὰ μὲν ἀχρωμάτιστα τὰ δὲ κεχρωματισμένα. χρωματίζουσι δὲ ἀμαράκινον ῥόδινον μεγαλεῖον, ἀχρωμάτιστα δὲ τῶν μὲν πολυτελῶν Αἰγύπτιον μήλινον κύπρος, τὰ δ' εὐτελῆ πάντα· ταῦτα δὲ ἀχρωμάτιστα διότι τὸ μὲν Αἰγύπτιον καὶ τὴν κύπρον λευκὰ εἶναι βούλονται, τῷ δὲ μηλίνῳ τὴν τῶν μήλων χρόαν, τοῖς δ' εὐτελέσιν οὐ λυσιτελεῖ τὸ χρῶμα προστιθέναι. χρωματίζουσι δὲ τὰ μὲν ἐρυθρὰ τῇ ἀγχούσῃ, τὸ δ' ἀμαράκινον τῷ καλουμένῳ χρώματι· τοῦτο δ' ἐστὶ ῥιζίον ὃ ἄγουσιν ἐκ τῆς Συρίας.

32 VII. Συνεργεῖν δὲ δοκοῦσι πρὸς τὰς γεύσεις οὐχ αἱ ὀδμαὶ μόνον ἀλλὰ καὶ αἱ δριμύτητες καὶ αἱ θερμότητες ἐνίων, διὸ καὶ τῶν οἴνων τισὶ τὰ τοιαῦτα μιγνύντες ὥσπερ κέντρον ἐμποιοῦσιν. ἔστι δὲ ἡ μὲν σμύρνη θερμὴ καὶ δηκτικὴ μετὰ

to make, since no others involve the mixture of so many and such costly ingredients. To make *megaleion*, they say, the oil is boiled for ten days and nights, and not till then do they put in the resin and the other things, since the oil is more receptive when it has been thoroughly boiled. The superior kind of sweet marjoram-perfume,[1] they say, is made of all the best spices except sweet marjoram: in fact this is the only spice which perfumers do not use for any perfume, and the name is a misnomer.

Some perfumes are made up colourless, some are given a colour. They give a colour to sweet marjoram-perfume, rose-perfume, and *megaleion*, while among expensive kinds the Egyptian, quince-perfume and *kypros* are colourless, as well as all the cheaper kinds. The reason why these are made without colour is that it is desired that the Egyptian and *kypros* should look white and that quince-perfume should have the colour of quinces, while it is not worth while to add colour to the cheaper sorts. The dye used for colouring red perfumes is alkanet; the sweet marjoram-perfume is dyed with the substance called *khroma* (dye), which is a root imported from Syria.

Of the properties of various spices.

VII. It is thought that not only the smells of perfumes contribute to a pleasant taste, but also the qualities of pungency and heat which are found in some of them: accordingly some of these perfumes are also mixed with certain wines to give, as it were, 'point' to them. Thus myrrh is hot and has a biting quality as well as being astringent, and it also

[1] Clearly distinct from that mentioned in 28.

στύψεως, ἔχει δὲ καὶ πικρίαν. τὸ δὲ κινάμωμον δριμύτητά τινα μετρίαν μετὰ θερμότητος. παραπλησίως δὲ καὶ τὸ κόστον. ἡ δὲ κασία τούτων ὑπερβάλλει θερμότητι καὶ δριμύτητι καὶ στύψει. θερμὴ δὲ καὶ στυπτικὴ καὶ ἡ ἶρις, καθ' ὑπερβολὴν δὲ καὶ πικρὰ νέα οὖσα καὶ τὸν χρῶτα τῶν ἐργαζομένων ἀφελκοῖ. δηκτικὸν δὲ καὶ τὸ καρδάμωμον μετὰ θερμότητος. τοῦ δὲ βαλσάμου ὁ μὲν ὀπὸς καὶ τὸ καρπίον ἀνδρικώτερα πρὸς ἀμφότερα ταῦτα, τὸ δὲ ξύλον ἀσθενέστερον. παραπλησίαν δ' ἔχει τούτῳ δύναμιν καὶ τὸ ἄμωμον.

33 Ὁ δὲ σχοῖνος δηκτικώτερον μὲν τοῦ καλάμου καὶ θερμότερον, στυπτικὰ δὲ ὁμοίως ἄμφω. τούτων δὲ στυπτικώτερον τὸ κύπειρον. στύφει δὲ καὶ ἡ ἀσπάλαθος ἡ εὐώδης. ἡ δὲ νάρδος δηκτικὴ μετὰ θερμότητος. τὸ δὲ μάρον καὶ τὸ χρῶμα τὸ εἰς τὸ ἀμαράκινον ἐμμιγνύμενον θερμαντικά. [συνεργεῖ δὲ καὶ τῆς ἀγχούσης τὸ ῥιζίον εἰς τὴν χρόαν τοῦ ῥοδίνου καὶ τῆς ἴριδος.]

34 Νέα μὲν οὖν ὄντα τῶν ἀρωμάτων ἔνια δυνάμεις μὲν εὐθὺς ἔχει βαρείας καὶ δριμείας, παλαιούμενα δὲ μέχρι τῆς ἀκμῆς γλυκαίνεται, εἶτ' ἀναλύεται πάλιν. οἷον ἡ ἶρις εἰς μὲν τὴν ἐργασίαν ἀκμάζει μετὰ τὴν συλλογὴν τρία ἔτη, καὶ διαμένει δὲ πλεῖστον χρόνον ἐξ ἔτη. τὸ δὲ μάρον ἔτη δύο. ἡ δὲ σμύρνα δέκα ἔτη διαμένει βελτίων γενομένη. παραπλησίως δὲ τούτοις ἡ τῆς ἀκμῆς διαμονὴ καὶ τοῦ κιναμώμου καὶ τοῦ κόστου καὶ τῆς κασίας. σχοῖνος δὲ καὶ κάλαμος παρακμάζει ταχύ. τῶν δ' ἀνθῶν τὰ μὲν εὐθὺς χλωρὰ ὄντα

---

[1] cf. Plin. 21. 42.   [2] cf. Index, σχοῖνος (2).
[3] cf. Index, κάλαμος ὁ εὐώδης.

CONCERNING ODOURS, 32–34

has a bitter quality. Cinnamon again has a fair amount of pungency as well as heat. So too is it with *koston*. Cassia exceeds both of these in heat pungency and astringency. Iris-perfume is hot and astringent, and excessively bitter when it is fresh, [1] in which state it causes sores on the skin of those that work on it. Cardamom has also a biting quality as well as heat. The juice and the fruit of balsam of Mecca are more active in producing both these qualities, while the wood is less so. Nepaul cardamom has also a property similar to this.

Ginger-grass [2] has a more biting quality than sweet-flag,[3] and is hotter; but both are equally astringent. *Kypeiron* is however more astringent than either. The sweet-scented *aspalathos* also has this quality. Spikenard has a biting quality as well as heat. *Maron* and the *khroma* which is mixed with sweet marjoram-perfume are heating. [4] The root of alkanet also contributes to the colour of rose-perfume and iris-perfume.

Now some spices when they are fresh have at first heavy and pungent qualities, but in course of time become sweet till they have reached their prime, and then lose their properties again. Thus the iris is at its prime for manufacturing the perfume for three years after it was gathered, and [5] lasts for six years at longest.[5] *Maron* lasts two years; myrrh ten, and improves with time. Cinnamon *koston* and cassia keep at their best for about the same periods as these. Ginger-grass and sweet-flag soon get past their prime. Of flowers some, like the rose, possess

---

[4] This sentence seems irrelevant here.

[5-5] καὶ—ἔτη. These words are omitted, apparently by accident, in both W.'s editions, though represented in his Latin translation.

τὰς δυνάμεις ἔχει, καθάπερ τὸ ῥόδον, τὰ δὲ ξηρανθέντα, καθάπερ ὁ κρόκος καὶ ὁ μελίλωτος· χλωρὰ γὰρ ὑγρότερα.

35 Τὰς μὲν οὖν φύσεις καὶ δυνάμεις τῶν ἀρωμάτων ἐκ τούτων θεωρητέον.

VIII. Δοκεῖ δὲ τὸ μεγαλεῖον ἀφλέγμαντον εἶναι παντὸς τραύματος· τὸ δὲ ῥόδινον ἄριστον πρὸς τὰ ὦτα. ταῦτα δ' οὐκ ἀλόγως. τοῦ μὲν γὰρ ἡ σύνθεσις ἐκ ῥητίνης κεκαυμένης, ὥσπερ ἐλέχθη, καὶ κασίας καὶ κιναμώμου καὶ σμύρνης, ἅπαντα δὲ ταῦτα στυπτικὰ καὶ ξηραντικά. τὸ δὲ ῥόδινον τοῖς ὠσὶν ἀγαθὸν ὅτι ἐν ἁλσὶν ἡ ποίησις· ἀναξηραίνει γὰρ καὶ ἐκθερμαίνει διὰ τοὺς ἅλας· διὸ καὶ ἡ ἁλοσάχνη ἀγαθόν. ἀλλὰ τὸ τῆς στραγγουρίας λόγου δεῖται· καὶ γὰρ ταύτῃ λέγουσι μάλιστα βοηθεῖν. αἴτιον δ' ἂν εἴη διότι πᾶν τὸ ὑπεξάγειν μέλλον ἀναλῦσαι δεῖ πρότερον τὸ ὑπεξαχθησόμενον· τοῦτο δὲ οἱ ἅλες ποιοῦσιν, ἡ δ' εὐωδία τὴν ὁρμὴν ἀπέδωκε.

36 Διὰ τί δὲ τὸ ἴρινον εὔοσμον μὲν οὐ ποιεῖ δὲ τὴν ὁρμήν; ἢ διότι στυπτικὸν καὶ συνάγει τοὺς πόρους, ὥστε συγκλείσει κωλύειν τὴν δίοδον; ἀλλὰ καὶ κοιλίας λυτικὴ διά τε τὴν θερμότητα καὶ διὰ τὸ ἀποστύφειν τοὺς ἐπὶ τὴν κύστιν πόρους· ἀποκλειομένων γὰρ τούτων εἰς τὴν κοιλίαν ἡ συρροή. τὸ δὲ ὅλον φαρμακῶδες καὶ

---

[1] cf. C.P. 6. 14. 8 and 11.

CONCERNING ODOURS, 34–36

their virtues from the first while they are still fresh, some only after they are dried, as crocus and *melilotos*,[1] these having a certain amount of moisture while they are fresh.

These examples may suffice for study of the characters and properties of spices.

Of the medicinal properties of certain perfumes.

VIII. *Megaleion* is believed to relieve the inflammation caused by any wound, and rose-perfume to be excellent for the ears. And this is probable enough. For the former is composed, as was said, of burnt resin cassia cinnamon and myrrh, and all these have astringent and drying properties: while the reason why rose-perfume is good for the ears is that salt is used in the manufacture of it: for it is by reason of the salt that it dries and warms (which is why 'sea-foam [2]' is also good for the ears). Its use against strangury however needs explanation: for it is said to be specially helpful against this. The explanation may be that anything which is to remove the difficulty must first dissolve that which is to be removed; and this is the effect of the salt, while the fragrance supplies the necessary stimulus.

Why however, it may be asked, though iris-perfume is fragrant, does it not give the stimulus? Perhaps it is because it is astringent and closes the passages, so that by shutting them it prevents free course. On the other hand this perfume acts as a laxative on the bowels because of its heating quality and because it astringes the passages leading to the bladder: for, when these are closed, the liquid collects in the bowels. In general iris-

[2] Said to be a zoophyte: *cf.* Arist. *H.A.* 9. 14. 2.

τὸ ἴρινον καὶ ἄλλα τῶν μύρων. ἡ δ' αἰτία πάντων ὡς καθόλου εἰπεῖν ἐν ταῖς δυνάμεσι ταῖς εἰρημέναις, ὅτι στυπτικὰ καὶ θερμαντικά· τὰ ἀρώματα γὰρ τὰ τοιαῦτα φαρμακώδη. ταῦτα μὲν οὖν ἔξω τῆς τέχνης.

37 IX. Κρᾶσις δὲ καὶ μῖξις οὐκ ἔστιν ὡρισμένη τῶν ἀρωμάτων, ὥστ' ἐκ τῶν αὐτῶν ἀεὶ χρηστὰ καὶ ὅμοια γίνεσθαι, ἀλλοῖα δὲ συμβαίνει διὰ τὴν ἀνωμαλίαν τῶν δυνάμεων τῶν ἐν τοῖς ἀρώμασι. τῆς δ' ἀνωμαλίας αἰτίαι πλείους. μία μέν, ἥπερ καὶ τοῖς ἄλλοις καρποῖς, ἡ τοῦ ἔτους κατάστασις· αὕτη γὰρ πολυχουστέρας ὁτὲ δ' ἀσθενεστέρας τὰς δυνάμεις <ποιεῖ>. ἑτέρα δὲ ἐν τῇ συλλογῇ, τὸ προτερῆσαι τῆς ἀκμῆς ἢ ὑστερῆσαι· καὶ γὰρ τοῦτο οὐ μικρὸν διαφέρει. τρίτη δ' ἡ μετὰ τὴν συλλογήν, ὅσα χρόνου δεῖται πρὸς τὴν ἀκμήν, ὥσπερ ἐλέχθη· καὶ γὰρ ἐνταῦθά ἐστι τὸ προτερεῖν καὶ ὑστερεῖν.

38 Τούτων δὲ τὸ μὲν τῶν ἐτῶν οὐκ ἐφ' ἡμῖν, πλὴν εἰς τὸ εἰδέναι τὰ ποῖα σφοδροτέρας καὶ ἀσθενεστέρας ἔχει τὰς δυνάμεις· τὰ δὲ κατὰ τὰς ἀκμὰς τῆς τε συλλογῆς καὶ μετὰ τὴν συλλογὴν ἐφ' ἡμῖν ἐστί, τῷ εἰδότι μᾶλλον τὸ ἐπιτυγχάνειν.

Ἡ μὲν οὖν γένεσις καὶ σύνθεσις τῶν μύρων ἐκ τούτων.

CONCERNING ODOURS, 36–38

perfume, as well as others, has medicinal properties. And the explanation in all cases, to put it generally, lies in the above-mentioned properties of astringency and heating; for it is spices possessing these properties that are medicinal. However these matters lie outside our subject of study.

Of rules for the mixture of spices, and of the storing of various perfumes.

IX. There is no fixed rule for the combination and mixture of spices in the sense that the same components will always produce a satisfactory and a uniform result: the result varies by reason of the varying quality of the virtues found in the spices. For this there are several reasons. One, which applies also to fruits, is the character of the season; this causes the virtues to be sometimes much more than ordinarily powerful, sometimes less so. Another is to be found in the time of collection, according as it is made before or after the spices are in their prime. A third cause operates after the collection, that is, in the case of those spices which need time to come to their prime, as was said [1]: for here too it is possible to be too soon or too late.

Of these causes that which depends on the seasons is not within our control, except so far as we can discover which spices in a particular season have powerful, and which weak virtues.[2] But we can control those which depend on collecting them when in their prime, or on keeping them after they are collected, that is, if we know pretty well how to hit the right moment.

So much for the origin and composition of perfumes.

[1] 34.   [2] *i.e.* and we can select accordingly.

Πολυχρονιώτατον δ' ἐστὶ τό τ' Αἰγύπτιον καὶ τὸ ἴρινον καὶ τὸ ἀμαράκινον καὶ τὸ νάρδινον, πάντων δὲ μάλιστα ἡ στακτή, διαμένει γὰρ ὁποσονοῦν χρόνον. μυροπώλης δέ τις ἔφη παρ' αὑτῷ μεμενηκέναι Αἰγύπτιον μὲν ὀκτὼ ἔτη, ἴρινον δὲ εἴκοσι, καὶ ἔτι διαμένειν βέλτιον ὂν τῶν ἀκμαζόντων. ἡ μὲν οὖν χρονιότης ἐν τούτοις.

39 Τὰ δ' ἄνθινα πάντα ἀσθενῆ. συμβαίνει δὲ τοῖς ἀνθίνοις ἀκμάζειν μὲν ὡς ἐπὶ τὸ πᾶν μετὰ δίμηνον, μεταβάλλειν δ' ἐπὶ τὸ χεῖρον ἐνιαυτοῦ προελθόντος καὶ περικαταλαβούσης τῆς ὥρας ἐν ᾗ τὴν ἀκμὴν λαμβάνει τὸ ἄνθος. ἀνὰ λόγον δὲ τῇ ἀσθενείᾳ καὶ τὸ εὐέπαντα εἶναι καὶ ὅλως εὐδιάπνευστα· τὰ δ' ἐκ τῶν ῥιζῶν καὶ τῶν λοιπῶν χρονιώτερα· πλείων γὰρ ἡ ὀσμὴ καὶ ἰσχυροτέρα καὶ σωματωδεστέρα.

40 Διαφθείρει δὲ τὰ μύρα καὶ ὥρα θερμὴ καὶ τόπος καὶ ὁ ἥλιος, ἂν τεθῶσι· διὸ καὶ οἱ μυροπῶλαι ζητοῦσι τὰς οἰκίας ὑπερῴους καὶ μὴ προσηλίους ἀλλ' ὅτι μάλιστα παλισκίους· ἀφαιρεῖται γὰρ τὰς ὀσμὰς ὁ ἥλιος καὶ τὸ θερμὸν καὶ ὅλως ἐξίστησι τῆς φύσεως μᾶλλον τοῦ ψυχροῦ· τὸ δὲ ψυχρὸν καὶ ὁ πάγος, εἰ καὶ ἀοσμότερον ποιεῖ διὰ τὸ συστέλλειν, ἀλλ' οὐκ ἀφαιρεῖταί γε τὴν δύναμιν τελέως. ἡ πονηρὰ γὰρ φθορά, καθάπερ τῶν οἴνων καὶ τῶν ἄλλων χυλῶν, τῷ

41 τὸ οἰκεῖον ἀφαιρεῖσθαι θερμόν. διὸ καὶ εἰς ἀγγεῖα μολυβδᾶ ἐγχέουσι καὶ τοὺς ἀλαβάστρους ζητοῦσι τοιούτου λίθου· ψυχρὸν γὰρ καὶ πυκνὸν καὶ ὁ

CONCERNING ODOURS, 38-41

Those which last longest are the Egyptian, the iris, the sweet marjoram and the spikenard perfumes: but myrrh-oil has the longest life of any; for it will keep any time. A certain perfumer said that he had had Egyptian perfume in his shop for eight years, and iris-perfume for twenty, and that it was still in good case, in fact better than fresh perfume. These are instances of perfumes which will keep a long time.

On the other hand all those made from flowers have little vigour. These are usually at their best after two months, but they deteriorate when a year has past and the season has come round again at which the flowers are at their best. Also, as these perfumes lack vigour, so also do they quickly mellow, and, in most cases, quickly evaporate. Those made from roots and the other parts of the plant last longer, their odour being fuller stronger and more substantial.

Perfumes are ruined by a hot season or place or by being put in the sun. This is why perfumers seek upper rooms which do not face the sun but are shaded as much as possible. For the sun or a hot place deprives the perfumes of their odour, and in general makes them lose their character more than cold treatment: while cold and frost, even if they make them less odorous by congealing them, yet do not altogether deprive them of their virtue. For the most destructive thing that can happen to them, as to wines and other savours, is that they should be deprived of their proper heat. This is why men put them into vessels of lead and try to secure phials of alabaster—a stone which has the required effect: for lead is cold and of close texture, and stone has

THEOPHRASTUS

μόλυβδος καὶ ὁ λίθος ὁ τοιοῦτος· καὶ ἄριστος τοῖς μύροις ὁ μάλιστα τοιοῦτος. ὥστε δι' ἄμφω τηροῦσι, καὶ τῷ ψυχρῷ καὶ τῷ πυκνῷ, μήτε διιέντες ἔξω τὴν ὀσμὴν μήθ' ὅλως ἐπιδεχόμενοι μηδέν. καὶ γὰρ ἡ ἀναπνοὴ φθείρει καὶ τὸ ἔξωθεν ἐπεισιὸν καὶ ἀλλότριον· ἐπεὶ καὶ τὰ πνεύματα φθείρει καὶ καταναλίσκει, καθάπερ ἐλέχθη, τὰς ὀσμάς, ἄλλως τε καὶ τὰς μὴ φυσικάς.

42 X. Κεφαλαλγῆ δὲ τῶν μὲν πολυτελῶν τὸ ἀμαράκινον καὶ τὸ νάρδινον καὶ μεγαλεῖον, τῶν δ' εὐτελῶν ὅλως μὲν τὰ πλεῖστα μάλιστα δὲ τὸ δάφνινον. ἐλαφρότατα δὲ τὸ ῥόδινον καὶ ἡ κύπρος, ἃ καὶ τοῖς ἀνδράσι μάλιστα ἁρμόττειν δοκεῖ, καὶ πρὸς τούτοις τὸ κρίνον· ταῖς δὲ γυναιξὶν ἡ στακτὴ καὶ τὸ μεγαλεῖον καὶ τὸ Αἰγύπτιον καὶ τὸ ἀμαράκινον καὶ τὸ νάρδινον· διὰ γὰρ τὴν ἰσχὺν καὶ τὸ πάχος οὐκ εὐαπόπνοα οὐδ' εὐαφαίρετα· ζητοῦσι <γὰρ> τὰ χρόνια.

43 Ἐπεὶ δὲ τὰ μὲν ἀσθενῆ τὰ δ' ἰσχυρά, καὶ ἰσχυρότερα τὰ ἀπὸ τῶν ῥιζῶν καὶ τὰ ἄλλα τὰ προειρημένα, διὰ τοῦτο τὰ μὲν ἄνθινα μὴ τριβόμενα εὐοσμότερα, τὰ δ' ἀπὸ τῶν ῥιζῶν καὶ τὰ λοιπὰ τριβόμενα· τὰ μὲν γὰρ διαπνεῖταί τε καὶ ἅμα διαθερμαινόμενα διὰ τὴν τρῖψιν ἐξίσταται καὶ ἀλλοιοῦται, τὰ δὲ διὰ τὴν ἰσχὺν

---

[1] *e.g.* alabaster, which here at least is spoken of as a kind of stone. [2] γὰρ ins. Sch.

the same character, that being the best for keeping perfumes which has it in the highest degree.[1] So that vessels made of these materials keep the perfume well for both reasons, their coolness and their closeness of texture: they neither let the odour pass away through them, nor do they take in anything else. For evaporation destroys the perfume, and so also does any foreign substance which finds its way in: for even draughts of air destroy odours and cause them to waste, as was said, especially those odours which do not belong to a thing's essential nature.

Of the properties of certain perfumes.

X. Headache is caused by sweet marjoram spikenard and *megaleion* among costly perfumes: most of the cheap ones have also this effect, notably that made from bay. The lightest are rose-perfume and *kypros,* which seem to be the best suited to men, as also is lily-perfume. The best for women are myrrh-oil, *megaleion,* the Egyptian, sweet marjoram, and spikenard: for these owing to their strength and substantial character do not easily evaporate and are not easily made to disperse, and [2] a lasting perfume is what women require.

Inasmuch however as some perfumes are stronger than others, the stronger being those made from roots and the others already mentioned, for this reason those derived from flowers are more fragrant if they are not bruised, while bruising improves those made from roots and the others. For the former kind evaporate and pass off as they are warmed by the bruising, thus losing their character, while the latter owing to their strength have, as it

365

ὥσπερ ἀνοιγομένων τινῶν πόρων ἐκ τῆς τρίψεως
44 ἐμφανεστέραν ποιεῖ τὴν ὀσμήν. ὃ καὶ ἐπ' αὐτῶν
τῶν ῥιζῶν καὶ ὅλως τῶν στερεῶν συμβαίνει,
καθάπερ ἐλέχθη. κατὰ δὲ τῶν ἀνθῶν ἐναντίως,
ὥστε ἠκολούθηκεν ἑκάτερα τῇ ἀρχῇ. τὰ δ' ἐκ
τῆς σμύρνης εὐλογώτατα δι' ἄμφω· καὶ γὰρ
μίγνυται μᾶλλον καὶ <ἡ> θερμότης ἡ τῆς τρίψεως
οὐκ ἀλλοτρία, μαλακή τις οὖσα· καὶ γὰρ <ἡ>
σμύρνα ζητεῖ τινα πύρωσιν. ἁπλῶς δὲ πᾶν τὸ
πολύοδμον ἄντ' εὐῶδες ἄντε κακῶδες ἄντε δριμὺ
ἄντ' ὀξὺ ἄντ' ὁποιονοῦν τυγχάνῃ, κινούμενον ἐμ-
φανέστερον· τότε γὰρ ὥσπερ ἐνεργείᾳ ἀναμίγνυται
μᾶλλον τῷ ἀέρι.

Τῶν δὲ μύρων τὸ Αἰγύπτιον καὶ ἡ στακτὴ καὶ
εἴ τι ἄλλο πολύοδμον [καὶ] μιγνύμενα τῷ οἴνῳ τῷ
εὐώδει ἡδίω· παραιρεῖται γὰρ ἡ βαρύτης αὐτῶν·
ἐπεὶ καὶ ἡ σμύρνη αὐτὴ πρὸς τὴν ἀναθυμίασιν
βρεχθεῖσα ἐν τῷ γλυκεῖ, καθάπερ ἐν τοῖς πρότερον
ἐλέχθη.

45 Πρὸς δὲ τὰς δυνάμεις σκοπουμένοις δόξειεν ἂν
ἄτοπον εἶναι τὸ συμβαῖνον ἐπὶ τοῦ ῥοδίνου· κου-
φότατον γὰρ ὂν καὶ ἀσθενέστατον ἀφανίζει τὰς
τῶν ἄλλων ὀσμάς ὅταν προμυρισθῶσι· διὸ καὶ
οἱ μυροπῶλαι τοὺς ἐπιδιστάζοντας καὶ μὴ ὠνου-
μένους παρ' αὐτῶν ἐπιμυρίζουσι τούτῳ πρὸς τὸ
μὴ αἰσθάνεσθαι τὰ παρὰ τῶν ἄλλων. αἴτιον
δ' ὅτι λεπτότατον ὂν καὶ προσφιλὲς τῇ αἰσθήσει
διὰ τὴν κουφότητα μάλιστα διικνεῖται καὶ συμ-

---

[1] ἡ ins. W. [2] ἡ ins. W.
[3] The words ἄντε δριμὺ are omitted in both W.'s editions, but represented in his Latin version.

were, certain passages opened by the bruising, and so their fragrance is made more obvious. This, as was said, also takes place in the case of the roots themselves and of the solid things in general; but the result in the case of flowers is just the opposite, so that both kinds behave according to their origin. That this should apply to the perfumes made from myrrh is quite natural for both reasons; they mingle more than others with the air, and the heat[1] due to the bruising is not prejudicial, since it is gentle, and myrrh[2] in fact requires a certain amount of heating. And in general any strong odour, whether it be pleasing or the reverse, whether it be pungent[3] or sharp, or whatever its character, becomes more pronounced with movement; for then it becomes, as it were, active and mingles more with the air.

The Egyptian perfume, myrrh-oil, and any others that have a strong odour become[4] sweeter if they are mixed with fragrant wine; for then their heavy quality is removed. In fact myrrh itself is made to exhale a more fragrant odour by being steeped in sweet wine, as was said[5] in a former treatise.

If one has regard to the virtues of the perfumes in question, one may well be surprised at what happens in the case of rose-perfume:—though it is lighter and less powerful than any other, if one has first been scented with it, it destroys the odour of the others. And this is why perfumers, if a purchaser hesitates and is not inclined to buy this perfume, scent him with it so that he is not able to smell the others. The explanation is that, being very delicate and acceptable to the sense of smell, by reason of its lightness it penetrates as no

---

[4] I have bracketed καί.   [5] *C.P.* 6. 17. 2.

πληροῖ τοὺς πόρους, ὥσθ' ἡ αἴσθησις κατειλημ-
μένη καὶ πλήρης οὖσα κρίνειν ἀδυνατεῖ. δύο
γάρ εἰσι τρόποι, τάχα δὲ τρεῖς, οἱ κωλύοντες τὴν
κρίσιν. εἷς μὲν ὁ νῦν εἰρημένος· ἄλλος δ' ὁ ἀπὸ
τῶν ἰσχυρῶν ὥσπερ μεθύσκων τὴν αἴσθησιν καὶ
καρηβαρᾶν ποιῶν· τρίτος δ' ὅταν προκαταληφθῇ
τῷ βελτίονι· τὸ γὰρ ἐπεισάγειν τὸ χεῖρον οὐ
ῥᾴδιον· οὐ δέχεται γὰρ ἡ αἴσθησις, ὥσπερ οὐδ'
ἐπὶ τῶν χυλῶν καὶ ὅλως τῶν κατὰ τὴν τροφήν.

47 Κατισχναίνειν δὲ δοκεῖ τὸ ῥόδον καὶ τὴν σύν-
θετον ὀσμήν· ὅταν γὰρ ἀκμάζῃ τὸ ἄνθος, ῥοδίζουσι
τὰς συνθέσεις, ἀνοιγόμεναι δ' ἐξόζουσι τούτου
μόνου καὶ μάλιστα. παύεται δὲ ταχὺ καὶ λήγει
διὰ τὴν ἀσθένειαν καὶ λεπτότητα, δι' ἣν καὶ ἐξόζει τῶν ἄλλων· λεπτὴ γὰρ οὖσα ἡ ἀναπνοὴ καὶ
ἠθροισμένη τῇ κατακλείσει προτερεῖ τε τῶν λοιπῶν καὶ διαδίδοται πανταχοῦ. διὰ ταὐτὸ δὲ
τοῦτο καὶ ἀπολήγει ταχὺ καὶ κατακρατεῖται
πάλιν· ἀσθενεῖ γὰρ τὸ λεπτὸν καὶ μαλακόν.

48 Ποιοῦσι δέ τινες τοῦτο καὶ τῶν οἴνων, ὥστε
προποθέντες ἀφανίζειν τὴν τῶν ἄλλων ἡδονήν.
ἔνιοι δ' ὥστε μὴ ἐπιδέχεσθαι ῥᾳδίως τοὺς ἄλλους,

---

[1] *cf.* 57, 58. σύνθετος ὀσμή or σύνθεσις seems to mean a kind of pot-pourri, which was from time to time renewed with fresh rose-petals. Sch. understands σύνθεσις to mean 'clothes in a wardrobe' (*cf.* Lat. *synthesis*), but it must surely have the same meaning here as σύνθετος ὀσμή: Sch.'s citation from 57 does not seem to prove his point, and μένουσι πολὺν χρόνον αἱ συνθέσεις in 58 is conclusive against him. *cf.* also 69.

other can and fills up the passages of the sense, so that being entirely taken up and filled with it, it is unable to judge of others. For the power of judging is inhibited in two, or possibly in three ways; one is that which has just been mentioned; another is that the sense of smell is, as it were, intoxicated with its powerful virtues and becomes stupefied: the third is that the sense may be preoccupied with the superior odour; for then it is not easy to introduce after it what is inferior, since the sense of smell refuses this—just as the sense of taste in like case refuses flavours and foods in general.

It is also thought that the rose even weakens the effect of compound perfume[1]; for, when the flower is at its best, they treat compound perfumes with it; and, when these come to be opened, they smell only or chiefly of rose. However this effect is only temporary and transient because of the weakness and delicacy of the rose-scent, (the very quality which also causes it to assert[2] itself over the scent of the other ingredients). For, as it is so delicate and is compressed by confinement, it is exhaled before the others and disperses in all directions. It is also for this reason that the rose-scent only asserts itself for a short time and then is overpowered again; for anything that is delicate and subtle must be lacking in vigour.

Certain wines have also a similar effect: if they are first drunk, there is no satisfaction in others. Some again make it even difficult to take others

---

[2] *i.e.* when the pot-pourri is first opened: the 'delicacy' of the rose-scent causes it to be given off quickly and so (1) to be the first scent perceived, (2) to be volatile. ἐξόζω in this passage is used with gen. in two distinct senses.

ὥσπερ ὁ Ἐρυθραίας ἁλυκός τις ὢν καὶ μαλακός. τὴν αἰτίαν <δὲ> πειρατέον ἐκ τῶν ὁμοίων λαμβάνειν· ἔχει δὲ τοῦτ' ἴδιον τὸ ῥόδινον, ὥσπερ σχεδὸν καὶ μικρῷ πρότερον εἴρηται· τὰ μὲν γὰρ ἄλλα πάντ' ἢ τὰ πλεῖστα κεφαλαλγῆ, τοῦτο δέ, ὥσπερ ἐλέχθη, λυτικὸν καὶ βάρους καὶ ἀλγηδόνος καὶ τῆς ἀπὸ τῶν μύρων.

49 Ἡ δ' αἰτία φανερὰ διὰ τῶν προειρημένων, εἴπερ ἐπικρατεῖ καὶ διαδύεται πανταχοῦ. τὰ μὲν γὰρ ἄλλ' ὅσα κεφαλαλγῆ βαρέα διὰ τὸ ἐκ τοιούτων συγκεῖσθαι τὰ μὲν ῥιζῶν τὰ δ' ὀπῶν· τοῦτο δὲ καὶ τῇ ὀσμῇ ἐλαφρὸν καὶ τῇ θερμότητι σύμμετρον εἰς τὸ συμπέψαι καὶ διανοῖξαι τοὺς πόρους. οἱ γὰρ δὴ πόνοι τῆς κεφαλῆς ἢ καθυγραινομένης ἢ πνευματουμένης τῷ ἐναπολαμβάνεσθαι, ὥστε τὸ μὲν ἐκκρῖναι δεῖ τὸ δὲ πέψαι ἢ ἀφελεῖν.

50 Πρὸς ἅπαντα δὲ ἡ θερμότης χρήσιμον, καὶ εἰς ἀφαίρεσιν καὶ ἔτι μᾶλλον εἰς τὸ πέττειν καὶ διανοίγειν τοὺς πόρους, εἰς ἃ συμβάλλεται τὸ ἐν τῷ ἁλὶ πεποιῆσθαι· καὶ γὰρ ἀναστομοῦσι καὶ διαθερμαίνουσιν οἱ ἅλες. ἡ δ' εὐοσμία καὶ ὁρμήν τινα ποιεῖ πρὸς τὴν κίνησιν. ἀγαθὸν δὲ καὶ δοκεῖ πρὸς τοὺς κόπους εἶναι τῇ θερμότητι σύμμετρον ὂν καὶ τῇ κουφότητι καὶ τῇ διαδύσει πρὸς τοὺς ἐντὸς πόρους· ὡς δέ τινές φασιν, οὐχ ἧττον ἡ κύπρος ἔτι τούτου· μαλακὴ γὰρ ἡ ὀσμὴ καὶ

---

[1] cf. 52.   [2] δὲ ins. W.
[3] i.e. the case is so far analogous to that of rose-perfume; but the comparison does not hold as to what follows.

after them; this is the effect for instance of wine of Erythrae,[1] which has a taste of brine and is subtle. The[2] explanation one must endeavour to find by comparing analogous cases.[3] However there is one peculiarity which as we have already more or less indicated, is possessed by rose-perfume only; while all or most of the others are heady, this, as was said, gives actual relief from heaviness and discomfort, even from that caused by other perfumes.

The reason for this is plain in view of what has been already said, seeing that this perfume overpowers others and penetrates everywhere. For the others that are heady are heavy because they are made of heavy substances, whether roots or juices; while this perfume is both light as to its scent and also by its heat well adapted to bring the passages to a suitable temperature and to open them. For pains in the head are due to an excess of moisture in it, or of air which gets confined in it, so that it is necessary to get rid of the one, and to raise the temperature of the other or to remove it.

And for all such purposes heat is useful, both for removing the moisture or air, and, still more, for raising the temperature of the passages and opening them: and to these ends it is helpful that the perfume should have been prepared with salt, since the effect of salt is to open the passages and to warm them thoroughly. Again the fragrance also supplies a stimulus to movement. This perfume is also considered to be good against lassitude, because its heat and its lightness make it suitable, and also because it penetrates to the inner passages. Some however say that *kypros* is quite as efficacious: for this too has a delicate scent which is grateful to the

προσφιλὴς τῷ χρωτὶ καὶ ἡ ταύτης. καὶ ταῦτα μὲν καὶ τὰ ὅμοια τούτοις ὥσπερ ἴδια ἂν εἴη.

51 XI. Τοῦ ῥοδίνου δὲ αἱ μίξεις καὶ ἐν ταῖς ὀσμαῖς καὶ ἐν τοῖς χυμοῖς, ἐὰν ἡρμοσμέναι τυγχάνωσιν, ἔχουσί τινα χρείαν, αἱ μὲν ἀφαιροῦσαι τὴν βαρύτητα καὶ τὴν ἰσχύν, αἱ δ' εὐοδμίαν τινὰ αἱ δὲ γλυκύτητα ἐμποιοῦσαι, καθάπερ καὶ ἐπὶ τῶν οἴνων. καὶ γὰρ ὁ ἐν Θάσῳ ὁ ἐν τῷ πρυτανείῳ διδόμενος, θαυμαστός τις ὡς ἔοικε τὴν ἡδονήν, ἠρτυμένος ἐστίν· ἐμβάλλουσι γὰρ εἰς τὸ κεράμιον σταῖς μέλιτι φυράσαντες, ὥστε τὴν μὲν ὀσμὴν ἀπ' αὐτοῦ, τὴν δὲ γλυκύτητα ἀπὸ τοῦ σταιτὸς λαμβάνειν τὸν οἶνον.

52 Συμβαίνειν δὲ τοῦτο καὶ κατὰ τὰς τῶν οἴνων μίξεις· οἷον ἐάν τις κεράσῃ σκληρὸν καὶ εὔοσμον μαλακῷ καὶ ἀόσμῳ, καθάπερ τὸν Ἡρακλεώτην καὶ τὸν Ἐρυθραῖον, τοῦ μὲν τὴν μαλακότητα τοῦ δὲ τὴν εὐοσμίαν παρεχομένου· συμπίπτει γὰρ ἅμα τὰ κακὰ ἀλλήλων ἀφανίζειν τῇ μαλακότητι θατέρου <καὶ τῇ εὐοσμίᾳ θατέρου>. πολλὰς δὲ καὶ ἄλλας οἱ ἔμπειροι λέγουσι καὶ ἴσασι μίξεις. ὃ καὶ ἐπὶ τῶν ὀσμῶν εὔλογον συμβαίνειν, καὶ ἐπὶ τῶν χρωμάτων ἄν τις λαμβάνῃ τὰς ἁρμοττούσας μίξεις. τοῦτο μὲν οὖν ἴδιον τοῦ ῥοδίνου.

53 Τὸ δὲ κοινὸν ἐπὶ πάντων ἀπόρημα, τί δή ποτε

---

[1] Quoted by Athen. 1. 58.   [2] cf. 48.

[3] This sentence must be defective : as it stands, the effect of only one wine is given, though the effect is said to be

CONCERNING ODOURS, 50-53

skin. These and similar properties may be considered peculiar to these particular perfumes.

Of other properties and peculiarities of perfumes.

XI. The admixture of rose-perfume, whether in scents or in flavours, if it be well blended, is beneficial, in the one case by removing the heaviness and strength of the scent, in the other by imparting a fragrant scent or a sweet taste to the flavour, as in the case of wines. [1] Thus the wine which is served in the town-hall of Thasos, which appears to be of wonderfully delightful quality, is thus flavoured. For they put into the jar a lump of dough which has been kneaded up with honey, so that the wine gets its fragrance from itself, but its sweet taste from the honeyed dough.

This result also follows, it is said, from the mixture of different wines,—for example, if a strong fragrant wine be mixed with one that is mild and without fragrance, (for instance, if wine of Heraclea be mixed with wine of Erythrae),[2] since the latter contributes its mildness and the former its fragrance: [3] for the effect is that they simultaneously destroy one another's inferior qualities through the mildness of the one and the fragrance of the other. There are many other such blends mentioned by and known to experts. And it is quite to be expected that such a result should follow from blending odours, as it does from blending colours, if one discovers the suitable combinations. This then is peculiar to rose-perfume.

However there is one question which applies to all perfumes, namely, why it is that they appear to

mutual. I have added καὶ τῇ εὐοσμίᾳ θατέρου after Sch.; his text however is συμπίπτει γὰρ ἅμα, καὶ τὰ κακὰ ἀλλήλων ἀφανίζει, τῇ μαλακότητι <καὶ τῇ εὐοσμίᾳ> θατέρου.

ἀπὸ τοῦ καρποῦ τῆς χειρὸς ἥδιστα φαίνεται, διὸ καὶ οἱ μυροπῶλαι τοῦτο μυρίζουσι τὸ μέρος. τὴν δ' αἰτίαν ἐκ τοῦ ἐναντίου ληπτέον, ὅτι τὸ θερμὸν ἐξίστησι καὶ ἀλλοιοῖ· ταχεῖα δ' ἤδη ἡ αἴσθησις τοῖς μύροις ἀναμιγνυμένοις τῷ χρωτί.

54 Ἀπορεῖται δὲ διότι οἱ μὴ εἰωθότες μυρίζεσθαι μᾶλλον ἐξόζουσι τῶν συνεχῶς μυριζομένων· εἴη μὲν γὰρ ἂν λέγειν καὶ ὅτι φαντασίαι καὶ οὐκ ἀλήθειαι διὰ τὸ μὴ εἰωθός· εἰ δ' οὖν καὶ ἀληθές, ἔοικε τὸ μὲν οἷον συναναμίγνυσθαι πλείοσιν ὀσμαῖς ἑτέραις ὑφ' ὧν ἀμαυροῦται, συγκαταμιγνυμένου καὶ τοῦ χρωτός, τὸ δὲ ὥσπερ ἀκέραιον δέχεσθαι τὸ μανὸν καὶ ἐκφαίνειν τῇ αἰσθήσει χρονιζόμενον. εἴη δ' ἂν καὶ ἐναντίως λαβεῖν ὡς ἧττον δεχομένου διὰ τὸ ἀσύνηθες, βραδύτερον δ' ἀναμιγνύμενα πλείω χρόνον ἐξόζειν. καὶ τοῦτο μὲν ἔλαττον καὶ οὐ φανερῶς ὁμολογούμενον.

55 Ἅπτεται δὲ μάλιστα τοῦ χρωτὸς καὶ κεφαλῆς καὶ τῶν ἄλλων καὶ πλεῖστον χρόνον ἐμμένει τὰ

---

[1] Sc. a part of the body which, not being fleshy, does not spoil the scent by its warmth. So Plin. 13. 19. appears to give the point—*experimentum* (*unguentorum*) *inversa manu capitur, ne carnosae partis calor vitiet*, though it may be questioned whether *inversa manu* represents καρποῦ. Pliny's

be sweetest when the scent comes from the wrist[1]; so that perfumers[2] apply the scent to this part. The explanation must be sought by observing what happens in the contrary case, inasmuch as heat changes or destroys the character of a scent, and the effect on the sense of smell is immediately perceived when perfumes are brought into close contact with the skin.

The question is also raised why those who do not habitually use perfumes smell of them more strongly, when they do so, than those who use them habitually. The suggestion might be made that this is an illusion due to the fact that the use is not habitual, and does not represent what really happens. If however it does, it would appear that in the one case the perfume becomes, as it were, confused with a number of other scents which weaken its force (the smell of the skin also becoming mixed with it), while in the other case the porous condition of the skin takes in the scent as it were uncontaminated, and so makes it perceptible by the sense of smell, because it lingers for some time. One might also make a suggestion of opposite character, that the skin takes in perfumes less readily because it is not used to them, and so, as the perfumes mingle with it more slowly, they preserve[3] their scent for a longer time. One may add that this is a small point and that all do not agree as to the fact.

Those perfumes whose scent is strongest get the best hold on the skin head and other parts of the body, and last for the longest time: such are

phrase presumably means the back of the hand, and suggests that ἀπὸ τοῦ καρποῦ may be corrupt.

[2] Sc. in offering samples for choice.
[3] Sc. it is not absorbed by the skin.

THEOPHRASTUS

ἰσχυρότατα ταῖς ὀσμαῖς, οἷον μεγαλεῖον, Αἰγύπτιον, ἀμαράκινον· τὰ δ' ἀσθενῆ καὶ <οὐ> πολύ-
οδμα, κούφην ἔχοντα τὴν ἀναπνοήν, ταχεῖαν ποιεῖ
καὶ τὴν ἀπόλειψιν, ὥσπερ τό τε ῥόδινον καὶ ἡ
κύπρος.

56 Ἔνια δὲ καὶ εἰς τὴν ὑστεραίαν οὐ χεῖρον ὄζει,
διαπεπνευκυίας εἴ τις ἐνῆν βαρύτης. τὰ δὲ καὶ
ὅλως ἔμμονα μᾶλλον, ὥσπερ ἡ νάρδος καὶ τὸ
ἴρινον, πάντων δὲ μάλιστα <τὰ>¹ ἰσχυρότατα.
καὶ τὰ μὲν ἔν τε τοῖς λουτροῖς καὶ τῇ ἀνέσει
διατηρεῖ πως τὴν ὀσμὴν ἢ οὐ συγκακύνει· τὰ δὲ
κακυνόμενα πλείω ποιεῖ δυσωδίαν αὐτῶν τῶν
ἱδρώτων, ὡς ἂν σήψεώς τινος ἢ διαφθορᾶς γινομένης.

Καὶ τὰ μὲν περὶ τῆς τῶν μύρων ποιήσεώς τε
καὶ δυνάμεως ἐπὶ τοσοῦτον εἰρήσθω.

57 XII. Τὰ δὲ περὶ τὴν τῶν ξηρῶν μίξιν, ἐξ ὧν
<τὰ> διαπάσματα καὶ αἱ συνθέσεις, οὐκ ἔτι ζητεῖ
μίξιν τῶνδέ τινων ὡρισμένων, ἀλλ' ὅσῳ ἄν τις
πλείω καὶ ποικιλώτερα μίξῃ, τοσούτῳ καὶ ἡ ὀσμὴ
λαμπροτέρα καὶ ἡδίων, ὥσπερ καὶ ἐξ αὐτῶν
τῶν ἀρωμάτων τῶν προχείρων· εἰς ταὐτὸ γὰρ
μιγνύντες ἅπαντα χρῶνται. ζητοῦσι δ' ἐν τούτοις καὶ σπεύδουσιν ὥστε μὴ ἑνὸς ἀλλὰ πάντων

¹ τὰ ins. Sch.

megaleion, Egyptian perfume and sweet marjoram-perfume. Those on the other hand which are weak and have not a powerful scent, since they are volatile and evaporate, also quickly come to an end: for instance rose-perfume and *kypros*.

There are some however whose scent is even better on the second day, when any heavy quality that they possessed has evaporated. Some again are altogether more permanent, as spikenard and iris-perfume, and the stronger[1] a perfume is, the longer it lasts. Again some perfumes for some reason keep their scent in the bath when the body is relaxed, or at least do not help to produce a disagreeable effect; while others become disagreeable and cause an even more unpleasant odour than the sweat, as though some sort of decomposition or decay took place.

Let this suffice for an account of the manufacture and properties of perfumes.

Of the making of perfume-powders and compound perfumes.

XII. As to the mixing of solid substances to make powders[1] and compound perfumes, we do not find it here necessary to mix certain specified ingredients: the more numerous and the more various the perfumes that are mixed, the more distinguished and the more grateful will be the scent—just as though one were mixing whatever spices themselves[2] were procurable. As a matter of fact the custom is to use a mixture made of all kinds. Again in perfumes of this class the aim and object is not to make the mixture smell of some one particular thing, but to

[2] Sc. the natural products from which the manufactured perfumes are made.

κοινήν τινα τὴν ὀσμὴν εἶναι. διὸ καὶ ἀνοίγοντες διά τινων ἡμερῶν τὸ ἐξόζον ἐξαιροῦσιν ἀεὶ καὶ τῶν ἰσχυρῶν ἐλάττω μιγνύουσιν, ὥσπερ . . . τὰ δ' ὅλως οὐ μιγνύουσιν, ὥσπερ τὸ ἐρυσίσκηπτρον, ὑπὲρ οὗ καὶ ἀρτίως ἐλέχθη.

58 Βρέχουσι δὲ συντιθέντες τῷ οἴνῳ τῷ εὐώδει· ἔοικε δ' οὖν χρήσιμος εἶναι πρὸς τὰς εὐοσμίας, εἴ γε καὶ οἱ μυρεψοὶ χρῶνται. μένουσι δὲ πολὺν χρόνον αἱ συνθέσεις. ἡ δὲ χρῆσις τούτων μὲν εἰς τὴν τῶν ἱματίων ὀσμήν, τῶν δὲ διαπασμάτων εἰς τὴν στρωμνήν, ὅπως πρὸς τὸν χρῶτα προσπίπτῃ· καὶ γὰρ ἅπτεται μᾶλλον καὶ ἐμμονώτερον τοῦτο, καὶ ὥσπερ ἀντ' ἐκείνου τοῦτο ποιοῦσιν. οἱ δὲ πρότερον ἐνέβαλον οἴνῳ καταβρέχοντες εὐώδει πρὸς τὸ παραιρεῖσθαι τὴν ὀσμήν, ἔνια δὲ καὶ μελικράτῳ καὶ οἴνῳ μιγνύντες ἀνέδευον, τὰ δὲ καὶ αὐτῷ τῷ μελικράτῳ. τὸ γὰρ ὅλον ἄμφω ταῦτα συνεργεῖ πρὸς εὐοσμίαν. διαμένουσι δὲ αἱ συνθέσεις. φανερὸν δ' ἐκ τούτων ὅπερ καὶ πρότερον ἐλέχθη, διότι τὰ ξηρὰ καὶ εὐοσμότερα πρὸς ἄλληλα <μιχθέντα> ταῖς ὀσμαῖς.

59 Εὐλόγως δὲ τὰ μύρα φαρμακώδη διὰ τὴν τῶν
(61) ἀρωμάτων δύναμιν· καὶ γὰρ τὰ ἀρώματα τοιαῦτα. δηλοῖ δὲ τά τε καταπλάσματα καὶ ἃ δή τινες

---

[1] The example is missing. Turn. supplies *costum et amomum* from Plin. 13. 16, which does not however certainly refer to this passage; see 69, where this passage seems to be repeated.
[2] The reference of ἐκείνου is obscure.
[3] μιχθέντα add. Turn.

produce a general scent derived from them all. This is why every few days they open the vessel and remove each time that perfume whose scent is overpowering the others, adding at the same time smaller quantities of the less powerful scents, such as . . .,[1] while some perfumes are never added, such as galingale, of which we spoke just now.

When they make compound perfumes, they moisten the spices with fragrant wine: and this certainly seems to be useful for producing fragrance, seeing that perfumers also use it. These compound perfumes last a long time. They are used to impart a pleasant odour to clothes, while the powders are used for bedding, so that they may come in contact with the skin: for this kind of preparation gets a better hold and is more lasting, so that men use it thus instead of scenting their bodies directly.[2] Some, before putting the powder in the bedding, soak it in fragrant wine, so that it may acquire its scent: and some powders they moisten by mixing them with mead and wine, or again simply with mead. For in general both these things help to give them fragrance. Compound perfumes also last well. From which what was said above becomes manifest, inasmuch as solid perfumes, when mixed [3] with one another, acquire a greater fragrance.

[4] It is to be expected that perfumes should have medicinal properties in view of the virtues of spices: for these too have such virtues. The effects of

[4] In W.'s text, which I have followed, there is some rearrangement (after Furlanus) of the order of sentences in this chapter and the next: *e.g.* part of §61 is transferred to §59. Both figures are retained for convenience of reference.

THEOPHRASTUS

μαλάγματα καλοῦσιν οἵας ἀποδείκνυται δυνάμεις τά τε φύματα καὶ τὰ ἀποστήματα διαχέοντα καὶ ἄλλα πλείω τῶν κατὰ τὸ σῶμα διαλλοιοῦντα, ἐπιπολῆς μὲν ἀλλὰ καὶ τὰ ἐν βάθει, οἷον, ἄν τις καταπλάσῃ τὰ ὑποχόνδρια καὶ τὸ στῆθος, εὐθὺς σὺν τοῖς ἐρυγμοῖς ἀποδίδωσιν εὐώδεις τὰς ὀσμάς. . . .

61 (59) XIII. Αἱ δὲ τῶν ζῴων ὀσμαὶ κατὰ τὰς ἰδίας γίνονται φύσεις· ἑκάστῳ γάρ ἐστί τις οἰκεία κατὰ τὴν κρᾶσιν. αὗται δ᾿ ἡδεῖαι μὲν καὶ καθαραὶ [καὶ] κατὰ τὰς ἀκμὰς καὶ ὅταν εὖ ἔχωσιν ἑαυτῶν, ἔτι δὲ ἡδίους ἁπαλῶν καὶ νέων ὄντων. πλεῖσται δὲ καὶ κακωδέσταται περὶ τὰς ὀχείας καὶ ὅλως συντηκομένων καὶ καμνόντων σωμάτων· διὸ καὶ οἱ τράγοι καὶ οἱ ἔλαφοι καὶ λαγοὶ καὶ τἆλλα τότε μάλιστα ὄζει.

62 (60) Θαυμαστὸν δὲ καὶ ἴδιον τὸ συμπάσχειν τὰς τραγέας, ὅταν ἡ ὥρα καθήκῃ τῆς ὁρμῆς. αἴτιον δὲ δηλονότι τὸ ὑπολείπεσθαί τινα ἐν τῷ δέρματι δύναμιν ἢ ὑγρότητα τοιαύτην ἀφ᾿ ἧς ἡ ὁρμὴ γίνεται καὶ ζώντων· κινουμένης οὖν καὶ διαθερμαινομένης ταύτης ὑπὸ τοῦ ἀέρος εὔλογον καὶ τὰ δέρματα καθ᾿ ὅσον ἐπιβάλλει. διὸ καὶ ὡς

---

[1] § 60 on some other medicinal effects of perfumes is omitted.   [2] καὶ bracketed by W.

plasters and of what some call 'poultices' prove what virtues they display, since they disperse tumours and abscesses and produce a distinct effect on various other parts of the body, on its surface, but also on the interior parts: for instance, if one lays a plaster on his abdomen and breast, the patient forthwith produces fragrant odours along with his eructations.[1]

Of the characteristic smells of animals, and of certain curious facts as to the smell of animal and vegetable products.

XIII. The smells of animals correspond to their several characters: each has a smell of its own according to its particular composition. These smells are pleasant and pure when[2] the animal is in its prime and in good condition, and even pleasanter when they[3] are young and tender. But the smell is strongest and least pleasant at the breeding season, and generally when the body is wasting or out of condition: wherefore goats stags hares and other animals have most smell at such times.

It is a remarkable fact and peculiar to the goat that goat-skins[4] are sympathetically affected when the breeding season comes round. The reason plainly is that there remains somehow in the hide the sort of virtue or moisture from which arises the breeding impulse when the animal is alive. It is natural therefore that, when this is excited and warmed by the air, the skin also should be excited[5] so far as it belongs to it to be so affected. Wherefore the original cause as it were of the

[3] ἑαυτῶν can hardly be sound: ? αὐτὰ (sc. τὰ ζῷα).
[4] *i.e.* the skin of a dead goat.
[5] W. adds κινεῖσθαι after ἐπιβάλλει.

πρῶτον αἴτιον ἡ διάθεσις· τότε γὰρ καὶ οἱ μὴ ὀχεύοντες ὄζουσι καὶ οἱ ἄγονοι καὶ αἱ αἶγες ὅλως. ἡ δ' ὀχεία τότε μὲν μεγάλην μερίδα συμβάλλεται, καθ' αὑτὴν δ' αἰτία γίνεται ἡ διάθεσις.

63 Συμβαίνει δὲ τρόπον τινὰ καὶ ἐν ἄλλοις ἡ τοιαύτη συμπάθεια· καὶ γὰρ ὁ οἶνος ἅμα τῇ σταφυλῇ δοκεῖ συνανθεῖν καὶ τὰ σκόροδα καὶ τὰ κρόμυα τότε δριμύτατον ὄζειν, ὅταν <τὰ> ἐν τῇ γῇ βλαστάνῃ· πλὴν τούτοις ἅμα συμβαίνει καὶ αὐτοῖς βλαστάνειν. ὅλως δὲ πάντα κινεῖται τὰ φλοιόριζα καὶ σαρκόριζα μὴ ἀπεξηραμμένα κατὰ τὰς βλαστητικὰς ὥρας· ἡ γὰρ ἐνυπάρχουσα δύναμις ἐν αὐτοῖς κινεῖται. θαυμασιώτατον δὲ τῶν τοιούτων τὸ ἐπὶ τοῦ στέατος τῆς ἄρκτου συμβαῖνον, εἴπερ ἅμα ταῖς φωλίαις ἐπαίρεται καὶ ἐκπληροῖ τὰ ἀγγεῖα.

64 XIV. Τί δή ποτε Δημόκριτος τοὺς μὲν χυμοὺς πρὸς τὴν γεῦσιν ἀποδίδωσι, τὰς δ' ὀσμὰς καὶ τὰς χρόας οὐχ ὁμοίως πρὸς τὰς ὑποκειμένας αἰσθήσεις; ἔδει γὰρ ἐκ τῶν σχημάτων. ἢ τοῦτό γε πρὸς ἅπαντας κοινόν; ἅπαντες γὰρ οἱ μὲν μόνης

---

[1] *i.e.* to form a 'crust.' [2] τὰ ins. Sch.
[3] *cf.* H.P. 1. 6. 7.
[4] *i.e.* when the fat of the living bear becomes abnormally developed. Plin. 8. 128. expresses his surprise at T.'s credulity

phenomenon is the special condition of the animal at such periods: for at these times even those males which are not breeding have the smell, and the sterile goats and the females in general. Indeed, though at that particular time the fact that animals are actually breeding is a powerful factor in producing the smell, yet their condition is in itself a cause.

Similar sympathetic behaviour is found in a manner in other things also. Thus wine appears to 'bloom'[1] at the same time as the growing grape, and stored garlic and onions appear to have the most pungent smell at the season when those[2] in the ground are sprouting: however in this case sprouting takes place in the stored vegetables also. And in general any plant whose root is in layers[3] or fleshy becomes active at the season of sprouting, unless it has been completely dried: for it is the force latent in such plants which is stirred into activity. But the most remarkable phenomenon of the kind is what occurs with bears' grease: it makes active growth at the time of the bear's winter sleep[4] and completely fills the vessels in which it is kept.

Of odours as compared with other sense-impressions.

XIV. What can be the reason why Democritus, though he assigns various flavours to the sense of taste, yet does not in like manner assign various smells and colours to the senses to which they belong? According to his system he should have done so. Perhaps the same criticism should apply to all who have dealt with the subject: for they all

in this matter: his version (*coctas ursorum carnes*) adds to the marvel.

THEOPHRASTUS

οἱ δὲ μάλιστα ταύτης τὰ πάθη λέγουσι καὶ τὰς
διαφοράς, ὡς ἐν χρώμασι λευκὸν καὶ μέλαν, καὶ
ἐν χυμοῖς γλυκὺ καὶ πικρόν, οὐχ οὕτω δ᾽ ἐν
ὀσμαῖς· οὐδὲν γὰρ πλὴν τό τ᾽ εὔοσμον καὶ τὸ
κάκοσμον. οὐδ᾽ ἐν ἁπτοῖς· πλείω γὰρ εὐθὺ τὰ
ὑποκείμενα, σκληρὸν μαλακὸν τραχὺ λεῖον.

65 Ἀλλὰ μᾶλλον ἐν φωναῖς, ὀξὺ καὶ βαρύ. ἔτι
δὲ τὰ μὲν μικτὰ τὰ δ᾽ ἄμικτα. ἄμικτοι χυλοὶ οἱ
μὲν τῷ μὴ καταμερίζεσθαι ὥστ᾽ ἐξ ἀμφοῖν, οἷον
ὕδωρ ἔλαιον φλέγμα αἷμα, ὅλως πᾶν τὸ ἐπινέον
ἢ τὸ διαιροῦν, ὥσπερ τὸ ὄξος καὶ τὸ γάλα. τὸ
γὰρ τῇ πιέσει καὶ τρίψει μιγνύμενον ἕτερον εἶδος.
ἄλλον δὲ τρόπον οἱ μὴ εὔμικτοι πρὸς τὴν χρείαν
ἢ καὶ λυμαινόμενοι ἀλλήλους, οἷον ἡ θάλαττα
καὶ τὰ νιτρώδη καὶ πικρὰ ὕδατα τοὺς οἴνους καὶ
τὰ πότιμα, ἐὰν μὴ εὐθὺς χρῆταί τις.

66 Ὀσμαὶ δὲ αἱ μὲν οὕτως ἄμικτοι πλείους καὶ
ὥστε καθόλου λαβεῖν αἱ κακώδεις ταῖς κακώδεσι.
ὡς δὲ βέλτιόν τι τὸ ἐξ ἀμφοῖν ἔργον εὑρεῖν εἰ

---

[1] *i.e.* taste.
[2] There seems to be some confusion here, as in the first sentence of the section T. complained that colours are *not* classified. The following passage is unusually elliptical, and the text is probably defective. T.'s complaint seems to be that sense-experiences *in general* have been inadequately classified (*cf.* 2); and in 66 foll. he specially mentions smells.

CONCERNING ODOURS, 64-66

either give the various qualities and distinguish the experiences of this sense [1] alone or at least comparatively neglect the others: thus with colours [2] they distinguish white and black, and with flavours sweet and bitter, yet they make no corresponding classification of smells, but merely class them as 'pleasant' or 'unpleasant.' So too they fail to distinguish different experiences of the sense of touch, whereas several belong immediately to this sense, as hardness, softness, roughness, smoothness.

In sounds still more are there differences, as that between shrill and deep. Again some sense-experiences are simple, some compound. Flavours are simple first in the sense that they cannot be resolved into two components [3]: instances are water oil phlegm blood,[4] and in general anything which floats, like milk,[5] or which causes separation, like vinegar. (Where mixture can be produced by pressure or crushing, it is quite a different matter). Secondly there are flavours which do not readily combine in another sense, namely for human use, or which even spoil one another if they are mixed, as sea-water, or water with soda in it or which has a bitter taste : these spoil wines or other things that are good to drink, unless they are taken at once.

Now the odours which in this sense do not combine are numerous, and, speaking generally, it is the pleasant odours which do not combine with the unpleasant ones. It would indeed be difficult, if not impossible, to find a case in which mixture is an

[3] Sch. and W. after Turn. add <ἐν γίνεσθαι> after ἀμφοῖν, which seems unnecessary.
[4] *i.e.* a liquid which, in one way or another, refuses to mix with another liquid.   [5] ? cream.

385

μὴ ἀδύνατον, εἰς τὴν τοιαύτην δὲ δύναμιν <οὐχ>
ἅπαν ὡς εἰπεῖν πρὸς πᾶν εὔοσμον. ἀλλ' ἔνθα
μὲν ἴσως χείρω ποιεῖ ἔνθα δὲ βελτίω, καθάπερ
ἐπὶ τῶν μύρων· τὰ μὲν γὰρ ἀφαιρεῖται τὸ ἄκρατον καὶ σκληρόν, τὰ δ' ἀποθηλύνει καὶ ὥσπερ
ἐξυδατοῖ τὰς ὀσμάς. ἐν δὲ τοῖς ξηροῖς ἅπασαι
πᾶσαις μικταί.

67 Τὰ γὰρ διαπάσματα ὅσῳ ἂν ᾖ πλειόνων
ἀμείνω. ποιεῖ δὲ καὶ ἡ τοῦ οἴνου κατάμιξις καὶ
μύρα ἔνια καὶ θυμιάματα εὐοσμότερα, καθάπερ
τὴν σμύρναν. δοκεῖ δὲ καὶ τὸ μύρον ἡδύνειν τοὺς
οἴνους, διὸ καὶ οἱ μὲν ἐν τῇ οἰνοποιΐᾳ μιγνύουσιν
οἱ δὲ οὕτως ἐπιχεόμενον πίνουσιν. οὐκ ἄλογον
δὲ συνέγγυς τὰς αἰσθήσεις οὔσας καὶ ἐν τοῖς
αὐτοῖς ὑποκειμένοις ἔχειν τινὰ ἐπικοινωνίαν· ὡς
γὰρ ἐπὶ τὸ πᾶν οὐδεὶς οὔτε χυλὸς ἄοσμος οὔτε
ὀσμὴ ἄχυλος· τοῦτο δὲ ὅτι οὐδεμία ἐκ μὴ ἔχοντος
χυλόν.

68 Συμβαίνει δὲ καὶ μεταβάλλειν τὰς ὀσμὰς ἅμα
τοῖς χυλοῖς, ὥσπερ ἐπί τε τοῦ οἴνου καὶ ἐπὶ
καρπῶν τινῶν· ἐνίων δὲ καὶ ἐν τῷ ἄνθει πρότερον,
ὥσπερ τῶν βοτρύων· ἡ δὲ τῶν μύρων εἰς ἀκμὴν
μόνον καὶ οἷον φθίσιν. μετακινοῦνται δ' ἐν ταῖς
ἐτείαις ὥραις πάνθ' ὡς εἰπεῖν, μάλιστα δὲ τὰ

---

[1] I have inserted οὐχ, suggested by Sch.
[2] Like Sch. I fail to see the relevance of this remark. The sense required is 'while the fruit is still on the tree':

CONCERNING ODOURS, 66-68

improvement to the odour: in fact one might say that not[1] every combination of one fragrant thing with another will produce such a quality, but though sometimes the effect of such mixture is an improvement, sometimes it may be the reverse, as in the case of perfumes: for while the effect of some admixtures is to remove excessive strength or harshness, in other cases the odour is enfeebled and made, as it were, insipid. With solids however all combinations are possible.

In fact powders are the better, the more ingredients they have. Also the admixture of wine makes some perfumes and things used for incense more fragrant, for instance myrrh. It appears also that perfume sweetens wines, wherefore some add it in the manufacture, some put it in at the time of drinking. Nor is it unnatural that between these senses, since they are akin and are affected by the same objects, there should be a sort of reciprocity: for, to speak generally, no taste is unaccompanied by smell and no smell is unaccompanied by taste, the reason being that a thing which has no taste produces no smell.

It is also the case that smells actually change along with tastes, for instance in wine and certain fruits. And in some cases, as with grapes, the change takes place earlier, during[2] the flowering period: while in perfumes it occurs only when they have reached their best and are about, as it were, to go off. Almost all perfumes undergo alteration at certain seasons of the year, and this applies specially to the weakest kinds: in the case of those made

possibly ἀνθεῖ has got in from below and we should read καρπῷ.

ἀσθενέστατα, τὰ δ' ἄνθινα καθ' ἣν ὥραν ἀνθεῖ τὸ ἄνθος.

69 [Τὰς συνθέσεις ποιοῦσιν ἐκ τῶν ἀρωμάτων· θραύσαντες πολλὰ καὶ μίξαντες εἰς ταὐτὸ κλείουσιν εἰς κιβώτιον, εἶτ' ἀνοίγοντες διά τινων ἡμερῶν ὅτι ἂν μάλιστα ὄζειν δοκῇ τοῦτ' αἴρουσι, καὶ πάλιν δὲ καὶ πάλιν διαλείποντες χρόνον, ὅπως ἂν μηδενὸς ἐξόζῃ. θαυμαστὴν δ' ὀσμὴν λαμβάνει τὰ ἱμάτια εἰς ταῦτα τιθέμενα.

Τὸ δὲ τῆς βαλάνου τῆς Αἰγυπτίας μύρον αὐτὸ μὲν οὐκ ἄγαν ἀναπνεῖ, μιγνύμενον δὲ ποιεῖ τἆλλα βελτίω μάλιστα δὲ τὴν ἶριν]. . . .

---

[1] cf. 63.
[2] cf. 57, of which this section seems to be a repetition.
[3] cf. 15.

from flowers this period is that at which the plants from which they are made are in bloom.¹

[Compound ² perfumes are made from spices: they bruise and mix a variety of these and shut them up together in a box. Then after a few days they open the box and take out the spice which seems to have the strongest smell: this treatment is repeated at intervals, so that the smell of no one ingredient may overpower the others. And clothes put away with such perfumes acquire a marvellous fragrance.

The perfume ³ made of the Egyptian *balanos*, though it has not much scent of its own, when mixed with others, especially iris-perfume, improves them]. . . .⁴

⁴ The remaining sentences (§§ 70, 71) seem to be disconnected scraps, which perhaps do not belong to this treatise at all. The text of them being defective, it seems not worth while to attempt translation.

ΠΕΡΙ ΣΗΜΕΙΩΝ ΥΔΑΤΩΝ ΚΑΙ ΠΝΕΥΜΑΤΩΝ ΚΑΙ ΧΕΙΜΩΝΩΝ ΚΑΙ ΕΥΔΙΩΝ

Ι. Σημεῖα ὑδάτων καὶ πνευμάτων καὶ χειμώνων καὶ εὐδιῶν ὧδε ἐγράψαμεν καθ' ὅσον ἦν ἐφικτόν, ἃ μὲν αὐτοὶ προσκοπήσαντες ἃ δὲ παρ' ἑτέρων οὐκ ἀδοκίμων λαβόντες.

Τὰ μὲν οὖν ἐπὶ τοῖς ἄστροις δυομένοις καὶ ἀνατέλλουσιν ἐκ τῶν ἀστρονομικῶν δεῖ λαμβάνειν. εἰσὶ δὲ δύσεις δισταί· οἵ τε γὰρ ἀφανισμοὶ δύσεις εἰσί· τοῦτο δέ ἐστιν ὅταν ἅμα συνδύνῃ τῷ ἡλίῳ τὸ ἄστρον, καὶ ὅταν ἀνατέλλοντος δύνῃ. ὁμοίως δὲ καὶ ἀνατολαὶ διτταί, αἱ μὲν ἑῷοι ὅταν προανατέλλῃ τοῦ ἡλίου τὸ ἄστρον, αἱ δ' ἀκρόνυχοι ὅταν ἅμα δυομένῳ ἀνατέλλῃ.

Αἱ μὲν οὖν τοῦ Ἀρκτούρου λεγόμεναι ἀνατολαὶ ἀμφοτέρως συμβαίνουσιν· ἡ μὲν γὰρ τοῦ χειμῶνος ἀκρόνυχός ἐστιν, ἡ δὲ μετοπωρινὴ ἑῴα. τῶν δ' ἄλλων αἱ πλεῖσται τῶν ὀνομαζομένων ἑῷαι, οἷον Πλειάδος καὶ Ὠρίωνος καὶ Κυνός.

Τῶν δὲ λοιπῶν σημείων ἔνια μὲν ἴδια κατὰ πάσας χώρας ἐστὶν ἐν ὅσαις ὄρη ὑψηλὰ καὶ αὐλῶνές εἰσι, μάλιστα δὲ ὅσα πρὸς θάλασσαν καθήκει τῶν ὑψηλῶν· τῶν τε γὰρ πνευμάτων ἀρχομένων τὰ νέφη προσπίπτει πρὸς τοὺς τοιούτους τόπους, καὶ μεθισταμένων εἰς τοὐναντίον

CONCERNING WEATHER SIGNS

Introductory: general principles.

I. THE signs of rain wind storm and fair weather we have described so far as was attainable, partly from our own observation, partly from the information of persons of credit.

Now those signs which belong to the setting or rising of the heavenly bodies must be learnt from astronomy.[1] Their settings are twofold, since they may be said to have set when they become invisible. And this occurs when the star sets along with the sun, and also when it sets at sunrise. In like manner their risings are twofold: there is the morning rising, when the star rises before the sun, and there is the rising at nightfall, when it rises at sunset.

Now what are called the risings of Arcturus occur at both times, his winter rising being at nightfall and his autumn rising at dawn. But the rising of most of the familiar constellations is at dawn, for instance, the Pleiad Orion and the Dog.

Of the remaining signs some belong specially to all such lands as contain high mountains and valleys, specially where such mountains extend down to the sea: for, when the winds begin to blow, the clouds are thrown against such places, and, when the winds

[1] Or, perhaps, 'from my astronomical works.'

THEOPHRASTUS

ἀντιμεθίστανται καὶ ὑγρότερα γινόμενα διὰ βάρος εἰς τὰ κοῖλα συγκαθίζει. διὸ δεῖ προσέχειν οὗ ἄν τις ἰδρυμένος ᾖ. ἔστι γὰρ ἀεί τινα λαβεῖν τοιοῦτον γνώμονα καὶ ἔστι σαφέστατα σημεῖα τὰ ἀπὸ τούτων.

4 Διὸ καὶ ἀγαθοὶ γεγένηνται κατὰ τόπους τινὰς ἀστρονόμοι ἔνιοι, οἷον Ματρικέτας ἐν Μηθύμνῃ ἀπὸ τοῦ Λεπετύμνου, καὶ Κλεόστρατος ἐν Τενέδῳ ἀπὸ τῆς Ἴδης, καὶ Φαεινὸς Ἀθήνησιν ἀπὸ τοῦ Λυκαβηττοῦ τὰ περὶ τὰς τροπὰς συνεῖδε, παρ' οὗ Μέτων ἀκούσας τὸν τοῦ ἑνὸς δέοντα εἴκοσιν ἐνιαυτῶν <κύκλον> συνέταξεν. ἦν δὲ ὁ μὲν Φαεινὸς μέτοικος Ἀθήνησιν ὁ δὲ Μέτων Ἀθηναῖος. καὶ ἄλλοι δὲ τὸν τρόπον τοῦτον ἠστρολόγησαν.

5 Ἄλλα δέ ἐστι σημεῖα ἃ λαμβάνεται ἀπό τε ζῴων τῶν κατ' οἰκίαν καὶ ἑτέρων τινῶν τόπων καὶ παθημάτων, μάλιστα δὲ κυριώτατα <τὰ> ἀπὸ τοῦ ἡλίου καὶ τῆς σελήνης· ἡ γὰρ σελήνη νυκτὸς οἷον ἥλιός ἐστι· διὸ καὶ αἱ σύνοδοι τῶν μηνῶν χειμέριοί εἰσιν, ὅτι ἀπολείπει τὸ φῶς τῆς σελήνης ἀπὸ τετράδος φθίνοντος μέχρι τετράδος ἱσταμένου. ὥσπερ οὖν ἡλίου ἀπόλειψις γίνεται κατὰ τὸν ὅμοιον τρόπον καὶ τῆς σελήνης ἔκλειψις.

6 δεῖ οὖν προσέχειν μάλιστα ταῖς ἀνατολαῖς ταῖς τούτων καὶ ταῖς δύσεσιν ὁποίας ἂν ποιῶνται τὸν βουλόμενον προγινώσκειν.

---

[1] ἀντιμεθίστανται. ? ἀντιμεθίσταται.
[2] Plin. 5. 140. Of Matriketas nothing is known.
[3] Said (Plin. 2. 31) to have first recognised the Ram and the Archer. Athen. (7. 278 b) connects him with Tenedos.

change, the clouds also change[1] and take a contrary direction, and, as they become laden with moisture, they settle down in the hollows because of their weight. Wherefore good heed must be taken to the local conditions of the region in which one is placed. It is indeed always possible to find such an observer, and the signs learnt from such persons are the most trustworthy.

Thus in some parts have been found good astronomers: for instance, [2] Matriketas at Methymna observed the solstices from Mount Lepetymnos, Cleostratus [3] in Tenedos from Mount Ida, Phaeinos at Athens from Mount Lycabettus: Meton, who made the cycle [4] of nineteen years, was the pupil of the last-named. Phaeinos was a resident alien at Athens, while Meton was an Athenian. Others also have made astronomical observations in like manner.

Again there are other signs which are taken from domestic animals or from certain other quarters and happenings. Most important of all are the [5] signs taken from the sun and moon: for the moon is as it were a nocturnal sun. Wherefore also the meetings of the months are stormy, because the moon's light fails from the fourth day from the end of one month to the fourth day from the beginning of the next: there is therefore a failure of the moon corresponding to the failure of the sun. Wherefore anyone who desires to forecast the weather must pay especial heed to the character of the risings and settings of these luminaries.

[4] Called 'the great year': cf. Aelian. *V.H.* 10. 7. τὸν τοῦ ἑνὸς δέοντα εἴκοσιν ἐνιαυτῶν <κύκλον> conj. Sch. ἐνιαυτὸν conj. W.

[5] τὰ seems necessary. ? κύρια τὰ.

Πρῶτον μὲν οὖν ληπτέον ὅτι αἱ διχοτομίαι διορίζουσι τὰς ὥρας, ὥστε ἐπὶ τούτων δεῖ ἀθρεῖν καὶ ἐνιαυτὸν καὶ μῆνα καὶ ἡμέραν. διχοτομεῖ δὲ τὸν μὲν ἐνιαυτὸν Πλειάς τε δυομένη καὶ ἀνατέλλουσα· ἀπὸ γὰρ δύσεως μέχρι ἀνατολῆς τὸ ἥμισυ τοῦ ἐνιαυτοῦ ἐστίν. ὥστε δίχα τέμνεται ὁ

7 πᾶς χρόνος. ὁμοίως δὲ καὶ αἱ τροπαὶ καὶ ἰσημερίαι ποιοῦσιν. οἵα τις ἂν οὖν ᾖ κατάστασις τοῦ ἀέρος Πλειάδος δυομένης, οὕτω ἔχει ὡς ἐπὶ τὸ πολὺ μέχρι τροπῶν, κἂν μεταβάλλῃ, μετὰ τροπάς· ἐὰν δὲ μὴ μεταβάλλῃ, διέχει ἕως ἰσημερίας, κἀκεῖθεν ὡσαύτως μέχρι Πλειάδος, καὶ ἀπὸ ταύτης μέχρι τροπῶν θερινῶν, καὶ ἐντεῦθεν μέχρι ἰσημερίας, καὶ ἀπὸ ἰσημερίας μέχρι Πλειάδος δύσεως.

8 Ὡς δ' αὕτως ἔχει καὶ περὶ τὸν μῆνα ἕκαστον· διχοτομοῦσι γὰρ αἵ τε πανσέληνοι καὶ αἱ ὀγδόαι καὶ αἱ τετράδες, ὥστε ἀπὸ νουμηνίας ὡς ἀπ' ἀρχῆς δεῖ σκοπεῖν. μεταβάλλει γὰρ ὡς ἐπὶ τὸ πολὺ ἐν τῇ τετράδι, ἐὰν δὲ μή, ἐν τῇ ὀγδόῃ, εἰ δὲ μή, πανσελήνῳ· ἀπὸ δὲ πανσελήνου εἰς ὀγδόην φθίνοντος, καὶ ἀπὸ ταύτης εἰς τετράδα, ἀπὸ δὲ τετράδος εἰς τὴν νουμηνίαν.

9 Ὡς δ' αὕτως καὶ ἐπὶ τῆς ἡμέρας ἔχουσιν αἱ μεταβολαὶ ὡς ἐπὶ τὸ πολύ. ἀνατολὴ γὰρ καὶ πρωῒ καὶ μεσημβρία καὶ δείλη καὶ δύσις, καὶ τὰ τῆς νυκτὸς μέρη τὰ ἀνάλογα ταὐτὸ ποιεῖ τοῖς εἰρημένοις περὶ πνευμάτων καὶ χειμῶνος καὶ εὐδίας. μάλιστα γὰρ ἐὰν μέλλῃ μεταβάλλειν, ἐν

Now the first point to be seized is that the various periods are all divided in half, so that one's study of the year the month or the day should take account of these divisions. The year is divided in half by the setting and rising of the Pleiad [1] : for from the setting to the rising is a half year. So that to begin with the whole period is divided into halves: and a like division is effected by the solstices and equinoxes. From which it follows that, whatever is the condition of the atmosphere when the Pleiad sets, that it continues in general to be till the winter solstice, and, if it does change, the change only takes place after the solstice: while, if it does not change, it continues the same till the spring equinox : the same principle holds good from that time to the rising of the Pleiad, from that again to the summer solstice, from that again to the autumnal equinox, and from that to the setting of the Pleiad.

So too is it with each month ; the full moon and the eighth [2] and the fourth days make divisions into halves, so that one should make the new moon the starting-point of one's survey. A change most often takes place on the fourth day, or, failing that, on the eighth, or, failing that, at the full moon ; after that the periods are from the full moon to the eighth day from the end of the month, from that to the fourth day from the end, and from that to the new moon.

The divisions of the day follow in general the same principle : there is the sunrise, the mid-morning, noon, mid-afternoon, and sunset; and the corresponding divisions of the night have like effects in the matter of winds storms and fair weather; that is to say, if there is to be a change, it will generally

[1] Plin. 18. 280.   [2] *cf.* Arat. 73 f.

ταῖς διχοτομίαις μεταβάλλει. καθόλου μὲν οὖν τὰς ὥρας οὕτω δεῖ παρατηρεῖν, καθ' ἕκαστα δὲ τῶν σημείων κατὰ τὸν ὑπογεγραμμένον τρόπον.[1]

10 Ὕδατος μὲν οὖν σημεῖα τὰ τοιαῦτα δοκεῖ εἶναι. ἐναργέστατον μὲν οὖν τὸ ἑωθινόν, ὅταν πρὸ ἡλίου ἀνατολῆς φαίνηται ἐπιφοινίσσον σημεῖον· ἢ γὰρ αὐθημερινὸν ἐπισημαίνει ἢ τριῶν ἡμερῶν ὡς ἐπὶ τὸ πολύ. δηλοῖ δὲ καὶ τὰ ἄλλα σημεῖα· ἐὰν γὰρ μὴ πρότερον, τριταῖα μάλιστα σημαίνει τὸ ἐπιφοινίσσον καὶ δύνοντος, ἧττον δὲ ἢ τὸ ἑωθινοῦ.

11 Καὶ ἐὰν δύνῃ χειμῶνος ἢ ἔαρος εἰς νεφέλιον, τριῶν ἡμερῶν ὡς τὰ πολλὰ ἐπισημαίνει. καὶ ἐὰν ῥάβδοι νοτόθεν, ταὐτὰ δὲ ταῦτα βορρᾶθεν γινόμενα ἀσθενέστερα. καὶ ἐὰν ἀνίσχων μέλαν σημεῖον ἴσχῃ, καὶ ἐὰν ἐκ νεφελῶν <ἂν>έχῃ, ὑδατικόν, καὶ ἐὰν ἀκτῖνες ἀνίσχοντος ἀνατείνωσι πρὶν ἀνατεῖλαι, κοινὸν ὕδατος σημεῖον καὶ ἀνέμου. καὶ ἐὰν καταφερομένου τοῦ ἡλίου ὑφίστηται νέφος, ὑφ' οὗ ἐὰν σχίζωνται αἱ ἀκτῖνες, χειμερινὸν τὸ σημεῖον. καὶ ὅταν καυματίας δύηται καὶ ἀνατέλλῃ, ἐὰν μὴ ἄνεμος γένηται, ὕδατος τὸ σημεῖον.

12 Τὰ αὐτὰ δὲ σημαίνει καὶ σελήνη πανσελήνῳ ἀνίσχουσα, ἀσθενέστερα δὲ ὁ μείς. ἐὰν μὲν ᾖ πυρώδης, πνευματώδη σημαίνει τὸν μῆνα, ἐὰν δὲ ζοφώδης, ὑδατώδη· σημαίνει δὲ ὅτι ἂν σημαίνῃ τριταῖος ὢν ὁ μείς.

---

[1] τὸν ὑπογ. τρόπον seems to mean the same as the Aristotelian τὸν ὑφηγημένον τρόπον, e.g. *Eth. Nic.* 2. 7. 9. The rendering 'the following method' would however suit the context.

occur at one of these divisions. In general therefore one should observe the periods in the way indicated, though as to particular signs we must follow the accepted method.[1]

The signs of rain.

Now the signs of rain appear to be as follows: most unmistakable is that which occurs at dawn, when the sky has a reddish appearance before sunrise; for this usually indicates rain within three days, if not on that very day. Other signs point the same way: thus a red sky at sunset indicates rain within three days, if not before, though less certainly than a red sky at dawn.

Again, if the sun sets in a cloud in winter or spring, this generally indicates rain within three days. So too, if there are streaks of light from the south, while, if these are seen in the north, it is a less certain sign. Again, if the sun when it rises has a black mark, or if it rises[2] out of clouds, it is a sign of rain; while, if at sunrise there are rays[3] shooting out before the actual rising, it is a sign of rain and also of wind. Again if, as the sun sinks, a cloud forms below it and this breaks up its rays, it is a sign of stormy weather. Again, if it sets or rises with a burning heat, and there is no wind, it is a sign of rain.

Moonrise gives similar indications, at the time of full moon: they are less certain when the moon is not full. If the moon looks fiery, it indicates breezy weather for that month, if dusky, wet weather; and, whatever indications the crescent moon gives, are given when it is three days old.

[2] ἀνέχῃ conj. Sch.   [3] Plin. 18. 344.

THEOPHRASTUS

13 Ἀστέρες πολλοὶ διάττοντες ὕδατος ἢ πνεύματος, καὶ ὅθεν ἂν διάττωσιν ἐντεῦθεν τὸ πνεῦμα ἢ τὸ ὕδωρ. καὶ ἐὰν ἀκτῖνες ἀθρόαι ἀνίσχωσιν ἀνιόντος ἢ δύνοντος, σημεῖον <ὕδατος>. καὶ ὅταν ἀνίσχοντος τοῦ ἡλίου αἱ αὐγαὶ οἷον ἐκλείποντος χρῶμα ἴσχωσιν, ὕδατος σημεῖον. καὶ ὅταν νεφέλαι πόκοις ἐρίων ὅμοιαι ὦσιν, ὕδωρ σημαίνει. [ὑετοῦ δὲ σημεῖα] πομφόλυγες ἀνιστάμεναι πλείους ἐπὶ τῶν ποταμῶν ὕδωρ σημαίνουσι πολύ. ὡς δ' ἐπὶ τὸ πολὺ ἶρις περὶ λύχνον ἢ διὰ λύχνου διαφαινομένη νότια σημαίνει ὕδατα.

14 Καὶ οἱ μύκητες ἐὰν νότια ᾖ, ὕδωρ σημαίνουσι, σημαίνουσι δὲ καὶ ἄνεμον κατὰ λόγον ὡς ἂν ἔχωσι πλήθους καὶ μεγέθους, σμικροὶ δὲ καὶ κεγχρώδεις καὶ λαμπροὶ ὕδωρ καὶ ἄνεμον. καὶ ὅταν χειμῶνος τὴν φλόγα <ὁ λύχνος> ἀπωθῇ διαλιπὼν οἷον πομφόλυγας, ὕδατος σημεῖον, καὶ ἐὰν πηδῶσιν αἱ ἀκτῖνες ἐπ' αὐτόν, καὶ ἐὰν σπινθῆρες ἐπιγένωνται.

15 Ὄρνιθες λουόμενοι μὴ ἐν ὕδατι βιοῦντες ὕδωρ ἢ χειμῶνας σημαίνουσι. καὶ φρύνη λουομένη καὶ βάτραχοι μᾶλλον ᾄδοντες σημαίνουσιν ὕδωρ. καὶ ἡ σαύρα φαινομένη, ἣν καλοῦσι σαλαμάνδραν, ἔτι δὲ καὶ χλωρὸς βάτραχος ἐπὶ δένδρου ᾄδων ὕδωρ σημαίνει. χελιδόνες τῇ γαστρὶ τύπτουσαι τὰς λίμνας ὕδωρ σημαίνουσι. βοῦς τὴν προσθίαν ὁπλὴν λείξας χειμῶνα ἢ ὕδωρ σημαίνει. ἐὰν

---

[1] cf. 37.   [2] ὕδατος ins. Furl.   [3] Plin. 18. 344.
[4] Plin. 18. 356.   [5] ὑετοῦ δὲ σημεῖα bracketed by Sch.
[6] cf. Arist. Meteor. 3. 4; Plut. Quaest. Nat. 1. 2.
[7] cf. 42.
[8] i.e. breaks up into small 'grains' (?). cf. 25, 42, 54.

Many shooting[1] stars are a sign of rain or wind, and the wind or rain will come from that quarter from which they appear. Again, if at sunrise or sunset the sun's rays appear massed together, it is a sign of rain.[2] Also it is a sign of rain when at sunrise the rays[3] are coloured as in an eclipse; and also when there are clouds[4] like a fleece of wool. The rising of bubbles[5] in large numbers on the surface of rivers is a sign of abundant rain. And in general, when a rainbow[6] is seen round or through a lamp, it signifies rain from the south.

Again, if the wind is from the south, the snuff[7] of the lamp-wick indicates rain; it also indicates wind in proportion to its bulk and size: while if the snuff is small, like millet-seed,[8] and of bright colour, it indicates rain as well as wind. Again, when in winter the lamp rejects[9] the flame but catches, as it were, here and there in spurts, it is a sign of rain: so also is it, if the rays of light leap up on the lamp, or if there are sparks.

It is a sign of rain or storm when birds which are not aquatic take a bath. It is a sign of rain when a toad takes a bath, and still more so when frogs are vocal. So too is the appearance of the lizard known as 'salamander,'[10] and still more the chirruping of the green frog in a tree. It is a sign of rain when swallows[11] hit the water of the lakes with their belly. It is a sign of storm or rain when the ox

---

[9] *i.e.* refuses to light properly. The appearance seems to be that described Verg. *Georg.* 1. 391 (*scintillare oleum*). In the same passage *putres concrescere fungos* perhaps illustrates the comparison of the snuff to millet-seed above.

[10] *cf. de igne* 60, where it is explained why the salamander puts fire out.

[11] Plin. 18. 363; Verg. *Georg.* 1. 377.

δὲ εἰς τὸν οὐρανὸν ἀνακύπτων ὀσφραίνηται, ὕδωρ σημαίνει.

16 Κορώνη ἐπὶ πέτρας κορυσσομένη ἣν κῦμα κατακλύζει ὕδωρ σημαίνει· καὶ κολυμβῶσα πολλάκις καὶ περιπετομένη ὕδωρ σημαίνει. κόραξ πολλὰς μεταβάλλειν εἰωθὼς φωνάς, τούτων ἐὰν ταχὺ δὶς φθέγξηται καὶ ἐπιρροιζήσῃ καὶ τινάξῃ τὰ πτέρα, ὕδωρ σημαίνει. καὶ ἐὰν ὑετῶν ὄντων πολλὰς μεταβάλλῃ φωνὰς καὶ ἐὰν φθειρίζηται ἐπ' ἐλαίας. καὶ ἐάν τε εὐδίας ἐάν τε ὕδατος ὄντος μιμῆται τῇ φωνῇ οἷον σταλαγμούς, ὕδωρ σημαίνει. ἐάν τε κόρακες ἐάν τε κολοιοὶ ἄνω πέτωνται καὶ ἱερακίζωσιν, ὕδωρ σημαίνουσι. καὶ ἐὰν κόραξ εὐδίας μὴ τὴν εἰωθυῖαν φωνὴν ἵῃ καὶ ἐπιρροιβδῇ, ὕδωρ σημαίνει.

17 Ἐὰν ἱέραξ ἐπὶ δένδρου καθεζόμενος καὶ εἴσω εἰσπετόμενος φθειρίζηται, ὕδωρ σημαίνει. καὶ θέρους ὅταν πολλοὶ ἀθρόοι φανῶσιν ὄρνιθες οἳ βιοτεύουσιν ἐν νήσῳ, ὕδωρ σημαίνουσιν· ἐὰν δὲ μέτριοι, ἀγαθὸν αἰξὶ καὶ βοτοῖς, ἐὰν δὲ πολλοὶ ὑπερβολῇ, αὐχμὸν ἰσχυρόν. ὅλως δὲ ὄρνιθες καὶ ἀλεκτρυόνες φθειριζόμενοι ὑδατικὸν σημεῖον, καὶ ὅταν μιμῶνται ὕδωρ ὡς ὗον.

18 Καὶ ἡ νῆττα ἥμερος <ἐὰν> ὑπιοῦσα ὑπὸ τὰ γεῖσα ἀποπτερυγίζηται, ὕδωρ σημαίνει, ὁμοίως δὲ καὶ κολοιοὶ καὶ ἀλεκτρυόνες, ἐάν τε ἐπὶ λίμνῃ ἢ θαλάττῃ ἀποπτερυγίζωνται, ὡς νῆττα ὕδωρ ση-

---

[1] Plin. 18. 364 ; Verg. *Georg.* 1. 375.

[2] ἐπιρροιζήσῃ. Sc. with his wings probably; not, as LS. 'croaks.' Plin. (18. 362) seems to have had a fuller text, or to have drawn also on some other authority.

[3] ὑετῶν ὄντων can hardly mean 'while it is raining.'

licks his fore-hoof; if he puts[1] his head up towards
the sky and snuffs the air, it is a sign of rain.

It is a sign of rain when a crow puts back its head
on a rock which is washed by waves, or when it often
dives or hovers over the water. It is a sign of rain
if the raven, who is accustomed to make many
different sounds, repeats one of these twice quickly
and makes a whirring[2] sound and shakes his wings.
So too if, during a rainy season,[3] he utters many
different sounds, or if he searches for lice perched on
an olive-tree. And if, whether in fair or wet weather,
he imitates, as it were, with his voice falling drops,
it is a sign of rain. So too is it if ravens or jackdaws fly high and scream[4] like hawks. And, if a
raven in fair weather does not utter his accustomed
note and makes a whirring with his wings,[5] it is a
sign of rain.

It is a sign of rain if a hawk perches on a tree, flies
right into it and proceeds to search for lice: also,
when in summer a number of birds living on an
island pack together: if a moderate number collect,
it is a good sign for goats and flocks, while if the
number is exceedingly large, it portends a severe
drought. And in general it is a sign of rain when
cocks and hens search for lice; as also when they
make a noise like that of falling rain.

Again it is a sign of rain when a tame[6] duck gets
under the eaves and flaps its wings. Also it is a sign
of rain when jackdaws and fowls flap their wings
whether on a lake or on the sea—like the duck. It

[4] ἱερακίζωσιν. ? 'hover like hawks.' However, Arat. 231
understood it to refer to the voice: so LS.
[5] ἐπιρροιβδῇ. Exact sense uncertain. cf. Soph. Ant. 1004.
[6] ἥμερος. ? ἡ ἥμερος.

THEOPHRASTUS

μαίνει. καὶ ἐρωδιὸς ὄρθριον φθεγγόμενος ὕδωρ ἢ πνεῦμα σημαίνει· καὶ ἐὰν ἐπὶ θάλατταν πετόμενος βοᾷ, μᾶλλον ὕδατος σημεῖον ἢ πνεύματος, καὶ ὅλως βοῶν ἀνεμῶδες.

19 Καὶ ὁ σπίνος ἐν οἰκίᾳ οἰκουμένῃ ἐὰν φθέγξηται ἕωθεν, ὕδωρ σημαίνει ἢ χειμῶνα. καὶ χύτρα σπινθηρίζουσα πᾶσα περίπλεως ὕδατος σημεῖον. καὶ ἴουλοι πολλοὶ πρὸς τοῖχον ἕρποντες ὑδατικόν. δελφὶς παρὰ γῆν κολυμβῶν καὶ ἀναδυόμενος πυκνὰ ὕδωρ ἢ χειμῶνα σημαίνει.

20 Ὑμηττὸς ἐλάττων, ἄνυδρος καλούμενος, ἐὰν τῷ κοίλῳ νεφέλιον ἔχῃ, ὕδατος σημεῖον· καὶ ἐὰν ὁ μέγας Ὑμηττὸς τοῦ θέρους ἔχῃ νεφέλας ἄνωθεν καὶ ἐκ πλαγίου, ὕδατος σημεῖον. καὶ ἐὰν ὁ ἄνυδρος Ὑμηττὸς λευκὰς ἔχῃ ἄνωθεν καὶ ἐκ πλαγίου. καὶ ἐὰν περὶ ἰσημερίαν λὶψ πνεύσῃ, ὕδωρ σημαίνει.

21 Αἱ δὲ βρονταὶ αἱ μὲν χειμεριναὶ καὶ ἑωθιναὶ μᾶλλον <ἄνεμον ἢ> ὕδωρ σημαίνουσιν· αἱ δὲ θεριναὶ μεσημβρίας καὶ ἑσπεριναὶ βρονταὶ ὑδατικὸν σημεῖον. ἀστραπαὶ δὲ ἐάν γε πανταχόθεν γένωνται, ὕδατος ἂν ἢ ἀνέμου σημεῖον, καὶ ἑσπεριναὶ ὡσαύτως. καὶ ἐὰν ἀκρωρίας νότου πνέοντος νοτόθεν ἀστράψῃ, ὕδωρ σημαίνει ἢ ἄνεμον. καὶ ζέφυρος ἀστράπτων πρὸς βορείου ἢ χειμῶνα ἢ ὕδωρ σημαίνει. καὶ θέρους αἱ ἑσπέριαι ἀστραπαὶ ὕδωρ αὐτίκα σημαίνουσιν ἢ

---

[1] Sch. cites Plin. 18. 364, *vermes terreni erumpentes*, as representing this, which seems doubtful.
[2] *cf.* Plin. 18. 361 ; Cic. *Div.* 2. 70.
[3] ἐὰν τῷ. ? ἐὰν ἐν τῷ.

CONCERNING WEATHER SIGNS, 18–21

is a sign of wind or rain when a heron utters his note at early morning: if, as he flies towards the sea, he utters his cry, it is a sign of rain rather than of wind, and in general, if he makes a loud cry, it portends wind.

It is a sign of rain or storm if a chaffinch kept in the house utters its note at dawn. It is also a sign if any pot filled with water causes sparks to fly when it is put on the fire. It is also a sign of rain when a number of millepedes[1] are seen crawling up a wall. A dolphin[2] diving near land and frequently reappearing indicates rain or storm.

If the lesser Mount Hymettus, which is called the Dry Hill, has cloud in[3] its hollows, it is a sign of rain : so also is it, if the greater Hymettus has clouds in summer on the top and on the sides : or if the Dry Hymettus has white clouds on the top and on the sides ; also if the south-west wind[4] blows at the equinox.

Thunder in winter and at dawn indicates wind[5] rather than rain ; thunder in summer at midday or in the evening is a sign of rain. If lightning is seen from all sides, it will be a sign of rain or wind, and also if it occurs in the evening. Again, if when the south wind[6] is blowing at early dawn,[7] there is lightning from the same quarter, it indicates rain or wind. When the west wind is accompanied by lightning from the north, it indicates either storm or rain. Lightning in the evening in summer time indicates rain within three days, if not immediately.

[4] *cf.* Arist. *Probl.* 26. 26.
[5] ἄνεμον ἢ add. Furl. from Plin. 18. 354.
[6] *cf.* Soph. *Aj.* 257 ; Arist. *Probl.* 26. 20.
[7] ἀκρωρίας. *cf.* 42. So Arat. 216 renders.

403

τριῶν ἡμερῶν. καὶ ὀπώρας βορρᾶθεν ἀστραπαὶ ὑδατικὸν σημεῖον.

22 Ἡ Εὔβοια ὅταν διαζωσθῇ μέσῃ, ὕδωρ διὰ ταχέων. καὶ ἐὰν ἐπὶ τὸ Πήλιον νεφέλη προσίζῃ, ὅθεν ἂν προσίζῃ, ἐντεῦθεν ὕδωρ ἢ ἄνεμον σημαίνει. ὅταν ἶρις γένηται, ἐπισημαίνει· ἐάν τε πολλαὶ ἴριδες γένωνται, σημαίνει ὕδωρ ἐπὶ πολύ. ἀλλὰ πολλάκις καὶ οἱ ὀξεῖς ἥλιοι, ὅταν ἐκ νεφέλης. μύρμηκες ἐν κοίλῳ χωρίῳ ἐὰν τὰ ᾠὰ ἐκφέρωσιν ἐκ τῆς μυρμηκιᾶς ἐπὶ τὸ ὑψηλὸν χωρίον, ὕδωρ σημαίνουσιν, ἐὰν δὲ καταφέρωσιν, εὐδίαν. ἐὰν παρήλιοι δύο γένωνται καὶ ὁ μὲν νοτόθεν ὁ δὲ βορρᾶθεν, καὶ ἄλως ἅμα ὕδωρ διὰ ταχέων σημαίνουσι. καὶ ἄλως αἱ μέλαιναι ὑδατικὸν καὶ μᾶλλον αἱ δείλης.

23 Ἐν τῷ Καρκίνῳ δύο ἀστέρες εἰσίν, οἱ καλούμενοι Ὄνοι, ὧν τὸ μεταξὺ τὸ νεφέλιον ἡ Φάτνη καλουμένη. τοῦτο ἐὰν ζοφῶδες γένηται, ὑδατικόν. ἐὰν μὴ ἐπὶ Κυνὶ ὕσῃ ἢ ἐπὶ Ἀρκτούρῳ, ὡς ἐπὶ τὸ πολὺ πρὸς ἰσημερίαν ὕδωρ ἢ ἄνεμος. καὶ τὸ δημόσιον τὸ περὶ τὰς μυίας λεγόμενον ἀληθές· ὅταν γὰρ δάκνωσι σφόδρα, ὕδατος σημεῖον. σπίνος φθεγγόμενος ἕωθεν μὲν ὕδωρ σημαίνει ἢ χειμῶνα, δείλης δὲ ὕδωρ.

24 Τῆς δὲ νυκτὸς ὅταν τὸν Ὑμηττὸν κάτωθεν τῶν ἄκρων νεφέλη διαζώσῃ λευκὴ καὶ μακρά, ὕδωρ γίνεται ὡς τὰ πολλὰ μετρίων ἡμερῶν. καὶ ἐὰν

---

[1] Evidently an Attic saying, of days when only the upper part of the Euboean mountains was visible.

CONCERNING WEATHER SIGNS, 21-24

Lightning from the north in late summer is a sign of rain.

[1] When Euboea has a girdle about it up to the waist, there will be rain in a short space. If cloud clings about Mount Pelion, it is an indication of rain or wind from the quarter to which it clings. When a rainbow appears, it is an indication of rain; if many rainbows appear, it is an indication of long-continued rain. So too is it often when the sun appears [2] suddenly out of cloud. It is a sign of rain if ants [3] in a hollow place carry their eggs up from the ant-hill to the high ground, a sign of fair weather if they carry them down. If two mock-suns [4] appear, one to the south, the other to the north, and there is at the same time a halo, these indicate that it will shortly rain. A dark halo round the sun indicates rain, especially if it occurs in the afternoon.

In the Crab are two stars called the Asses, and the nebulous space between them is called the Manger [5]; if this appears dark, it is a sign of rain. If there is no rain at the rising of the Dog or of Arcturus, there will generally be rain or wind towards the equinox. Also the popular saying about flies is true; when they bite excessively, it is a sign of rain. If a chaffinch [6] utters its note at dawn, it is a sign of rain or storm, if in the afternoon, of rain.

When at night a long stretch of white cloud encompasses Hymettus below the peaks, there will generally be rain in a few days. If cloud settles on

---

[2] *cf. H.P.* 8. 10. 3.
[3] Plin. 18. 364; Verg. *Georg.* 1. 379.
[4] *cf.* 29. [5] *cf.* 43, 51.
[6] *cf.* 19, of which this seems to be in part a repetition.

ἐν Αἰγίνῃ [καὶ][1] ἐπὶ τοῦ Διὸς τοῦ Ἑλλανίου
νεφέλη καθίζηται, ὡς τὰ πολλὰ ὕδωρ γίνεται.
ἐὰν ὕδατα πολλὰ γίνηται χειμερινά, τὸ ἔαρ ὡς
τὰ πολλὰ γίνεται αὐχμηρόν· ἐὰν δ' αὐχμηρὸς
ὁ χειμών, τὸ ἔαρ ὑδατῶδες. ὅταν χιόνες πολλαὶ
γίνωνται, ὡς τὰ πολλὰ εὐετηρία γίνεται.

25 Φασὶ δέ τινες καὶ εἰ ἐν ἄνθραξι λαμπρὰ χάλαζα
ἐπιφαίνηται, χάλαζαν προσημαίνειν ὡς τὰ πολλά·
ἐὰν δὲ ὥσπερ κέγχροι μικροὶ λαμπροὶ πολλοί,
ἀνέμου μὲν ὄντος εὐδίαν, μὴ ἀνέμου δὲ ὕδωρ ἢ
ἄνεμον. ἔστι δ' ἄμεινον πρῶτον γίνεσθαι βόρειον
ὕδωρ νοτίου καὶ τοῖς φυομένοις καὶ τοῖς ζῴοις· δεῖ
δὲ γλυκὺ εἶναι καὶ μὴ ἁλμυρὸν τοῖς γενομένοις.
καὶ ὅλως ἔτος βέλτιον νοτίου βόρειον καὶ ὑγιει-
νότερον. καὶ ὅταν <πάλιν> ὀχεύωνται πρόβατα
ἢ αἶγες, χειμῶνος μακροῦ σημεῖον.[2]

26 II. Ὕδατος μὲν οὖν ταῦτα λέγεται σημεῖα·
ἀνέμου δὲ καὶ πνευμάτων τάδε. ἀνατέλλων ὁ
ἥλιος καυματίας, κἂν μὴ ἀποστίλβῃ, ἀνεμῶδες
τὸ σημεῖον· καὶ ἐὰν κοῖλος φαίνηται ὁ ἥλιος,
ἀνέμου ἢ ὕδατος τὸ σημεῖον. καὶ ἐὰν ἐπὶ πολλὰς
ἡμέρας καυματίας, αὐχμοὺς καὶ ἀνέμους πολυ-
χρονίους σημαίνει. ἐὰν αἱ ἀκτῖνες αἱ μὲν πρὸς
βορρᾶν αἱ δὲ πρὸς νότον σχίζωνται τούτου μέσου

---

[1] So called also by Pind. *Nem.* 5. 19.· Paus. 2. 30. 3 calls it the temple of Ζεὺς Πανελλήνιος. καὶ bracketed by Sch.
[2] *cf. C.P.* 2. 2.

the temple of Zeus Hellanios[1] in Aegina, usually rain follows. If a great deal of rain falls in winter, the spring is usually dry; if the winter has been dry, the spring is usually wet. When there is much[2] snow in winter, a good season generally follows.

Some say that, if in the embers[3] there is an appearance as of shining hail-stones, it generally prognosticates hail; while, if the appearance is like a number of small shining millet-seeds,[4] it portends fair weather, if there is wind at the time, but, if there is no wind, rain or wind. It is better both for plants and for animals that rain should come from the north before it comes from the south; it must however be fresh and not briny to the taste. And in general a season[5] in which a north wind prevails is better and healthier than one in which southerly winds prevail. It is a sign of a long winter when sheep or goats have a second[6] breeding season.

The signs of wind.

II. Such then are said to be the signs of rain. The following are signs of wind and breezes. [7] If the sun rises with a burning heat but does not shine brilliantly, it is a sign of wind. If the sun has a hollow appearance, it is a sign of wind or rain. If it blazes with a burning heat for several days, it portends long-continued drought or wind. If at dawn its rays are parted, some pointing to the north and some

[3] ἄνθραξι conj. Sch., supported by Plin. 18. 358; Arat. 309. ἀστράσι MSS.
[4] *cf.* 14, 42, 54.   [5] *cf. C.P.* 2. 2.
[6] πάλιν ins. Sch.; text probably defective.
[7] Plin. 18. 342.

THEOPHRASTUS

ὄντος κατ' ὄρθρον, κοινὸν ὕδατος καὶ ἀνέμου σημεῖόν ἐστιν.

27 Ἔστι δὲ σημεῖα ἐν ἡλίῳ καὶ σελήνῃ, τὰ μὲν μέλανα ὕδατος τὰ δ' ἐρυθρὰ πνεύματος. ἐὰν δὲ καὶ ὁ μεὶς βορείου ὄντος ὀρθὸς εἰστήκῃ, ζέφυροι εἰώθασιν ἐπιπνεῖν καὶ ὁ μὴν χειμερινὸς διατελεῖ. ὅταν μὲν ἡ κεραία <ἡ ἄνω> τοῦ μηνὸς ἐπικύπτῃ, βόρειος ὁ μείς· ὅταν δ' ἡ κάτωθεν, νότιος· ἐὰν δ' ὀρθὸς καὶ μὴ καλῶς ἐγκεκλιμένος μέχρι τετράδος καὶ εὔκυκλος, εἴωθε χειμάζειν μέχρι διχομηνίας. σημαίνει ζοφώδης μὲν ὢν ὕδωρ πυρώδης δὲ πνεῦμα.

28 Αἴθυιαι καὶ νῆτται [πτερυγίζουσαι] καὶ ἄγριαι καὶ τιθασσαὶ ὕδωρ μὲν σημαίνουσι δυόμεναι, πτερυγίζουσαι δὲ ἄνεμον. οἱ κέπφοι εὐδίας οὔσης ὅποι ἂν πέτωνται ἄνεμον προσημαίνουσι. στρουθοὶ χειμῶνος ἀφ' ἑσπέρας θορυβοῦντες ἢ ἀνέμου μεταβολὴν σημαίνουσιν ἢ ὕδωρ ὑέτιον. ἐρωδιὸς ἀπὸ θαλάσσης πετόμενος καὶ βοῶν πνεύματος σημεῖόν ἐστι· καὶ ὅλως βοῶν μέγα ἀνεμώδης.

29 Κύων κυλινδούμενος χαμαὶ μέγεθος ἀνέμου σημαίνει. ἀράχνια πολλὰ φερόμενα πνεῦμα ἢ χειμῶνα σημαίνει. ἡ ἄμπωτις βόρειον πνεῦμα σημαίνει, πλημμύρα δὲ νότιον. ἐὰν μὲν γὰρ ἐκ βορείων πλημμύρα ἥκῃ, εἰς νότιον μεταβάλλει, ἐὰν δ' ἐκ νοτίων ἄμπωτις γίνηται, εἰς βόρειον

---

[1] Plin. 18. 343 suggests that this is the meaning: text perhaps defective. cf. Verg. *Georg.* 1. 445.
[2] cf. 38.
[3] Lit. 'the crescent moon has a northerly character.' ἡ ἄνω add. Furl.

to the south, while the orb itself is[1] clearly seen between, it is a sign of rain and wind.

Also black spots on the sun or moon indicate rain, red spots wind. Again, if, while a north wind blows, the horns[2] of the crescent moon stand out straight, westerly winds will generally succeed, and the rest of the month will be stormy. When the upper horn of the crescent moon is bent, northerly winds[3] will prevail for that part of the month: when the lower horn is bent, southerly winds will prevail. [4] If however the horns up to the fourth day point straight and have not a graceful bend inwards but round to a circle, it will generally be stormy till the middle of the month. If the moon is dusky, it indicates rain, if fiery, it indicates wind.

It is a sign of rain when gulls and ducks, whether wild or tame, plunge under water, a sign of wind when they flap their wings. Wherever the bird called *kepphos* flies during a calm, it is a sign of coming wind. If sparrows in winter begin to be clamorous at evening, it is a sign of a coming change or of a fall of rain. A heron flying from the sea and screaming is a sign that a breeze is coming: so is it in general a sign of wind when he screams loudly.

A dog rolling on the ground is a sign of violent wind. A number of cobwebs[5] in motion portends wind or storm. The ebb-tide indicates a north wind, the flowing tide a wind from the south. For, if the flowing tide sets from the north, there is a change to the south, and if an ebb-tide comes from the south, there is a change to the north. It is

[4] *cf.* 38; Plin. 18. 347; Verg. *Georg.* i. 428; the English sign, 'the young moon with the old moon in her arm.'
[5] Plin. 11. 84; Arist. *Probl.* 26. 61.

THEOPHRASTUS

μεταβάλλει. θάλασσα οἰδοῦσα καὶ ἀκταὶ βοῶσαι καὶ αἰγιαλὸς ἠχῶν ἀνεμώδης. καὶ ὁ μὲν βορέας λήγων ἐλάττων ὁ δὲ νότος ἀρχόμενος. παρήλιος ὁπόθεν ἂν ᾖ ὕδωρ ἢ ἄνεμον σημαίνει.

30 Ἡ πέμπτη καὶ δεκάτη ἀπὸ τροπῶν τῶν χειμερινῶν ὡς τὰ πολλὰ νότιος. βορείων δὲ γινομένων ξηραίνει πάντα, νοτίων δὲ ὑγραίνει. ἐὰν δὲ νοτίων ὄντων ψοφῇ <τι> τῶν κεκολλημένων, εἰς τὰ νότια σημαίνει τὴν μεταβολήν· ἐὰν δὲ πόδες οἰδῶσι, νοτία ἡ μεταβολή. τὸ δὲ αὐτὸ σημεῖον καὶ ἐκνεφίου. καὶ ὀδαξῶν τὸν δεξιόν. ἐχῖνος ὁ χερσαῖος σημαντικόν· ποιεῖται δὲ δύο ὀπὰς ὅπου ἂν οἰκῇ, τὴν μὲν πρὸς βορρᾶν τὴν δὲ νοτόθεν· ὁποτέραν δ' ἂν ἀποφράττῃ, ἐντεῦθεν πνεῦμα σημαίνει, ἐὰν δ' ἀμφοτέρας, ἀνέμου μέγεθος.

31 Ἐὰν ὄρος ..., πρὸς βορρᾷ ἄνεμον προσημαίνει. ἐὰν ἐν θαλάττῃ ἐξαίφνης πνεύματος γαλήνη γίνηται, μεταβολὴν πνεύματος ἢ ἐπίδοσιν. ἐὰν ἄκραι μετέωροι φαίνωνται ἢ καὶ νῆσοι ἐκ μιᾶς πλείους, νοτίαν μεταβολὴν σημαίνει· γῆ τε μέλαινα ὑποφαινομένη <βόρειον>, λευκὴ δὲ νότιον. αἱ ἅλωνες περὶ τὴν σελήνην πνευματώδεις μᾶλλον ἢ περὶ ἥλιον· σημαίνουσι δὲ πνεῦμα ῥαγεῖσαι περὶ ἄμφω, καὶ ᾗ ἂν ῥαγῇ ταύτῃ πνεῦμα. ἐπι-

---

[1] cf. 40; Plin. 18. 359; Verg. Georg. 1. 356.

[2] cf. Arist. Probl. 26. 12 ad fin.

[3] ξηραίνει, ὑγραίνει seem to be used quasi-impersonally; but the text is perhaps defective.

[4] νότια MSS.; βόρεια conj. Furl., surely with good reason. cf. Arist. Probl. 1. 24.

[5] After δεξιόν Sch. and W. mark a lacuna, which does not seem necessary. [6] cf. Arist. H.A. 9. 6 ad fin.

a sign of wind when the sea [1] has a swell or promontories moan or there is loud noise on the beach. Now the north wind has less force as it ceases to blow, the south wind as it begins. A mock sun, in whatever quarter it appears, indicates rain or wind.

The fifteenth [2] day after the winter solstice is generally marked by southerly winds. If there is a northerly wind, everything gets dried [3] up, if a southerly, there is abundant moisture. If, while a south wind is blowing, glued articles make a cracking sound, it indicates a change to a south [4] wind. If the feet swell, there will be a change to a south wind. This also sometimes indicates a hurricane. So too does it, if a man has a shooting pain in the right foot. [5] The behaviour [6] of the hedgehog is also significant: this animal makes two holes wherever he lives, one towards the north, the other towards the south: now whichever hole he blocks up, it indicates wind from that quarter, and, if he closes both, it indicates violent wind.

If a mountain . . . , [7] it indicates wind from the north. If at sea during a wind there is a sudden calm, it indicates a change or an increase of wind. If promontories [8] seem to stand high out of the sea, or a single island looks like several, it indicates a change to south wind. If the land looks black from the sea, it indicates a north wind, [9] if white, a south wind. A halo [10] about the moon signifies wind more certainly than a halo about the sun: but in either case, if there is a break in the halo, it indicates wind, which will come from the quarter in which the break is. If the sky is overcast in whatever quarter

[7] I have marked a lacuna after ὅρος. Furl. renders *si mons versus aquilonem extenditur, venti signum est*, with what meaning I cannot see. [8] *cf.* Arist. *Meteor.* 3. 4 *ad init.*
[9] βόρειον add. Furl. [10] *cf.* 51.

νεφέλων ὅθεν ἂν ἀνατέλληται, ἐντεῦθεν ἄνεμος. αἱ κηλάδες νεφέλαι θέρους ἄνεμον σημαίνουσι.

82 Ἐὰν ἀστραπὴ πανταχόθεν γίνηται, ὕδωρ σημαίνει, καὶ ὅθεν ἂν αἱ ἀστραπαὶ πυκναὶ γίνωνται, ἐντεῦθεν πνεύματα γίνεται. θέρους ὅθεν ἂν ἀστραπαὶ καὶ βρονταὶ γίνωνται, ἐντεῦθεν πνεύματα γίνεται ἰσχυρά· ἐὰν μὲν σφόδρα καὶ ἰσχυρὸν ἀστράπτῃ, θᾶττον καὶ σφοδρότερον πνεύσουσιν, ἐὰν δ' ἠρέμα καὶ μανῶς, κατ' ὀλίγον. τοῦ δὲ χειμῶνος καὶ φθινοπώρου τοὐναντίον· παύουσι γὰρ τὰ πνεύματα αἱ ἀστραπαί· καὶ ὅσῳ ἂν ἰσχυρότεραι γίνωνται ἀστραπαὶ καὶ βρονταί, τοσούτῳ μᾶλλον παύονται· τοῦ δ' ἔαρος ἧττον ἂν ταὐτὰ σημεῖα λέγω, ὥσπερ καὶ χειμῶνος.

33 Ἐὰν νότου πνέοντος βορρᾶθεν ἀστράπτῃ, παύεται· ἐὰν ἕωθεν ἀστράπτῃ εἴωθε παύεσθαι τριταῖος, οἱ δὲ ἄλλοι πεμπταῖοι ἑβδομαῖοι ἐνναταῖοι, οἱ δὲ δειλινοὶ ταχὺ παύονται. οἱ βορέαι παύονται ὡς ἐπὶ τὸ πολὺ ἐν περιτταῖς οἱ δὲ νότοι ἐν ἀρτίαις. ἄνεμοι αἴρονται ἅμ' ἡλίῳ ἀνατέλλοντι καὶ σελήνῃ. ἐὰν ἀνατέλλων ὁ ἥλιος καὶ σελήνη παύσωσιν, ἐπιτείνει τὰ πνεύματα· χρονιώτερα δὲ καὶ ἰσχυρότερα τὰ πνεύματα γίνεται τὰ ἡμέρας ἢ νύκτωρ ἀρχόμενα.

34 Ἐὰν ἐτησίαι πολὺν χρόνον πνεύσωσι καὶ μετόπωρον γένηται ἀνεμῶδες, ὁ χειμὼν νήνεμος γίνεται, ἐὰν δ' ἐναντίως, καὶ ὁ χειμὼν ἐναντίος.

---

[1] κηλάδες, i.e. a 'mackerel sky' (?) The word seems to occur nowhere else except in Hesych., who renders ἄνυδρος: derivation obscure. It should probably be read in § 51 for κοιλάδες. [2] Plin. 18. 354.

[3] ἄν. Sc. εἶναι, which perhaps should be added.

the sun is first seen, there will be wind from that quarter. Light[1] clouds in summer-time indicate wind.

If lightning comes from all sides, it indicates rain, and from any quarter from which the flashes come in quick succession there will be wind. In summer[2] from whatever quarter lightning and thunder come, there will be violent winds: if the flashes are brilliant and startling, the wind will come sooner and be more violent; if they are of gentler character and come at longer intervals, the wind will get up gradually. In winter and autumn however the reverse happens, for the lightning causes the wind to cease: and, the more violent the lightning and thunder are, the more will the wind be reduced. In spring I consider that the indications would[3] not so invariably have the same meaning,—and this is also true of winter.

If, while a south wind is blowing, there comes lightning from the north, the wind ceases. If there is lightning at dawn, the wind generally ceases on the third day: other winds than a south wind however do not cease till the fifth seventh or ninth day, though a wind which got up in the afternoon will cease sooner. A north[4] wind generally ceases in an odd, a south wind in an even number of days. Winds get up at sunrise or moonrise. If the rising sun or moon have caused the wind to cease, presently[5] it gets up again with more force, and winds which begin to blow in the day-time last longer and are stronger than those which begin at night.

If periodic winds have been blowing for a long time, and a windy autumn follows, the winter is windless: if however the contrary happens, the character

[4] Plin. 2. 129.
[5] So Furl. renders: W. inserts μὴ after σελήνη.

πρὸς κορυφῆς ὄρους ὁπόθεν ἂν νεφέλη μηκύνηται, ταύτῃ ἄνεμος πνευσεῖται. αἱ νεφέλαι ἐκ τῶν ὄπισθεν προσίζουσαι καὶ ὄπισθεν πνευσοῦνται. Ἄθως μέσος διεζευγμένος νότιος, καὶ ὅλως τὰ ὄρη διεζωσμένα νότια ὡς τὰ πολλά. οἱ κομῆται ἀστέρες ὡς τὰ πολλὰ πνεύματα σημαίνουσιν, ἐὰν δὲ πολλοί, καὶ αὐχμόν. μετὰ χιόνα νότος, μετὰ πάχνην βορέας εἴωθε πνεῖν. μύκητες ἐπὶ λύχνου νότιον πνεῦμα ἢ ὕδωρ σημαίνουσιν.

35 Αἱ δὲ στάσεις τῶν πνευμάτων οὕτως ἔχουσιν ὡς ἐν τῷ γράμματι διώρισται. τῶν δ᾽ ἀνέμων ἔτι πνέουσι τοῖς ἄλλοις ἐπιπίπτουσι μάλιστα ἀπαρκτίας θρακίας ἀργέστης. ὅταν δὲ μὴ ὑπ᾽ ἀλλήλων διαλύωνται τὰ πνεύματα, ἀλλ᾽ αὐτὰ καταμαρανθῶσι, μεταβάλλουσιν εἰς τοὺς ἐχομέ-

---

[1] cf. 22. [2] cf. 57.
[3] cf. de Ventis 50; Arist. Probl. 26. 3. [4] cf. 14, 25, 42, 54.
[5] The 'figure' (giving points of the compass) has not been preserved. Arist. Meteor. 2. 6. describes such a figure (ὑπογραφή), which may be reconstructed thus :—

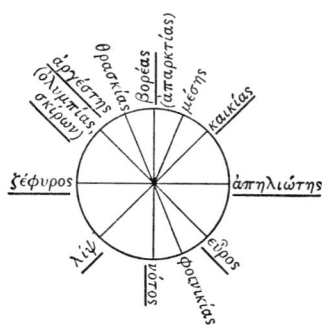

CONCERNING WEATHER SIGNS, 34-35

of winter is also reversed. From whatever quarter cloud streams out from a mountain peak, wind will blow in the direction thus indicated. Clouds which cling to the back of the mountain will also produce wind from the back of it. If there is a girdle[1] of cloud half way up Mount Athos, and if mountains in general wear such a girdle, there will generally follow a southerly wind. Comets[2] usually indicate wind, and, if there are many of them, drought is also indicated. After snow[3] a south wind, after hoar-frost a north wind generally blows. Snuff[4] in a lamp indicates wind or rain from the south.

The points from which the winds come are as they are given in the figure.[5] The winds which most often come on the top of other winds while these are still blowing are the north wind (*aparktias*),[6] the north-north-east and the north-west. When however the winds are not dispersed by one another but die down of their own accord, they change[7] to the next winds on the figure, reckoning from left

Arist. does not seem to distinguish βορέας and ἀπαρκτίας: his θρασκίας is T.'s θρακίας: his eight principal winds (underlined in diagram) correspond to those represented on the famous Tower of the Winds at Athens, built about two hundred years later.

[6] *cf.* Arist. *l.c.*   [7] Plin. 2. 128.

νους ἐπὶ δεξιά, ὥσπερ ἡ τοῦ ἡλίου ἔχει φορά. ὁ νότος ἀρχόμενος ξηρὸς τελευτῶν δὲ ὑγρός. καὶ ὁ εὖρος. ὁ δ' ἀπηλιώτης ἀπὸ ἀνατολῆς ἰσημερινῆς ὑδατώδης· διὰ λεπτῶν δὲ ἄγει τὰ ὕδατα.

36 Ὑγροὶ δὲ μάλιστα ὅ τε καικίας καὶ λίψ· χαλαζώδης δ' ἀπαρκτίας καὶ θρακίας καὶ ἀργέστης· νιφετώδης δὲ ὅ τε μέσης [ἢ βορέας] καὶ ἀπαρκτίας· καυματώδης δὲ νότος καὶ ζέφυρος καὶ εὖρος· οἱ μὲν οἷς ἂν ἐκ πελάγους προσπίπτωσιν, οἱ δὲ οἷς ἂν διὰ γῆς. δασύνουσι δ' οὐρανὸν νέφεσι καὶ καλύπτουσι καικίας μάλιστα εἶτα λίψ. καὶ οἱ μὲν ἄλλοι ἄνεμοι ἀφ' ἑαυτῶν τὰ νέφη ὠθοῦσι, καικίας δὲ μόνος πνέων εἰς ἑαυτόν. αἴθριοι δὲ μάλιστα θρακίας καὶ ἀργέστης καὶ τῶν λοιπῶν ἀπαρκτίας· ἐκνεφίαι δὲ μάλιστα ὅ τε ἀπαρκτίας καὶ ὁ θρακίας καὶ ὁ ἀργέστης.

37 Γίνονται δὲ ἐκνεφίαι ὅταν εἰς ἀλλήλους ἐμπίπτωσι πνέοντες μάλιστα μὲν μετοπώρου τῶν δὲ λοιπῶν ἔαρος. ἀστραπαῖος δὲ θρακίας καὶ ἀργέστης καὶ ἀπαρκτίας καὶ μέσης. ἐὰν ἐν τῇ θαλάττῃ πάπποι φέρωνται πολλοὶ οἱ γινόμενοι ἀπὸ τῶν ἀκανθῶν, ἄνεμον σημαίνουσιν ἔσεσθαι μέγαν. ὅθεν ἂν ἀστέρες διᾴττωσι πολλοί, ἄνεμον

---

[1] I have bracketed ἢ βορέας as probably a gloss on ἀπαρκτίας; ἢ is difficult to account for otherwise. See diagram.
[2] Plin. 18. 360.

to right according to the course of the sun. When the south wind begins to blow, it is dry, but it becomes wet before it ceases: so too does the south-east wind. The east wind, coming from the quarter where the sun rises at the equinox, is wet: but it brings the rain in light showers.

The north-east and south-west are the wettest winds; the north the north-north-east and the north-east bring hail; snow comes with the north-north-east[1] and north. The south, the west, and the south-east winds bring heat. Some of these have their effect on places which they strike as they come from the sea, others on places which they visit as they come over land. The winds which more than any others make the sky thick with cloud and completely cover it are the north-east and the south-west, especially the former. While the other winds repel the clouds from themselves, the north-east alone attracts them as it blows. Those winds which chiefly bring a clear sky are the north-north-west and the north-west, and next after them the north. Those which most have the character of a hurricane are the north the north-north-west and the north-west.

They acquire this character when they fall upon one another as they blow, especially in autumn, but to some extent in spring. Those which are accompanied by lightning are the north-north-west the north-west the north and the north-north-east. If at sea[2] a quantity of down is seen blown along, which has come from thistles, it indicates that there will be a great wind. Wind[3] may be expected from any quarter in which a number of shooting stars are

[3] *cf.* 13; Plin. 18. 352; Verg. *Georg.* 1. 365.

THEOPHRASTUS

ἐντεῦθεν· ἐὰν δὲ πανταχόθεν ὁμοίως, πολλὰ πνεύματα σημαίνουσι.

Πνευμάτων μὲν οὖν σημεῖα ταῦτα.

38 III. Χειμῶνος δὲ τάδε. ἥλιος δυόμενος εἰς μὴ καθαρόν. καὶ ὡς ἂν μερισθῇ δυόμενος, οὕτως αἱ ἡμέραι ἐπιτελοῦνται. οἷον εἰ τὸ τρίτον μέρος ἀπολειφθείη ἢ τὸ ἥμισυ. τὸ σελήνιον ἐὰν ὀρθὸν ᾖ μέχρι τετράδος καὶ εἰ εὔκυκλον, χειμάσει μέχρι διχοτόμου· γέρανοι ἐὰν πρωῒ πέτωνται καὶ ἀθρόοι, πρωῒ χειμάσει, ἐὰν δὲ ὀψὲ καὶ πολὺν χρόνον, ὀψὲ χειμάσει. καὶ ἐὰν ὑποστραφῶσι πετόμενοι, χειμῶνα σημαίνουσι.

39 Χῆνες βοῶντες μᾶλλον ἢ περὶ σῖτον μαχόμενοι χειμέριον. σπίνος στρουθὸς σπίζων ἕωθεν χειμέριον. ὄρχιλος [ὡς] εἰσιὼν καὶ εἰσδυόμενος εἰς ὀπὰς χειμῶνα σημαίνουσι καὶ ἐριθεὺς ὡσαύτως. κορώνη ἐὰν ταχὺ δὶς κρώζῃ καὶ τρίτον, χειμερία. καὶ κορώνη καὶ κόραξ καὶ κολοιὸς ὀψὲ ᾄδοντες χειμέριοι. στρουθὸς ἐὰν λευκὸς ᾖ χελιδὼν ἢ ἄλλο τι τῶν μὴ εἰωθότων λευκῶν, χειμῶνα μέγαν σημαίνουσιν, ὥσπερ καί, μέλανες ἐὰν πολλοὶ φανῶσιν, ὕδωρ.

40 Καὶ ἐὰν ἐκ πελάγους ὄρνιθες φεύγωσι, χειμῶνα σημαίνουσι. καὶ σπίνος ἐν οἰκίᾳ οἰκουμένῃ φθεγγόμενος χειμέριον. ὅσα ὕδωρ σημαίνει, χειμῶνα ἄγει, ἐὰν μὴ ὕδωρ, χιόνα καὶ χειμῶνα.

---

[1] *i.e.* and the succeeding day will be more or less stormy in proportion. ἀπολειφθείη. ? ἀποληφθείη = 'may be obscured.'

[2] *cf.* 27. *i.e.* it is possible, more or less, to see the whole circle.

seen. If these appear in every quarter alike, it indicates many winds.

Such then are the signs of winds.

III. The following are signs of storm. The sun becoming obscured as it sinks indicates storm. And, according as its orb is divided as it sets, so the succeeding days turn out; for instance, a third or a half of the orb may remain visible.[1] If the horns [2] of the moon point straight up till the fourth day, and if it rounds to a circle, it will be stormy till the middle of the month. If cranes fly early and in flocks, it will be an early winter[3]; if they fly late and for a long time, it will be a late winter; and, if they wheel as they fly, it indicates stormy weather.

[4] It is a sign of storm when geese make more clamour than usual or fight for their food; so too is it when a sparrow or chaffinch twitters at dawn. It indicates a storm when the goldcrest[5] goes into holes and hides itself; so also when the redbreast does the same. It is a sign of storm when the crow caws twice in quick succession and then a third time; also when the crow or raven or jackdaw makes its call late. It is a sign of a great storm when a white sparrow or swallow is seen, or a white specimen of any other bird which is not usually white, even as the appearance of a large number of such birds of a dark colour signifies rain.

It is also an indication of storm when birds flee from the sea. A chaffinch uttering its note in an inhabited house is a sign of storm. All the signs which indicate rain bring stormy weather, that is to say, snow and storm, if not rain. If the raven utters

[3] So Arat. 343 f. interprets.   [4] Plin. 18. 363.
[5] ὡς bracketed by Sch.

κόραξ φωνὰς πολλὰς μεταβάλλων χειμῶνος χειμέριον. κολοιοὶ ἐκ τοῦ νότου πετόμενοι καὶ τευθίδες χειμέριαι. φωνὴ ἐν λιμένι ἀποψοφοῦσα καὶ πολύπλοκον ἠχοῦσα χειμέριον. καὶ οἱ πνεύμονες οἱ θαλάττιοι ἐὰν πολλοὶ φαίνωνται ἐν τῷ πελάγει, χειμερινοῦ ἔτους σημεῖον. πρόβατα ἐὰν πρωῒ ὀχεύηται, πρώϊον χειμῶνα σημαίνουσι.

41 Μετοπώρῳ ἐὰν πρόβατα ἢ βόες ὀρύττωσι καὶ κοιμῶνται ἁθρόοι πρὸς ἀλλήλους ἔχοντες τὰς κεφαλάς, τὸν χειμῶνα χειμέριον σημαίνει. ἐν δὲ τῷ Πόντῳ φασίν, ὅταν Ἀρκτοῦρος ἀνατείλῃ θᾶττον, ἐναντίους τῷ βορρᾷ νέμεσθαι. βόες μᾶλλον ἐσθίοντες τοῦ εἰωθότος καὶ ἐπὶ τὸ δεξιὸν κατακλινόμενοι χειμέριον. καὶ ὦτα κρούων ὄνος χειμέριον· καὶ μαχόμενα πρόβατα καὶ ὄρνιθες περὶ σίτου παρὰ τὸ ἔθος· προπαρασκευάζονται γάρ· καὶ μύες τρίζοντες καὶ ὀρχούμενοι χειμέριον.

42 Καὶ κύων τοῖς ποσὶν ὀρύττουσα καὶ ὀλολυγὼν ᾄδουσα μόνη ἀκρωρίας χειμέριον. γῆς ἔντερα πολλὰ φαινόμενα χειμῶνα σημαίνει. καὶ ἐὰν πῦρ μὴ θέλῃ ἅπτεσθαι, χειμέριον· καὶ ἐὰν λύχνος ἅπτεσθαι μὴ ἐθέλῃ, χειμῶνα σημαίνει· καὶ τέφρα πηγνυμένη νιφετόν. λύχνος εὐδίας ἡσυχαῖος καιόμενος χειμῶνα σημαίνει· καὶ ἐὰν χειμῶνος ὄντος μύκαι μέλαιναι ἐπιγίνωνται, χειμῶνα σημαίνει· καὶ ἐὰν ὥσπερ κέγχροις πολλοῖς κατά-

---

[1] τευθίδες. The word is perhaps corrupt and conceals the name of a bird.
[2] cf. 21, 29. πολύπλοκον is Furlanus' conj. for Vulg. πολύποδον.
[3] πνεύμονες. Plin. 18. 359. pulmones: cf. 9. 154.

a great variety of sounds in winter, it is a sign of storm. Jackdaws flying from the south are a sign of storm, and so are cuttle-fish.[1] It is a sign of storm when a loud[2] voice is heard in harbour, which is re-echoed many times. It is a sign of a stormy season when a number of jelly-fish[3] appear in the sea. It indicates an early winter when the breeding season of sheep begins early.

If in autumn sheep or oxen dig holes and lie keeping their heads close to one another, it indicates a severe winter. They say that in Pontus when Arcturus rises, (the cattle[4]) face northwards as they graze. It is a sign of storm when cattle eat more than usual and lie down on their right sides.[5] So is it when the ass shakes[6] his ears, or when sheep or birds fight for their food more than usual, since they are then trying to secure a store against bad weather: also when mice squeak and dance.

A bitch digging holes with her paws and a tree-frog croaking alone at early dawn[7] are signs of storm: it indicates storm when a number of the worms[8] called 'the earth's entrails' appear. It is a sign of storm if the fire refuses to catch, or if a lamp refuses to light: while, if much ash is formed, it is a sign of snow. If a lamp burns steadily in fine weather, it is a sign of storm: so is it if in winter-time dark snuff[9] forms: if it is, as it were, full of numerous millet-seeds, there will be stormy weather;

[4] θᾶττον is clearly corrupt, and words indicating what the sign portends are missing.   [5] *cf.* 54.
[6] ὦτα κρούων doubtful. Sch. suggests οὖδας for ὦτα.
[7] ἀκρωρίας. *cf.* 21.
[8] γῆς ἔντερα. So Arat. 225 explains. One might guess 'worm-casts.'
[9] *cf.* 14, 25, 34, 54.

πλεως ᾖ, χειμερίσει· καὶ ἐὰν κύκλῳ περὶ τὸ λαμπρὸν ὦσιν εὐδίας οὔσης, χιονικόν.

43 Ἡ τοῦ ὄνου Φάτνη¹ εἰ συνίσταται καὶ ζοφερὰ γίνεται, χειμῶνα σημαίνει. καὶ ἐὰν ἀστραπὴ λαμπρὰ μὴ ἐν τῷ αὐτῷ μένῃ, χειμέριον. ἐπὶ Πλειάδι δυομένῃ ἐὰν λάμψῃ κατὰ Πάρνηθα καὶ Βρίληττον καὶ Ὑμηττον, ἐὰν μὲν ἅπαντα καταλάμψῃ, μέγαν χειμῶνα σημαίνει, ἐὰν δὲ τὰ δύο, ἐλάττω, ἐὰν δὲ Πάρνηθα μόνον, εὐδιεινόν· καὶ ἐὰν χειμῶνος ὄντος νεφέλη μακρὰ ἐπὶ τὸν Ὑμηττὸν ᾖ, χειμῶνος ἐπίτασιν σημαίνει. Ἄθως καὶ Ὄλυμπος καὶ ὅλως ὀρέων κορυφαὶ κατεχόμεναι ὑπὸ νεφελῶν χειμέριον. ἐὰν εὐδίας γινομένης νεφέλιον φαίνηται ἐν τῷ ἀέρι παρατεταμένον καὶ τιλλόμενον, οὔπω παύεται ὁ χειμών.

44 Ἐὰν τὸ μετόπωρον εὐδιεινὸν παρὰ τὸ εἰκὸς γένηται, τὸ ἔαρ γίνεται ψυχρὸν ὡς τὰ πολλά. ἐὰν πρωὶ χειμάζειν ἄρξηται, πρωὶ παύεται καὶ ἔαρ καλόν, ἐὰν δὲ τοὐναντίον, καὶ ἔαρ ὄψιον ἔσται. ἐὰν χειμὼν ὑέτιος, τὸ ἔαρ αὐχμηρόν, ἐὰν δ' αὐχμηρὸς ὁ χειμών, τὸ ἔαρ καλόν. ἐὰν ἡ ὀπώρα γίνηται ἐπιεικής, τὰ πολλὰ γίνεται τοῖς προβάτοις λιμός. ἐὰν τὸ ἔαρ καὶ τὸ θέρος ψυχρὰ γίνηται, ἥ τε ὀπώρα γίνεται καὶ <τὸ> μετόπωρον πνιγηρὸν καὶ οὐκ ἀνεμῶδες.

45 Οἱ πρῖνοι ἐὰν εὐκαρπῶσι, χειμῶνες πολλοὶ σφόδρα γίνονται. ἐὰν ἐπὶ κορυφῆς ὄρους νέφος ὀρθὸν στῇ, χειμῶνα σημαίνει, ὅθεν καὶ Ἀρχίλοχος ἐποίησε "Γλαῦχ' ὅρα· βαθὺς γὰρ ἤδη κύμασιν

---

¹ ὄνου φάτνη. cf. 23, 51. See LS, s.v. ὄνος; Theocr. 22. 21. Plin. 18. 353, sunt in signo Cancri duae stellae parvae

and if these in fine weather appear in a circle round the flame, it is a sign of snow.

If the 'Ass's Manger[1]' shrinks in size and becomes dark, it is a sign of storm; also if there is vivid lightning which does not remain in the same quarter. If at the setting of the Pleiad there is lightning over Parnes Brilessus and Hymettus—when it appears over all three mountains, it indicates a great storm; when over the two lesser heights, a less violent storm; when over Parnes alone, fine weather. Again, if during a storm a long cloud stretches over Hymettus, it signifies that the storm will increase in force. It is a sign of storm when Athos Olympus and mountain-peaks in general are covered with clouds. If during fine weather a cloud appears in the sky stretching a long way and torn to shreds, stormy weather will continue.

If the autumn is unusually fine, the succeeding spring is generally cold. If winter begins early, it ends early and there is a fair spring; if the reverse, spring will also be late. If the winter is wet, the spring will be dry, if the winter is dry, the spring will be fair. If the late summer is satisfactory, the sheep will generally suffer from hunger. If the spring and summer are cold, the late summer and autumn[2] will be stifling hot and windless.

If the kermes-oak[3] fruits well, there follows a long succession of storms. If a cloud stands upright on a mountain-peak, it indicates storm; whence Archilochus' lines " Mark you,[4] Glaucus; deep ocean

aselli appellatae, exiguum inter illas spatium obtinente nubecula, quam praesepia appellant.   [2] τὸ add. Sch.   [3] *cf.* 49.

[4] A comparison of war to stormy weather. Quoted also by Plut. *de Superstitione*, 72, and by Heraclides, *Allegoriae Homericae*, 4. In both citations the Greek is corrupt.

423

ταράσσεται Πόντος ἀμφὶ δ' ἄκρα <Γυρῶν> ὀρθὸν ἵσταται νέφος Σῆμα χειμῶνος." ἐὰν δ' ὁμόχρων ᾖ ὑμένι λευκῷ, χειμέριον. ὅταν ἑστώτων νεφῶν ἕτερα ἐπιφέρηται τὰ δ' ἠρεμῇ, χειμέρια.

46 <Ὁ ἥλιος> ἐὰν χειμῶνος διαλάμψας πάλιν ἀποκρυφθῇ καὶ τοῦτο ποιήσῃ δὶς ἢ τρίς, ἡμέρα χειμέριος δίεισιν. ὁ τοῦ Ἑρμοῦ ἀστὴρ χειμῶνος μὲν φαινόμενος ψύχη σημαίνει θέρους δὲ καῦμα. ὅταν μέλιτται μὴ ἀποπέτωνται μακρὰν ἀλλ' αὐτοῦ ἐν τῇ εὐδίᾳ πέτωνται, χειμῶνα ἐσόμενον σημαίνει. λύκος ὠρυόμενος χειμῶνα σημαίνει διὰ τριῶν ἡμερῶν. λύκος ὅταν πρὸς τὰ ἔργα ὁρμᾷ ἢ εἴσω χειμῶνος ὥρᾳ, χειμῶνα σημαίνει εὐθύς.

47 Ἔστι δὲ σημεῖον χειμώνων μεγάλων καὶ ὄμβρων καὶ ὅταν γένωνται ἐν τῷ μετοπώρῳ πολλοὶ σφῆκες, καὶ ὅταν ὄρνιθες λευκοὶ πρὸς τὰ ἐργάσιμα πλησιάζωσι, καὶ ὅλως τὰ ἄγρια θηρία ἐὰν πρὸς τὰ ἐργάσιμα, βόρειον καὶ χειμῶνος μέγεθος σημαίνει. τῆς Πάρνηθος ἐὰν τὰ πρὸς ζέφυρον ἄνεμον καὶ τὰ πρὸς Φύλης φράττηται νέφεσι βορείων ὄντων, χειμέριον τὸ σημεῖον.

48 Ὅταν πνίγη γίνηται ἰσχυρά, ὡς τὰ πολλὰ ἀνταποδίδωσι καὶ γίνεται χειμὼν ἰσχυρός. ἐὰν ὕδατα ἐαρινὰ πολλὰ γένηται, καύματα ἰσχυρὰ ἐν τοῖς πεδινοῖς καὶ κοίλοις γίνεται. δεῖ οὖν τὴν ἀρχὴν ὁρᾶν. ἐὰν τὸ μετόπωρον εὐδιεινὸν γίνηται

---

[1] Γυρῶν. γυροῦν W. Heraclides gives γυρεὸν, Plut. γυρεῦον; but one MS. of Plut. gives γύρων with a marginal gloss 'sc. πετρῶν,' which suggests that the word is a proper name. *Od.* 4. 500 mentions the Γύραι (*i.e.* the 'round-backed rocks')

is now stirred up with waves, and about the heights of the Gyrae[1] there rises a cloud erect, the sign of storm." If the clouds are of uniform colour, like[2] a white membrane, it is a sign of storm. When, as some clouds are motionless, others move towards them while they remain at rest, it is a sign of storm.

If the sun in winter after gleaming out is again obscured, and this is repeated two or three times, it will be stormy all day. If the star Hermes appears in winter, it indicates cold, if in summer, heat. When in fine weather bees do not fly[3] long distances, but fly about where they are, it indicates that there will be a storm. The howling of a wolf indicates a storm within three days. When a wolf approaches or enters cultivated ground in the season of winter, it indicates that a storm will come immediately.

It is also a sign of great storms and heavy rain when many wasps appear in autumn, or when white birds[4] approach cultivated lands; and in general when wild creatures approach such lands, it indicates a north wind and a severe storm. If the western side of Parnes and the side towards Phyle are blocked with clouds during a north wind, it is a sign of storm.

When there is severe heat, generally there is compensation and a severe winter follows. If there is much rain in spring, it is followed by severe heat in low-lying districts and valleys; so that one should mark how the season begins. If the autumn is

where Aias Oileus perished. The word is missing in the MSS. of T.

[2] ὅμοιον has perhaps dropped out after ὁμόχρων ᾖ; the adjective seems to agree with νέφος.

[3] *cf.* Arist. *H.A.* 9. 40 *ad fin.*

[4] Plin. 18. 363: presumably gulls, etc.

σφόδρα, τὸ ἔαρ ὡς τὰ πολλὰ γίνεται ψυχρόν· ἐὰν δὲ τὸ ἔαρ ὄψιον γένηται καὶ ψυχρόν, ἡ ὀπώρα ὀψία γίνεται καὶ <τὸ> μετόπωρον ὡς τὰ πολλὰ πνιγηρόν.

49 Οἱ πρῖνοι ὅταν εὐκαρπῶσι σφόδρα, ὡς μὲν τὰ πολλὰ χειμῶνα ἰσχυρὸν σημαίνουσιν, ἐνίοτε δὲ καὶ αὐχμούς φασι γίνεσθαι. καὶ ἐάν τις σπάλακα λαβὼν ὑποπάσας ἄργιλον εἰς πιθάκνην θῇ, σημαίνει ταῖς φωναῖς αἷς ἀφίησιν ἄνεμον καὶ εὐδίαν. καὶ τὸ πανταχοῦ δὲ λεγόμενον σημεῖον δημόσιον χειμέριον, ὅταν μύες περὶ φορυτοῦ μάχωνται καὶ φέρωσιν.

50 IV. Εὐδίας δὲ σημεῖα τάδε. ἥλιος μὲν ἀνιὼν λαμπρὸς καὶ μὴ καυματίας καὶ μὴ ἔχων σημεῖον μηδὲν ἐν ἑαυτῷ εὐδίαν σημαίνει. ὡς δ' αὕτως σελήνη πανσελήνῳ. καὶ δυόμενος ἥλιος χειμῶνος εἰς καθαρὸν εὐδιεινός, ἐὰν μὴ ταῖς προτέραις ἡμέραις εἰς μὴ καθαρὸν δεδυκὼς ᾖ ἐξ εὐδιῶν· οὕτω δὲ ἄδηλον. καὶ ἐὰν χειμάζοντος ἡ δύσις γένηται εἰς καθαρόν, εὐδιεινόν· καὶ ἐὰν δύνων χειμῶνος ὠχρὸς ᾖ, εὐδίαν σημαίνει.

51 Καὶ ὁ μεὶς ἐὰν τριταῖος ὢν λαμπρὸς ᾖ, εὐδιεινόν. καὶ ἡ τοῦ ὄνου Φάτνη ὅτε ἂν καθαρὰ καὶ λαμπρὰ φαίνηται, εὐδιεινόν. ἅλως δὲ ἐὰν ὁμαλῶς

---

[1] τὸ add. Sch.   [2] cf. 45.
[3] σπάλακα Vulg.; σπάκα Bas. Ald.; σκολόπακα (woodcock?) conj. Furl.
[4] *i.e.* (reading σκολόπακα) for the bird to find worms in with its long beak (Sch.). It is hard to say, without illus-

exceedingly fine, generally the spring is cold: if the spring is late and cold, the summer goes on late and the [1] autumn is usually scorching hot.

When the kermes-oak [2] fruits exceedingly well, it generally indicates a severe winter, and sometimes they say that this sign is followed by droughts. If one takes a mole [3] and puts it in a tub, the bottom [4] of which has been covered with clay, it indicates by the sounds which it utters wind or fine weather. There is also the sign of storm which is popularly recognized everywhere, namely when mice fight for the possession of chaff and carry it about.

The signs of fair weather.

IV. The following are signs of fair weather. [5] If the sun rises brilliant but without scorching heat and without showing any special sign in his orb, it indicates fair weather. The same may be said of the moon when it is full. If in winter that part of the sky into which the sun goes down is clear, it is a sign of fair weather, unless on the preceding days that part has not been clear, though it was clear above the horizon: in that case the prospect is uncertain. It is also a sign of fair weather, if during stormy conditions that part of the sky into which the sun sets is clear; and also if, in winter at the time of setting, the sun has a pale colour.

Again, it indicates fair weather if the outline of the moon on the third day is bright; also if the 'Ass's [6] Manger' is clear and bright. If the halo [7] forms and disappears evenly, it is a sign of fair

tration, which is the more convincing of the creatures suggested.   [5] Plin. 18. 342.   [6] *cf.* 23, 43.
[7] *cf.* 22, 31; Plin. 18. 345; Arist. *Meteor.* 3. 3.

παγῇ καὶ μαρανθῇ, εὐδίαν σημαίνει. αἱ κηλάδες[1] νεφέλαι χειμῶνος εὐδιειναί. Ὄλυμπος δὲ καὶ Ἄθως καὶ ὅλως τὰ ὄρη τὰ σημαντικὰ ὅταν τὰς κορυφὰς καθαρὰς ἔχωσιν, εὐδίαν σημαίνει. καὶ ὅταν τὰ νέφη πρὸς τὴν θάλασσαν αὐτὴν παραζωννύῃ, εὐδιεινόν. καὶ ὅταν ὕσαντος πρὸς δυσμὰς χαλκῶδες τὰ νέφη χρῶμα ἔχῃ· εὐδία γὰρ ὡς τὰ πολλὰ τῇ ὑστεραίᾳ.

52 Ὅταν δὲ ὀμίχλη γένηται, ὕδωρ οὐ γίνεται ἢ ἔλαττον. ὅταν γέρανοι πέτωνται καὶ μὴ ἀνακάμπτωσιν, εὐδίαν σημαίνει· οὐ γὰρ πέτονται πρὶν ἢ ἂν πετόμενοι καθαρὰ ἴδωσι. γλαῦξ ἡσυχαῖον φθεγγομένη ἐν χειμῶνι[5] εὐδίαν προσημαίνει· καὶ νύκτωρ χειμῶνος ἡσυχαῖον ᾄδουσα. θαλαττία δὲ γλαῦξ ᾄδουσα χειμῶνος μὲν εὐδίαν σημαίνει, εὐδίας δὲ χειμῶνα. καὶ κόραξ δὲ μόνος μὲν ἡσυχαῖον κράζων, καὶ ἐὰν τρὶς κράξῃ μετὰ τοῦτο πολλάκις κράξῃ, εὐδιεινός. . . .

53 Καὶ κορώνη ἕωθεν εὐθὺς ἐὰν κράξῃ τρίς, εὐδίαν σημαίνει, καὶ ἑσπέρας χειμῶνος ἡσυχαῖον ᾄδουσα. καὶ ὄρχιλος ἐξ ὀπῆς ἐκπετόμενος καὶ ἐξ ἑρκίων καὶ ἐξ οἰκίας ἔξωθεν εὐδίαν σημαίνει. καὶ ἐὰν χειμῶνος βορεύοντος βορρᾶθεν ὑπόλαμψις γένηται λευκή, νοτόθεν δὲ ἐναντία τεταγμένη ᾖ νεφέλη ὀγκώδης, ὡς ἐπὶ τὸ πολὺ εἰς εὐδίαν σημαίνει μεταβολήν. καὶ ὅταν βορέας νεφέλας πολλὰς κινῇ ἐκπνέων μέγας, εὐδίαν σημαίνει.

---

[1] κηλάδες I conj. cf. 31, to which this statement answers. κοιλάδες MSS.
[2] Plin. 18. 356. [3] Plin. 18. 357. cf. Verg. Georg. 1. 401.
[4] Plin. 18. 362.
[5] ἐν χειμῶνι. ? 'in winter.' The same ambiguity occurs in many places: the sense seems fixed here by the next sentence.

weather. Light[1] clouds in winter are a sign of fine weather. It is a sign of fine weather when Olympus Athos and in general the mountains which give signs have their tops[2] clear : so too is it, when clouds encompass them at the sea-level.[3] Also when after rain the clouds have a bronze colour towards sundown : in that case there will generally be fine weather the next day.

When there is mist, little or no rain follows. When cranes[4] take flight and do not come back, it is a sign of fair weather : for they do not do so till they see a clear sky before them as they fly. It is a sign of fair weather when during a storm[5] an owl makes a low hoot, or at night during a storm it utters a low sound. If the sea-owl utters its note during a storm, it indicates fair weather, if during fair weather, it indicates a storm. It is a sign of fair weather if a solitary raven makes a low croak, and, after croaking three times, repeats the sound again and again. . . .[6]

If the crow caws thrice directly the dawn appears, it indicates fair weather, as also if it makes a low note in the evening during a storm. It is a sign of fair weather if a goldcrest flies out abroad from a hole or from a hedge or from its nest. Again, if during a storm from the north there is a white gleam from that quarter, while in the south a solid mass of cloud has formed, it generally signifies a change to fair weather. Again when the north wind (Boreas) as it begins to blow violently stirs up a number of clouds, it indicates fair weather.

[6] I have marked a lacuna : the answer to μὲν is missing, presumably a statement about the significance of *more than one* raven. *cf.* Verg. *Georg.* 1. 410.

THEOPHRASTUS

54 Πρόβατα ὀψὲ ὀχευόμενα εὐδιεινὸν ἀποτελοῦσι τὸ σημεῖον. καὶ βοῦς ἐπὶ τὸ ἀρίστερον ἰσχίον κατακλινόμενος εὐδίαν σημαίνει· καὶ κύων ὡσαύτως· ἐπὶ δεξιὸν δὲ χειμῶνα. τέττιγες πολλοὶ γινόμενοι νοσῶδες τὸ ἔτος σημαίνουσι. λύχνος χειμῶνος καιόμενος ἡσυχαῖος εὐδίαν σημαίνει. καὶ ἐὰν ἐπ' ἄκρῳ οἷον κέγχρους ἔχῃ λαμπράς· καὶ ἐὰν ἐν κύκλῳ τὴν μύξαν περιγράφῃ λαμπρὰ γραμμή.

55 Ὁ τῆς σχίνου καρπὸς σημαίνει τοὺς ἀρότους· ἔχει δὲ τρία μέρη καὶ ἔστιν ὁ πρῶτος τοῦ πρώτου ἀρότου σημεῖον, ὁ δεύτερος τοῦ δευτέρου, ὁ τρίτος τοῦ τρίτου· καὶ ὡς ἂν τούτων ἐκβαίνῃ κάλλιστα καὶ γένηται ἀδρότατος, οὕτως ἕξει καὶ ὁ κατὰ τοῦτον ἄροτος.

56 Λέγεται δὲ καὶ τοιάδε σημεῖα ὅλων τε τῶν ἐνιαυτῶν γίνεσθαι καὶ τῶν μορίων. ἐὰν ἀρχομένου τοῦ χειμῶνος ζόφος ᾖ καὶ καύματα γίνηται καὶ ταῦτα ἄνευ ὕδατος ὑπ' ἀνέμων διαλυθῇ, πρὸς τὸ ἔαρ σημαίνει χάλαζαν ἐσομένην. καὶ ἐὰν μετὰ τὴν ἐαρινὴν ἰσημερίαν ὁμίχλαι πίπτωσι, πνεύματα καὶ ἀνέμους σημαίνουσιν εἰς ἕβδομον μῆνα ἀμφοτέρων ἀριθμουμένων. ὅσαι μὲν ἅμα μηνοειδεῖ τῇ σελήνῃ πίπτουσιν, αὗται μὲν πνεύματα σημαίνουσιν εἰς ἐκεῖνον τὸν χρόνον, ὅσαι δ' ἀμφικύρτου οὔσης τῆς σελήνης ὕδατα. καὶ ὅσῳ

---

[1] cf. 41.   [2] cf. 14, 25, 42.
[3] H.P. 7. 13. 6 the same is said of σκίλλα.

When sheep begin to breed late, it is a sign which fulfils itself in fair weather. So is it when an ox lies[1] on his left side, and also when a dog does the same: if they lie on the right side, it indicates storm. The appearance of a number of cicadas indicates that the season will be unhealthy. If a lamp burns quietly during a storm, it indicates fair weather. So also if it has on the surface an appearance like shining millet-seeds:[2] also if a bright line surrounds the lamp-nozzle.

The fruiting of the mastich[3] gives signs as to the seasons of sowing:[4] it takes place at three several periods, which indicate respectively the time for the first the second and the third sowing: and according as one or other of these fruiting-times turns out[5] best and produces the most abundant fruit, so too will be the success of the corresponding time of sowing.

Miscellaneous signs.

The following signs are said to affect either the whole year or whole periods[6] of it. If at the beginning of winter there is dull weather followed by heat, and these conditions are dispersed by wind without rain, it indicates that towards the spring there will be hail. Again, if after the spring equinox mists come down, it is an indication of breezes and winds by the seventh month, reckoning inclusively. Those mists which come down when the moon is in its first quarter indicate breezes for that period, those which come down when the moon is in its third quarter indicate rain. And the more mists

[4] *cf. H.P.* 7. 1. 1 foll.
[5] ἐκβαίνῃ I conj.: *cf. H.P.* 7. 13. 6; κλίνῃ MSS.
[6] *cf.* 6.

ἂν μᾶλλον ἐφ' ἑκατέρῳ τῷ σχήματι ὁμίχλαι πίπτωσι, μᾶλλον τὰ εἰρημένα σημαίνει.

57 Σημαίνει δὲ καὶ τὰ πνεύματα ἅμα ταῖς ὁμίχλαις ἐπιπιπτούσαις γινόμενα· καὶ ἐὰν μὲν ἀπ' ἠοῦς καὶ μεσημβρίας γίνηται τὰ πνεύματα, ὕδατα σημαίνει· ἐὰν δ' ἀφ' ἑσπέρας καὶ ἀπὸ τῆς ἄρκτου πνεύματα καὶ ψύχη. οὓς δὲ κομήτας Αἰγύπτιοι λέγουσιν οὐ μόνον τὰ προειρημένα σημαίνουσιν ὅταν φαίνωνται ἀλλὰ καὶ ψύχη· ἐπὶ δὲ τοῖς ἄστροις εἴωθεν ὡς ἐπὶ τὸ πολὺ σημαίνειν καὶ ταῖς ἰσημερίαις καὶ τροπαῖς, οὐκ ἐπ' αὐταῖς ἀλλ' ἢ πρὸ αὐτῶν ἢ ὕστερον μικρῷ.

---

[1] cf. 34; Arist. Meteor. 1. 6.
[2] Text seems doubtful, as cold weather was included above.

come down when the moon is assuming either shape, the more certainly is the result just mentioned indicated.

Also the winds which accompany the falling of the mists are significant: if the breezes come from the east or south, rain is indicated; if from the west or north, breezes and cold weather. And the stars which the Egyptians[1] call 'comets' indicate not only the conditions just mentioned but also cold[2] weather. [3] In the case of the rising of the stars the indication, as in the case also of the equinoxes and solstices, is given not at the actual time but a little earlier or later.

---

[3] The text of this sentence can hardly be sound. σημαίνειν has no subject and ταῖς ἰσημερίαις καὶ τροπαῖς no construction.

INDEX OF PLANTS

NOTE TO THE INDEX OF PLANTS

Sprengel made the first comprehensive attempt to determine in modern nomenclature the plants mentioned by Theophrastus: Wimmer gives the result in the Introduction to his 1842 edition. Sprengel adopted the most probable identifications of earlier botanists, supplemented by his own conjectures and Sibthorp's exploration of the Greek flora. The ambitious but uncritical *Conspectus Florae Classicae* of Fraas did not add much to our knowledge, which throughout had been vitiated by failure to recognise the fact that the Mediterranean flora differed from that of Western and Central Europe. Halacsy's *Conspectus Florae Graecae* now gives us a scientific enumeration of the native plants of Greece; a Greek plant-name can be wedded to a plant which at any rate is Greek. Incidentally much has been cleared up by special research at the hands of De Candolle, Hanbury, Yule, Schweinfurth, Bretzl, and others.

The identifications in the following Index are drawn from various sources; for their selection in view of the botanical data available I am indebted to Sir William Thiselton-Dyer. A considerable number may be accepted as certain, many are probable, some no more than possible.

INDEX OF PLANTS

comp. = compared.
= denotes a synonym. Where a reference is added (see *e.g.* ἀτρακτυλίς), it indicates that Theophrastus himself states that the names are synonymous.

ἀβρότονον, southernwood, *Artemisia arborescens*
 1. 9. 4. evergreen; 6. 1. 1. in list of under-shrubs; 6. 3. 6. an unnamed plant comp.: see App. (23): 6. 7. 3. propagation; 6. 7. 4. much seed: roots described.

ἄγνος (=οἴσος), chaste-tree, *Vitex Agnus-castus*
 1. 3. 2. a shrub which becomes tree-like; 1. 14. 2. bears fruit at the top; 3. 12. 1. growth of κράνεια comp.; 3. 12. 2. roots of θηλυκράνεια comp.; 4. 10. 2. ἐλαίαγνος comp.; 9. 5. 1. size of κινάμωνον and κασία comp.

ἀγριέλαιος (? = κότινος), wild olive, *Olea Oleaster*
 2. 2. 5. comes from seed of ἐλάα.

ἄγρωστις, dog's tooth grass, *Cynodon Dactylon*
 1. 6. 7. root jointed; 1. 6. 10. roots large and numerous; 2. 2. 1. propagation; 4. 6. 6. φῦκος (6) comp.; 4. 10. 5-6 root described; 4.11.13. an unnamed form of κάλαμος comp.: root of κ. ὁ Ἰνδικός comp.; 9. 13. 6. habit of ἐρευθέδανον comp.

ἄγχουσα, alkanet, *Anchusa tinctoria*
 7. 8. 3. leaves 'on the ground': 7. 9. 3. roots red.

ἀδίαντον, maiden-hair, *Adiantum Capillus-Veneris*, etc.
 7. 10. 5. evergreen; 7. 14. 1. leaf cannot be wetted: two kinds (*see below*): medicinal use: grows in damp places.

ἀδίαντον τὸ λευκόν (= τριχομανές 7. 14. 1), English maiden-hair, *Asplenium Trichomanes*
 7. 14. 1. described by comparison with ἀ. τὸ μέλαν: medicinal use: likes shady places.

ἀδίαντον τὸ μέλαν, maiden-hair, *Adiantum Capillus-Veneris*
 7. 14. 1. comp. with ἀ. τὸ λευκόν.

ἀδράφαξυς, orach, *Atriplex rosea*
 1. 14. 2. bears fruit both on top and at sides; 3. 10. 5. seeds of φίλυρα comp.; 7. 1. 2-3. time of sowing and of germination; 7. 2. 6. root described; 7. 2. 7-8. root of βλίτον comp.; 7. 2. 8. root: 7.3.2. seeds; 7. 3. 4. seed borne both at top and at side; 7. 4. 1. only one kind; 7. 5. 5. seed does not keep well.

ἀείζωον, house-leek, *Sempervivum tectorum*
 1. 10. 4. leaves fleshy; 7. 15. 2. always moist and green: habitat.

ἀθραγένη, traveller's joy, *Clematis Vitalba*
 5. 9. 6. wood makes good firesticks: described; 5. 9. 7. the stationary piece should be made of this or κιττός.

αἴγειρος, black poplar, *Populus nigra*
 1. 2. 7. bark; 1. 5. 2. bark fleshy; 2 2.10. Cretan form bears fruit; 3. 1. 1. propagation; 3. 3. 1. tree of mountain and plain; 3. 3. 4. a question if it bears fruit; etc.;

437

INDEX OF PLANTS

3. 4. 2. time of budding ; 3. 6. 1. quick growing; 3.14.2. described; 4. 1. 1. likes wet ground ; 4. 7. 4. size of unnamed Arabian tree, see App. (12a), comp.; 4. 13. 2. shorter-lived by water; 5. 9. 4. wood makes an evil smoke when burnt for charcoal.

αἰγίλωψ (1) (= ἄσπρις), Turkey oak, *Quercus Cerris*
  3. 8. 2. one of the five (Idaean) kinds of oak: fruit ; 3. 8. 4. habit and timber ; 3. 8. 6. galls : φάσκος (q.v.).

αἰγίλωψ (2) (grass), *Aegilops ovata*
  7. 13. 5. seed sometimes takes two years to germinate; 8. 7. 1. comp. with αἶρα ; 8. 8. 3. grows specially among κριθαί; 8. 9. 2. like a wild plant; 8. 9. 3. greatly exhausts the soil; 8. 11. 8-9. peculiarities about seed.

αἱμόδωρον, broom-rape, *Orobanche cruenta*
  8.8.5. parasitic on βουκέρας (only) : described.

αἶρα, darnel, *Lolium temulentum*
  1.5.2. 'bark' in one layer; 2.4.1. πυρός turns into ἀ.; 4. 4. 10. ὄρυζον comp.; 8. 4. 6. does not infest certain kinds of πυρός: contrasted with μελάμπυρον; 8. 7. 1. κριθή and especially πυρός said to change into ἀ. under certain conditions : described: λίνον also said to change into ἀ.: comp. with αἰγίλωψ (2); 8. 8. 3. produced possibly by degeneration of κριθή and πυρός, or else specially affects such crops; 8. 9. 3. altogether a wild plant.

ἀκαλύφη, nettle, *Urtica urens*
  7. 7. 2. a λάχανον ; needs cooking.

ἄκανθα (1) ἡ Αἰγυπτία, acacia, *Acacia arabica* (and *albida*)
  4. 2. 1. peculiar to Egypt ; 4. 2. 8. described : two kinds (ἡ λευκή and ἡ μέλαινα) distinguished (see below); 9. 1. 2. sap gummy.

ἄκανθα (ἡ Αἰγυπτία) ἡ λευκή, acacia, *Acacia albida*
  4. 2. 3. distinguished from ἀ. ἡ μέλαινα.

ἄκανθα (ἡ Αἰγυπτία) ἡ μέλαινα, acacia, *Acacia arabica*
  4. 2. 8. distinguished from ἀ. ἡ λευκή.

ἄκανθα (2) ἡ ἀκανώδης (see 4.10.6. n.), corn-thistle, *Carduus arvensis*
  4. 10. 6. root etc. described.

ἄκανθα (3) ἡ διψάς, *Acacia tortilis*
  4. 7. 1. the only tree which grows on part of the 'Red Sea' coast.

ἄκανθα (4) ἡ 'Ινδική (see App. (9)), *Balsamodendron Mukul*
  9. 1. 2. sap gummy: gum like σμύρνα.

ἄκανθα (5) ἡ λευκὴ Ἡρακλέους (= ἄκανθα (6)), *Euphorbia antiquorum*
  4. 4. 12. described : uses of wood.

ἄκανθα (6) (peculiar to Gedrosia), =ἄκανθα (5), *Euphorbia antiquorum*
  4. 4. 13. described : has a blinding juice.

ἄκανθα (7) τις, gum arabic, *Acantha arabica*
  9. 18. 1. said to have the property of thickening water.

ἄκανθα (?) (8) (= ἄκανος = ἰξία (2) = ἰξίνη = χαμαιλέων ὁ λευκός 9. 12. 1.), pine-thistle, *Atractylis gummifera*

ἄκανος (=ἄκανθα(8)=ἰξία(2)=ἰξίνη= χαμαιλέων ὁ λευκός), pine-thistle, *Atractylis gummifera*
  1. 10. 6. spinous-leaved ; 1. 13. 3. flower attached above each seed ; 6. 1. 3. has spines on the leaves : a wild under-shrub; 6. 4. 4. many stalks and sidegrowths ; 6. 4. 5. one form only ; 6. 4. 8. root of σόγκος contrasted: χαμαιλέων comp.; 6. 4. 11. fruitcase of κάκτος (1) comp.; 6. 6. 6. seed of ῥόδον comp.; 9. 12. 1. 'head' of χαμαιλέων ὁ λευκός comp.: another name for χαμαιλέων (?); 9. 12. 2. leaf of χαμαιλέων ὁ μέλας comp.

ἀκόνιτον (= θηλυφόνον = μυόφονον = σκορπιός (3)), wolf's bane, *Aconitum Anthora*
  9. 16. 4. localities : described: habitat : eaten by no animal; 9. 16. 5. difficulty of compounding drug : effects : has no anti-

INDEX OF PLANTS

dote; 9. 16. 7. use requires expert knowledge: legal restrictions: proportion between times of gathering and of administering.

ἄκορνα, *Cnicus Acarna*
1. 10. 6. spinous-leaved; 7. 4. 3. a 'thistle-like' plant; 6. 4. 6. described.

ἀκτέος (?) (= ἀκτῆ), elder, *Sambucus nigra*
3. 4. 2. time of budding.

ἀκτῆ (=ἀκτέος), elder, *Sambucus nigra*
1. 5. 4. wood without knots; 1. 6. 4. core fleshy: has no core, according to some: 1. 8. 1. few knots; 4. 13. 2. shorter-lived by water; 5. 3. 3. character of wood.

ἀλθαία (= μαλάχη ἡ ἀγρία 9. 15. 5.), marsh-mallow, *Althaea officinalis*
9. 15. 5. a drug, called in Arcadia μαλάχη ἡ ἀγρία; 9.18.1. root said to thicken water: described: medicinal use.

ἄλιμον, *Atriplex Halimus*
4. 16. 5. very dangerous to trees.

ἀλίφλοιος (δρῦς), see δρῦς (3).

ἀλσίνη, *Parietaria cretica*
9. 13. 3. leaf of ἀριστολοχία comp.

ἀλωπέκουρος, *Polypogon monspeliensis*
7. 11. 2. flowers in a spike: described.

ἀμάρακον (ἀμάρακος), sweet marjoram, *Origanum Majorana*
1. 9. 4. evergreen; 6. 1. 1. in list of under-shrubs; 6. 7. 4. propagation: roots described; 6.8.3. flowering time; 9. 7. 3. in list of ἀρώματα.

ἄμπελος (1) (leaf οἰναρον 9, 13. 5.), vine, *Vitis vinifera*
1. 2. 1. has tendrils; 1. 2. 7. bark; 1. 3. 1. a typical 'tree'; 1. 3. 5. evergreen at Elephantine; 1.5.2. bark cracked and fibrous: bark in layers; 1. 6. 1. core fleshy; 1. 6. 3. roots thin; 1. 6. 5. roots branching upwards; 1. 8. 5. highest shoots 'roughest': 'eye' analogous to knot in other trees; 1. 9. 1. effect of pruning; 1. 10. 4. leaves broad; 1. 10. 5. leaf divided; 1. 10. 7. long leaf-stalk: attachment of leaf-stalk; 1. 10. 8. leaves made of 'bark' and flesh; 1. 11. 4. seeds all together in a single case; 1.11.5. each grape separately attached; 1. 12. 1. taste of fruit; 1. 12. 2. taste of sap; 1. 13. 1. flower 'downy'; 1. 13. 3. flower surrounds fruit; 1. 13. 4. some kinds sterile; 1. 14. 1. bears on new shoots; 1. 14. 4. many cultivated forms; 2. 1. 3. propagation; 2. 2. 4. degenerates from seed; 2. 3. 1. sometimes spontaneously changes character; 2. 3. 2. *a.* ὁ κάπνειος varies in colour of grapes on same bunch; 2. 3. 3. sometimes bears fruit on the stem; etc.; 2. 5. 3. propagation: cannot be grafted: 2. 5. 4. propagation; 2. 5. 7. low ground suitable: great variety of kinds according to soil; 2. 6. 12. cuttings set upside down; 2.7.1. water-loving; 2. 7. 2. needs much pruning; 2. 7. 5. use of dust; 2. 7. 6. root-pruning; 3. 5. 4. autumn bud-ding; 3. 17. 3. bark of κολοιτία (2) comp.; 3.18.5. flower and fruit of ῥοῦς comp.; 3. 18. 12. cluster of berries of σμῖλαξ (2) comp.; 4. 4. 8. unnamed Indian tree (cotton-plant) planted in rows like *a.*; 4. 4. 11. in India confined to hill-country; 4. 5. 4. grows on Mount Tmolus and Mysian Olympus; 4. 7. 7. leaf of δένδρον τὸ ἐριόφορον comp.; 4. 7. 8. occurs on island of Tylos; 4. 13. 2. some kinds short-lived; 4. 13. 4–6. said to be longest-lived of trees: reason: method of prolonging life artificially; 4. 14. 2. young plants liable to 'sun-scorch'; 4. 14. 6. other diseases; 4. 14. 7. effects of injury to roots; 4. 14. 8. effect of rain on fruiting; 4. 14. 9. a special pest at Miletus; 4. 14. 10. effect of hot winds; 4.14.13. effects of frost; 4. 15. 1. outer bark can be stripped; 4. 16. 1. survives

439

INDEX OF PLANTS

splitting of stem; 4. 16. 6. natural antipathy of ἀ. to ῥάφανος; 5. 3. 4. character of wood; 5. 4. 1. the less fruitful trees produce more solid wood; 5.9.4. wood, if damp, makes an evil smell when burnt for charcoal; 5. 9. 6. ἀθραγένη comp.; 8. 2. 8. ἀ. in Melos; 9. 1. 6. time of tapping; 9. 13. 5. leaf and time of growth of πεντακετές comp.; 9. 18. 11. peculiar properties of certain local kinds.

ἄμπελος (2) (Mt. Ida), currant grape, *Vitis vinifera*, var. *corinthiaca*
3. 17. 4. a local Idaean kind; 3. 17. 6. do. described.

ἄμπελος (3) ἡ ποντία, *Fucus spiralis*
4. 6. 2. peculiar to certain waters; 4. 6. 9. described.

ἄμπελος (4), ἡ ἀγρία (= μήλωθρον), bryony, *Bryonia cretica*
3. 18. 12. fruit of σμίλαξ (2) comp. 9. 14, 1. how long drug prepared from it will keep; 9. 20. 3. properties of root: medicinal use.

ἀμυγδαλῆ, almond, *Prunus Amygdalus*
1. 6. 3. large central root; 1. 9. 6. leaves produced early, but not shed early; 1. 11. 1. seed immediately within envelope; 1. 11. 3. seed in a woody shell; 1. 12. 1. taste of fruit; 1. 13. 1. flower 'leafy': flower of some kinds reddish; 1. 14. 1. bears on last year's wood; 2. 2. 5. degenerates from seed; etc.; 2. 2. 9. effects of cultivation; 2. 2. 11. do.: effect of tapping gum; 2. 5. 6. trees should be planted far apart; 2. 7. 6. 'punishing' the tree; 2. 7. 7. tapping the gum; 2. 8. 1. apt to shed immature fruit; 3. 11. 4. fruit of μελία comp.; 3. 12. 1. leaf of κράνεια comp.; 4. 4. 7. fruit of τέρμινθος ἡ Ἰνδική comp.; 4.7.5. fruit of unnamed Persian tree (see App. (13)), comp., 4.14.12. uninjured by special winds; 5.9.5. wood-ashes make pungent smoke; 7. 13. 6. flower appears before leaves and (new growth of) stem; 8. 2. 2. germination described; 9. 1. 2. sap gummy; 9. 1. 3. gum scentless; 9. 1. 5. gum useless; 9. 19. 1. leaf of ὀνοθήρας comp.

ἄμωμον, Nepaul cardamom, *Amomum subulatum*
9. 7. 2. an ἄρωμα, Median or Indian.

ἀνδράχλη, andrachne, *Arbutus Andrachne*
1. 5. 2. bark readily drops off; 1. 9. 3. evergreen; 3. 3. 1. a mountain tree; 3. 3. 3. evergreen; 3. 4. 2. time of budding; 3. 4. 4. time of fruiting; 3. 4. 6. do.; 3. 6. 1. slow growing (?); 3. 16. 5. described; 3. 16. 6. leaf of κοκκυγέα comp.; 4. 4. 2. leaf of μηλέα ἡ Περσική comp.; 4.7.5. an unnamed Persian tree (see App. (14)), comp.; 4. 15. 1. does not perish if bark is stripped; 4. 15. 2, bark cracks; 5. 7. 6. wood used for parts of loom; 9. 4. 3. bark of σμύρνα comp.

ἀνδράχνη, purslane, *Portulaca oleracea*
7. 1. 2–3. time of sowing and germination; 7.2.9. root described.

ἀνεμώνη, anemone, *Anemone* spp.
7. 8. 3. leaves 'on the ground.'

ἀνεμώνη, anemone, *Anemone coronaria*
7. 7. 3. puts forth flower soon after season of growth begins; 7. 10. 2. flowers in winter.

ἀνεμώνη ἡ λειμωνία, *Anemone pavonina*
6. 8. 1. flowering-time.

ἀνεμώνη ἡ ὀρεία, *Anemone blanda*
6. 8. 1. flowering-time.

ἄνηθον (= ἄνητος), dill, *Anethum graveolens*
1. 11. 2. seeds naked; 1. 12. 2. taste of sap; 6. 2. 8. fruit of νάρθηξ and ναρθηκία comp.; also setting of flowers and fruit; 7. 1. 2–3. time of sowing and germination; 7. 2. 8. root described; 7. 3. 2. seeds described; 7. 4. 1. only one kind; 7. 6. 4. fruit of ὀρειοσέλινον comp.

ἄνθεμον, *Anthemis chia*, etc. (see below)
1. 13. 3. flower attached above

440

INDEX OF PLANTS

each seed; 7.14.2. flowering begins at top: 'flower and fruit: several kinds (*see below*).

ἄνθεμον τὸ ἀφύλλανθες, wild chamomile, *Matricaria Chamomilla* 7.8.3. leaves 'on the ground.'

ἄνθεμον τὸ φυλλῶδες, *Anthemis chia* 7.8.3. leaves on the stem.

ἀνθέρικος, *see* ἀσφόδελος.

ἄννησον, anise, *Pimpinella Anisum* 1.12.1. scent.

ἄννητος (= ἄνηθον.) dill, *Anethum graveolens* 9.7.3. in list of ἀρώματα.

ἀντίρρινον, snapdragon, *Antirrhinum Orontium* 9.19.2. alleged magic properties: described.

ἀπάπη, dandelion, *Taraxacum officinale* 6.4.8. (?) flower of χαμαιλέων comp.; 7.7.1. a λάχανον: classed as 'chicory-like' from its leaves; 7.7.3. season of growing; 7.7.4. prolonged flowering-time; 7.8.3. leaves 'on the ground'; 7.10.2. (?) flowers in winter, earliest of all; 7.10.3. flowers borne in succession; 7.11.3. flowering-time; 7.11.4. inedible: growth described.

ἀπαργία, hawk's beard, *Crepis Columnae* 7.8.3. leaves 'on the ground.'

ἀπαρίνη, bedstraw, *Galium Aparine* 7.8.1. stem 'clasping,' but, for want of support, 'on the ground'; 7.14.3. clings to clothes: peculiar setting of flower described; 8.8.4. grows specially among φακοί: growth described; 9.19.2. ἀντίρρινον comp.

ἄπιος (1), pear, *Pyrus communis*, var. *sativa* 1.2.7. bark; 1.3.3. a tree whose stem is not single; 1.8.2. has less knots than ἀχράς; 1.10.5. leaves round; 1.11.4. seeds all together in a single case; 1.11.5. seeds in a membrane; 1.12.2. taste of sap; 1.13.1. flower 'leafy'; 1.13.3. flower above fruit-case; 1.14.1.

bears on last year's wood; 1.14.4. a cultivated form of ἀχράς; many cultivated forms; 2.1.2. propagation; 2.2.4. degenerates from seed; 2.2.5. seed produces wild form; 2.2.12. cannot be made out of ἀχράς by cultivation; 2.5.3. grafting; 2.5.6. trees should be planted rather far apart; 2.7.7. 'punishing' the tree; 2.8.1. apt to shed immature fruit; 3.2.1. produces less fruit than ἀχράς, but ripens more; 3.3.2. has better fruit and timber in lowlands; 3.4.2. time of budding; 3.6.2. formation of buds; 3.11.5. mountain and lowland forms comp.; 3.12.8. fruit of ὄη comp. as to keeping; 3.14.1. leaf of πτελέα comp.; 3.14.3. leaf of κλήθρα comp.; 3.18.7. does not differ in kind from ἀχράς; 4.2.5. περσέα comp.; 4.3.1. size of λωτός (4) comp.; 4.4.2. thorns of μηλέα ἡ Περσική comp.; 4.5.3. abundant in Pontus; 4.13.1. shorter-lived than ἀχράς; 4.14.2. apt to get wormeaten; 4.14.10. fruit gets wormeaten; 4.14.12. uninjured by special winds; 5.3.2. leaf of an unnamed tree comp. (*see* App. (20)); 9.4.2. leaf of λιβανωτός comp.

ἄπιος (2) (= ἰσχάς 9.9.5. = ῥάφανος ἡ ὀρεία), spurge, *Euphorbia Apios* 9.9.5. medicinal use; 9.9.6. described.

ἄρακος, *Vicia Sibthorpii* 1.6.12. an unnamed plant (*see* App. (1)) comp.; 8.8.3. ('the rough hard kind') grows specially among φακοί.

ἀράχιδνα, *Lathyrus amphicarpus* 1.1.7. fruit underground; 1.6.12. root like a second fruit.

ἀρία (= ἴψος = φελλόδρυς 3.16.3.), holm-oak, *Quercus Ilex* var. *agrifolia* 3.3.8. doubt whether it has a flower; 3.4.2. time of budding; 3.4.4. time of fruiting; 3.16.3. Dorian name for φελλόδρυς;

INDEX OF PLANTS

3. 17. 1. acorn of φελλός comp.; 4. 7. 2. (?) leaf of δίφνη (6) comp.; 5. 1. 1. time of cutting timber; 5. 3. 3. character of wood; 5.4.2. wood proof against decay; 5. 5. 1. wood hard to work; 5. 9. 1. wood makes good charcoal.

ἀριστολοχία, birthwort, *Aristolochia rotunda*
9. 13. 2. described: medicinal use; 9. 14. 1. how long drug will keep; 9.15.5. grows in Arcadia; 9. 20. 4. *cf.* 9. 13. 2.

ἄρκευθος (= κέδρας (3)), Phoenician cedar, *Juniperus phoenicea*
1. 9. 3. evergreen; 3. 3. 1. a mountain tree; 3. 3. 3. evergreen; 3. 3. 8. doubt whether it has a flower; 3. 4. 1. takes a year to ripen fruit; 3. 4. 5. time of fruiting; etc.; 3. 4. 6. do.; 3. 6. 1. slow-growing(?); 3.6.5. shallow-rooting according to Arcadians; 3.12.3–4. described: distinguished from κέδρος (1): 4. 1. 3. grows high on mountains, but not tall; 5. 7. 4. use of wood in house-building; 5. 7. 6. other uses of wood: does not decay; 9. 1. 2. sap gummy.

ἀρνόγλωσσον (= στελέφουρος 7. 11. 2., according to some, = ὄρτυξ 7. 11. 2., according to some), plantain, *Plantago maior*
7. 8. 3. leaves 'on the ground'; 7. 10. 3. flowers borne in succession; 7. 11. 2. flowers in a spike: described by comparison with ἀλωπέκουρος.

ἄρον, cuckoo-pint, *Arum italicum*
1. 6. 7. root fleshy; 1. 6. 8. has a stout root and also fibrous roots: roots not tapering; 1. 6. 10. cultivation; 1. 16. 10.(?) flower made of flesh; 7. 2. 1. propagation; 7. 9. 4. root described; 7.12.2. root and leaves edible: use in surgery: special treatment to promote growth of root: one kind inedible (*see* δρακόντιον); 7. 13. 1. leaves described; 7. 13. 2. no stem or flower.

ἀρρενόγονον (= θηλύγονον), dog mercury, *Mercurialis perennis*
9. 19. 5. properties: described.

ἀσπάλαθος, *Calycotome villosa*
9. 7. 3. in list of ἀρώματα.

ἄσπρις (= αἰγίλωψ (1)), Turkey oak, *Quercus Cerris*
3. 8. 2. one of the four Macedonian kinds of oak: acorns and timber.

ἀστέρισκος, Michaelmas daisy, *Aster Amellus*
4. 12. 2. seed of μελαγκρανίς comp.

ἀσταφίς, *Delphinium Staphisagria*
9. 12. 1. medicinal use.

ἀσφάραγος, asparagus, *Asparagus acutifolius*
1. 10. 6. spines for leaves; 6. 1. 3. do.; a wild under-shrub; 6. 4. 1. one of very few plants which are altogether spinous; 6. 4. 2. described.

ἀσφόδελος (stem ἀνθέρικος), (= πόθος (2)), asphodel, *Asphodelus ramosus*
1. 4. 3. belongs to 'ferula-like' plants; 1. 10. 7. attachment of leaves; 6. 6. 9. leaves of νάρκισσος (1) comp.; 7. 9. 4. root acorn-shaped; 7. 12. 1. root edible; 7.13.1. leaves described; 7. 13. 2–3. stem of ἶρις comp.: largest stem of herbaceous plants: fruit inflorescence etc. described; worm which infests it: uses for food of stem and roots; 7. 13. 4. grown from seed; 9. 9. 6. leaf of ἰσχας comp.; 9.10.1. stem of ἐλλέβορος comp. by some.

ἄσχιον, puff-ball, *Lycoperdon giganteum*
1. 6. 9. not a root, though underground.

ἀτρακτυλίς (= φόνος 6. 4. 6.), distaff-thistle, *Carthamus lanatus*
6. 4. 3. a 'thistle-like' plant; 6. 4. 6. described: also called φόνος: reason; 9. 1. 1. juice blood-coloured.

ἀφάκη, tare, *Vicia sativa* var. *angustifolia*
8. 1. 4. (a pulse) sown late; 8. 5. 3. shape of pod; 8. 8. 3. πελεκῖνος

442

INDEX OF PLANTS

grows specially among ἄ.; 8.11.1. seed does not keep.

ἀφάρκη (a natural hybrid between ἀνδράχλη and κόμαρος), hybrid arbutus, *Arbutus hybrida*
1. 9. 3. evergreen; 3. 3. 1. a mountain tree; 3. 3. 3. evergreen; 3. 4. 2. time of budding; 3. 4. 4. time of fruiting; 5. 7. 7. uses of wood.

ἀφία, lesser celandine, *Ranunculus Ficaria*
7. 7. 3. puts forth flowers at season of growth.

ἀχράς, wild pear, *Pyrus amygdaliformis*
1. 4. 1. more fruitful than cultivated kind; 1. 8. 2. has more knots than ἄπιος; 1. 9. 7. time of shedding leaves; 1. 14. 4. a wild form of ἄπιος; 2. 2. 5. produced from seed of ἄπιος; 2.2.12. cannot be made into ἄπιος by cultivation; 3. 2 1. produces more fruit than ἄπιος, but ripens less; 3. 3. 1. a tree of mountain and plain; 3. 3. 2. has better fruit and timber in lowlands; 3. 4. 2. time of budding; 3. 4. 4. time of fruiting; 3. 6. 1. slow growing (?); 3. 11. 5. mountain and lowland forms comp.; 3. 12. 8. fruit of ὄη comp. as to keeping; 3. 14. 2. bark of λεύκη comp.; 3. 18. 7. does not differ in kind ·from ἄπιος; 4. 13. 1. longer lived than ἄπιος; 5. 5. 1. cobblers' strops made of the wood.

ἀψίνθιον, wormwood, *Artemisia Absinthium*
1. 12. 1. taste of fruit; 4. 5. 1. seeks cold regions; 7. 9. 5. leaves and stem bitter, yet wholesome; 9. 17. 4. said to become by use non-poisonous to sheep.

βάλανος, *Balanites aegyptiaca*
4. 2. 1. peculiar to Egypt; 4. 2. 6. described.

βάλσαμον (gum ὑποβάλσαμον),balsam of Mecca, *Balsamodendron Opobalsamum*
9. 1. 2. sap gummy; 9. 1. 7. time of tapping; 9. 4. 1. collection of gum; 9. 6. 1–4. described : habitat : method of collection : nowhere found wild; 9. 7. 3. in list of ἀρώματα.

βάτος, bramble, *Rubus ulmifolius*
1. 3. 1. a typical 'shrub'; 1. 5 3. thorns on wood; 1. 9. 4. evergreen; 1. 10. 6. leaf with spinous projections; 1. 10. 7. stem presently spinous; 3. 18. 3. grows in wet and dry places alike; 3. 18. 4. kinds distinguished; 3. 18. 12. cluster of berries of σμῖλαξ (2) comp; 4. 8. 1. to some extent grows in marshes; 4. 12. 4. to some extent aquatic; 6. 1. 3. has spines on the shoots.

βληχώ, pennyroyal, *Mentha Pulegium*
9. 16. 1. leaf etc. of δίκταμνον comp.

βλίτον, blite, *Amaranthus Blitum*
1. 14. 2. bears fruit both on top and at sides; 7. 1. 2–3. time of sowing and of germination; 7. 2. 7–8. root described; 7. 3. 2. seeds described; 7. 3. 4. seed borne both on top and at side; 7. 4. 1. only one kind.

βολβίνη, star-flower, *Ornithogalum umbellatum*
7. 13. 9. belongs to τὰ βολβώδη.

βολβός, purse-tassels, *Muscari comosum* etc. (*see below*).
1. 6. 7. root in scales; 1. 6. 8. root not tapering; 1. 6. 9. no side roots : (part of) stem underground; 1. 10. 7. no leaf-stalk : attachment of leaves; 6. 8. 1. flowering time : used as a coronary plant; 7. 2. 1. propagation; 7. 2. 2. root makes offsets; 7. 2. 3. offsets specially numerous; 7. 4. 12. formation of roots of κρόμυον comp.; 7. 9. 4., *cf*. 1. 6. 7.; 7. 12. 1. example of an edible root; 7. 12. 2. special treatment to promote growth of root; 7. 13. 1. leaves described; 7. 13. 2. flower-stem not the only stem; 7. 13. 4–5. grown from seed : seed sometimes takes two years to germinate; 7. 13. 7. root of νάρκισσος (1) comp.; 7. 13. 8.

443

INDEX OF PLANTS

several kinds: 7. 13. 9. roots of various plants comp.; 8. 8. 3. grows specially among πυρός.

βολβὸς ὁ ἐριόφορος, *Pancratium maritimum*
7. 13. 8. grows on beach: described: uses for food and clothing.

βουκέρας, (=τῆλις), fenugreek. *Trigonella Foenum-Graecum*
4. 4. 10. an Indian plant (see App. (8)), comp. 8. 8. 5. αἱμόδωρον parasitic on β.

βουμέλιος, ash, *Fraxinus excelsior*
3. 11. 4–5. described; 4. 8. 2. common in Egypt.

βούπρηστις, ?
7. 7. 3. season of growing.

βούτομος, sedge, *Carex riparia*
1. 5. 3. stem very smooth; 1. 10. 5. leaves end in a point: further described; 4. 8. 1. in list of τὰ λοχμώδη; 4. 10. 4. described; 4. 10. 6. grows both on land and in water: grows on the floating islands of Lake Copais; 4. 10. 7. part used for food; 4.11.12. foliage of some κάλαμοι comp.

βρόμος, oats, *Avena sativa*
8. 4. 1. seed has more coats than other cereals; 8. 9. 2. exhausts the soil: reason: like a wild plant.

βρύον, oyster-green, *Ulva Lactuca*
4. 6. 2. occurs generally in Greek waters; 4. 6. 6. described.

γήθυον, long onion, *Allium Cepa* var.
1. 6. 9. part of stem underground; 7. 1. 2–3. time of sowing and of germination; 7. 1. 6. germination; 7. 1. 7. bears fruit in second year: has single stem; 7. 2. 2–3. root makes offsets; 7. 5. 1. likes water; 7. 5. 3. bears transplanting; 7. 5. 5 seed does not keep well; 7. 9. 4. root in scales; 7.12.3. root of φάσγανον comp.; 9. 11. 6. 'head' of στρύχνος ὁ μανικός comp.

γήτειον (Attic for γήθυον), hornonion, *Allium Cepa* var.
1. 10. 8. leaves hollow; 7. 4. 10.

described: cultivation (classed as a form of κρόμυον).

γλεῖνος, *Acer creticum*
3. 3. 1. name for lowland form of σφένδαμνος; 3. 11. 2. timber.

γλυκεῖα (sc. ῥίζα) (=ῥίζα Σκυθική q.v.), 9. 13. 2., liquorice, *Glycyrrhiza glabra*

γλυκυσίδη (=παιωνία q.v.), 9. 8. 6., peony, *Paeonia officinalis*.

γογγυλίς, turnip, *Brassica Rapa*
1. 6. 6. root fleshy; 1. 6. 7. root of bark and flesh; 7. 1. 2. time of sowing; 7, 1. 7. germination; 7. 2. 5. survives and increases in size under a heap of soil: root described; 7. 2. 8. do.; 7. 3. 2. seeds described; 7. 3. 4. seed borne at side; 7. 4. 3. doubtful if more than one kind: seed, method of sowing, effect of weather; 7.5. 3. bears transplanting; 7. 6. 2. wild form distinguished; 7. 9. 4. root has 'bark.'

δαῦκον (1), carrot, *Daucus Carota*
9. 15. 5. Arcadian drug: described (see note).

δαῦκον (2), *Malabaila aurea*
9. 15. 8. grows about Patrai: properties: root black; 9.20. 2. *cf.* 9. 15. 8.

δάφνη (1) (ἡ ἥμερος), (berry δαφνίς, 1. 11. 3.), sweet bay, *Laurus nobilis*
1. 5. 2. bark thin; 1. 6. 2. roots both stout and fine; 1.6.4. roots crooked; etc.; 1.8.1. few knots; 1. 9. 3. evergreen (cultivated and wild forms, see *below*); 1. 11. 3. fleshy seed in a shell (δαφνίς); 1. 12. 1. taste of fruit; 1.14. 4. many cultivated forms; 2. 1. 3. propagation; 2. 2. 6. sometimes improves from seed; 2. 5. 6. trees should be planted close together; 3. 3. 3. evergreen; 3. 4. 2. time of budding; 3. 7. 3. (one kind) produces a cluster; 3. 11. 3. leaves of μελία comp.; 3.11.4. winter-buds of μελία comp.; 3. 12. 7. leaf of ὄη comp.to that of δ. ἡ λεπτόφυλλος; 3. 13. 5. leaflet of ἀκτή comp. to

INDEX OF PLANTS

leaf of δ. ἡ πλατύφυλλος; 3.14.3. flower (?) of κλήθρα comp.; 3.15.4. leaf of τέρμινθος comp·; 3.16.4. leaf of κόμαρος comp.; 3.17.3. leaf of κολοίτια (2) comp. to δ. ἡ πλατύφυλλος; 4.4.12. leaf of an unnamed Arian shrub comp. (see App. (10)); 4.4.13. leaf of an unnamed Gedrosian tree comp. (see App. (11)); 4.5.3. does not thrive in cold regions; 4.5.4. grows in Propontis; 4.7.1. a class of marine Atlantic plants comp.; 4.7.4. leaf of an unnamed Arabian tree (see App. (12b)) comp.; 4.13.3. after decaying shoots again from same stock; 4.16.6. spoils flavour of grape; 5.3.3–4. character of wood; 5.7.7. wood used for walking-sticks; 5.8.3. grows in lowland parts of Latium: abundant on Circeian promontory; 5.9.7. fire-drills made of the wood, because it does not wear away; 9.4.2. bark of λιβανωτός comp.; 9.4.3. leaf of λιβανωτός comp. (by some); 9.4.9. do; 9.10.1. leaf of ἐλλέβορος ὁ μέλας comp. (by some); 9.15.5. δαῦκον (1) comp.; 9.20.1. one kind of πέπερι (fruit) comp.

δάφνη (2) ἡ ἀγρία (= ὀνοθήρας), oleander, *Nerium Oleander*

1.9.3. distinguished from δ. ἡ ἥμερος.

δάφνη (3) ἡ Ἀλεξανδρεία, Alexandrian laurel, *Ruscus Hypophyllum*

1.10.8. bears fruit on leaves; 3.17.4. do.

δάφνη (4) ἡ λεπτόφυλλος, sweet bay, *Laurus nobilis*

3.12.7. (see under δάφνη).

δάφνη (5) ἡ πλατύφυλλος, sweet bay, *Laurus nobilis*

3.11.3., 3.13.5., 3.17.3. (see under δάφνη).

δάφνη (6) (=ἐλάα (3) = App. (14)), white mangrove, *Avicennia officinalis*

4.7.1. grows in 'Red Sea'; 4.7.2. described: produces a drug for stanching blood.

δίκταμνον, dittany, *Origanum Dictamnus*

9.16. 1–2. described: medicinal use: popular belief about its use to goats: comp. with ψευδοδίκταμνον; 9.16.3. habitat.

δίκταμνον (ἕτερον), *Ballota Pseudodictamnus*

9.16.3. Cretan: has nothing in common with true δ. except the name: described: properties different.

διόσανθος, carnation, *Dianthus inodorus*

6.1.1. in list of under-shrubs; 6.6.2. a cultivated under-shrub: a coronary plant: scentless; 6.6.11. grown from seed: woody; 6.8.3. flowering time.

διοσβάλανος (fruit κάρυον κασταναϊκόν), 4.8.11., chestnut, *Castanea vesca*

1.12.1. taste of fruit; 3.2.3. evidence that it is really wild; 3.3.1. a mountain tree; 3.3.8. doubt whether it has a flower; 3.4.2. time of budding; 3.4.4. time of fruiting; 3.5.5. winterbuds; 4.5.1. in list of northern trees; 4.5.4. abundant on Mount Tmolus and Mysian Olympus; 4.8.11. bark of root of λωτός (2) comp. to shell of chestnut.

διόσπυρος, *Diospyros Lotus*

3.13. fruit of κέρασος comp.

δόλιχος, calavance, *Vigna sinensis*

8.3.2. stem; 8.11.1. seed does not keep.

δόναξ (=κάλαμος ὁ Λακωνικός = κ. ὁ αὐλητικός = κ. ὁ συριγγίας = κ. ὁ τοξικός),pole-reed,*Arundo Donax*

4.11.11. a kind of κάλαμος: habit and habitat.

δρακόντιον, edderwort, *Dracunculus vulgaris*

7.12.2. an inedible and poisonous kind of ἄρον; 9.20.3. medicinal use: described.

δρυπίς, *Drypis spinosa*

1.10.6. spinous-leaved.

δρῦς (1), oak, *Quercus Robur*

1.2.1. has galls (κηκίς); 1.2.7. bark; 1.5.2. bark thick: bark fleshy; 1.5.3. wood fleshy; 1.5.5.

445

INDEX OF PLANTS

wood heavy because it contains mineral matter; 1.6.1. core hard and close; 1.6.2. core called 'oak-black': core large and conspicuous; 1.6.3. roots many and long; 1.6.4. roots fleshy: deep-rooting; 1.8.5. diseased formation (κραδή); 1.9.5. an evergreen specimen; 1.10.6. leaves notched: leaves with spinous projections; 1.10.7. attachment of leaves; 1.11.3. seed in a leathery shell; 2.2.3. propagation; 2.2.6. deteriorates from seed; 3.3.1. tree of mountain and plain; 3.3.3. evergreen in some places; 3.3.8. doubt whether it has a flower (βρύον); 3.4.2. time of budding; 3.4.4. time of fruiting; 3.5.1. periods of budding; etc.; 3.5.2. galls; 3.5.5. winter-buds; 3.6.1. quick growing; 3.6.5. instance of a deep-rooting tree; 3.7.4–6. various galls; etc.; 3.8.2. four or five kinds, viz. ἡμερίς or ἐτυμόδρυς, αἰγίλωψ, πλατύφυλλος, φηγός, ἁλίφλοιος or εὐθύφλοιος (five recognised by inhabitants of Mt. Ida); 3.16.1. leaf growth and bark of πρῖνος comp.; 3.16.3. φελλόδρυς δρῦς and πρῖνος comp.; 4.2.8. common in Thebaid; 4.5.1. in list of northern trees; 4.5.3. grows in Pontus; 4.14.10. infested by knips; 4.15.2. survives stripping of bark for some time; 4.15.3. effect of stripping bark in winter; 5.1.2. time of cutting timber: reasons; 5.1.4. do.; 5.3.1. core very close and heavy; 5.3.3. character of wood; 5.4.1. wood hard and heavy; 5.4.2. wood proof against decay; 5.4.3. wood does not decay if buried or soaked in water: rots in seawater; 5.4.8. effect of salt water on different parts; 5.5.1. wood hard to work; 5.6.1. wood contains mineral matter and so gives under weight: apt to split; 5.7.2. used for keel of triremes and for merchantmen to make extra keel for hauling: does not glue well on to ἐλάτη or πεύκη; 5.7.4. use of wood in housebuilding; 5.8.3. grows in Latium on Circeian promontory; 5.9.1. wood makes good charcoal, but inferior to ἀρία and κόμαρος; 5.9.2. charcoal of this wood less esteemed by smiths than that of πεύκη; 8.2.2. germination from acorn described; 9.9.5. leaf of χαμαίδρυς comp.

δρῦς (2) ἡ ἀγρία (=φηγός 3.8.2.), Valonia oak, *Quercus Aegilops*
1.5.2. rough bark; 3.8.2. *see under* δρῦς.

δρῦς (3) ἡ ἀλίφλοιος (=δ. ἡ εὐθύφλοιος 3.8.2.), sea-bark oak, *Quercus Pseudo-Robur*
3.8.2. one of the five kinds of oak (Mt. Ida): =δ. ἡ εὐθύφλοιος; 3.8.3–4. acorns; 3.8.5. habit and timber; 3.8.6. φάσκος (q.v.) grows on it; 3.8.7. timber; 5.1.2. time of cutting timber.

δρῦς (4) ἡ εὐθύφλοιος (=δ. ἡ ἀλίφλοιος 3.8.2.), sea-bark oak, *Quercus Pseudo-Robur*
3.8.2. one of the five kinds of oak (Mt. Ida).

δρῦς (5) ἡ ἥμερος (= ἐτυμόδρυς q.v. = ἡμερίς (2)), true oak, *Quercus Robur*
3.8.2. one of the five kinds of oak (Mt. Ida).

δρῦς (6) ἡ πλατύφυλλος, broad-leaved oak (scrub oak), *Quercus lanuginosa*
3.8.2. one of the five kinds of oak (Mt. Ida): fruit; 3.8.5. habit and timber; 3.8.6. galls: one of the four Macedonian oaks: has bitter acorns.

δρῦς (7) (φῦκος), *Cystoseira ericoides*
4.6.2. peculiar to certain waters; 4.6.7–8. described.

δρῦς (8) (ποντία), *Sargassum vulgare*
4.6.9. distinguished from δρῦς (7); has a useful βάλανος.

ἐβένη (ἔβενος) (1), ebony, *Diospyros Ebenum*
1.5.4. wood heavy; 1.5.5. do. because of close grain; 1.6.1.

INDEX OF PLANTS

core hard and close; 4. 4. 6. described: two kinds distinguished (*see below*); 5. 3. 1. wood very close and heavy, especially the core; 5. 3. 2. colour of wood of τέρμινθος comp.: wood of an unnamed tree (*see* App. (20)) comp. to a variegated ἐ.; 5. 4. 2. wood proof against decay; 9. 20. 4. colour and medicinal use of wood.

ἐβένη (2), *Diospyros melanoxylon*
4. 4. 6. a kind with inferior wood.

εἰλετίας, *see* κάλαμος ὁ εἰλετίας

ἔκτομον τὸ μελαμπόδιον *see* ἑλλέβορος ὁ μέλας

ἐλάα, olive, *Olea Europea*
1. 3. 1. a typical 'tree'; 1. 5. 4. wood easily broken, not split: wood has many knots; 1. 5. 5. wood easily broken, because tough and not of straight grain; 1. 6. 2. core not conspicuous; 1. 6. 3. roots both stout and thin; 1. 6. 4. roots branching; etc.: shallow rooting; roots crooked; etc.; 1. 8. 2. has less knots than κότινος; 1. 8. 6. liable to excrescences: etc.; 1. 9. 3. evergreen; 1. 10. 1. leaves inverted in summer; 1.10.2. colour of leaves; 1. 10. 4. leaves narrow; 1. 10. 7. leaf-stalk short; etc.; 1. 11. 1. seed enveloped in flesh and stone; 1. 11. 3. fleshy seed in a stone; 1. 11. 4. effect on fruit of rich feeding; 1. 12. 1. taste of fruit; 1. 13. 2. flower consists of one 'leaf' only partly divided; 1. 13. 3. flower surrounds fruit; etc; 1.14.1. bears on last year's wood; 1. 14. 2. bears fruit both on top and at side; 1.14.4. a cultivated form of κότινος; 2.1.2. propagation; 2.1.4. do.; 2.2.5. seed produces wild form; 2. 2. 12. cannot be made out of κότινος by cultivation; 2. 3. 1. sometimes changes to κότινος spontaneously; etc.; 2. 5. 3. grafting; 2. 5. 4. propagation; 2. 5. 6. do.: trees should be planted far apart; 2. 5. 7. low ground suitable; 2. 7. 2. needs much pruning; 2. 7. 3. requires pungent manure and much water; 3. 2. 1. produces less fruit than κότινος but ripens more; 3. 12. 2. flower and fruit of θηλυκράνεια comp.; 3. 17. 5. size of fruit of συκῆ ἡ Ἰδαία comp.; 4. 2. 8. common in Thebaid; 4. 2. 9. character in Thebaid; 4. 3. 1. grows and bears well in Cyrenaica; 4. 4. 1. (?) distribution in Asia; 4. 7. 2. leaf and fruit of ἐλάα (3) comp.; 4. 7. 4. size of fruit of unnamed Arabian tree comp. (*see* App. (12*b*)); 4. 13. 1. shorter-lived than κότινος; 4. 13. 2. story of a very old tree at Athens; 4. 13. 5. explanation of longevity; 4.14. 2. diseases; 4. 14. 8. effect of rain on fruiting; 4. 14. 9. specially apt to shed fruit at Taras; 4. 14. 10. other diseases: effect of hot winds; 4. 4. 11. suffers much from special winds; 4.16.1. improved by lopping branches; 5.3.3. character of wood; 5.3.7. images made from the root; 5. 4. 2. wood proof against decay; 5. 4. 4. wood not eaten by *teredon*; 5. 5. 2. core not obvious: wherefore wood not apt to 'draw'; 5. 5. 3. core not obvious but exists; 5.6.1. wood apt to split under pressure; 5. 9. 6. wood good for kindling furnaces; 5. 9. 7. wood not suitable for fire-sticks: reason; 5. 9. 8. articles made of the wood have been known to produce shoots: instances; 6. 2. 1. leaf of κνέωρος ὁ λευκός comp.; 6. 2. 4. will not grow more than a short distance from the sea; 8. 2. 8. abundant in Melos; 9.18. 5. fruit of θηλύγονον comp. to βρύον of ἐ.: fruit of ἀρρενόγονον comp. to undeveloped olive.

ἐλ 'α (2), *Olea cuspidata*
4. 4. 11. Indian (in hill-country only).

'ἐλάα' (3) (= δάφνη (6) = App. (14)), white mangrove, *Avicennia officinalis*
4.7.1. grows in 'Red Sea'; 4.7.2. described.

447

INDEX OF PLANTS

ἐλαίαγνος (properly ἐλέαγνος), goat willow, *Salix Caprea*
4. 10. 1–2. in list of plants of L. Copais: described.

ἐλάτη (1), silver-fir, *Abies cephalonica*
1. 1. 8. branches opposite; 1. 3. 6. refuses cultivation; 1. 5. 1. erect and tall; 1. 5. 2. bark in layers; 1. 5. 3. wood fibrous; 1. 5. 4. wood easily split; 1.5.5. do. because of straight grain; 1. 6. 3. root single; 1. 6. 4. roots fibrous; 1.6.5. roots not branching; 1.8.1. many knots; 1. 8. 2. 'male' has more knots than 'female'; 1. 8. 3. branches at right angles; 1. 9. 1. growth chiefly upwards; 1.9. 2. growth affected by position; 1. 9. 3. evergreen; 1. 10. 5. leaves described; 1. 12. 1. taste of fruit; 1. 12. 2. taste of sap; 1. 13. 1 flower yellow; 2. 2. 2. propagated only by seed; 2. 7. 3. requires pungent manure; 3. 1. 2. grows only from seed; 3.3.1. a mountain tree; 3. 3. 3. evergreen; 3. 4. 5. time of flowering and fruiting; 3. 5. 1. period of budding; 3. 5. 3. do.; 3. 5. 5. winter-buds; 3. 6. 1. quick growing: even young tree fruits; 3. 6. 2. formation of buds; 3. 6. 4. not deep-rooting; 3. 6. 5. deep-rooting according to Arcadians; etc.; 3. 7. 1–2. dies if topped: formation of *callus*; 3. 9. 5. timber compared with πεύκη: etc.; 3. 9. 6. differences between 'male' and 'female': described; 3. 9. 7. further comparison with πεύκη: produces λούσσον; 3. 9. 8. do.: core and callus; 4. 1. 1. likes shade; 4. 1. 2. grows tall in shade but has inferior timber; 4. 1. 3. grows high on mountains, but not tall; 4. 4. 1. (?) distribution in Asia; 4. 5. 1. in list of Northern trees; 4. 5. 3. does not grow in Pontus; 4. 15. 3. effects of stripping bark at various seasons; 4. 16. 1. topping fatal; 4. 16. 1–2. not injured by cutting for resin; 4. 16. 4. said to perish if entirely deprived of its heart-wood; 5. 1. 1. time when timber is of best colour; 5. 1. 2. time of cutting timber; 5. 1. 4. do.; 5. 1. 5–6. timber comp. with πίτυς; 5. 1. 7. uses of timber; 5. 1. 8. growth and character; 5. 1. 9–10. methods of cleaving; 5. 3. 3. character of wood; 5. 3. 5. used for doors; 5. 4. 4. less eaten by *teredon* than πεύκη; 5. 4. 6. wood, if barked just before time of budding, does not decay in water: story in proof of this; 5. 5. 1. knotty parts of wood hard to work; 5. 5. 2. core most obvious in ἐ.; 5. 6. 1. wood good for struts: behaviour under pressure; 5. 6. 2. strongest of all woods; 5. 7. 1–2. use of wood in ship-building; 5.7. 4–5. uses of wood in house-building and crafts: the most generally useful of woods: more so than πεύκη; 5. 9. 8. wood has a peculiar exudation; 9.1.2. sap gummy; 9. 2. 1. production of resin (ῥητίνη); 9. 2. 2. quality of resin.

ἐλάτη (2), silver-fir, *Abies pectinata*
5. 8. 1. grows to great size in Latium, but finer still in Corsica; 5.8. 3. grows in hill-country of Latium.

' ἐλάτη' (3), 'sea-fir,' *Cystoseira Abies-marina*
4. 6. 2. peculiar to certain waters; 4. 6. 7–8. described.

ἐλατήριον, see σίκυος ὁ ἄγριος
4. 5. 1. in list of northern plants.

ἐλειοσέλινον (= σέλινον τὸ ἕλειον), marsh celery, *Apium graveolens*
7. 6. 3. comp. with σέλινον: medicinal use.

ἐλειόχρυσος, gold-flower, *Helichrysum siculum*
6. 8. 1. flowering time; 9. 19. 3. alleged magic properties: described: medicinal use.

ἐλελίσφακος, salvia, *Salvia triloba*
6. 1. 4. a spineless wild undershrub; 6. 2. 5. like wild σφάκος: leaf described.

INDEX OF PLANTS

ἐλένιον, calamint, *Calamintha incana*
2. 1. 3. propagation; 6. 1. 1. in list of under-shrubs; 6. 6. 2. a cultivated under-shrub: a coronary plant: the whole plant scented; 6. 6. 3. woody: only one form; 6. 7. 2. said by some to have no fruit; 6. 7. 4. roots described.

ἐλίκη, crack willow, *Salix fragilis*
3.13.7. Arcadian name for willow.

ἔλιξ, ivy, *Hedera Helix*
3. 18. 7–8. described: does not develop into κιττός; 3. 18. 8. kinds; 7. 8. 1. stem 'clasping.'

ἔλιξ ἡ λευκή, white-berried ivy, *Hedera Helix*
3. 18. 8. several kinds.

ἔλιξ ἡ ποικίλη (= ἐ. ἡ Θρακία, 3.18.8.), ivy, *Hedera Helix*
3. 18. 8. several kinds.

ἔλιξ ἡ χλωερά, ivy, *Hedera Helix*
3. 18. 8. described.

ἐλλεβορίνη, rupture-wort, *Herniaria glabra*
9. 10. 2. seed mixed with ἐλλέβορος ὁ λευκός to make an emetic.

ἐλλέβορος, hellebore, *Helleborus cyclophyllus* and *Veratrum album*
4. 5. 1. seeks cold regions; 6. 2. 9. belongs to 'ferula-like' plants: has a hollow stem; 9.8.4. what part of root cut for medicinal use: 'bulbous' part a purge for dogs; 9. 8. 6. poisonous effect on those who dig it; precautions; 9. 9. 2. medicinal use; 9. 10. 1–4. kinds distinguished (*see below*); 9. 14. 1. how long drug will keep; 9. 17. 1–3. the drug can be made ineffectual by use: instances.

ἐλλέβορος ὁ λευκός, white hellebore, *Veratrum album*
9. 10. 1. has nothing in common with ἐ. ὁ μέλας except the name: divergent accounts given of the resemblances between the two plants: described; 9. 10. 2. not poisonous to sheep; when in season: distribution; 9. 10. 3–4. very local: local varieties, Οἰταῖος, Ποντικός, Ἐλεάτης, Μαλιώτης, Παρνάσιος, Αἰτωλικός:
Οἰταῖος the best: properties of Ἐλεάτης; 9. 15. 5. grows in Arcadia; 9. 18. 2. restores scorpion to life when it has been killed with σκορπίος (3).

ἐλλέβορος ὁ μέλας (drug σησαμοειδές 9. 14. 4.), hellebore, *Helleborus cyclophyllus*
9. 8. 8. superstition as to gathering; 9. 10. 1. (*see* under ἐ. ὁ λευκός): described; 9. 10. 2. poisonous to animals; 9. 10. 3. grows everywhere: some localities specified; 9. 10. 4. called by some ἔκτομον τὸ μελαμπόδιον: uses for purification and as charm; 9. 14. 4. use of fruit in medicine; 9. 15. 5. grows in Arcadia; 9.16.6. leaf of ἐφήμερον comp.

ἔλυμος, Italian millet, *Setaria italica*
4. 4. 10. inflorescence of ὄρυζον comp.; 8.1.1. in list of 'summer crops' distinct from cereals and pulses; 8. 11. 1. seed keeps well.

ἔνθρυσκον, chervil, *Scandix australis*
7. 7. 1. a λάχανον.

ἐπετίνη (?) (? πιτυΐνη), *Ajuga Iva*
7. 8. 1. stem 'clasping,' but, for want of support, 'on the ground.'

Ἐπιμενίδειος, *see* σκίλλα ἡ Ἐ.

ἐπίπετρον, stone-crop, *Sedum anopetalum*
7. 7. 4. flowerless.

ἐρέβινθος, chick-pea, *Cicer arietinum*
2. 4. 2. seed soaked before sowing; 2. 6. 6. size of some dates comp.; 4.4.4. size of fruit of συκῆ ἡ Ἰνδική comp.; 4. 4. 9. not found in India; 6. 5. 3. leaf of a kind of τρίβολος comp.; 8. 1. 1. in list of pulses; 8. 1. 4. sown both early and late; 8. 2. 1. germination described; 8. 2. 3. comes up with several leaves: deep-rooting; 8. 2. 5. flowering time; 8. 2. 6. time of maturing seed; 8. 3. 2. stem; 8. 5. 1. several kinds: three mentioned, κριοί, ὀροβιαῖοι, οἱ ἀνὰ μέσον: white forms sweetest; 8. 5. 2. pod round: seeds comparatively few; 8.5.4. attachment of seed;

449

INDEX OF PLANTS

8. 6. 5. rain hurtful when ἐ. is in flower: three kinds mentioned, μέλας, πυρρός, λευκός; 8. 7. 2. comp. with other pulses: destroys weeds: suitable soil: grows well after κύαμος; 8. 9. 1. exhausts the soil most of pulses; 8. 10. 1. diseases and pests; 8.10. 5. infested by caterpillars; 8. 11. 2. only seed which does not engender 'worms' etc.: seed keeps well; 8. 11. 6. do. especially in hill country.

ἐρείκη, heath, *Erica arborea*
1. 14. 2. bears fruit on the top; 9. 11. 11. λιβανωτὶς ἡ ἄκαρπος ἐ. is abundant.

ἐρευθεδανόν, madder, *Rubia tinctorum*
6. 1. 4. a spineless wild undershrub; 7.9.3. roots red; 9.13.4. do.; 9. 13. 6. described: habit: habitat: medicinal use.

ἐρινεός, wild fig, *Ficus Carica*
1. 8. 2. has more knots than συκῆ; 1. 14. 4. wild form of συκῆ; 2. 2. 12. cannot be made into συκῆ by cultivation; 2. 3. 1. sometimes changes to συκῆ spontaneously; 3. 3. 1. a mountain tree; 3. 4. 2. time of budding; 4. 2. 3. fruit of συκῆ ἡ Κυπρία comp.; 4. 13. 1. longerlived than συκῆ; 4. 14. 4. not liable to diseases of συκῆ; 5. 6. 2. wood tough and easy to bend: uses; 5. 9. 5. wood makes pungent smoke.

(δένδρον τὸ) ἐριόφορον, cotton-plant, *Gossypium arboreum*
4. 4. 8. (not named) clothes made from it; 4. 7. 7–8. described.

ἕρπυλλος (1) (ἐ. ὁ ἥμερος), tufted thyme, *Thymus Sibthorpii*
1. 9. 4. evergreen; 2. 1. 3. propagation; 6. 1. 1. in list of undershrubs; 6. 6. 2. a cultivated under-shrub: a coronary plant: the whole plant scented: 6. 6. 3. woody: only one form; 6. 7. 2. said by some to have no fruit: 6. 7. 4. roots described; 6. 7. 5. growth peculiar: wild forms (see ἕρπυλλος (2)); 6. 7. 6. cultivation.

ἕρπυλλος (2) ὁ ἄγριος, Attic thyme, *Thymus atticus*
6. 7. 2. produces seeds, unlike ἐ.(1); brought from Hymettus; sometimes quite like θύμος; 6. 7. 5. has various forms.

ἐρύσιμον, *Sisymbrium polyceratium*
8. 1. 4. sown later than cereals and pulses. a 'summer crop'; 8. 3. 1. leaf; 8. 3. 3. flower; 8. 6. 1. rain not beneficial after sowing; 8. 7. 3. doubtful if eaten green by animals: described.

ἐρυσίβη (cf. ἐρυσιβᾶν, ἐρυσιβώδης 8. 3. 2.), wheat-rust, *Puccinia graminis*
8. 10. 1. a pest common to all crops (cereals, pulses etc.).

ἐτυμόδρυς (= ἡμερίς (2) 3. 8. 2. = δρῦς ἡ ἥμερος), true oak, *Quercus Robur*
3. 8. 2. one of the five kinds of oak (Mt. Ida): = ἡμερίς: fruit; 3. 8. 7. one of the four Macedonian kinds: has sweet acorns.

Εὐβοϊκόν, see καρύα ἡ Εὐβοϊκή.

εὔζωμον, rocket, *Eruca sativa*
1. 6. 6. root woody; 7. 1. 2–3. time of sowing and germination; 7. 2. 8. root described; 7. 4. 1. only one kind; 7. 4. 2. leaf of a kind of ῥαφανίς comp.; 7. 5. 5. seed keeps well; 9. 11. 6. leaf of στρυχνός ὁ μανικός comp.

εὐθύφλοιος (δρῦς), see δρῦς (4).

εὐώνυμος, spindle-tree, *Euonymus europaeus*
[3. 18. 13. described].

ἐφήμερον (= σπάλαξ (?)), meadow saffron, *Colchicum parnassicum*
9. 16. 6. a poison which has an antidote: described: effects.

ζειά, rice-wheat, *Triticum dicoccum*
2. 4. 1. seed, unless bruised, produces πυρός; 4. 4. 10. ὄρυζον comp.; 8. 1. 1. in list of cereals; 8. 1. 2. sown early; 8. 3. 3. ζ. and τίφη only plants which can change into something quite different (cf. 2. 4. 1.); 8. 9. 2. exhausts the soil: reason: likes rich soil: ζ. and τίφη the cereals most like πυρός.

INDEX OF PLANTS

ζυγία, maple, *Acer campestre*
3. 3. 1. a mountain tree: name for mountain form of σφένδαμνος; 3. 4. 2. time of budding; 3. 6. 1. slow growing (?); 3. 11. 1–2. described; 5. 1. 2. time of cutting timber; 5. 1. 4. do.; 5. 3. 3. character of wood; 5. 7. 6. uses of wood.
ζωστήρ, see φῦκος (1) τὸ πλατύφυλλον.

ἡδύοσμον, (= μίνθη), green mint, *Mentha viridis*
7. 7. 1. a λάχανον.
ἡλιοτρόπον, *Heliotropium villosum*
7. 3. 1. length of flowering season of ὤκιμον comp.; 7. 8. 1. stem 'on the ground'; 7. 9. 2. long in flower; 7. 10. 5. evergreen; 7. 15. 1. flowering depends on the heavenly bodies.
ἡμερίς (1), gall-oak, *Quercus infectoria*
3. 8. 2. one of the five 'Idaean' kinds of oak: fruit; 3. 8. 4. habit and timber; 3. 8. 6. galls.
ἡμερίς (2), (so-called by some) (= δρῦς ἡ ἥμερος = ἐτυμόδρυς 3.8.2.), true oak, *Quercus Robur*
3. 8. 2. bears sweet fruit.
ἡμεροκαλλές, Martagon lily, *Lilium Martagon*
6. 1. 1. in list of under-shrubs (see note); 6. 6. 11. grown from seed: a coronary plant.
ἡμιόνιον, milt-waste, *Asplenium Ceterach*
9. 18. 7. properties of leaf: described: habitat: mules fond of it.
ἡρακλεία (=μήκων ἡ Ἡρακλεία), *Silene venosa*
9. 15. 5. an Arcadian drug.
ἡρακλεωτική (καρύα), see καρύα ἡ Ἡρακλεωτική.
ἡριγέρων, groundsel, *Senecio vulgaris*
7. 7. 1. a λάχανον: classed as 'chicory-like' from its leaves: 7. 7. 4. prolonged flowering-time; 7.10.2. flowers in winter.
ἠρύγγιον, eryngo, *Eryngium campestre*
6. 1. 3. has spines on the leaves: a wild under-shrub.

θαψία, *Thapsia garganica*
9. 8. 3. most powerful juice from root; 9. 8. 5. superstition as to method of cutting; 9. 9. 1. root and juice used; 9. 9. 5. medicinal use; 9. 9. 6. described; 9. 11. 2. leaf of πάνακες τὸ Ἀσκληπίειον comp.; 9. 20. 3. medicinal use: grows specially in Attica: properties; effect on foreign and native cattle.
θέρμος, lupin, *Lupinus alba*
1. 3. 6. refuses cultivation; 1.7.3. seed roots through undergrowth; 3. 2. 1. fruits better in wild state; 4. 7. 5. fruit of an unnamed Arabian tree (see App. (13)) comp.; 4. 7. 6. fruit of an unnamed Persian tree (see App. (13)) comp.; 4. 7. 7. fruit of a tree of the island of Tylos (see App. (13)) comp.; 8. 1. 3. sown early; 8. 2. 1. germination described; 8. 5. 2. seeds in compartments; 8. 5. 4. attachment of seed; 8. 7. 3. not eaten green by any animal; 8. 11. 2. seed keeps well: like a wild plant; 8. 11. 6. seed keeps specially well in hill country; 8. 11. 8. peculiarities about sowing seed.
θηλύγονον (= ἀρρενόγονον), dog-mercury, *Mercurialis perennis*
9. 18. 5. properties: described.
θηλυκράνεια, cornel, *Cornus sanguinea*
1. 8. 2. has less knots than κράνεια; 3. 3. 1. tree of mountain and plain; 3. 4. 2. time of budding; 3. 4. 3. time of fruiting; 3. 4. 6. fruit inedible; 3. 12. 1–2. described; 5. 4. 1. less fruitful than κράνεια.
θηλύπτερις, bracken, *Pteris aquilina*
9. 18. 5. properties: distinguished from πτερίς.
θηλύφονον (= ἀκόνιτον = μυόφονον = σκορπίος (3) 9. 18. 2.), wolf's bane, *Aconitum Anthora*
9. 18. 2. properties: habit: fatal to the scorpion.
θήσειον, *Corydalis densiflora*
7. 12. 3. root bitter: medicinal use.

451

INDEX OF PLANTS

θραύπαλος, joint-fir, *Ephedra campylopoda*
3. 6. 4. very shallow-rooting: many roots; 4.1.3. likes shade

θριδακίνη (properly, but not always, distinguished from θρῖδαξ), wild lettuce, *Lactuca scariola*
1. 10. 7. time of leaf-growth: stem presently spinous; 1.12.2. taste of sap; 7. 1. 2-3. time of sowing and of germination; 7. 3. 2. seeds; 7. 4. 1. several kinds; 7. 4. 5. do. viz. λευκή, πλατύκαυλος. στρογγυλόκαυλος, Λακωνική: differences; 7. 5. 4. pests; 7. 6. 2. wild form distinguished: medicinal use; 9. 8. 2. juice of stalk collected, with a piece of wool; 9. 11. 10. leaf of λιβανωτὶς ἡ ἄκαρπος comp. to ὁ ἡ πικρά.

θρῖδαξ, lettuce, *Lactuca sativa*
7. 2. 4. grows again when stem is cut: effect on flavour; 7. 2. 9. root described; 7. 5. 3. bears transplanting.

θρυαλλίς, *Plantago crassifolia*
7. 11. 12. flowers more or less in a 'spike.'

θρύον, (a grass), *Imperata arundinacea*
4. 11. 12. foliage of some κάλαμοι comp.

θρύορον (?) (= στρυχνὸς ὁ μανικός 9. 11. 6.), thorn-apple, *Datura Stramonium*.

θυία (θύεια), odorous cedar, *Juniperus foetidissima*
1. 9. 3. evergreen; 3. 4. 2. time of budding; 3. 4. 6. time of fruiting; 4. 1. 3. grows on hill-tops.

(θῦμα, ? a madrepore
4. 7. 1. grows in Atlantic: turns to stone).

θύμβρα (θύμβρον), savory, *Satureia Thymbra*
1. 3. 1. (?) a typical under-shrub; 1. 12. 1. taste of fruit; 1. 12. 2. taste of sap; 6. 1. 4. a spineless wild under-shrub; 6. 2. 3. seed conspicuous: not, like θύμος, particular as to situation; 6.7.5. a wild form of ἕρπυλλος comp.; 7. 1. 2-3. time of sowing and of germination; 7. 1. 6. germination; 7. 5. 5. seed keeps well; 7. 6. 1. wild form distinguished.

θύμον (1) (θύμος), Cretan thyme, *Thymbra capitata*
1. 12. 2. taste of sap; 3. 1. 3. reproduces itself without seed; 6. 2. 3. two forms, black and white: seed inconspicuous; 6. 2. 4. requires sea-breezes.

('θύμον' (2), ? a madrepore
4. 7. 2. a marine plant which turns to stone: described).

θύον (θύα), thyine-wood, *Callitris quadrivalvis*
5. 3. 7. described: character and use of wood; 5.4.2. wood proof against decay.

ἰασιώνη, bindweed, *Convolvulus sepium*
1. 13. 2. flower consists of one 'leaf.'

ἴκμη, ? duckweed, *Lemna minor*
4.10. 1-2. in list of plants of Lake Copais; 4.10.4. requires further investigation

ἰξία (1). oak-mistletoe, *Loranthus europaeus*
3. 7. 6. grows on oak and other trees; 3. 16. 1. grows on πρῖνος.

ἰξία (2) (= ἄκανθα (9) = ἄκανος = ἰξίνη = χαμαιλέων ὁ λευκός), pine-thistle, *Atractylis gummifera*
9. 1. 3. Cretan: produces a gum.

ἰξίνη (gum (ἀκανθική) μαστίχη 6. 4. 9., 9. 1. 2.) (= ἄκανθα (9) = ἄκανος = ἰξία (2) = χαμαιλέων ὁ λευκός), pine-thistle, *Atractylis gummifera*
6. 4. 3. a 'thistle-like' plant; 6. 4. 4. time of growing; 6. 4. 9. described; 9. 1. 2. produces a gum called μαστίχη.

ἴον (= ἰωνία = ἴον τὸ λευκόν), gilliflower, *Matthiola incana*
1. 9. 4. evergreen; 2. 1. 3. propagation; 4. 7. 4. colour and scent of unnamed Arabian tree (*see* App. (12a)) comp.; 6. 1. 1. in list of under-shrubs; 6. 6. 1. a cultivated under-shrub: a coronary plant; sweet-scented; 6.6.5. sweetest-scented at Cyrene; 6. 6. 11. grows from

452

INDEX OF PLANTS

seed; woody; 6. 8. 5. position and climate important for fragrance: flowers very early in Egypt; 6. 8. 6. on mountains blooms well, but has inferior scent; 7. 6. 4. wild form quite distinct, alike only in leaf.

ἴον τὸ λευκόν (=λευκόϊον (1) = ἰωνία ἡ λευκή), gilliflower, *Matthiola incana*

3. 18. 13. flower of εὐώνυμος comp.; 4. 7. 8. flower of an Arabian tree (*see* App. (15)) comp; 6. 6. 3. several colour forms; 6. 6. 7. distinguished from ἰ. τὸ μέλαν; 6. 8. 1-2. flowering-time; 6. 8. 5. plant lives three years at most: degenerates with age: 7. 8. 3. leaves 'on the stem.'

ἴον τὸ μέλαν (=ἰωνία ἡ μέλαινα), violet, *Viola odorata*

1. 13. 2. has a 'twofold' flower; 6. 6. 3. only one form; 6. 6. 7. distinguished from ἰ. τὸ λευκόν; 6. 8. 1-2. a coronary plant: flowering time.

ἴπνον, ? marestail, *Hippuris vulgaris*

4. 10. 1-2. in list of plants of Lake Copais; 4.10.4. requires further investigation.

ἱππομάραθον (=μαγύδαρις), *Prangos ferulacea*

6. 1. 4. a spineless wild undershrub: belongs to 'ferula-like' plants.

ἱπποσέλινον, Alexanders, *Smyrnium Olusatrum*

1. 9. 4. evergreen; 2. 2. 1. propagation; 7. 2. 6. root of τεύτλιον comp.; 7. 2. 8. root; 7. 6. 3. comp. with ἐλειοσέλινον: medicinal use; 9. 1. 3. root produces a gum: which is like σμύρνα; 9. 1. 4. propagated from a δάκρυον: a popular error about ἑ. and σμύρνα; 9. 15. 1. grows in Arcadia.

ἱπποφαές, *see* τιθύμαλλος.

ἱππόφεως, spurge, *Euphorbia acanthathamnos*

6. 5. 1. in list of spinous plants which have leaves as well as spines; 6. 5. 2. has no spines on the leaves.

ἶρις, iris, *Iris pallida*, etc.

1. 7. 2. root fragrant; 4. 5. 2. grows best in Illyria on shores of Adriatic; 6. 8. 3. a coronary plant: flowering time; 7. 13. 1. leaves described; 7. 13. 2. flower-stem not the only stem: stem comp. with ἀσφόδελος: 9.7.3. in list of ἀρώματα; 9.7.4. only European ἄρωμα: best in Illyria: preparation; 9. 9. 2. perfume.

ἴσχαιμος, *Andropogon Ischaemum*

9. 15. 3. Thracian: properties.

ἰσχάς (=ἄπιος (2) 9. 9. 5. = ῥάφανος ἡ ὀρεία), spurge, *Euphorbia Apios*.

ἰτέα, willow, *Salix* spp.

1. 4. 2. lives near water; 1. 4. 3. 'amphibious'; 1. 5. 1. crooked and low; 1. 5. 4. wood light; 3. 1. 1. propagation; 3. 1. 2. seems to have no fruit, yet reproduces itself: instance; 3.1.3. sheds its fruit unripened; 3.3.1. tree of mountain and plain; 3.3.4. a question if it bears fruit; 3. 4. 2. time of budding; 3. 6. 1. quick or slow grower?; 3. 13. 7. described: kinds (*see below*): called in Arcadia ἑλίκη; 3. 14. 4. leaf of κολυτέα comp.; 4. 1. 1. likes wet ground; 4. 5. 7. common in some Mediterranean regions; 4. 8. 1. grows partially in water; 4. 10. 1. in list of plants of Lake Copais; 4. 10. 6. grows both on land and in water; 4. 13. 2. shorter-lived by water; 4. 16. 2. grows again after being cut or blown down; 4. 16. 3. instance of a tree which survived the lopping of its branches; 5. 3. 4. character of wood; 5. 7. 7. uses of wood; 5. 9. 4. wood makes an evil smoke when burnt for charcoal.

ἰτέα ἡ λευκή, white willow, *Salix alba*

3. 13. 7. described.

ἰτέα ἡ μέλαινα, *Salix amplexicaulis*

3. 13. 7. described.

ἴφυον, spike-lavender, *Lavandula Spica*

6. 6. 11. a coronary plant: grown

453

INDEX OF PLANTS

from seed; 6. 8. 3. flowering time.
ἴψος, ? cork-oak, *Quercus Suber* (G. from Plin. 16. 98. Hesych. has ἴψος = κισσός)
3. 4. 2. time of budding.
ἰωνία (=ἴον q.v. = ἰωνία ἡ λευκή = ἴον τὸ λευκόν = λευκόϊον (1)), gillillower, *Matthiola incana*.
ἰωνία ἡ λευκή (= ἰωνία = ἴον q.v.), gillillower, *Matthiola incana*.
ἰωνία ἡ μέλαινα (= ἴον τὸ μέλαν q.v.), violet, *Viola odorata*.

κάκτος (1), cardoon, *Cynara Cardunculus*
6. 4. 10–11. a 'thistle-like' plant: described: peculiar to Sicily.
κάκτος (2), artichoke, *Cynara Scolymus*
6. 4. 11. has erect 'stalk' called πτέρνιξ: described; edible; base of receptacle called σκαλίας.
κάλαμος, reed
1. 5. 2. bark fibrous: bark in one layer; 1. 5. 3. stem jointed; 1.6.2. core membranous; 1. 6. 7. root jointed; 1. 6. 10. roots large and numerous; 1. 8. 3. joints regular; 1.8.5. joints analogous to 'knots'; 1. 9. 4. evergreen; 1. 10. 5. leaves end in a point; further described; 1.10.9. leaves made of fibre: leaf-stalk made of fibre; 2. 2. 1. (a kind of) propagation; 4. 8. 1. in list of τὰ λοχμώδη; 4. 8. 7. κύαμος ὁ Αἰγύπτιος comp.; 4. 8. 8. thickness of root of κύαμος ὁ Αἰγύπτιος comp.; 4. 9. 1. class of rivers in which κ. grows; 4. 9. 3. has 'side-growths'; 4. 10. 1. in list of plants of Lake Copais; 4.10.6. grows both on land and in water; 4.10.7. effect of drought; 4. 11. 1. distinguished from κ. ὁ αὐλητικός (*see below*): a stout and a slender form (ὁ χαρακίας and ὁ πλόκιμος) (*see below*); 4. 11. 10–13. other forms; 6. 2. 8. setting of leaves of νάρθηξ and ναρθηκία comp.; 9.16.1. δίκταμνον kept ἐν καλάμῳ.
κάλαμος ὁ αὐλητικός (= κ. ὁ Λακωνικός = κ. ὁ συριγγίας = κ. ὁ τοξικός = κ. ὁ χαρακίας = δόναξ, pole-reed, *Arundo Donax*
4. 10. 1. in list of plants of Lake Copais; 4. 10. 6. grows only in water; 4. 11. 1. distinguished from the ordinary form of κ.; 4. 11. 2. not true that it takes nine years to grow; 4. 11. 3. conditions of growth; 4. 11. 4. described by contrast with other κάλαμοι; 4. 11. 4–7. manufacture of the mouthpieces of pipes; 4. 11. 8–9. distribution in region of Lake Copais.
κάλαμος ὁ εἰλετίας, *Ammophila arundinacea*
4. 11. 13. the 'male kind' of κ. ἐπίγειος, so called by some.
κάλαμος (ἐπίγειος), bush-grass, *Calamogrostis Epigeios*
4.11.13. described: growth comp. to ἄγρωστις.
κάλαμος ὁ εὐώδης, sweet flag, *Acorus Calamus*
4. 8. 3. grows in a Syrian lake; 9.7.1. habitat (east of Lebanon): described: fragrance; 9. 7. 3. in list of ἀρώματα.
κάλαμος ὁ Ἰνδικός, bamboo, *Bambusa arundinacea*
4. 11. 13. described.
κάλαμος ὁ Ἰνδικός ('male'), Male bamboo, *Dendrocalamus strictus*
4. 11. 13. distinguished as solid.
κάλαμος ὁ Λακωνικός (= κ. ὁ αὐλητικός = κ. ὁ συριγγίας = κ. ὁ τοξικός = κ. ὁ χαρακίας = δόναξ), pole-reed, *Arundo Donax*
4. 11. 12. colour.
κάλαμος ὁ πλόκιμος, spear-grass, *Phragmites communis*
4. 11. 1. pliant reed; compared with κ. ὁ χαρακίας: grows on floating islands of Lake Copais.
κάλαμος ὁ συριγγίας (= κ. ὁ αὐλητικός = κ. ὁ Λακωνικός = κ. ὁ τοξικός = κ. ὁ χαρακίας = δόναξ), pole-reed, *Arundo Donax*
4. 11. 10 described.
κάλαμος ὁ τοξικός (Κρητικός) (= κ. ὁ αὐλητικός = κ. ὁ Λακωνικός = κ. ὁ συριγγίας = κ. ὁ χαρακίας = δόναξ), pole-reed, *Arundo Donax*
4. 11. 11. described.

INDEX OF PLANTS

κάλαμος ὁ χαρακίας (= κ. ὁ Λακωνικός etc.), pole-reed, *Arundo Donax*
4. 11. 1. stout form: described: grows in reed-beds of Lake Copais.

κάλαμος (other kinds)
4. 11. 10. briefly described.

κάππαρις, caper, *Capparis spinosa*
1. 3. 6. refuses cultivation; 3.2.1. fruits better in wild state; 4. 2. 6. fruit of βάλανος comp.; 6. 1. 3. has spines on the shoots; 6. 4. 1. has spines on leaves as well as on stem; 6. 5. 2. described; 7. 8. 1. stem 'on the ground'; 7. 10. 1. grows and flowers entirely in summer.

κάρδαμον, cress, *Lepidium sativum*
1. 12. 1. taste of fruit; 7. 1. 2–3. time of sowing and germination; 7. 1. 6. germination; 7. 4. 1. only one kind; 7. 5. 5. seed keeps well.

καρδάμωμον, cardamom, *Elettaria Cardamomum*
9. 7. 2. an ἄρωμα, Median or Indian; 9 7. 3. in list of ἀρώματα.

καρύα (fruit κάρυον), hazel, *Corylus avellana*
1. 12. 1. taste of fruit; 3. 2. 3. evidence that it is really wild; 3. 3. 1. a mountain tree; 3.4.2. time of budding; 3. 4. 4. time of fruiting; 4. 5. 4. abundant on Tmolus and Mysian Olympus; 8. 2. 2. germination described.

καρύα ἡ Εὐβοϊκή, sweet chestnut, *Castanea vesca* var. (improved form)
1. 11. 3. seed in a leathery shell; 4. 5. 4. common in Euboea and Magnesia; 5. 4. 2. wood proof against decay; 5. 4. 4. wood does not decay in water (?) 5. 6. 1. wood makes a noise when about to split: instance; 5. 7. 7. uses of wood: does not rapidly decay; 5. 9. 2. charcoal of this wood used in iron-mines.

καρύα ἡ Ἡρακλεωτική (Ἡρακλεῶτις) (fruit κάρυον), filbert, *Corylus avellana* vars.
1. 3. 3. effect of not pruning; 1. 10. 6. leaves notched; 1.11.1. seed immediately within envelope; 1. 11. 3. seed in a woody shell; 3. 3. 8. doubt whether it has a flower (ἴουλος); 3. 5. 5–6. catkins; 3. 6. 2. formation of buds; 3. 6. 5. deep-rooting according to Arcadians: etc.; 3. 7. 3. catkins; 3. 15. 1–2. described: kinds.

καρύα ἡ Περσική, walnut, *Juglans regia*
3.6.2. formation of buds; 3.14.4. leaf of σημύδα comp.

κασία, cassia, *Cinnamomum iners*
4. 4. 14. in list of oriental aromatic plants; 9. 4. 2. Arabian; 9. 5. 1. and 3. described: method of collection; 9. 7. 2. Arabian; 9. 7. 3. in list of ἀρώματα.

καυκαλίς, *Tordylium apulum*
7. 7. 1. a λάχανον.

κέγχρος, millet, *Panicum miliaceum*
1. 11. 2. seeds in a husk; 4. 4. 10. inflorescence of ὄρυζον comp.; 4. 8. 10. fruit of λωτός (2) comp.; 4. 10. 3. size of sceds of σίδη comp.; 8. 1. 1. in list of 'summer crops' distinct from cereals and pulses; 8. 1. 4. sown later than cereals and pulses; 8. 2. 6. time of maturing seed; 8. 3. 2. stem; 8. 3. 3. flower; 8. 3. 4. seed abundant; 8. 7. 3. needs little water: comp. with μέλινος; 8. 9. 3. reasons why it might have been expected to exhaust the soil: contrasted with pulses as to 'lightness'; 8.11.1. seed keeps well; 8.11.6. do. specially in hill-country; 9.18. 6. fruit of κραταιγών comp.

κεδρίς, juniper, *Juniperus communis*
1. 9. 4. evergreen: a dwarf kind (cf. κέδρος 3. 13. 7.); 1. 10. 6. leaf spinous at tip; 1. 12. 1. taste of fruit.

κέδρος (1) (= ὀξύκεδρος 3. 12. 3.), prickly cedar, *Juniperus Oxycedrus*
1. 5. 3. wood not fleshy; 1. 10. 6. leaf spinous at tip; 3. 6. 5. shallow-rooting according to Arcadians; 3.10.2. μίλος comp.; 3.12. 3–4. described: two kinds, ἡ Λυκίη and ἡ Φοινίκη (? Φοινι-

455

INDEX OF PLANTS

κική) (see κέδρος (3)); distinguished from ἄρκευθος; 3. 13. 7. has a dwarf form (? κεδρίς, cf. 1. 9. 4.); 4. 3. 3. size of fruit of παλίουρος ὁ 'Αιγύπτιος comp.; 4.5.2. grows on Thracian and Phrygian mountains; 4. 16. 1. some think topping fatal; 5.3.7. images made from the wood; 5. 4. 2. wood proof against decay; 5. 9. 8. wood exudes moisture : hence 'sweating' statues; 9. 1. 2. sap gummy.

κέδρος (2), Syrian cedar, *Juniperus excelsa*
3. 2. 6. characteristic of mountains of Cilicia and Syria; 4. 5. 5. grows in Syria and is used for ships; 5. 7. 1–2. use of wood in ship-building; 5. 7. 4. use of wood in house-building; 5. 8. 1. remarkably fine in some regions *e.g.* Syria.

κέδρος (3), ἡ Φοινικική (= ἄρκευθος), Phoenician cedar, *Juniperus phoenicea*
3. 12. 3. see κέδρος (1); 9. 2. 3. said to be burnt for pitch in Syria.

κέδρος (4) ἡ Λυκίη
3. 12. 3. a kind so distinguished by some from κέδρος (3).

κενταύριον, centaury, *Centaurea salonitana*
1. 12. 1. taste of fruit; 3. 3. 6. only bears fruit in hill country; 4.5.1. seeks cold regions; 7.9.5. leaves and stems bitter, yet wholesome ; 9. 1. 1. juice bloodred; 9. 11. 6. juice mixed with στρύχνος ὁ μανικός to make a potion.

κενταυρίς, feverwort, *Erythraea Centaurium*
9. 8. 7. superstition as to gathering; 9. 14. 1. how long drug will keep.

κεντρομυρρίνη (= μυάκανθος), butcher's broom, *Ruscus aculeatus*
3. 17. 4. bears fruit on its leaves.

κεραΐς (= ῥάφανος ἡ ἀγρία 9. 15. 5.), charlock, *Raphanus Raphanistrum*.

κέρασος (= λακάρη), bird-cherry, *Prunus avium*
3. 13. 1–3. described; 4. 15. 1. bark can be stripped; 9. 1. 2. sap gummy.

κεραύνιον, 'thunder-truffle,' *Tuber aestivum*
1. 6. 5. has no roots.

κερκίς (1), Judas-tree, *Cercis Siliquastrum*
1. 11. 2. seeds in a pod.

κερκίς (2), aspen, *Populus tremula*
3. 14. 3. described.

κερωνία (= συκῆ ἡ Αἰγυπτία 1. 11. 2.), carob, *Ceratonia Siliqua*
1. 11. 2. seeds in a pod; 1. 13. 2. bears on stem and branches; 4. 2. 4. described.

κήλαστρος (κήλαστρον), holly, *Ilex Aquifolium*
1. 3. 6. refuses cultivation; 1. 9. 3. evergreen; 3. 3. 1. tree of mountain and plain; 3. 3. 3. evergreen; 3. 4. 5–6. time of fruiting ; said to lose fruit in winter ; 4. 1. 3. grows in very cold positions; 5. 6. 2. colour of wood of φιλύκη comp.; 5. 7. 7. wood used for walking-sticks.

κινάμωμον, cinnamon, *Cinnamomum Cassia*
4. 4. 14. in list of oriental aromatic plants; 9.4.2. Arabian; 9.5.1–2. two kinds, white and black, described : habitat: method of collection : a story; 9. 7. 2. in list of ἀρώματα.

κίσθος, rock-rose, *Cistus* spp.
6. 1. 4. a spineless wild undershrub; 6. 2. 2. described : two forms (see below).

κίσθος ὁ ἄρρην, *Cistus villosus*
6. 2. 1. described.

κίσθος ὁ θῆλυς, *Cistus salvifolius*
6. 2. 1. described.

κιττός, ivy, *Hedera Helix*
1. 3. 2. a shrub which becomes tree-like; 1. 9. 4. evergreen; 1. 10. 1. leaves change shape with age of plant; 1. 10. 7. long leaf-stalk; 1. 13. 1. flower 'downy'; 1. 13. 4. attachment of flower; 3. 4. 6. time of fruiting; 3. 10. 5. fruit and leaf of φίλυρα comp.; 3. 14. 2. leaf of κερκίς (2) comp.; 3. 18. 6. kinds distinguished (see below); 3. 18. 7. distinguished from

INDEX OF PLANTS

ἕλιξ; 3. 18. 9–10. described: habit etc.; 3. 18. 11. cluster of berries of σμῖλαξ (2) comp.: described; 4. 4. 1. distribution in Asia; 4. 16. 5. overgrowth of κ. dangerous to trees; 5. 3. 4. character of wood; 5. 9. 6. wood said to make best fire-sticks; 5. 9. 7. the stationary fire-stick should be made of κ. or ἀθραγένη; 9. 13. 6. leaf of ἐρευθεδανόν comp.

κιττὸς ὁ ἕλιξ, see ἕλιξ

κιττὸς ὁ λευκός, white-berried ivy, *Hedera Helix*
3. 18. 6. described : several kinds: one=κορυμβίας, one=κ. ὁ Ἀχαρνικός; 3. 18. 9. roots; 3. 18. 10. fruit; 9. 18. 5. properties of fruit.

κιττὸς ὁ μέλας, black-berried ivy, *Hedera Helix*
3. 18. 6. several kinds; 3. 18. 9. roots; 3. 18. 10. fruit.

κιχόριον (κιχόρη), chicory, *Cichorium Intybus*
1. 10. 7. attachment of leaves; 7. 7. 1. a λάχανον; a class of plants called 'chicory-like' from their leaves; 7. 7. 3. season of growing; 7. 8. 3. leaves 'on the ground' and 'on the stem'; etc.; 7. 9. 2. long in flower; 7. 10. 3. flowers borne in succession; 7. 11. 3. root, inflorescence and seed-vessel described; 9. 12. 4. μήκων ἡ ῥοιάς comp. to κ. τὸ ἄγριον; 9. 16. 4. leaf of ἀκόνιτον comp.

κλήθρα, alder, *Alnus glutinosa*
1. 4. 3. 'amphibious'; 3. 3. 1. tree of mountain and plain; 3. 3. 6. does not always fruit; 3. 4. 2. time of budding; 3. 4. 4. time of fruiting; 3. 6. 1. slow growing (?); 3. 6. 5. roots slender and 'plain,' according to Arcadians; 3. 14. 3. described; 3.15. 1. leaf of καρύα ἡ Ἡρακλεωτική comp.; 4. 8. 1. grows partially in water.

κλινότροχος (?), *Acer Pseudo-platanus*
3. 11. 1. a form of σφένδαμνος.

κλύμενον, honeysuckle, *Lonicera etrusca*
9. 8. 5. superstition as to time of cutting; 9. 18. 6–7. properties of fruit.

κνέωρον (berry Κνίδιος κόκκος), *Daphne Gnidium*
6. 1. 4. a spineless wild undershrub; 9. 20. 2. berry described; medicinal use and properties.

κνέωρος ὁ λευκός, *Daphne oleoides*
6. 2. 2. distinguished from κ. ὁ μέλας; use of root.

κνέωρος ὁ μέλας, *Thymelaea hirsuta*
1. 10. 4. leaves fleshy; 6. 2. 2. see κ. ὁ λευκός.

κνῆκος (= κ. ὁ ἥμερος = κρόκος ὁ ἀκανθώδης), safflower, *Carthamus tinctorius* etc. (*see below*)
1. 13. 3. flowers attached above each seed; 6. 1. 3. a wild undershrub : has spines on the leaves; 6. 4. 3. a 'thistle-like' plant; 6. 4. 4. no side-growths; 6. 4. 5. three forms distinguished and described, one cultivated (*see below*); 6. 6. 6. seed of ῥόδον comp.

κνῆκος ἡ ἀγρία, *Carthamus leucocaulos*
6. 4. 5. distinguished from κ. ἡ ἥμερος.

κνῆκος ἡ ἀγρία (ἑτέρα), *Cnicus benedictus*
6. 4. 5. described.

κνῆκος ἡ ἥμερος, *Carthamus tinctorius*
6. 4 5. distinguished from wild kinds.

Κνίδιος κόκκος, see κνέωρον

κοΐξ (= κουκιόφορον), doum-palm, *Hyphaene thebaica*
1. 10. 5. reedy leaves; 2. 6. 10. a shrubby palm : Ethiopian.

κοκκυγέα, wig-tree, *Rhus Cotinus*
3. 16. 6. described.

κοκκυμηλέα (fruit κοκκύμηλον), plumtree, *Prunus domestica*
1. 10. 10. fruit made of flesh and fibre; 1. 11. 1. seed enveloped in flesh and stone; 1. 12. 1. taste of fruit; 1. 13. 1. flower 'leafy'; 1. 13. 3. flower above fruit-case; 3. 6. 4. very shallow-rooting: few roots; 3. 6. 5. deep-rooting according to Idaeans; etc.; 4. 2. 3. size of fruit of συκῆ ἡ Κυπρία comp.; 4.2.5. fruit-stone of περσέα comp.

457

INDEX OF PLANTS

κοκκυμηλέα (ἡ Αἰγυπτία) (sebesten), *Cordia Myxa*
4. 2. 10. described.

κολοιτία (1) (κολουτέα 3. 17. 2.: cf. 3. 17. 3. n.), *Cytisus aeolicus*
1. 11. 2. tree of Lipari islands: seeds in a pod ; 3. 17. 2. described.

κολοιτία (2), *Salix cinerea*
3. 17. 3. Idaean : described.

κολοκύντη, gourd, *Cucurbita maxima*
1. 11. 4. seeds in a row : 1. 12. 2. taste of sap ; 1. 13. 3. flower attached above fruit : 2. 7. 5. use of dust ; 7. 1. 2–3. time of sowing and of germination; 7.1.6. germination ; 7. 2. 9. root described ; 7. 4. 1. several kinds; 7. 5. 5. seed does not keep well.

κολυτέα, bladder-senna, *Colutea arborescens*
[3. 14. 4. described.]

κόμαρος (fruit μεμαίκυλον 3. 16. 4.), arbutus, *Arbutus Unedo*
1. 5. 2. bark readily drops off ; 1. 9. 3. evergreen ; 3. 16. 4. described ; 3. 16. 6. leaf of κοκκυγέα comp.; 5.9.1. wood makes good charcoal.

κόμη = τραγοπώγων 7. 7. 1. *q.v.*

κόνυζα, *Inula* spp.
6. 1. 4. a spineless wild undershrub ; 6. 2. 5. two kinds described and compared ('male' and 'female') (*see below*); 7. 10. 1. grows and flowers entirely in summer.

κόνυζα ἡ ἄρρην, *Inula viscosa*
6. 2. 5.

κόνυζα ἡ θήλεια, *Inula graveolens*
6. 2. 5.

κορίαννον, coriander, *Coriandrum sativum*
1. 11. 2. seeds naked ; 7. 1. 2–3. time of sowing and germination; 7. 1. 6. germination ; 7. 2. 8. root described ; 7. 3. 2. seeds described ; 7. 4. 1. only one kind ; 7. 5. 4. effect of hot weather ; 7.5.5. seed keeps well.

κορυμβίας, *see* κιττὸς ὁ λευκός.

κόρχορος, blue pimpernel, *Anagallis caerulea*
7. 7. 2. a λάχανον : proverbial for bitterness.

κορωνόπους, hartshorn, *Plantago Coronopus*
7. 8. 3. leaves 'on the ground.'

κόστος, *Saussurea Lappa*
9. 7. 3. in list of ἀρώματα

κότινος (? = ἀγριέλαιος), wild olive, *Olea Oleaster*
1. 4. 1. more fruitful than ἐλάα ; 1. 8. 1. many knots ; 1. 8. 2. more knots than ἐλάα ; 1. 8. 3. knots regular : knots opposite ; 1. 8. 6. liable to excrescences; 1.14.4. wild form of ἐλάα ; 2.2.11. cannot be made into ἐλάα by cultivation: effect of transplanting and removing top-growth; 2. 3. 1. occasionally changes to ἐλάα spontaneously ; 3. 2. 1. produces more fruit than ἐλάα but ripens less ; 3. 6. 2. knots opposite ; 3. 15. 6. size of fruit of κράταιγος comp.; 4. 4. 11. Indian olive between κ. and ἐλάα ; 4. 13. 1. longer-lived than ἐλάα ; 4. 13. 2. story of a very old κ. at Olympia ; 4. 14. 12. suffers less than ἐλάα from special winds ; 5. 2. 4. story of a tree at Megara ; 5. 3. 3. character of wood ; 5. 4. 2. wood proof against decay ; 5. 4. 4. wood not eaten by *teredon*; 5. 7. 8. uses of wood for carpenter's tools.

κουκιόφορον (= κοῖξ), doum-palm, *Hyphaene thebaica*
2. 6. 9. (not named) described ; 4. 2. 7. described.

κράνεια (fruit κράνεον 4. 4. 5.), cornelian cherry, *Cornus mas*
1.6.1. core hard and close ; 1.8.2. has more knots than θηλυκράνεια ; 3. 2. 1. fruit sweeter and better ripened in wild than in cultivated form ; 3. 3. 1. tree of mountain and plain ; 3. 4. 2. time of budding ; 3. 4. 3. time of fruiting ; 3. 6. 1. slow-growing (?) ; 3. 12. 1–2. described ; 4. 4. 5. root of an unnamed Indian tree (*see* App. (6)) comp.; 5. 4. 1. more fruitful than θηλυκράνεια ; 5. 6. 4. wood very strong.

INDEX OF PLANTS

κραταίγονος, willow-weed, *Polygonum Persicaria*
9. 18. 6. properties: described.
κράταιγος (= κραταιγών 3. 15. 6.), *Crataegus Heldreichii*
3. 15. 6. described: perhaps a wild form of μεσπίλη.
κρηπίς, ox-tongue, *Helminthia echioeides*
7. 8. 3. leaves on the stalk.
κριθή, barley, *Hordeum sativum*
1. 6 5. roots numerous; 1. 6. 6. do.; 1. 11. 5. each seed separately attached; 2. 2. 9. said to turn sometimes into wheat; 2. 4. 1. wild κ. turns into cultivated with cultivation; 4. 4. 9. India has a corresponding cereal and a wild form of κ.; 8. 1. 1. in list of cereals; 8. 1. 3. sown early, before πυρός; 8. 1. 5–6. time of germination in Hellas (and in Egypt?); 8. 2. 1. germination described; 8. 2. 3. single leaf first appears: roots described; 8. 2. 6. time of maturing seed; 8. 2. 7. time of harvest in Hellas and in Egypt; 8. 2. 9. crop very early in island of Chalkia; 8.3.2. stem; 8.4.1–2. comp. in detail with other cereals: kinds distinguished (*see below*); 8.6.1. conditions for sowing; 8. 6. 4. suitable soil; 8. 6. 5–6. rain hurtful when κ. is in flower: and when it is ripe; 8. 7. 1. said to change into αἶρα under certain conditions; 8.7.5. in many places comes up again next year; etc.; 8. 8. 2. favourable localities; 8.8.3. αἰγίλωψ (2) grows specially among κ.; 8.9.1. exhausts the soil, but less than πυρός: reason; 8. 10. 2. wheat-rust; 8. 10. 3. effects of weather; 8. 11. 1. seed keeps less well than πυρός; 8. 11. 3. grain stored without drying; 8. 11. 7. at Babylon grain jumps on the threshing-floor: reason; 9. 11. 9. τιθύμαλλος ὁ μυρτίτης gathered at time of barley-harvest; 9. 12. 4. μήκων ἡ ῥοιάς grows in fields of κ.

κριθαὶ αἱ ἀγρίαι (Indian), *Sorghum halepense*
4. 4. 9. can be used for bread.
κριθαὶ αἱ Ἀχιλλεῖαι, barley, *Hordeum sativum* var.
8. 4. 2. ear close to leaf; 8. 10. 2. specially liable to wheat-rust.
κριθαὶ αἱ Ἰνδικαί, barley, *Hordeum sativum* var.
8 4.2. branching.
κριθῶν γένος τρίμηνον, barley, *Hordeum sativum* var.
8. 1. 4. sown late.
κρίνον (= κρινωνία, *cf.* λείριον (1)), lily, *Lilium candidum* etc.
1. 13. 2. has a 'twofold' flower; 2.2.1. propagation from exudation; etc.; 4. 8. 6. an unnamed Egyptian plant (*see* App. (18)) comp.; 4.8.9. petals of flower of λωτός (2) comp.; 6. 6. 3. several colour forms; 6. 6. 8. do.; a coronary plant: described: propagation; 6. 6. 9. leaves of νάρκισσος (1) comp.; 6. 8. 3. flowering time; 9. 1. 4. *cf.* 2. 2. 1.

κρίνον τὸ πορφυροῦν, Turk's cap lily, *Lilium chalcedonicum*
6. 6. 3. (*see* κρίνον).
κριοί, *see* ἐρέβινθος.
κρόκος, crocus, *Crocus* spp. etc. (*see below*)
1. 6. 6. root fleshy; 1. 6. 7. do. 1. 6. 11. large fleshy root; 7.7.1. leaf of τραγοπώγων comp.; 7. 7. 4. flowering time short: three kinds mentioned, εὔοσμος, λευκός, ἀκανθώδης (*see below*). 7.9.4. root acorn-shaped; 7.10 2. flowers in winter; 7.13.1. leaves described; 7. 13. 2. no stem except flower-stem.
κρόκος ὁ ἀκανθώδης (= κνῆκος = κ. ἡ ἥμερος), safflower, *Carthamus tinctorius*
7. 7. 4. (*see* κρόκος).
κρόκος ὁ εὔοσμος, saffron crocus, *Crocus sativus*
4. 3. 1. abundant in Cyrenaica; 6. 6. 5. sweetest-scented at Cyrene; 6. 6. 10. a coronary plant: described: propagation; 6. 8. 3. flowering time: a wild (scentless) and a cultivated

INDEX OF PLANTS

kind; 7. 7. 4. *see* κρόκος; 9. 7. 3. in list of ἀρώματα.

κρόκος ὁ λευκός, crocus, *crocus cancellatus*
7. 7. 4: 7. 10. 2. (*see* κρόκος).

κρομυογήτειον, onion, *Allium Cepa* var.
4. 6. 2. root of φῦκος τὸ πλατύ comp.

κρόμυον, onion, *Allium Cepa*
1. 5. 2. 'bark' in layers; 1. 6. 7. root in scales; 1. 6. 9. no side roots; 1. 10. 7. attachment of leaves; 1. 10. 8. leaves hollow; 7. 1. 7. stem single; 7. 2. 1. propagation; 7. 2. 3. growth of γήθυον and πράσον comp.: offsets specially numerous; 7. 3. 4. seed borne at top; 7. 4. 7–10. kinds distinguished, Σάρδιον, Κνίδιον, Σαμοθράκιον, σητάνιον, σχιστόν, Ἀσκαλώνιον: cultivation and special points of σχιστόν (*see below*), Ἀσκαλώνιον: further local varieties; 7. 4. 12 formation of roots of σκόρoδον contrasted; 7. 5. 1. likes water; 7 5. 2. said to dislike rainwater; 7. 8. 2. stem smooth, not branched; 7. 9. 4. *cf.* 1. 6. 7; 7. 13. 4. grows in colonies because of offsets; 9. 15. 7. root of μῶλυ comp.

κρόμυον τὸ σχιστόν, shallot, *Allium Cepa* var.
7. 4. 7–10. distinguished from other varieties of κρόμυον: cultivation.

κρότων, castor-oil plant, *Ricinus communis*
1. 10. 1. leaves change shape with age of plant; 3. 18. 7. do.

κύαμος, bean, *Vicia Faba*
3. 3. 3. size of fruit of κέρασος comp.; 3. 15. 3. fruit of τέρμινθος comp.; 3. 17. 6 size of berry of ἄμπελος (2) comp.; 4. 3. 1. size of fruit of λωτός (4) comp.; 7. 3. 1. length of flowering of ὤκιμον comp.; 8. 1. 1. in list of pulses; 8. 1. 3–4, sown early, but can be sown late; 8. 1. 5. time of germination: very slow; 8. 2. 1. germination described; 8. 2. 3. comes up with several leaves : roots and side-growths contrasted with other pulses; 8. 2. 5. flowering time; 8. 2. 6. time of maturing seed; 8. 3. 1. leaf; 8. 3. 2. stem; 8. 5. 1. more than one kind: white form sweetest; 8. 5. 4. attachment of seed; 8. 6. 1. rain not beneficial after sowing; 8. 6. 5. likes water when in flower, but not later; 8. 7. 2. makes ground fertile for ἐρέβινθος; 8. 8. 6 causes etc. of κ. becoming 'cookable' or 'uncookable'; 8. 9. 1 improves the soil (*cf.* 8. 7. 2.); 8. 10. 5. infested by ἶνες; 8. 11. 1. seed does not keep; 8. 11. 3. seed keeps well in some localities.

κύαμος (ὁ Αἰγύπτιος), *Nelumbium speciosum*
4. 8. 7–8 described; 4. 8. 9. stalk, leaves and growth of fruit of λωτός (2) comp.

κυδώνιος (fruit μῆλον κυδώνιον) (= στρουθίον (1)), quince, *Cydonia vulgaris*
2. 2. 5. produced from seed of στρουθίον; 4. 8. 11. size of root of λωτός (2) comp. to μῆλον κυδώνιον.

κυϊξ, ?
7. 13. 9. (in defective sentence): belongs to τὰ βολβώδη.

κυκλάμινος, cyclamen, *Cyclamen graecum*
7. 9. 4. root has 'bark'; 9. 9. 1. root and juice used; 9. 9. 3. use in medicine and as charm; 9. 18. 2. leaf of σκορπίος (3) comp.

κύμινον, cummin, *Cuminum Cyminum*
1. 11. 2. seeds naked; 7. 3. 2–3. seeds described: popular belief about sowing; 7. 4. 1. several kinds; 8. 3. 5. seed very abundant and small; 8. 6. 1 rain not beneficial after sowing; 8. 8. 5. a plant parasitic on root (*see* App. (25)); 8. 10. 1. diseases; 9. 8. 8. *cf.* 7. 3. 2–3.

κυνόρροδον, dog-rose, *Rosa canina*
4. 4. 8. an unnamed Indian tree (cotton plant) comp.

INDEX OF PLANTS

κυνόσβατος, wild rose, *Rosa sempervirens*
3. 18. 4. described; 9. 8. 5. superstition as to method of cutting.
κύνωψ, rib-grass, *Plantago lanceolata*
7.7.3. time of growing: 7.11.2. (?) flowers in a spike.
κυπάριττος, cypress, *Cupressus sempervirens*
1. 5. 1. erect and tall; 1. 5. 3. wood not fleshy; 1. 6. 4. shallow rooting; 1. 6. 5. roots not branching; 1. 8. 2. ' male ' has more knots than ' female '; 1. 9. 1. growth chiefly upwards; 1. 9. 3. evergreen; 1. 10. 4. leaves fleshy; 2. 2. 2. propagation; 2. 2. 6. seed of ' female ' produces ' male ' trees; 2. 7. 1. dislikes manure and water; 3. 1. 6. comes up spontaneously in Crete; 3. 2. 3. evidence that it is really wild (at least ' male ' form); 3. 2. 6. characteristic of the Cretan Ida; 3. 12 4. bark of κέδρος (1) comp.; 4.1.3. grows very high on Cretan mountains; 4. 3. 1. grows in Cyrenaica; 4. 5. 2. abounds in Crete, Lycia, Rhodes; 4. 18. 12. beer (βρυτός) made from κ. in Egypt; 4,16.1. some think topping fatal; 5.3.7. θύον comp.: images made from the wood; 5. 4. 1. the ' male ' form the more fruitful; 5. 4. 2. wood proof against decay: an instance: takes a fine polish; 5. 7. 4. use of wood in housebuilding.
κύπειρον, *Cyperus rotundus*
9. 7. 3. in list of ἀρώματα.
κύπειρος, galingale, *Cyperus longus*
1. 5. 3.. stem very smooth; 1. 6. 8. a stout root and also fibrous roots; 1.8.1. no knots; 1.10.5. leaves end in a point; further described; 4. 8. 1. in list of τὰ λοχμώδη; 4. 8. 12. leaves of μαλιναθάλλη comp.; 4.ι10. 1. in list of plants of Lake Copais; 4. 10. 5. described; 4. 10. 6. grows both on land and in water: grows on the floating islands of Lake Copais;

4. 11. 12. foliage of some κάλαμοι comp.
κύτινος, see ῥόα.
κύτισος (1), laburnum, *Laburnum vulgare*
1. 6. 1. core hard and close; 4. 4. 6. habit of ἐβένη comp.; 5. 3. 1. wood of the core very close and heavy.
κύτισος (2), tree-medick, *Medicago arborea*
4. 16. 5. dangerous to trees.
κώμακον, *Ailanthus malabarica*
9. 7. 2. an Arabian ἄρωμα (i.e. imported through Arabia: mixed with other ἀρώματα:

[name also given to fruit of a different plant].

κώνειον, hemlock, *Conium maculatum*
1. 5. 3. stem fleshy; 6. 2. 9. belongs to ' ferula-like ' plants : has a hollow stem; 7. 6. 4. leaf of ὀρεισσέλινον comp.; 9. 8. 3. most powerful juice from root; 9.15. 8. localities; 9. 16. 8. medical experience; 9. 16. 9. treatment in Keos; 9. 20. 1. πέπερι an antidote to κ.
κωνόφορος, see [πεύκη ἡ] κωνόφορος

λάθυρος, *Lathyrus sativus*
8. 3. 1. leaf; 8. 3. 2. stem; 8. 10. 5. infested by ' worms.'
λακάρη (λάκαρα) (= κέρασος ? Macedonian name), bird-cherry, *Prunus avium*
3. 3. 1. a tree of mountain and plain; 3. 6. 1. slow-growing (?).
λάπαθος (λάπαθον), monk's rhubarb, *Rumex Patientia*
1. 6. 6. root single; 7. 1. 2. time of sowing; 7. 2. 2. 7–8. root described; 7. 4. 1. only one kind; 7. 6. 1. wild form distinguished (*see below*); 9. 11. 1. leaf of πάνακες τὸ Χειρώνειον comp.
λάπαθον τὸ ἄγριον, dock, *Rumex conglomeratus*
7. 6. 1. distinguished from λ. τὸ ἥμερον; 7.7.2. a λάχανον; needs cooking.

461

INDEX OF PLANTS

λειμωνία, (? = σκόλυμος 6. 4. 3.), golden thistle, *Scolymus hispanicus*
  6. 4. 3. a 'thistle-like' plant: leaves spinous.

λείριον (1) (= κρίνον q.v.) Madonna lily, *Lilium candidum*
  3. 13. 6. flower of ἀκτῆ has the heavy scent of λ.; 3. 18. 11. scent of flower of σμίλαξ (2) comp.; 9.16.6.(?) leaf of ἐφήμερον comp.

λείριον (2), narcissus, *Narcissus* spp. (see below)
  1. 13. 2. flower consists of one 'leaf' only partly divided.

λείριον, polyanthus narcissus, *Narcissus Tazetta*
  6. 8. 1. flowering time; 7. 13. 4. grown from seed.

λείριον (τὸ ἕτερον 6.8.3.) (= νάρκισσος (1) 6.6.9.), narcissus, *Narcissus serotinus*
  6. 6. 9. a coronary plant: described; flowering time.

λέμνα, water chickweed, *Callitriche verna*
  4. 10. 1. in list of plants of Lake Copais.

λευκάκανθα, milk-thistle, *Silybum marianum*
  6. 4. 3. a 'thistle-like' plant.

λεύκη, abele, *Populus alba*
  1. 10. 1. leaves change shape with age of tree: leaves inverted in summer; 3. 1. 1. propagation; 3. 3. 1. tree of mountain and plain; 3. 4. 2. time of budding; 3. 6. 1. quick-growing; 3. 14. 2. described; 3. 18. 7. cf. 1. 10. 1.; 4. 1. 1. likes wet ground; 4. 2. 3 stem of συκῆ ἡ Κυπρία comp.; 4. 8. 1. grows partially in water; 4. 8. 2. scarce on Nile; 4. 10. 2. flower of ἐλαίαγνος comp.; 4.13. 2. shorter-lived by water; 4. 16. 3. instance of a tree which grew again after falling down; 5. 9. 4. wood makes an evil smoke when burnt for charcoal.

λευκόϊον (1) (= ἴον τὸ λευκόν = ἰωνία ἡ λευκή), gilliflower, *Matthiola incana*

λευκόϊον (2), snowdrop, *Galanthus nivalis*
  7. 13. 9. (in defective sentence) belongs to τὰ βολβώδη.

λιβανωτίς
  9.9 5. medicinal use; 9.11.10–11. two kinds (see below).

λιβανωτὶς ἡ ἄκαρπος, *Lactuca graeca*
  9. 11. 10–11. described: medicinal use: habitat.

λιβανωτὶς ἡ κάρπιμος, (fruit κάχρυ 9. 11. 10.), *Lecokia cretica*
  9. 11. 10. described: medicinal use: habitat: prevents moth in clothes.

λιβανωτός, (gum λίβανος, frankincense: also λιβανωτός 9. 4. 4–9. etc.), frankincense-tree, *Boswellia Carteri*
  4. 4. 14. in list of oriental ἀρώματα; 9. 1. 6. time of tapping; 9. 4. 1. collection of gum; 9. 4. 2. Arabian: described: habitat; 9.4.3. another account; 9.4.4–10. accounts of travellers; 9. 11. 3 scent of πάνακες τὸ Ἡράκλειον comp.; 9. 11. 10. scent of root of λιβανωτὶς ἡ κάρπιμος comp.; 9. 20. 1. an antidote to κώνειον.

λίνον, flax, *Linum usitatissimum*
  3. 18. 3. seeds oily; 8. 7. 1. said to change into αἷρα.

λίνον πύρινον ?
  9. 18. 6. growth of κραταίγονος comp.

λινόσπαρτον, Spanish broom, *Spartium junceum*
  1. 5. 2. bark in layers.

λυχνίς, rose-campion, *Lychnis coronaria*
  6.8. 3. a coronary plant: flowering time.

(λωτός)
  7. 15. 3. many plants called by this name which have nothing in common but the name.

λωτός (1), nettle-tree, *Celtis australis*
  1. 5. 3. wood not fleshy; 1. 6. 1. core hard and close; 1. 8. 2. few knots; 4. 2. 5. colour of wood of περσέα comp.; 4. 2. 9. wood of olive of Thebaid comp.; 4. 2. 12. wood of an unnamed tree (? κοκκυμηλέα ἡ Αἰγυπτία)

INDEX OF PLANTS

comp.; 5. 3. 1. wood very close and heavy; 5. 3. 7. images made from the wood; 5. 4. 2. wood proof against decay: turns black when old; 5. 5. 4. core not obvious but exists; 5. 5. 6. treatment of core for making door-hinges; 5. 8. 1. grows in some places exceptionally fine.

λωτός (2) (aquatic) (root κόρσιον), Nile water-lily, *Nymphaea stellata*
4. 8. 9–11. described.

λωτός (3) (herb), trefoil, *Trifolium fragiferum*
7. 8. 3. leaves 'on the stem'; 7. 13. 5. seed sometimes takes two years to germinate.

λωτός (4) (Libyan tree), *Zizyphus Lotus*
4 3.1. common in Libya; 4.3.1–2 described; 4. 3. 4. further described.

λωτός (5) (aromatic) (= μελίλωτος), *Trigonella graeca*
9. 7. 3. in list of ἀρώματα.

μαγύδαρις (= ἱππομάραθον), *Prangos ferulacea*
1. 6. 12. root most characteristic part; 6. 3. 7. distinct from σίλφιον: described: distribution.

[6. 3. 4. name also given to seed of σίλφιον].

μαδωνάϊς (=νυμφαία 9.13.1.), yellow water-lily. *Nuphar luteum*

μαλάχη (1), mallow, *Lavatera arborea*
1. 3. 2. a herb which becomes tree-like under cultivation; 1. 9. 2. do.; 4. 15. 1. outer bark can be stripped; 9. 18. 1. leaf fruit and taste of that of ἀλθαία comp.

μαλάχη (2), cheese-flower, *Malva silvestris*
7. 7. 2. a λάχανον; needs cooking; 7. 8. 1. stem 'on the ground'

μαλάχη (3) ἡ ἀγρία (= ἀλθαία 9.15.5.), marsh-mallow, *Althaea officinalis*

μαλιναθάλλη (= μνάσιον), *Cyperus esculentus*
4. 8. 12. described.

μανδραγόρας (1), mandrake, *Mandragora officinarum*
9. 8. 8. superstition as to gathering; 9. 9. 1. root and juice used: medicinal use of leaf and root.

μανδραγόρας (2), ? deadly nightshade, *Atropa Belladonna*
6. 2. 9. belongs to 'ferula-like' plants: has hollow stem: fruit described.

μάραθον, fennel, *Foeniculum vulgare*
1. 11. 2. seeds naked; 1. 12. 2. taste of sap; 4. 6. 3. φῦκος τὸ τριχόφυλλον comp.; 6.1.4. a spineless wild under-shrub: belongs to 'ferula-like' plants; 6. 2. 9. do.: has a fibrous stem; 7. 3. 2. seeds described; 9. 9. 6. leaf of θαψία comp.

μάσπετον, see σίλφιον

μελαγκρανίς (= σχοῖνος ὁ κάρπιμος 4. 12. 1.), bog-rush, *Schoenus nigricans*

μελάμπυρον (μελάμπυρος), *Neslia paniculata*
8. 4. 6. infests πυρός ὁ Σικελός: contrasted with αἶρα; 8. 8. 3. (μελάμπυρος ὁ Ποντικός), specially affects crops of πυρός.

μελία, manna-ash, *Fraxinus Ornus*
3. 3. 1. tree of mountain and plain; 3. 4. 4. time of fruiting; 3. 6. 1. slow growing (?); 3. 6. 5. roots numerous matted and run deep, according to Arcadians; 3. 11. 3–4. described: two kinds, see βουμέλιος; 3.17.1. leaf of φελλός comp.; 4. 5. 3. grows in Pontus; 4. 8. 2. common on Nile; 5. 1. 2. time of cutting timber; 5. 6. 4. wood 'moist': used for elastic bedsteads; 5. 7. 3. wood used for bent-wood work: use in shipbuilding; 5. 7. 8. uses of wood for carpenter's tools.

μελίλωτος (= λωτός (5)), *Trigonella graeca*
7. 15. 3. one of the many diverse plants called λωτός.

μέλινος, (in other authors μελίνη:

463

INDEX OF PLANTS

see 8. 1. 1. *n*.), Italian millet, *Setaria italica*
8. 1. 4. sown later than cereals and pulses; 8. 2. 6. time of maturing seed; 8. 3. 2. stem; 8. 3. 3. flower; 8. 7. 3. needs little water : comp with κέγχρος.

μελισσόφυλλον, balm, *Melissa officinalis*
6. 1. 4. a spineless wild undershrub.

μεμαίκυλον, see κόμαρος

μεσπίλη (fruit μέσπιλον) (= μ. ἡ σατάνειος), medlar, *Mespilus germanica*
3. 12. 5–6. described : three kinds (Idaean account, *see below*); 3. 13. 1. leaf of κέρασος comp ; 3. 15. 6. leaf, bark and taste of fruit of κράταιγος comp.; 3.17.5. flower of συκῆ ἡ Ἰδαία comp. also taste of fruit; 4. 2. 10, fruit of κοκκυμηλέα comp.; 4 8. 12. μαλιναθάλλη comp.; 4. 14. 10. fruit gets worm-eaten.

μεσπίλη ἡ ἀνθηδονοειδής, hawthorn, *Crataegus Oxyacantha*
3. 12. 5. described.

μεσπίλη ἡ ἀνθήδων, oriental thorn, *Crataegus orientalis*
3. 12. 5. described.

μεσπίλη ἡ σατάνειος, medlar, *Mespilus germanica*
3. 12. 5. described.

Μηδικὴ (πόα), lucerne, *Medicago sativa*
8. 7. 7. destroyed by sheep sleeping on it.

μήκων, poppy etc., *Papaver* spp. etc. (*see below*)
1. 9. 4. evergreen ; 1. 11. 2. seeds in a vessel; 4. 8. 7. size of flower of κύαμος ὁ Αἰγύπτιος comp.; 4. 8. 10. size of 'head' of λωτός (2) comp.; 4.10.3. σίδη comp.(?); 9.8. 2. juice of 'head' collected ; 9. 12. 3–5. kinds (*see below*) having nothing in common but the name; 9. 16. 9. medical experience ; 9. 20. 1. seeds of one kind of πέπερι comp.

μήκων ἡ Ἡρακλεία (= Ἡρακλεία), *Silene venosa*
9. 12. 5. described : medicinal use.

μήκων ἡ κερατῖτις, horned poppy, *Glaucium flavum* var. *Serpierii*
9. 12. 3. described : medicinal use ; habitat.

μήκων ἡ μέλαινα, *Papaver Rhoeas*
9. 11. 9. mixed with τιθύμαλλος ὁ μυρτίτης to make a medicine.

μήκων (ἡ ὀπώδης), opium poppy, *Papaver somniferum*
1. 12. 2. juice.

μήκων ἡ ῥοιάς, *Papaver hybridum*
9. 12. 4. described : edible : habitat : medicinal use.

μηκώνιον (= τιθύμαλλος 9. 8. 2.), spurge, *Euphorbia Peplus*
9. 8. 2. collection of juice.

μηλέα (fruit μῆλον), apple, *Pyrus Malus*
1. 3. 3. a tree whose stem is not single ; 1. 5. 2. bark smooth : bark readily drops off; 1. 6. 1. core fleshy ; 1. 6. 3. few roots ; 1. 6. 4. shallow rooting ; 1. 8. 4. knots peculiar ; 1. 9. 1. trunk divides low down ; 1. 10. 4. (?) leaves fleshy ; 1. 10. 5. leaves oblong; 1. 11. 4. seeds all together in a single case; 1. 11. 5. seeds in a membrane; 1. 12. 1. taste of fruit ; 1. 12. 2. taste of sap ; 1. 13. 1. flower 'leafy'; 1. 13. 3. flower above fruit-case; 1. 14. 1. bears on last year's wood : some kinds bear also on new wood ; 1.14. 4. many cultivated forms; 2. 1. 2. propagation ; 2. 2. 4. degenerates from seed ; 2. 2. 5. seed produces wild form ; 2. 5. 3. grafting ; 2. 5. 6. trees should be planted fairly close together; 2. 6. 6. some dates round like μῆλα ; 2. 8. 1. apt to shed immature fruit; 3. 3. 1. tree of mountain and plain ; 3. 3. 2. has better fruit and timber in lowlands; 3. 4. 2. time of budding ; 3. 4. 4. time of fruiting ; 3. 11. 5. mountain and lowland forms compared ; 4. 5. 3. abundant in Pontus ; 4. 5. 4. grows on Mount Tmolus and Mysian Olympus ; 4. 7. 7. size of fruit of δένδρον τὸ ἐριόφορον comp.; 4. 10. 2. leaf of ἐλαίαγνος comp. (?); 4. 10. 3.

INDEX OF PLANTS

size of flower of σίδη comp.;
4. 13. 2. short-lived, especially
certain kinds; 4. 13. 3. after
decaying shoots again from
same stock; 4. 14. 2. apt to get
worm-eaten; 4. 14. 10 fruit
gets worm-eaten; 4. 14. 12. uninjured by special winds; 4.16.1.
survives splitting of stem; 5.3.3.
character of wood; 5. 4. 1. the
less fruitful trees produce more
solid wood; 6. 4. 9 'head' of
ἰξίνη comp. to μῆλον.

μηλέα ἡ γλυκεῖα, *Pyrus Malus* var.?
4. 13. 2. specially short-lived;
4. 14. 7. has specially weak constitution; a form of μ. ἡ ἐαρινή;
9. 11. 5. leaf of στρύχνος ὁ ὑπνώδης comp.

μηλέα ἡ ἐαρινή, *Pyrus Malus* var.?
2. 1. 3. propagation; 4. 7. 7. size
of cotton-bearing vessel comp.;
4. 13. 2. specially short-lived;
4. 14. 7. has weak constitution;
(*cf.* μ. ἡ γλυκεῖα).

μηλέα ἡ ὀξεῖα, *Pyrus Malus* var.?
4. 13. 2. comparatively long-lived.

μηλέα ἡ Περσική (Μηδική) citron, *Cit-
rus Medica*
1. 11. 4. seeds in a row; 1. 13. 4.
only pistillate flower fruitful;
4. 4. 2. peculiar to Media and
Persia: described.

μῆλον τὸ Κυδώνιον, *see* Κυδώνιος

μήλωθρον (= ἄμπελος (4)), bryony,
Bryonia cretica
3. 18. 11. fruit of σμῖλαξ (2) comp.;
6. 1. 4. a spineless wild under-
shrub.

μήνανθος, *Limnanthemum nymphoi-
des*
4. 10. 1-2. in list of plants of
Lake Copais; 4. 10. 4. requires
further investigation.

μῖλαξ (= σμῖλαξ (2)), smilax, *Smilax
aspera*
1. 10. 5. leaf described; 1. 10. 6.
leaf with spinous projections;
6. 8. 3. flower used in garlands.

μῖλος, yew, *Taxus baccata*
1. 9. 3. evergreen; 3. 3. 1. a
mountain tree; 3. 3. 3. evergreen; 3. 4. 2. time of budding;
3. 4. 5. time of flowering and
fruiting; 3. 4. 6. time of fruiting; 3. 6. 1. slow growing (?)
3. 10. 2. described; 4. 1. 3. likes
shade; 5. 7. 6. uses of wood.

μίνθη (μίνθα) (= ἡδύοσμον), green
mint, *Mentha viridis*
2. 4. 1. σισύμβριον turns into μ.
unless often transplanted; 6.7.2.
said by some to have no fruit.

μνάσιον (= μαλιναθάλλη), *Cyperus
esculentus*
4. 8. 2. used for food in Egypt;
4. 8. 6. described.

μυάκανθος (= κεντρομυρρίνη), butcher's broom, *Ruscus aculeatus*
6. 5. 1. in list of spinous plants
which have leaves as well as
spines.

μύκης, mushroom etc., *Fungus*
1. 1. 11. has not all the 'parts' of
a plant; 1. 5. 3. stem very
smooth; 1. 6. 5. no roots;
3. 7. 6. grows on roots of trees.

[4. 7. 2. marine growths which
turn to stone];
[4. 14. 3. name given to a disease
of ἐλάα].

μυόφονον (= ἀκόνιτον = θηλύφονον =
σκορπίος (3)), wolf's bane, *Aconitum Anthora*
6. 1. 4. a spineless wild under-
shrub: belongs to 'ferula-like'
plants; 6. 2. 9. do.; has a
fibrous stem.

μυρίκη (1), tamarisk, *Tamarix te-
trandra*
1. 4. 3. 'amphibious'; 1. 9. 3.
evergreen; 1. 10. 4. leaves
fleshy; 3.3. 1. tree of mountain
and plain; 3. 3. 3. evergreen;
3. 16. 4. bark of κόμαρος comp.;
4. 2. 6. (?) leaf of βαλανος comp.;
4. 6. 7. leaf of δρῦς (7) comp;
6. 2. 1. leaf of κνέωρος ὁ μέλας
comp.; 6. 4. 8. flower of χαμαιλέων comp.

μυρίκη (2), tamarisk, *Tamarix artic-
ulata*
5.4.8. Arabian: wood very strong.

μυρρίνη (μύρρινος, μύρτος) (fruit μύρ-
τον), myrtle, *Myrtus communis*
1. 3. 3. effect of not pruning;
1. 9. 3. evergreen; 1. 10. 2.
leaves close-set and opposite;
1. 10. 4. leaves narrow; 1. 10. 8.

INDEX OF PLANTS

leaves regular; 1. 12. 1. taste of fruit; 1. 13. 3. flower above fruit-case; 1. 14. 1. bears on last year's wood: flowers borne on new wood not fertile; 1.14.4. many cultivated forms; 2. 1. 4. propagation; 2. 2. 6. sometimes improves from seed; 2. 5. 6. propagation: trees should be planted close together; 2. 7 2. needs much pruning; 2. 7. 3. requires pungent manure and much water; 3. 6 2. formation of buds; 3. 12. 4. fruit of κέδρος (1) comp.; 3. 15. 5. leaf of πύξος comp.; 3. 16. 4. flower of κόμαρος comp.; 4. 2. 6. (?) leaf of βάλανος comp.; 4. 3. 1. arrangement of fruit of λωτός (4) comp.; 4. 5. 3. does not thrive in cold regions; 4. 5. 4. grows in Propontis; 5. 8. 3. grows in lowland parts of Latium: and on Circeian promontory (a dwarf kind); 6. 8. 5. very fragrant in Egypt; 9. 11. 9. leaf of τιθύμαλλος ὁ μυρτίτης comp.

μῶλυ, moly, *Allium nigrum*
9. 15. 7. localities in Arcadia: said to be like the μ. of Homer: described: use as charm.

ναῖρον ?
9. 7. 3. in list of ἀρώματα.

νᾶπυ, white mustard, *Brassica alba*
1. 12. 1. taste of fruit; 7. 1. 2–3; time of sowing and of germination; 7. 3. 2. seeds described; 7. 5. 5. seed keeps well.

νάρδον, spikenard, *Nardostachys Jatamansi*
9. 7. 2. an Indian ἄρωμα; 9. 7. 3. in list of ἀρώματα; 9. 7. 4. an unnamed Thracian plant (*see* App. (25)) comp.

ναρθηκία (= νάρθηξ *see* 6.2.7.), ferula, *Ferula communis*
6.1.4 spineless: belongs to 'ferulalike, plants'; 6. 2. 7. perhaps differs only in size from νάρθηξ; 6. 2. 8. described.

νάρθηξ (= ναρθηκία *see* 6. 2. 7.), ferula, *Ferula communis*
1. 2. 7. flesh turns to wood; 1.6.1. core fleshy; 1. 6. 2. core membranous; 6.2.7. perhaps differs only in size from ναρθηκία; 6. 2. 8. described; 6. 3. 1. stalk of σίλφιον comp.; 9. 9. 6. stem of θαψία comp.; 9. 10. 1. leaf of both ἐλλέβοροι comp. by some; 9. 16. 2. δίκταμνον kept ἐν νάρθηκι.

νάρκισσος (1) (= λείριον (2) 6. 6. 9.), narcissus, *Narcissus serotinus*
6.6.9. a coronary plant: described; 7.13.1. leaves described; 7.13.2. no stem except the flower-stem; 7. 13. 5–7. stem appears before leaves, viz. flower-stem: sequence described and comp. with σκίλλα.

νάρκισσος (2), pheasant's eye narcissus, *Narcissus poeticus*
6. 8. 1. flowering time.

νάρτη ?
9. 7. 3. in list of ἀρώματα.

(νηπενθές) = μήκων ἡ ὑπώδης, opium poppy, *Papaver somniferum*.
9. 15. 1. mythical: supposed effects.

νυμφαία (= μαδωνάϊς 9.13.1.), yellow water-lily, *Nuphar luteum*
9. 13. 1. fragrant: habitat and localities: leaf described: medicinal use: called μαδωνάϊς in Boeotia.

ξίρις, gladwyn, *Iris foetidissima*
9.8.7. superstition as to gathering.
ξίφιον (= ξίφος 7.13.1. = φάσγανον), corn-flag, *Gladiolus segetum*
6. 8. 1. flowering time; 7. 13. 2. flower-stem not the only stem.
ξίφος (= ξίφιον 7. 13. 1. = φάσγανον), corn-flag, *Gladiolus segetum*.

ὄγχνη, wild pear, *Pyrus communis* var. *Pyraster*
2. 5. 6. trees should be planted rather far apart.

ὄη (οἴη), sorb, *Sorbus domestica*
2. 2 10. becomes sterile in a warm place; 2. 7. 7. 'correcting' the tree; 3. 2. 1. fruit sweeter and better ripened in wild than in cultivated form; 3. 5. 5. winterbuds; 3. 6. 5. roots shallow but strong: thick according to Arcadians; 3 11. 3. leaf of μελία

INDEX OF PLANTS

comp.; 3. 12. 6–9. described; 3. 15. 4. leaf of τέρμινθος comp.

οἰνάνθη (1), drop-wort, *Spiraea Filipendula*
6. 6. 11. a coronary plant: grown from seed; 6. 8. 1–2. flowering time: flower described.

οἰνάνθη (2) ἡ ἀγρία, wild vine, *Vitis silvestris*
5. 9. 6. ἀθραγένη comp.

οἶσος (= ἄγνος), withy, *Vitex Agnus-castus*
3. 18. 1–2. has two forms, 'white' and 'black'; 6. 2 2. used for tying-up.

ὀλόσχοινος, see σχοῖνος ὁ ὁλ.

ὀλύρα, (cultural variety of ζειά), rice-wheat, *Triticum dicoccum*
8. 1. 3. sown early; 8. 4. 1. comp. in detail with other cereals; 8. 9 2. does not exhaust the soil much: reason.

ὀνοθήρας (= δάφνη ἡ ἀγρία), oleander, *Nerium Oleander*
9. 19. 1. effect on 'mind': described.

ὀνόπυξος, *Onopordon illyricum*
6. 4. 3. a 'thistle-like' plant.

ὀνοχειλές, bugloss, *Echium diffusum*
7. 10. 3. flowers borne in succession.

ὀνωνίς, rest-harrow, *Ononis antiquorum*
6. 1. 3. has leaves as well as spines: a wild under-shrub; 6. 5. 1. in list of spinous plants which have leaves as well as spines; 6. 5. 3–4. described: troublesome to farmers.

ὀξυάκανθος, cotoneaster, *Cotoneaster Pyracantha*
1. 9. 3. evergreen; 3. 3. 1. tree of mountain and plain; 3. 3. 3. evergreen; 3. 4. 2. time of budding; 3. 4. 4. time of fruiting; 4.4.2. thorns of μηλέα ἡ Περσική comp.; 6. 8 3. fruit used in garlands.

ὀξύη (ὀξύα), beech, *Fagus silvatica*
3. 3. 8. doubt whether it has a flower; 3. 6. 5. roots few slender and 'plain' according to Arcadians: shallow - rooting; 3. 10. 1. described; 3. 11. 5. mountain and lowland forms compared; 5. 1. 2. time of cutting timber; 5. 1. 4. do.; 5. 4. 4. wood does not decay in water; 5. 6. 4. wood 'moist': used for elastic bedsteads; 5. 7. 2. wood used for keel etc. of small vessels; 5. 7. 6. other uses of wood; 5. 8. 3. grows very fine in lowland part of Latium.

ὀξύκεδρος (= κέδρος (1) 3. 12. 3.), prickly cedar, *Juniperus Oxycedrus*
3. 12. 3. some, who call ἄρκευθος a κέδρος, distinguish κέδρος (1) as ὀξύκεδρος.

ὑπιτίων, ?
7. 13. 9. (in defective sentence) belongs to τὰ βολβώδη.

ὑποβάλσαμον, see βάλσαμον.

ὀρεοσέλινον, parsley, *Petroselinum sativum*
7. 6. 3–4. distinguished from other forms of σέλινον: medicinal use.

ὀρειπτελέα, wych-elm, *Ulmus montana*
3.14.1. distinguished from πτελέα.

ὀρίγανον (ὀρίγανος) (= ὁ. ἡ μέλαινα), marjoram, *Origanum viride* etc.
1. 9. 4. evergreen (partly); 1. 12. 1. taste of fruit; 6. 1. 4. a spineless wild under-shrub; 6. 2. 3. two forms, 'black' and 'white' (*see below*): seed conspicuous: not, like θύμος, particular as to situation; 7. 1. 3. time of germination; 7. 1. 6. germination; 7. 2. 1. propagation; 7. 6 1. wild form distinguished.

ὀρίγανος ἡ λευκή, marjoram, *Origanum heracleoticum*
6. 2. 3. distinguished from ὀ. ἡ μέλαινα.

ὀρίγανος ἡ μέλαινα (= ὀρίγανον), marjoram, *Origanum viride*
6. 2. 3. distinguished from ὀ. ἡ λευκή.

ὅρμινον, *Salvia Horminum*
8. 1. 4. sown later than cereals and pulses; 8. 7. 3. doubtful if eaten green by animals: described: sown at same time as σησάμη.

ὀροβάγχη, dodder, *Cuscuta europaea*
8. 8. 4. grows specially among ὄροβοι: reason: ἀπαρίνη comp.

467

INDEX OF PLANTS

ὄροβος, bitter vetch, *Ervum Ervilia*
2. 4. 2. more digestible if sown in spring; 7. 5. 4. used to prevent ψύλλαι in ῥαφανίς; 7.6.3. size of fruit of ἱπποσέλινον comp.: 8.1.4. sown both early and late; 8.2.5. flowering time; 8. 3. 2. stem; 8. 5. 1. more than one kind: white form sweetest; 8. 5. 2. seeds not in compartments; 8. 5. 3. shape of pod; 8. 8. 4. ὀροβάγχη grows specially among ὀ.; 8. 10. 1. a pest; 8. 11. 2. seed keeps well; 8. 11. 6. do. specially in hill-country; 9.20.1. shape of one kind of πέπερι (fruit) comp.

ὄρτυξ (= στελέφουρος according to some, 7. 11. 2.), plantain, *Plantago Lagopus*.

ὄρυζον, rice, *Oryza sativa*
4. 4. 10. described.

ὄρχις (1) (μέγας), orchis, *Orchis papilionacea*
9.18.3. properties: leaf and stalk.

ὄρχις (2) (μικρός), orchis, *Orchis longicruris*
9.18.3. properties: leaf and stalk.

ὀστρύα (ὀστρυίς) (ὀστρυίς = ὄστρυς 3.10.3.), hop-hornbeam, *Ostrya carpinifolia*
1. 8. 2. ' male ' has more knots than ' female '; 3. 3. 1. tree of mountain and plain; 3. 6. 1. slow-growing (?); 3. 10. 3. described.

ὄστρυς (= ὀστρύα 3.10.3.), hop-hornbeam, *Ostrya carpinifolia*
3. 10 3. described.

οὔιγγον, *Colocasia antiquorum*
1.1.7. 'fruit' underground; 1.6.9. grows underground; 1. 6. 11. described.

πάδος (? = πηδός (?)), *Prunus Mahaleb*
4. 1. 3. likes shade.

παιωνία (= γλυκυσίδη 9.8. 6.), peony, *Paeonia officinalis*
9. 8. 6. superstition as to time of digging.

παλίουρος (?), Christ's thorn, *Paliurus australis*
1. 3. 1. a typical 'shrub'; 1. 3. 2. becomes tree-like; 1.5.3. thorns on wood; 1. 10. 6. leaf with spinous projections; 1. 10. 7. stem presently spinous; 3. 3. 1. tree of mountain and plain; 3. 4. 2. time of budding; 3. 4. 4. time of fruiting; 3. 11. 2. fruit of σφένδαμνος comp.; 3. 18. 3. kinds: described; 4. 8. 1. to some extent grows in marshes; 4.12.4. to some extent aquatic; 6.1.3. has spines on the shoots.

παλίουρος (2) (ὁ Αἰγύπτιος), *Zizyphus Spina-Christi*
4. 3. 1-2. common in Libya; 4. 3. 3. described: distinguished from π. of Hellas.

πανάκεια (= πάνακες τὸ Ἡράκλειον), *Opopanax hispidus*
9. 15. 7. localities.

πάνακες (τὸ Σύριον ? 9. 7. 2: 9. 10. 1.), (juice χαλβάνη (?) 9.7.2: 9.9.2., *see* note), all-heal, *Ferulago galbanifera*
9. 1. 2. in list of plants whose juice is a gum; 9. 7. 2. Syrian: χαλβάνη made from π.; 9. 7. 3. in list of ἀρώματα; 9. 9. 1. root fruit and juice used; 9. 9. 2. uses for medicine and perfume; 9. 11. 1. kinds (*see below*); 9. 11. 4. two further kinds, one fine-leaved, the other not: medicinal use.

πάνακες τὸ Ἀσκληπίειον, *Ferula nodosa*
9. 8. 7. superstition as to gathering; 9. 11. 1. described: medicinal use.

πάνακες τὸ Ἡράκλειον (= πανάκεια), *Opopanax hispidus*
9. 11. 1. in list of kinds of π.; 9. 11. 3. described: medicinal use.

πάνακες τὸ Χειρώνειον, elecampane, *Inula Helenium*
9. 11. 1. described: habitat: medicinal use.

παντάδουσα, star-thistle, *Centaurea Calcitrapa*
6. 5. 1. in list of spinous plants which have leaves as well as spines.

πάπυρος (stalk πάπυρος), papyrus, *Cyperus Papyrus*
4. 8. 2. useful for food in Egypt; 4. 8. 3-4. described: uses;

INDEX OF PLANTS

4. 8. 5. stem of σάρι comp; 6. 3. 1. belongs to 'ferula-like' plants.

παρθένιον, bachelor's buttons, *Pyrethrum Parthenium*
7. 7. 2. a λάχανον: needs cooking

πέζις, bullfist, *Lycoperdon Bovista*
1. 6. 5. no roots.

πελεκῖνος, axe-weed, *Securigera Coronilla*
8. 8. 3. grows specially among ἀφάκη: name explained.

πεντατετές (=πεντάφυλλον 9.13. 5.), cinquefoil, *Potentilla reptans*
9. 13. 5. described.

πεντάφυλλον (=πεντατετές 9. 13. 5.), cinquefoil, *Potentilla reptans*.

(πέπερι), pepper, *Piper nigrum*
9. 20. 1. a fruit: two forms: described: properties: antidote to Κνίδιος κόκκος comp.

περδίκιον, 'partridge-plant,' *Polygonum maritimum*
1. 6. 11. large fleshy roots.

περιττός (? στρύχνος ὁ περιττός) (= στρύχνος ὁ μανικός 9. 11. 6.), thorn-apple, *Datura Stramonium*.

περσέα (=πέρσιον), *Mimusops Schimperi*
3. 3. 5. not fruitful everywhere; 4. 2. 1. peculiar to Egypt; 4. 2. 5. described; 4. 2. 8. common in Thebaid.

πέρσιον (=περσέα), *Mimusops Schimperi*
2. 2. 10. effects of climate.

πευκέδανον, sulphur-wort, *Peucedanum officinale*
9. 14. 1. how long drug will keep; 9. 15. 1. grows in Arcadia; 9. 20. 2. properties of root: use in medicine: grows in Arcadia.

πεύκη, fir, *Pinus* spp.
1. 3. 6. refuses cultivation; 1.5. 1. erect and tall; 1. 5. 4. wood has many knots; 1.6.1. core woody; 1. 6. 3. root single; 1. 6. 5. roots not branching; 1. 8. 1. many knots; 1. 9. 3. evergreen (the wild and one cultivated kind); 1. 10. 4. leaves like teeth of comb (?); 1.10.6. leaf spinous at tip; 1. 12. 1. taste of fruit; 1. 12. 2. taste of sap; 2. 2. 2. propagated only by seed; 2.5.2. instance of very long roots; 3. 1. 2. grows only from seed: 3. 2. 3. evidence that it is really wild; 3. 3. 1. a mountain tree; 3. 3. 3. evergreen; 3. 3. 8. doubt whether it has a flower; 3. 4. 5. time of budding and fruiting; 3. 4. 6. time of fruiting; 3. 5. 1. periods of budding; 3. 5. 3. do.; 3. 5. 5. winter-buds; 3. 5. 6. cone; 3. 6. 1. quick growing: even young tree fruits; 3. 6. 4. not deep-rooting; 3. 7. 1. dies if topped; 3. 7. 3. produces a 'tuft' (κύτταρος); 3. 9. 1–8. kinds according to various authorities (*see below*): distinction from πίτυς; 3. 9. 4. timber, foliage; 3. 9. 5. further distinction from πίτυς: the disease 'pitch-glut'; 3.9.7. comparison with ἐλάτη; 3. 9. 8. do.: core and callus; 4. 1. 1. likes sun; 4. 1. 2. in shade has inferior timber; 4.5.1. in list of Northern trees; 4. 5. 3. does not grow in Pontus; 4. 15. 3. effects of stripping bark at various seasons; 4. 16. 1. topping fatal; 4.16. 1–2. not injured by cutting for tar; 4. 16. 4. said to perish if entirely deprived of its heartwood; 5. 1. 2. time of cutting timber; 5. 1. 4. do.; 5. 1. 9–10. methods of cleaving; 5. 4. 2. wood (when resinous) proof against decay; 5. 4. 4. more eaten by *teredon* than ἐλάτη: 5. 4. 8. effect of salt water on different parts; 5. 5. 1. knotty parts of wood hard to work; 5. 6. 1. wood good for struts: behaviour under pressure; 5.6.2. takes glue best of all woods; 5. 7. 1–2. uses of wood in shipbuilding; 5. 7. 4–5. uses in house-building and crafts; 5.8.1. grows to great size in Latium, but finer still in Corsica; 5. 8. 3. grows in hill-country of Latium; 5. 9. 3. charcoal of this wood preferred by smiths to that of δρῦς; 9. 1. 2. sap gummy; 9.1.6. time of tapping; 9. 2. 1. pro-

469

INDEX OF PLANTS

ductive of resin (ῥητίνη); 9.2.2; quality of resin; 9.2.3-4. Macedonians only burn the 'male' for pitch (πίττα), and the roots of the 'female'; aspect etc. required for production of good pitch; 9.2.5. Idaean account different (see π. ἡ Ἰδαία and π. ἡ παραλία); 9.2.6. filling up the holes; 9.2.7. do. Idaean account; 9.2.8. further rules for collecting pitch: age of tree; etc.; 9.3.1–3. method of preparing pitch; 9.1.4. does not grow in Syria.

πεύκη ἡ ἄκαρπος (= π. ἡ θήλεια = π. ἡ Ἰδαία), Corsican pine, *Pinus Laricio*
3.9.2. described; 3.9.4. one of three wild kinds (Idaean account).

πεύκη ἡ ἄκαρπος ('male'), Corsican pine, *Pinus Laricio*
3.9.2. comp. with 'female.'

πεύκη ἡ ἄκαρπος ('female'), Aleppo pine, *Pinus halepensis*
3.9.2. comp. with 'male.'

πεύκη ἡ ἄρρην (= π. ἡ παραλία = πίτυς in 3.9.5.), Aleppo pine, *Pinus halepensis*
3.9.3. timber: produces συκῆ (Mt. Ida); 3.9.4. one of three wild kinds (Idaean account).

πεύκη ἡ ἥμερος (= [π. ἡ] κωνόφορος), stone pine, *Pinus Pinea*
3.9.1. distinguished from other kinds; 3.9.4. Arcadians say it is a πίτυς: timber, foliage, resin.

πεύκη ἡ θήλεια (= π. ἡ ἄκαρπος = π. ἡ Ἰδαία), Corsican pine, *Pinus Laricio*
3.9.3. timber: contains αἰγίς; 3.9.4. one of three wild kinds (Idaean account).

πεύκη ἡ Ἰδαία (= π. ἡ ἄκαρπος = π. ἡ θήλεια), Corsican pine, *Pinus Laricio*
3.9.1-2, described; 9.2.5. Idaean account of pitch (πίττα).

[πεύκη ἡ] κωνόφορος (= π. ἡ ἥμερος), stone pine, *Pinus pinea*
2.2.6. seeds true; 3.9.4. foliage: pitch (πίττα): Arcadians say it is a πίτυς.

πεύκη ἡ παραλία (= π. ἡ ἄρρην = πίτυς in 3.9.5), Aleppo pine, *Pinus halepensis*
3.9.1. described; 9.2.5. Idaean account of pitch (πίττα).

πήγανον (πηγάνιον), rue, *Ruta graveolens*
1.3.1. a typical under-shrub; 1.3.4. becomes tree-like; 1.9.4. evergreen; 1.10.4. leaves fleshy; 2.1.3. propagation; 6.1.1. may be classed as an under-shrub; 6.5.3. leaf of ὀνωνίς comp.; 6.7.3. strong plants of ἀβρότονον comp.; 7.2.1. propagation: seed slow to germinate; 7.4.1. only one kind; 7.5.1. dislikes manure; 7.6.1. wild form distinguished; 9.4.2. colour of leaf of λιβανωτός comp.; 9.5.1 leaf of βάλσαμον comp.; 9.9.6. leaf of ἰσχάς (ἄπιος (2)) comp.

πηδός (?) (? = πάδος 4.1.3.), *Prunus Mahaleb*
5.7.6. uses of wood.

πικρίς, *Urospermum picroeides*
7.11.4. inedible: flowers in spring, but also throughout winter and summer.

πῖλος, *Polyporus igniarius* (?)
3.7.4. produced by δρῦς; described.

πισός, pea, *Pisum sativum*
8.1.1. in list of pulses; 8.1.4 sown late; 8.2.3. comes up with several leaves; 8.3.1. leaf; 8.3.2. stem; 8.5.2. seeds not in compartments; 8.5.3. shape of pod; 8.10.5. infested by 'worms.'

πίτυς (= πίτυς ἡ ἀγρία = πεύκη ἡ ἄρρην in 3.9.5. = πεύκη ἡ παραλία in 3.9.5.), Aleppo pine, *Pinus halepensis*
1.6.1. core woody; 1.10.4. leaves like teeth of comb; 1.10.6. leaf spinous at tip; 1.12.1. taste of fruit; 2.2.2. propagated only by seed; 3.1.2. do.; 3.3.3. evergreen; 3.4.5. time of budding and fruiting; 3.5.5. winterbuds; 3.6.1. slow growing (?);

470

INDEX OF PLANTS

even young tree fruits; 3.9 4–8.
πεύκη and πίτυς; 3 9. 6. distinction from πεύκη; 3. 11. 1.
bark of σφένδαμνος comp.; 3.17.1.
bark of φελλός comp.; 4. 5. 3.
does not grow in Pontus:
4. 14. 8. if topped, becomes
barren, but is not destroyed;
4. 16. 1. topping fatal; 5. 1. 2.
time of cutting timber; 5. 1. 4.
do.; 5. 1. 5–6. timber comp.
with ἐλάτη; 5 7. 1. used in
Cyprus for ship-building instead
of πεύκη; 5. 7. 3. wood used for
bent-wood work in triremes;
5. 7. 5. use of wood in ship-
building and house-building:
soon rots; 5. 7. 8. use of wood
for carpenter's tools; 5. 9. 2.
charcoal of this wood used in
silver mines; 9.1.2. sap gummy;
9. 2. 1. production of resin
(ῥητίνη); 9. 2. 2. quality of
resin; 9. 2. 3. said to be burnt
for pitch (πίττα) in Syria.

πίτυς ἡ ἀγρία (= πίτυς = πεύκη ἡ
ἄρρην = πεύκη ἡ παραλία in 3.9.5.)
Pinus halepensis (mountain
form)
1. 9. 3 evergreen; 3. 3. 1. a
mountain tree (Macedonian).

πίτυς ἡ φθειροποιός, *Pinus brutia*
2. 2. 6. seeds come true.

πλάτανος, plane, *Platanus orientalis*
1. 4. 2. lives near water; 1. 6. 3.
roots many and long; 1 7. 1.
example of long roots; 1. 8. 5.
diseased formation (κραδή);
1. 9. 5. evergreen specimens;
1. 10. 4. leaves broad; 1.10. 7.
attachment of leaf-stalk; 3.1.1.
propagation; 3. 1 3. produces
seed and seedlings; 3.3. 3. evergreen
in some places; 3 4 2
time of budding; 3. 6. 1. quick
growing; 3. 11 1. leaf of σφένδαμνος
comp.; 3. 11. 4. has a
sort of winter-bud like that
of μελία; 4. 5. 6. found at
only one place on Adriatic
coast: rare in Italy; 4. 5. 7.
common in some Mediterranean
regions; 4. 7. 4. size of unnamed
Arabian tree (see App. 12a)
comp.; 4. 8. 1. grows partially

in water; not common on Nile;
4. 13. 2; trees said to have been
planted by Agamemnon; 4.15.2.
bark cracks; 4. 16. 2. grows
again after being cut or blown
down: instances; 5. 3. 4. character
of wood; 5. 7. 3. wood
used for bent-wood work: soon
decays; 5. 9. 4. wood makes an
evil smoke when burnt for charcoal;
9. 11. 6. 'head' of στρύχνος ὁ μανικός comp. to fruit of π.

πλατύφυλλος (δρῦς) see δρῦς (6).

πόα, grass
7. 8. 3. leaves 'on the ground.'

πόα ἡ Μηδική, see Μηδική.

πόθος (1), larkspur, *Delphinium orientale*
6. 8. 3. a coronary plant: flowers
in summer: flower like ὑάκινθος.

πόθος (2) (= ἀσφόδελος), asphodel,
Asphodelus ramosus
6. 8. 3. a coronary plant: flowers
in summer: flower white: used
in connexion with funerals.

πόλιον, hulwort, *Teucrium Polium*
1. 10. 4. leaves fleshy: prevents
moth in clothes; 2. 8. 3. used
for caprification; 7 10. 5. evergreen.

πολυάκανθος, *Carduus acanthoides*
6 4. 3. a 'thistle-like' plant.

πολυπόδιον, polypody, *Polypodium vulgare*
9.13.6. peculiar shape: described:
named from cuttle-fish (πολύπους),
and used as charm to
prevent polyp (πολύπους): other
medicinal use: habitat; 9.20.4.
comes up after rain: has no
seed.

πράσιον, *Marrubium* spp.
6. 1. 4. a spineless wild undershrub;
6. 2. 5. two kinds;
see below.

πράσιον (1), horehound, *Marrubium peregrinum*
6. 2. 5. leaf described: used by
druggists.

πράσιον (2), horehound, *Marrubium vulgare*
6. 2. 5. leaf described.

πράσον (1), leek, *Allium Porrum*
7. 1. 2–3. time of sowing and of
germination; 7. 1. 6. germina-

471

INDEX OF PLANTS

tion; 7. 1. 7. bears fruit in second year: stem single; 7. 2. 2 3. root makes offsets; 7. 3. 4. seed borne at top: method of sowing; 7. 4. 10. coat often like γήτειον; 7. 4. 11. size of 'head' of one year old σκόροδον comp.; 7. 5. 3. improved by transplanting; 7. 5. 4. pests; 7. 5. 5. seed keeps well; 7. 8. 2. stem smooth, not branched; 9. 10. 1. leaf of ἐλλέβορος ὁ λευκός comp. by some.

πράσον (2) (= ζώστηρ 4. 6. 2. = φῦκος (1)); grass-wrack, *Posidonia oceanica*
4. 6. 2. = ζώστηρ, *q.v.*

πράσον (3) (=φῦκος (2)), riband-weed, *Laminaria saccharina*
4. 6. 4. carried by current from Atlantic into Mediterranean: described; 4.7.1. refers to 4 6.4.

πρῖνος, kermes-oak, *Quercus coccifera*
1. 6. 1. core hard and close; 1.6.2. core large and conspicuous; 1. 9. 3. evergreen; 1. 10. 6. leaves with spinous projections; 3. 3. 1. a mountain tree; 3.3.3. evergreen; 3. 3. 6. does not always fruit; 3. 4. 1. takes a year to ripen fruit; 3. 4. 4–6; time of fruiting; 3. 6. 4. deep rooting; 3. 7. 3. produces a scarlet 'berry'; 3. 16. 1. described; 3. 16. 2. σμῖλαξ (1) comp.; 3. 16. 3. φελλόδρυς δρῦς and π. comp.; 3. 16. 4. leaf of κόμαρος comp.; 4. 3. 1. leaf of λωτός (4) comp.; effect of stripping bark in winter; 5. 4. 8. wood of μυρίκη (2) comp. for strength; 5. 5. 4. core not obvious, but exists; 5. 7. 6. uses of wood; 5. 9. 7. wood used for fire-sticks; 9. 4. 3. leaf of σμύρνα comp. by some.

προύμνη (= σποδίας), bullace, *Prunus insititia*
9. 1. 2. sap gummy.

πτελέα, elm, *Ulmus glabra*
1.8.5. diseased formation (κραδή); 1. 10. 1. leaves inverted in summer; 1. 10. 6. leaves notched; 3. 1. 1. propagation; 3. 1. 2. seems to have no fruit yet re-

produces itself: instance; 3.1.3. do.: proof; 3. 3. 1. tree of mountain and plain; 3. 3. 4. a question if it bears fruit; 3.4.2. time of budding; 3. 6. 1. quick growing; 3. 7. 3. produces a βότρυς and leaf-galls; 3. 11. 5. mountain and lowland forms comp.; 3. 14. 1. description: kinds; 3. 15. 4. leaf-galls of τέρμινθος comp.; 3. 17. 5. leaf of Idaean κολοιτία comp.; 3. 17. 5. leaf of συκῆ ἡ Ἰδαία comp.; 3.18.5. leaf of ῥοῦς comp.; 4. 2. 3. leaf of συκῆ ἡ Κυπρία comp.; 4. 5. 3. grows in Pontus; 4. 5. 7. common in some Mediterranean regions; 4. 9 2. leaf of τρίβολος (3) comp.; 4. 15. 2. survives stripping of bark; 5. 1. 2. time of cutting timber; 5. 3. 4. character of wood; 5. 3. 5. method of making door-hinges of the wood; 5. 4. 3. wood does not decay if exposed to air; 5. 6. 4. wood strong: used for door-hinges; 5. 7. 3. wood used for bent-wood work: use in shipbuilding; 5. 7. 6. other uses of wood; 5. 7. 8. uses of wood for carpenter's tools; 9. 1. 2. sap gummy: gum contained in the 'bag'; 9. 4. 3. leaf of σμύρνα comp. (by some).

πτερίς, fern, *Nephrodium Filix-mas*
1. 10. 5. frond described; 4. 2. 11. leaf of an unnamed Memphian shrub (*see* App. (2)) comp.; 8.7.7. destroyed by manure or by sheep sleeping on it; 9. 13. 6. leaf of πολυπόδιον comp. to π. ἡ μεγάλη; 9. 18. 8. distinguished from θηλύπτερις; 9. 20. 5. root only of use: medicinal use: time of gathering.

πύξος, box, *Buxus sempervirens*
1. 5. 4. wood heavy; 1. 5. 5. do. because of close grain; 1. 6. 2. core not conspicuous; 1. 8. 2. few knots; 1. 9. 3. evergreen; 3. 3. 1. a mountain tree; 3. 3. 3. evergreen; 3. 4 6. time of fruiting: fruit inedible; 3. 6. 1. slow growing (?); 3.15.5. described; 4. 4. 1. hard to grow in Baby-

INDEX OF PLANTS

lon; 4. 5. 1. in list of Northern trees; 5. 3. 1. wood very close and heavy; 5. 3. 7. images made from the wood; 5. 4. 1. wood hard and heavy; 5. 4. 2. wood proof against decay; 5. 4. 5. wood not attacked by σκώληξ; 5. 5. 2. core not obvious: wherefore wood not apt to 'draw'; 5. 5. 4. core not obvious but exists; 5. 7. 7. uses of wood: that grown on Mount Olympus useless; 5. 7. 8. uses of wood for carpenter's tools; 9. 20. 4. wood of ἔβενος comp.

πυρός, wheat, *Triticum vulgare*
1.5.2. 'bark' fibrous; 1.6.5. roots numerous; 1. 6. 6. do.; 1. 11. 2. seeds in a husk; 1 11. 5. each seed separately attached; 2.2.9. said to turn sometimes into κριθή; 2.4.1. turns into αἶρα: wild π. turns into cultivated with cultivation; 4.4.9. πυροί grow in India; 4. 10. 3. taste of seeds of σίδη comp.; 7.11.2. inflorescence and general appearance of στελέφουρος comp.; 8. 1. 1. in list of cereals; 8. 1. 3. sown early, but after κριθή; 8. 1. 4. one kind sown late; 8. 1. 5. time of germination; 8. 2. 1. germination described; 8. 2. 3. single leaf first appears: roots described; 8. 2. 6. time of maturing seed; 8. 2. 7. time of harvest in Hellas and in Egypt; 8. 3. 2. stem; 8. 4. 1-2. compared in detail with other cereals; 8. 4. 3-5. kinds distinguished, local and other (*see below*); 8. 6. 1. conditions for sowing; 8. 6. 4. suitable soil; 8. 6. 5-6. rain hurtful when π. is in flower: and when it is ripe, but less so than to κριθή; 8. 7. 1. said to change into αἶρα under certain conditions; 8. 7. 4. effect of cutting down or grazing young crop in Thessaly and in Babylon; 8.7.5. in many places comes up again next year; etc.; 8.8.2. favourable localities; 8. 8. 3. degenerates into αἶρα,—or else αἶρα is specially apt to grow among π.;
8. 9. 1. exhausts the soil most of cereals; 8. 10. 1. a pest of π.; 8. 10. 2. wheat-rust; 8. 10. 3. effects of weather; 8.10.4. effects of 'worms' in various localities; 8. 11. 1. seed keeps better than κριθή; 8. 11. 3. grain stored without drying; 8. 11. 7. effect of mixing earth with the grain in some places: at Babylon grain jumps on the threshing-floor: reason.

πυρὸς ὁ Αἰγύπτιος, *Triticum vulgare* var.
8. 4. 3. in list of varieties of π.; 8. 4. 6. escapes αἶρα.

πυρὸς ὁ Ἀλεξάνδρειος, *Triticum vulgare* var.
8. 4. 3. in list of varieties.

πυρὸς ὁ Ἀσσύριος, *Triticum vulgare* var.
8. 4. 3. in list of varieties.

πυρὸς ὁ Βοιώτιος, *Triticum vulgare* var.
8. 4. 5. heaviest grain.

πυρὸς ὁ Θράκιος, *Triticum vulgare* var.
8. 4. 3. grain has many coats.

πυρὸς ὁ καχρυδίας
8. 4. 3. thick stem.

πυρὸς ὁ κριθανίας
8. 2. 3. branching.

πυρὸς ὁ Λακωνικός
8. 4. 5. grain light.

πυρὸς ὁ Λιβυκός
8. 4. 3. grain not long in husk thick stem.

πυρὸς ὁ Ποντικός
8. 4. 3.-4. lightest grain; 8. 4. 5. variation in grain; 8.4.6. escapes αἶρα.

πυρὸς ὁ σιτανίας
8. 2. 3. branching.

πυρὸς ὁ Σικελός
8. 4. 3., 8. 4. 5. heaviest grain of kinds imported to Hellas; 8.4.6. fairly free from αἶρα, especially that of Akragas: infested with μελάμπυρον.

πυρὸς ὁ στλεγγύς
8. 4. 3. in list of varieties, *see note*

ῥάμνος, buckthorn, *Rhamnus* spp.
1. 5. 3. stem fleshy; 1. 9. 4. evergreen; 3. 18. 12. cluster of ber-

473

INDEX OF PLANTS

ries of σμῖλαξ (2) comp.; 5. 9. 7. wood used for fire-sticks, especially for the stationary piece.

ῥάμνος ἡ λευκή, buckthorn, *Rhamnus graeca*
3. 18. 2. distinguished from ῥ. ἡ μέλαινα.

ῥάμνος ἡ μέλαινα, buckthorn, *Rhamnus oleoides*
3. 18. 2. distinguished from ῥ. ἡ λευκή.

ῥαφανίς, radish, *Raphanus sativus*
1. 2. 7. flesh of root turns to wood; 1. 6. 6. root fleshy; 1. 6. 7. root of 'bark' and flesh; 7. 1. 2–3. time of sowing and of germination; 7.1. 5. do.; 7. 1. 7. germination; 7. 2. 5. survives and increases in size under a heap of soil; 7. 2. 5–6. root described; 7. 2. 8. do.; 7. 3. 2. seeds described; 7. 3. 4. seed borne at side; 7. 4. 1–2. several kinds (described) Κορινθία, Κλεωναία, λειοθασία, (or Θρᾳκία), ἀμωρέα (see below), Βοιωτία, and one with leaf like εὔζωμον; 7.4.3. effects of weather; 7. 5. 3. improved by transplanting; 7.5.4. pests; 7. 6. 2. root of wild γογγυλίς comp.; 7. 6. 3. root of ἱπποσέλινον comp.; 7. 8. 2. stem branched; 9. 9. 1. method of cutting root of μανδραγόρας (1) comp.; 9. 12. 1. method of cutting up χαμαιλέων ὁ λευκός for medicinal use comp.

ῥαφανίς ἡ ἀμωρέα, horse-radish (?)
7. 4. 2. in list of varieties of ῥ.

ῥάφανος, cabbage, *Brassica cretica*
1. 3. 4. becomes tree-like; 1. 6. 6. root single; 1. 9. 4. evergreen; 1. 10. 4. leaves fleshy; 1. 14. 2. bears fruit on top and at side; 4.4.12. size of an unnamed Asian shrub (see App. (10)) comp.; 4.16.6. spoils flavour of grape: vine-shoot turns away from ῥ; whence use of ῥ. as cure for effects of wine; 6. 1. 2. may be classed as an under-shrub; 7. 1. 2–3. time of sowing and of germination; 7. 2. 1. propagation; 7. 2. 4. grows again when stem is cut; effect on flavour; 7. 4. 1. several kinds; 7. 4. 4. three kinds distinguished, οὐλόφυλλος, λειόφυλλος, ἀγρία (see below); 7. 5. 3. bears transplanting; 7.5.4. pests; 7.6.1–2. wild form distinguished.

ῥάφανος ἡ ἀγρία (= κεράϊς 9. 15. 5.), charlock, *Raphanus Raphanistrum*
7. 4. 4. see ῥάφανος; 7. 6. 1–2. see ῥάφανος; 9. 15. 5. Arcadian: a drug: also called κεράϊς.

ῥάφανος ἡ ὀρεία (= ἄπιος (2) = ἰσχάς), spurge, *Euphorbia Apios*
9. 12. 1. used to kill a pig, mixed with χαμαιλέων ὁ λευκός.

ῥοά (ῥοιά), (flower κύτινος), pomegranate, *Punica Granatum*
1. 3. 3. a tree which has not however a single stem; 1. 5. 1. crooked and low; 1. 6. 1. core fleshy; 1. 6. 3. few roots; 1. 6. 4. shallow rooting; 1. 6. 5. roots branching upwards; 1. 9. 1. much branched; 1. 10. 4. leaves narrow; 1. 10. 10. fruit made of fibre and skin; 1. 11. 4. seeds all together in a single case; 1. 11. 5. each fruit separately attached (?); 1. 11. 6. arrangement of stones; 1. 12. 1. taste of fruit; 1. 13. 1. flower red; 1.13.3. flower above fruit-case; 1. 13. 4. some kinds sterile; 1.13.5. flower described; 1.14.1. bears on last year's wood; 1. 14. 4. many cultivated forms; 2. 1. 2–3. propagation; 2. 2. 4. degenerates from seed; 2. 2. 5. do. details; 2. 2. 7. in some places improves from seed; 2. 2. 9. effects of cultivation; 2. 2. 10. ref. to 2. 2. 9.; 2. 2. 11. effect of good cultivation; 2.3.1. sometimes changes character; 2. 3. 2. ref. to 2. 2. 7.; 2. 3. 3. sometimes bears fruit on the stem; 2. 5. 5. propagation; 2. 5. 6. trees should be planted close together; 2. 6. 8. size of fruit of a kind of φοῖνιξ (1) comp.; 2. 6. 12. cuttings set upside down; 2.7.1. water-loving; 2.7.3. requires pungent manure and much water; 2. 8. 1. apt to

474

INDEX OF PLANTS

shed immature fruit; 3. 5. 4. autumn budding; 3. 6. 2. formation of buds; 3. 18. 4. fruit and growth of κυνόσβατος comp.; 3.18.13. size and leaf of ‘εὐώνυμος comp.; 4. 3. 3. stones eaten with fruit; 4. 5. 3. grows well in Pontus with shelter; 4. 5. 4. grows on Mt. Tmolus and Mysian Olympus; 4. 10. 3. flower of σίδη comp.: seeds of σίδη contrasted; 4. 13. 2. short-lived, especially the stoneless form (*see below*); 4. 13. 3. after decaying shoots again from same stock; 4. 14. 10. fruit gets worm-eaten; 4. 14. 12. uninjured by special winds; 4. 16. 1. survives splitting of stem; 6.1.3. has spines on the shoots; 7.13.4. fruit kept by inserting stalk in bulb of σκίλλα; 9. 5. 2. size of βάλσαμον comp.

ῥοὰ ἡ ἀπύρηνος, *Punica Granatum* var.

4. 13. 2. specially short-lived.

ῥοδωνία (flower ῥόδον, fruit μῆλον 6. 6. 6.), rose, *Rosa centifolia* etc.

1. 9. 4. evergreen; 1. 13. 1. colour alluded to; 1. 13. 2. has a ‘twofold’ flower; 1. 13. 3. flower above fruit-case; 1.13.5. flower of ῥοά comp.; 2. 2. 1. propagation; 4. 8. 7. colour of flower of κύαμος ὁ Αἰγύπτιος comp.; 4.10.3. sepals of σίδη comp.; 6. 1. 1. in list of under-shrubs; 6.1.3. has spines on the shoots; 6. 6. 4–6. a cultivated under-shrub: a coronary plant: many kinds: localities: propagation and cultivation; 6.8.2. flowering time; 6. 8. 5. bush lives five years and then degenerates unless pruned: position and climate important for fragrance: flowers very early in Egypt; 6. 8. 6. blooms well on mountains, but has inferior scent; 9. 19. 1. colour of flower of ὀνοθήρας comp.

ῥόδον τὸ ἄγριον, wild rose, *Rosa dumetorum*

6. 2. 1. flower of κίσθος comp.

ῥοῦς (drug ῥοῦς 3. 18. 5.), sumach *Rhus Coriaria*

3. 18. 1. has more than one form (*see below*); 3. 18. 5. ‘male’ and ‘female’ forms: described: used for dyeing: produces a drug called ῥοῦς.

ῥοῦς ἡ λευκή

3. 18. 2. distinguished from ῥ. ἡ μέλαινα.

ῥοῦς ἡ μέλαινα

3. 18. 2. distinguished from ῥ. ἡ λευκή

ῥύτρος, globe-thistle, *Echinops spinosus*

6. 4. 4. a ‘thistle-like’ plant: branches from the top.

σάρι (stalk σάρι), *Cyperus auricomus*

4. 8. 2. useful for food in Egypt; 4. 8. 5. described.

σέλινον, celery, *Apium graveolens*

1.2.2. takes two years to mature; 1. 6. 6. root single, but with large side-growths; 1.9.4. evergreen (partly); 1. 10. 7. time of leaf-growth; 1. 12. 2. taste of sap; 2. 4. 3. effect of trampling and rolling in seed; 6. 3. 1. leaf of σίλφιον comp.; 7. 1. 2–3. time of sowing and germination; 7. 1. 6. germination; 7. 1. 7. bears fruit in second year; 7. 2. 2. root makes offsets; 7. 2. 5. root described; 7. 2. 8. do.; 7. 3. 4–5. methods of sowing and transplanting; 7. 4. 6. kinds distinguished; 7. 5. 3. bears transplanting; 7.6.3. wild forms (ἱπποσέλινον, ἑλειοσέλινον, ὀρειοσέλινον) distinguished.

σέλινον τὸ ἕλειον (= ἑλειοσέλινον), marsh celery, *Apium graveolens*

4. 8. 1. in list of marsh plants; 9. 11. 1. leaf of λιβανωτὶς ἡ κάρπιμος comp.

σέσελι, hartwort, *Tordylium officinale*

9. 15. 5. an Arcadian drug.

σημύδα (?), Judas-tree, *Cercis Siliquastrum*

3. 14. ¹4. described; 5. 7. 7. (?) wood used for walking-sticks.

475

INDEX OF PLANTS

σήσαμον (seed σήσαμη), sesame, *Sesamum indicum*
1.11.2. seed-vessel; 3.13.6. seeds of berry of ἀκτῆ comp.; 3.18.13. fruit of εὐώνυμος comp.; 4.8.14. size of fruit of an unnamed Egyptian plant (*see* App. (20)) comp.; 6. 5. 3. seed of a kind of τρίβολος comp.; 8.1.1. in list of 'summer crops' distinct from cereals and pulses; 8. 1. 4. sown later than cereals and pulses; 8 2. 6. time of maturing seed; 8. 3. 1. leaf; 8. 3. 2. stem; 8.3.3. flower; 8. 3. 4. seed abundant; 8. 5. 1. a white kind, which is the sweetest; 8. 5. 2. seeds in compartments; 8. 6. 1. rain not beneficial after sowing; 8. 7. 3. not eaten green by any animal: ἐρύσιμον comp.: sown at same time as ὄρμινον; 8. 9. 3. exhausts the soil; 9. 9. 2. fruit of ἐλλέβορος comp.; 9. 14. 4. do.

σίδη, waterlily, *Nymphaea alba*
4. 10. 1–2. in list of plants of Lake Copais; 4. 10. 3–4. described: size of fruit of βούτομος comp. (to seed of σ.); 4. 10. 6. grows only in water; 4. 10. 7. part used for food.

σικύα, bottle-gourd, *Lagenaria vulgaris*
1. 11. 4. seeds in a row; 1. 13. 3. flower attached above fruit; 7. 2. 9. root described; 7. 3. 5. takes shape of vessel in which it is grown.

σίκυος (σίκυον), cucumber, *Cucumis sativus*
1. 10. 10. fruit made of flesh and fibre; 1. 12. 2. taste of sap; 1. 13. 3. flower attached above fruit; 1. 13. 4. some flowers sterile; 2. 7. 5. use of dust; 7. 1. 2–3. time of sowing and germination; 7. 1. 6. germination; 7. 2. 9. root described; 7. 3. 1. long in flower; flower described; 7. 3. 5. effect of soaking seed in milk; 7. 4. 1. several kinds; 7.4. 6. do. viz. Λακωνικός, σκυταλίας, Βοιώτιος; 7. 5. 2. said to dislike rain-water; 7. 5 3. bears trans-planting; 7. 5. 5. seed does not keep well; 7. 5. 6. seed not liable to pests; 7. 13. 1. leaf of ἄρον comp.

σίκυος ὁ ἄγριος, (drug ἐλατήριον 9. 9. 4.), squirting cucumber, *Ecballium Elaterium*
4. 5. 1. in list of Northern plants; 7. 6. 4. quite distinct from cultivated σ.; 7. 8. 1. stem 'on the ground'; 9. 9. 4. medicinal use: ἐλατήριον made from seed; 9. 4. 1–2. how long drug will keep: conditions; 9. 15. 6. grows in Arcadia.

σίλφιον (leaf μάσπετον 6. 3. 1), (seed φύλλον, μαγύδαρις 6. 3. 4), silphium, *Ferula tingitana*
1. 6. 12. root most characteristic part; 3. 1. 6. comes up spontaneously; 3. 2. 1. fruits better in wild state; 4. 3. 1. grows in Cyrenaica; 4. 3. 7. consideration postponed; 6. 3. 1–2. described: belongs to 'ferula-like' plants: rules as to cutting and treatment; 6. 3. 3. distribution; 6. 3. 4–6. another account, inconsistent in some points; 6.5.2. grows in mountain country; 7. 3. 2. seeds of ἀδράφαξυς comp.; 9. 1. 3. stem and root produce a gum; 9. 1. 4. gum pungent; 9. 1. 7. time of tapping: details about juices of stem and root.

σισύμβριον, bergamot-mint, *Mentha aquatica*
1.3.1. (?) a typical 'under-shrub'; 2. 1. 3. propagation; 2. 4. 1. turns into μίνθη, unless often transplanted; 6. 1. 1. in list of under-shrubs; 6. 6. 2. a cultivated under-shrub; a coronary plant: the whole plant scented; 6. 6. 3. woody: only one form; 6. 7. 2. said by some to have no fruit: but the wild form certainly seeds; 6. 7. 4. roots described; 6. 7. 6. cultivation; 9.16.3. leaf of δίκταμνον (ἕτερον) comp.

σισυρίγχιον, Barbary nut, *Iris Sisyrinchium*
1. 10. 7. attachment of leaves;

476

INDEX OF PLANTS

7. 13. 9 (in defective sentence) belongs to τὰ βολβώδη: peculiar growth of root: upper part edible.

σκαλίας, see κάκτος (2).

σκαμμωνία, scammony, *Convolvulus Scammonia*
4. 5. 1. seeks cold regions; 9.1.3. root produces a gum; 9. 1. 4. gum has medicinal properties; 9. 9. 1. root and juice used; 9. 20. 5. juice only used.

σκάνδιξ, wild chervil, *Scandix Pecten-Veneris*
7.7.1. a λάχανον; a class of plants called σκανδικώδη; 7. 8. 1. stem 'on the ground.'

σκίλλα, squill, *Urginea maritima*
1. 6. 7. root in scales; 1. 6. 8. root fleshy and bark-like: root not tapering; 1. 6. 9. no side roots; 1. 10. 7. no leaf-stalk: attachment of leaves; 2. 5. 5. cuttings of συκῆ etc. set in the bulb. of σ.; 7. 2. 2. root makes offsets; 7. 4. 12. formation of roots of κρόμυον comp.; 7. 9. 4., cf. 1.6.7.; 7.12.1. root edible (of the kind called ἡ Ἐπιμενίδειος) (see below) leaves described: 7. 13. 2. flower-stem not the only stem; 7. 13. 3. 'successive' flowering of ἀσφόδελος comp.; 7. 13. 4. very tenacious of life: hence various uses: use as a charm; 7.13.5–7. stem appears before leaves: sequence described and comp. with that of νάρκισσος (1); 9. 18. 3. leaf of ὄρχις comp.

σκίλλα ἡ Ἐπιμενίδειος, French sparrow-grass, *Ornithogalum pyrenaicum*
7. 12. 1. see above.

σκόλυμος (= ? λειμωνία 6.4.3.), golden thistle, *Scolymus hispanicus*
6. 4. 3. a 'thistle-like' plant: leaves spinous; 6. 4. 4. time of flowering; 6. 4. 7. described; 7. 4. 5. leaf of θριδακίνη ἡ Λακωνική comp.; 7. 10. 1. grows and flowers entirely in summer; 7. 15. 1. flowering depends on the heavenly bodies; 9. 12. 1. leaf of χαμαιλέων ὁ λευκός comp.;

9. 13. 4. an unnamed plant of Tegea comp.

σκολόπενδρον, hart's tongue, *Scolopendrium vulgare*
9. 18. 7. leaf of ἡμιόνιον comp.

σκόρδον (σκόρδον), garlic, *Allium sativum*
1. 6. 9. no side-roots; 1. 10. 7. attachment of leaves; 7. 1. 7. stem single; 7. 2. 1. propagation; 7. 2. 3. offsets specially numerous; 7. 4. 1. several kinds; 7. 4. 7. do.; 7. 4. 11. do.; e.g. τὸ Κύπριον: cultivation etc.; 7. 4. 12. formation of roots of κρόμυον contrasted; 7. 8. 2. stem smooth, not branched; 7. 13. 4. grows in colonies because of offsets; 9. 8. 6. eaten as precaution by diggers of ἐλλέβορος.

σκορπίος (1), *Genista acanthoclada*
6. 1. 3. has spines for leaves; 6. 4. 1. one of very few plants which are altogether spinous; 6. 4. 2. described.

σκορπίος (2), leopard's bane, *Doronicum cordatum*
9. 13. 6. peculiar shape, resembles scorpion, and is useful against its sting.

σκορπίος (3) (? σκορπίον) (= ἀκόνιτον = θηλύφονον 9.18.2.= μυόφονον), wolf's bane, *Aconitum Anthora*
9. 18. 2. properties: habitat: fatal to scorpion.

σκυθική (= γλυκεία, sc. ῥίζα 9.13.2.), liquorice, *Glycyrrhiza glabra*
9. 13. 2. fragrant: grows on Lake Maiotis: medicinal use: use against thirst.

σμῖλαξ (1), holm-oak, *Quercus Ilex typica*
3. 16. 2. described.

σμῖλαξ (?) (2) (= μῖλαξ), smilax, *Smilax aspera*
3. 18. 11–12. described; 7. 8. 1. stem clasping.

σμύρνα (gum σμύρνα 9. 1. 2.), myrrh, *Balsamodendron Myrrha*
4.4.12. gum of an Arian ἄκανθα (see App. (9)) comp.; 4. 4. 14. in list of Oriental ἀρώματα; 9.1.2. sap gummy (called σμύρνα); 9. 1. 6 time of tapping; 9. 4. 1. collec-

477

INDEX OF PLANTS

tion of gum; 9. 4. 2. Arabian: habitat; 9. 4. 3. described (two accounts); 9. 4. 4–10. accounts of travellers; 9. 7. 3. in list of ἀρώματα.

σόγκος, sow-thistle, *Sonchus Nymani* 4.6.10. growth of φοῖνιξ (2) comp.; 6. 4. 3. a 'thistle-like' plant, but has not the characteristic 'head' of such plants; 6. 4. 5. stalk of a kind of ἄκανος comp.; 6. 4. 8. root.

σπάλαξ (?) (= ἐφήμερον), meadow saffron, *Colchicum parnassicum* 1. 6. 11. large fleshy roots.

σπειραία, privet, *Ligustrum vulgare* 1. 14. 2. bears fruit at top; 6. 1. 4. a spineless wild under-shrub.

(σπογγιά, sponge 4. 6. 5. found on North Coast of Crete; 4. 6. 10. distinguished from 'plants.')

σποδιάς (= προύμνη), bullace, *Prunus insititia* 3. 6. 4. very shallow rooting: few roots.

στελέφουρος (= ἀρνόγλωσσον 7. 11. 2. according to some) (= ὄρτυξ 7. 11. 2. according to some), plantain, *Plantago Lagopus*

στοιβή (= φλέως 6. 1. 3.), *Poterium spinosum* 1. 10. 4. leaves fleshy; 6. 1. 3. has leaves as well as spines: a wild under-shrub; 6. 5. 1. in list of such plants; 6. 5. 2. has no spines on the leaves.

στρουθίον (1) (= κυδώνιος), quince, *Cydonia vulgaris* 2. 2. 5. seed produces κυδώνιος.

στρουθίον (2) (= στρουθός), soap-wort, *Saponaria officinalis* 6. 4. 3. a 'thistle-like' plant, but has not the characteristic 'head' of such plants; 6. 8. 3. a coronary plant: flowering tree: scentless.

στρουθός (= στρουθίον (2)), soap-wort, *Saponaria officinalis* 9. 12. 5. leaf of μήκων ἡ Ἡρακλεία comp.

στρύχνος 7. 15. 4. several plants called by this name, which have nothing in common but the name: three mentioned (*see below*); 9.11. 5–6. kinds (*see below*); 9. 15. 5. two kinds grow in Arcadia.

στρύχνος ὁ ἐδώδιμος, garden night-shade, *Solanum nigrum* 3. 18. 11. fruit of σμῖλαξ (2) comp.; 7. 7. 2. a λάχανον: can be eaten raw; 7. 15. 4. more or less cultivated: has berries.

στρύχνος ὁ μανικός (= θρύορος 9.11 6. = περιττός 9.11.6) thorn-apple, *Datura Stramonium* 7. 15. 4. causes madness, or, in a large dose, death; 9. 11. 6. described: medicinal use; 9. 19. 1. effect on 'mind.'

στρύχνος ὁ ὑπνώδης, *Withania somnifera* 7. 15. 4. induces sleep; 9. 11. 5. described: medicinal use: habitat.

στύραξ, storax, *Storax officinalis* 9. 7. 3. in list of ἀρώματα.

συκάμινος, mulberry, *Morus nigra* 1. 6. 1. core hard and close; 1. 9. 7. time of leafing; 1. 10. 10. fruit made of fibre and skin; 1. 12. 1. taste of fruit; 1. 13. 1. flower 'downy'; 1.13.4. attachment of flower; 5. 3. 4. character of wood; 5. 4. 2. do.: wood little liable to decay: turns black when old; 5. 6. 2. wood tough and easy to bend: uses; 5. 7. 3. wood used for bent-wood work: use in ship-building.

συκάμινος ἡ Αἰγυπτία, sycamore, *Ficus Sycamorus* 1. 1. 7. position of fruit; 1. 13. 2. bears on stem; 4. 1. 5. barren in uncongenial climate; 4.2.1–2. peculiar to Egypt: described; 4. 2. 4. κερωνία distinguished.

συκῆ (1) (fruit σῦκον), fig, *Ficus Carica* 1. 3. 1. a typical 'tree'; 1. 3. 5. evergreen at Elephantine; 1.5.1. crooked and low; 1. 5. 2. bark smooth: bark in one layer; 1. 5. 3. wood fleshy; 1. 5. 3. wood not fibrous; 1. 6. 1. core fleshy; 1. 6. 3. roots many and long; 1. 6. 4. roots crooked; 1. 7 2. roots very long; 1. 8. 1.

INDEX OF PLANTS

no knots; 1. 8. 2. has less knots than ἐρινεός; 1. 8. 5. young branches 'roughest'; 1.9.7. time of shedding leaves; 1. 10. 4. leaves broad; 1. 10. 5. leaf divided: etc.; 1. 10. 8. leaves made of 'bark' and flesh; 1. 11. 4. seeds all together in a single case; 1. 11. 6. arrangement of seeds; 1. 12. 1. taste of fruit; 1. 12. 2. taste of sap; 1. 14. 1. bears on new shoots: sometimes also on old wood (?); 1. 14. 4. cultivated form of ἐρινεός: many cultivated forms; 2. 1. 2. propagation; 2. 2. 4. degenerates from seed: etc.; 2. 2. 12. cannot be made out of ἐρινεός by cultivation; 2. 3. 1. sometimes changes to ἐρινεός spontaneously; 2. 3. 3. sometimes bears fruit from behind the leaves: other anomalies; 2. 5. 3. grafting; 2. 5. 4₁ propagation; 2. 5. 5. cuttings set in a bulb of σκίλλη; 2. 5. 6. easily propagated: trees should be planted far apart; 2. 5. 7. low ground suitable; 2. 6. 6. dates said to vary as much as figs in colour etc.; 2. 6. 12. cuttings set upside down; 2. 7. 1. effects of watering; 2. 7. 5. use of dust; 2. 7. 6. root-pruning etc.; 2. 8 1. apt to shed immature fruit: caprification; 2.8.2–4. do. and pseudo-caprification; 3. 3. 8. sheds ἐρινά; 3. 4. 2. time of budding; 3.5.4. autumn budding; 3. 6. 2. formation of buds; 3. 7. 3. produces ἐρινά and ὄλυνθοι; [3. 17. 4. a local Idaean kind (*see below*); 3. 17. 5. do. described;] 4. 2. 3. taste of fruit of σ. ἡ Κυπρία comp.; 4. 4. 4. fruit of ἡ συκῆ Ἰνδική comp.; 4. 5. 3. grows well in Pontus with shelter; 4. 7. 7. size of marine trees of island of Tylos comp.; 4.13.1. shorter - lived than ἐρινεός; 4. 13. 2. short-lived; 4. 14. 2. apt to get worm-eaten: young plants liable to 'sunscorch'; 4. 14. 3. destroyed by 'worms'

which breed in it: gets scab in some regions; 4. 14. 4–5. other diseases; 4. 14. 8. effects on fruit of rain and drought; 4. 14. 10. infested by *knips*; 4. 14. 12. suffers most from special winds; 4. 15. 2. survives stripping of bark for some time; 4. 15. 2. instance of bark growing again; 4. 16. 1. survives splitting of stem; 5. 3. 3. character of wood; 5. 6. 1. wood strong only against a vertical strain; 5. 9. 5. wood makes pungent smoke; 5. 9. 6. wood good for kindling furnaces; 7. 13. 2. root of ἀσφόδελος eaten with figs.

συκῆ (2) ἡ Αἰγυπτία (= κερωνία 1.11.2.), carob, *Ceratonia Siliqua*

συκῆ (3) ἡ Ἰδαία (fruit σῦκον), *Amelanchier vulgaris*

3. 17. 4–5. described.

συκῆ (4) ἡ Ἰνδική, banyan, *Ficus bengalensis*

1. 7. 3. roots again from branches; 4. 4. 4–5. described.

συκῆ (5) ἡ Κυπρία, sycamore, *Ficus Sycamorus* var.

4. 2. 3. described.

συκῆ (6) ἡ Λακωνική, *Ficus Carica* var.

2. 7. 1. water-loving; 2. 8. 1. caprification not used.

συκῆ (7) (ἡ Ἀραβική), *Ficus Carica* var.?

4. 7. 8. an evergreen Arabian kind.

(συκῆ (8) (? an alcyonidian polyp)

4. 6. 2. peculiar to certain waters; 4. 6. 9. described).

σφάκος, sage, *Salvia calycina*

6. 1. 4. a spineless wild undershrub; 6. 2. 5. like cultivated ἐλελίσφακος: leaf of one kind of πράσιον comp.

σφένδαμνος, maple, *Acer monspessulanum*

3. 3. 1. a tree of mountain and plain; 3. 3. 8 doubt whether it has a flower; 3. 4. 4. time of fruiting; 3. 6. 1. slow-growing (?); 3. 6. 5. roots shallow and few according to Arcadians; 3. 11. 1–2. described; 5. 1. 2. time of cutting timber; 5. 1. 4.

479

INDEX OF PLANTS

do.; 5. 3. 3. character of timber; 5. 7. 6. uses of wood.

σχῖνος (fruit σχινίς 9.4.7.), mastich, *Pistacia Lentiscus*
9. 1. 2. produces a gum; 9. 4. 7. λιβανωτός comp. by some.

σχοῖνος (1), rush, *Juncus* spp. etc.
1. 5. 3. not jointed; 1. 8. 1. no knots; 4. 8. 1. in list of τὰ λοχμώδη; 4. 12. 1–3. kinds distinguished.

σχοῖνος (2) ὁ εὐώδης ?), ginger-grass, *Cymbopogon Schoenanthus*
9. 7. 1. habitat (E. of Lebanon): described: fragrance; 9. 7. 3. in list of ἀρώματα.

σχοῖνος (3) ὁ κάρπιμος, (= μελαγκρανίς, 4.12.1.), bog-rush, *Schoenus nigricans*
4. 12. 1–3. described.

σχοῖνος (4) ὁ ὁλόσχοινος, *Scirpus Holoschoenus*
4. 12. 2. described; 9 12. 1. used for stringing pieces of χαμαιλέων ὁ λευκός.

σχοῖνος (5) ὁ ὀξύς, *Iuncus acutus*
4. 12. 1–2. described.

σχοῖνος (6)
4. 7. 3. stone ('σχοῖνοι' in 'Red Sea.')

τέρμινθος (τερέβινθος), terebinth, *Pistacia Terebinthus*
1. 9. 3. evergreen (wild form); 3. 2. 6. characteristic of Syria; 3. 3. 1. a mountain tree; 3. 3. 3. evergreen; 3. 4. 2. time of budding; 3. 4. 4. time of fruiting; 3. 15. 3–4. described; 4. 4. 7. τ. ἡ Ἰνδική comp.; 4. 16. 1–2. not injured by cutting for resin; 5.3.2. character of wood: uses; 5. 7. 7. fruit and resin useful, wood not used in crafts; 9. 1. 2. sap gummy; 9. 1. 6. time of tapping; 9. 2. 1. method of tapping; 9. 2. 2. produces best resin (ῥητίνη); 9. 2. 2. said to be burnt for pitch (πίττα) in Syria: *cf*. 3. 2. 6.; 9. 3. 4. method of preparing pitch in Syria; 9. 4. 7. σμύρνα comp. by some; 9. 4. 8. some say σμύρνα = τ.; 9. 5. 1. fruit of βάλσαμον comp.

τετραγωνία, *Euonymus latifolius*
3. 4. 2. time of budding; 3. 4. 6; time of fruiting.

τετράλιξ, yellow star-thistle, *Centaurea solstitialis*
6. 4. 4. a 'thistle-like' plant: time of growing.

τεῦτλις (= τεῦτλον)
7. 7. 2. a λάχανον; needs cooking.

τεῦτλον (τεύτλιον) = τεῦτλις, beet, *Beta maritima*
1. 3. 2. becomes large in cultivation; 1.5.3. stem fleshy; 1.6.6. root single, but with large sidegrowths; 1. 6. 7. root fleshy; 1. 9. 2., *cf*. 1. 3. 2; 1. 10. 4. leaves fleshy; 7. 1. 2–3. time of sowing and germination; 7.1.5. do.; 7. 1. 6. germination; 7.2.2 root makes offsets; 7. 2. 5–6. root described; 7. 2. 7. root of λάπαθον comp.; 7. 2. 8. root; 7. 3. 2. seeds; 7. 4. 1. several kinds; 7. 4. 4. two kinds, τὸ λευκόν (Σικελικόν) and τὸ μέλαν; 7. 5. 5. seed keeps well.

τῆλις (= βουκέρας), fenugreek, *Trigonella Foenum-graecum*
3. 17. 2. leaf of κολουτέα (κολοιτία (1) comp.

τιθύμαλλος (produces ἱπποφαές ? 9. 15. 6. *see note*) (= μηκώνιον 9. 8. 2.), spurge, *Euphorbia Peplus* etc.
9. 8. 2. juice of stalk, how collected; 9. 11. 1. several kinds; 9. 11. 5. do.; leaf of στρύχνος ὁ ὑπνώδης comp.; 9. 11. 7–9. kinds (*see below*); 9. 15. 6. Arcadian: localities.

τιθύμαλλος ὁ ἄρρην, *Euphorbia Sibthorpii*
9. 11. 8. described: medicinal use.

τιθύμαλλος ὁ μυρτίτης (fruit κάρυον 9. 11 9.), *Euphorbia Myrsinites*
9. 11. 9. described: habitat: time of gathering: medicinal use.

τιθύμαλλος ὁ παράλιος, sea-spurge, *Euphorbia Paralias*
9. 11. 7. described: medicinal use.

τίφη, one-seeded wheat, *Triticum monococcum*
1. 6. 5. roots numerous; 2. 4. 1.

INDEX OF PLANTS

seed, unless bruised, produces πυρός; 8. 1. 1. in list of cereals; 8. 1. 3. sown early; 8. 2. 1. germination described; 8. 2. 6. time of ripening seed; 8. 4. 1. comp. in detail with other cereals; 8. 8. 3. τ. and ζειά only plants which can change into something quite different (cf. 2. 4. 1.); 8. 9 2. exhausts the soil less than any other cereal: reason: likes light soil: τ. and ζειά the cereals most like πυρός.

τίφυον, autumn squill, *Scilla autumnalis*
7. 13. 7. flower appears before leaves and stem.

τραγάκανθα (1), tragacanth, *Astragalus creticus*
9. 1. 3. produces a gum: now known to grow best only in Crete (see τ. (2)); 9. 8. 2. no cutting needed to collect gum.

τραγάκανθα (2), tragacanth, *Astragalus Parnassi*
9. 15. 8. abundant in Achaia and not inferior to the τ. of Crete.

τραγοπώγων (= κόμη 7. 7. 1.) goat's beard, *Tragopogon porrifolius*
7. 7. 1. described: a λάχανον.

τρίβολος (1), caltrop, *Tribulus terrestris*
3. 1. 6 comes up spontaneously in damp places; 6. 1. 3. has leaves as well as spines: has spines on the fruit-vessel: 6. 5. 3. distinguished from τρίβολος (2); 7. 8. 1. stem 'on the ground'; 8. 7. 2. (as a weed) destroyed by ἐρέβινθος.

τρίβολος (2), caltrop, *Fagonia cretica*
6. 1. 3. has leaves as well as spines; 6. 4. 1. do.; 6. 5. 1. in list of such plants; 6. 5. 3. distinguished from τρίβολος (1); grows near enclosures.

τρίβολος (3), water chestnut, *Trapa natans*
4. 9. 1–3. described.

τριπόλιον (?), *Aster Tripolium*
9. 19. 2. use as charm.

τριχομανές (? = ἀδίαντον τὸ λευκόν) 7. 14. 1., English maidenhair, *Asplenium Trichomanes*

τύφη, bulrush, *Typha angustata*
1. 5. 3. not jointed; 1. 8. 1. no knots; 4. 10. 1. in list of plants of Lake Copais; 4. 10. 5. described; 4. 10. 6. grows both on land and in water: some doubt this; 4.10.7. part used for food.

ὑάκινθος ἡ ἀγρία, *Scilla bifolia*
6. 8. 1–2, flowering time.

ὑάκινθος ἡ σπαρτή, larkspur, *Delphinium Ajacis*
6. 8. 2. flowering time: flower of πόθος (1) comp.

ὕδνον, truffle, *Tuber cibarium*
1. 1. 11. has not all the 'parts' of a plant; 1. 6. 5. no roots; 1. 6. 9. underground.

ὑποχοιρίς, cat's ear, *Hypochoeris radicata*
7. 7. 1. a λάχανον; classed as 'chicory-like' from its leaves; 7. 11. 4. growth contrasted with χόνδρυλλα.

ὕφεαρ, mistletoe, *Viscum album*
3. 16. 1. grows on ποῖνος.

φακός, lentil, *Ervum Lens*
2. 4. 2. seed sown in dung; 3. 15. 3. fruit of τέρμινθος comp.; 3. 17. 2. size of fruit of κολοιτία (1) comp.; 3. 18. 5. arrangement of fruit of ῥοῦς comp.; 4. 4. 9. not found in India; 4. 4. 10. a so-called φ. in India; 8. 1. 4. sown late; 8.3.2. stem; 8. 3. 4. seed; 8. 5. 1. several kinds; white form sweetest; 8. 5. 2. seeds comparatively few; 8. 5. 3. shape of pod; 8. 8. 3. ἄρακος grows specially among φ.; 8. 8. 4. so also ἀπαρίνη; 8. 8. 6. causes etc. of φ. becoming 'cookable' or 'uncookable.'

φάσγανον (= ξίφιον = ξίφος 7. 13. 1.), corn-flag, *Gladiolus segetum*
7. 12. 3. use of root in food: root described; 7. 13. 1. leaves described; 7. 13. 4. grown from seed.

φάσκος, tree-moss, *Usnea barbata*
3. 8. 6. borne only on αἰγίλωψ (1).

φελλόδρυς (= ἀρία 3. 16. 3.), holm-oak, *Quercus Ilex* var. *agrifolia*
1. 9. 3. evergreen; 3. 3. 3. do.;

481

INDEX OF PLANTS

3. 16. 3. described: called ἀρία by Dorians.
φελλός, (?= ἴψος cf. Plin. 16. 98.), cork-oak, *Quercus Suber*
1. 2. 7. bark; 1. 5. 2. bark rough and fleshy; 1. 5. 4. wood light; 3. 17. 1. grows in Tyrrhenia: described; 4. 15. 1. is the stronger for having its bark stripped; 5. 3. 6. wood of φοῖνιξ (1) comp.
φέως (= στοιβή 6. 1. 3.), *Poterium spinosum*.
φηγός (= δρῦς ἡ ἀγρία), Valonia oak, *Quercus Aegilops*
3. 3. 1. a mountain tree; 3. 4. 2. time of budding; 3. 6. 1. slow-growing (?); 3. 8. 2. one of the five 'Idaean' kinds of oak: described: fruit; 3. 8. 3–4. acorns; 3. 8. 4. timber; 3. 8. 7. one of the four 'Macedonian' kinds of oak; 4. 13. 2. ancient trees at Ilium; 5. 1. 2. time of cutting timber.
φιλύκη, alaternus, *Rhamnus Alaternus*
1. 9. 3. evergreen; 3. 3. 1. a mountain tree; 3. 3. 3. evergreen; 3. 4. 2. time of budding; 3. 4. 4. time of fruiting; 5. 6. 2. easiest wood for turning.
φίλυρα (= φιλύρα ἡ θήλεια), lime (or silver lime), *Tilia platyphyllos*, (or *tomentosa*)
1. 5. 2. bark thin: bark in layers; 1. 5. 5. wood pliable; 1. 10. 1. leaves inverted in summer; 1. 12. 4. leaves, but not fruit, eaten by animals; 3. 3. 1. a mountain tree; 3. 4. 2. time of budding; 3. 4. 6. time of fruiting: fruit inedible; 3. 5. 5–6. winter-buds; 3. 10. 4–5. distinguished: 'male' and 'female' forms distinguished (*see below*); 3. 11. 1. bark of σφένδαμνος comp.; 3. 13. 1. bark of κέρασος comp.; 3. 13. 3. grows where κέρασος grows; 3. 17. 5. leaf of συκή ἡ Ἰδαία comp.; 4.4.1. hard to grow in Babylon; 4. 5. 1. in list of Northern trees; 4. 8. 1. grows partially in water; 4.15.1. outer bark can be stripped;

4. 15. 2. survives stripping of bark for some time; 5. 1. 2. time of cutting timber; 5. 1. 4. do.; 5. 3. 3. character of wood; 5. 5. 1. wood easy to work; blunts tools; 5. 6. 2. wood soft and easy to work; 5.7.5. various uses of wood and bark; 5. 9. 7. wood used for fire-sticks.
φίλυρα ἡ ἄρρην (= φιλυρέα), mock-privet, *Phillyrea media*
3. 10. 4–5. distinguished from φ. ἡ θήλεια.
φίλυρα ἡ θήλεια (= φίλυρα), silver-lime, *Tilia tomentosa*
3. 10. 4–5. distinguished from φ. ἡ ἄρρην.
φιλυρέα, mock-privet, *Phillyrea media*
1. 9. 3. evergreen; 3. 4. 6. (?) time of fruiting.
φλεώ (φλεώς), *Erianthus Ravennae*
4.8.1. in list of τὰ λοχμώδη; 4.10.1. in list of plants of Lake Copais; 4. 10. 4. described; 4. 10. 6. grows both on land and in water; grows on the floating islands of Lake Copais; 4. 10. 7. part used for food; 4. 11. 12. foliage of some κάλαμοι comp.
φλόγινον (= φλόξ), wall-flower, *Cheiranthus Cheiri*
6. 8. 1–2. a coronary plant: flowering time.
φλόμος ἡ μέλαινα, mullein, *Verbascum sinuatum*
9. 12. 3. leaf of μήκων ἡ κερατῖτις comp.
φλόξ (= φλόγινον), wall-flower, *Cheiranthus Cheiri*
6. 6. 2. a cultivated under-shrub; a coronary plant: scentless; 6. 6. 11. grown from seed.
φοῖνιξ (1), date-palm, *Phoenix dactylifera*
1. 2. 7. 'flesh' turns to wood; 1. 4. 3. (?) tolerant of sea-water; 1. 5. 1. few branches; 1. 5. 2. rough bark; 1. 5. 3 wood fibrous; 1. 6. 2. core not distinguishable; 1. 9. 1. growth chiefly upwards; 1. 9. 3. evergreen; 1. 10. 5. reedy leaves; 1.11.1. seed immediately within envelope: envelope not single;

INDEX OF PLANTS

1. 11. 3. seed solid and 'dry' throughout; 1. 13 5. 'male' only flowers; 1.14.2. bears fruit at top; 2. 2. 2. propagation; 2. 2. 6 seeds come true; 2. 2. 8. effect of locality; 2.2.10. effects of climate; 2. 6. 1. propagation from fruit; 2. 6. 2. propagation from 'head'; 2. 6. 3. cultivation; 2. 6. 4. do.; 2. 6. 5. cultivation in Syria; 2. 6. 8. various kinds; [2. 6. 9. branching kind: *see* κουκιόφορον]; [2. 6 10. shrubby kind: *see* κοίξ]; 2.6.12. cuttings set upside down; 2. 8. 1. apt to shed immature fruit; 2. 8. 4. artificial fertilisation; 3. 3. 5. not fruitful wherever it grows; 3. 13. 7. dwarf form (? φ. ὁ χαμαιρριφής); 4.1.5. *cf.* 2. 2. 10; 4. 2. 7. κουκιόφορον comp.; 4. 3. 1. grows in parts of Libya; 4. 3. 5. grows well in waterless Libya: salt in soil, water supply; 4. 3. 7. kept alive by dew in dry regions; 4. 4. 3. sown in pots; 4.4.13. dangerous to eat unripe dates (in Gedrosia); 4. 7. 8. occurs on the island of Tylos; 4. 13. 2. story of the very old φ. on Delos; 4. 14. 8. if topped, becomes barren, but is not destroyed; 4. 15. 2. survives stripping of bark; 5. 3. 6. character of wood: used for images; 5. 6. 1. wood strong: behaviour under pressure: 5. 9. 4. wood makes a very evil smoke when burnt for charcoal; 6. 4. 11. seed-vessel of κάκτος (1), when stripped of seeds, comp. to 'brain' of φ.; 9. 4. 4. mats in Arabia made of leaves of φ.

φοῖνιξ (2), *Nannorhops ritchiana*
4. 4. 8. Bactrian.

φοῖνιξ (3), ὁ χαμαιρριφής, dwarf palm, *Chamaerops humilis*
2. 6. 11. described; 3. 13. 7. (?).

φοῖνιξ (4), *Callophyllis laciniata*
4. 6. 2. peculiar to certain waters; 4. 6. 10. described.

φόνος (=ἀτρακτυλίς 6. 4. 6.), distaff-thistle, *Carthamus lanatus*
6. 4. 6. reason for name

φῦκος (1) τὸ πλατύφυλλον (= ζωστηρ 4. 6. 2. = πράσον (2)), grass-wrack, *Posidonia oceanica*
4. 6. 2. occurs generally in Greek waters: root described.

φῦκος (2) θαυμαστὸν τὸ μέγεθος, riband-weed, (= πράσον (3)), *Laminaria saccharina*
4.6. 4. described: grows in Atlantic: washed into Mediterranean; 4. 7. 1. refers to 4. 6. 4.

φῦκος (3) τὸ πόντιον
4.6.4. collected by sponge-fishers.

φῦκος (4) τὸ τριχόφυλλον, *Cystoseira foeniculosa*
4. 6. 3. described.

φῦκος (5), litmus, *Roccella tinctoria*
4. 6. 5. Cretan: dye described.

φῦκος (6), grass-wrack, *Cymodocea nodosa* (and *Zostera marina*)
4. 6. 6. described: comp. to ἄγρωστις.

χαλβάνη, see πάνακες (τὸ Σύριον).
χάλκειος, *Carlina corymbosa*
6. 4. 3. a 'thistle-like' plant.

χαμαίβατος (=βάτος), *Rubus ulmifolius*
3. 18. 4. described.

χαμαιδάφνη, periwinkle, *Vinca herbacea*
3. 18. 13. leaf of εὐώνυμος comp.

χαμαίδρυς, germander, *Teucrium Chamaedrys*
9. 9. 5. medicinal use: described.

χαμαιλέων, chamaeleon
6. 4. 3. a 'thistle-like' plant, but leaves not spinous (*see* n. on 6. 4. 8.); 6.4.8. flower turns into 'down'; 9. 12. 1–2, kinds (*see below*).

χαμαιλέων ὁ λευκός (= ἄκανθα (9) 9.12.1. = ἄκανος = ἰξία (2) = ἰξίνη), pine-thistle, *Atractylis gummifera*
9.12.1. described: medicinal use: fatal to dogs and pigs: how administered: grows everywhere.

χαμαιλέων ὁ μέλας, *Cardopatium corymbosum*
9. 12. 2. described: medicinal use: habitat: fatal to dogs; 9. 14. 1. how long drug will keep.

483

INDEX OF PLANTS

χελιδόνιον, greater celandine, *Chelidonium maius*
  7. 15. 1. flowering depends on the heavenly bodies.

χόνδρυλλα, *Chondrilla juncea*
  7. 7. 1. a λάχανον: classed as 'chicory-like' from its leaves; 7. 11. 4. growth of ὑποχοιρίς contrasted.

ψευδοδίκταμνον, *Ballota acetabulosa*
  9. 16. 2. comp with δίκταμνον as to appearance and properties: said by some to be only a degenerate form of δίκταμνον: method of keeping.

ὤκιμον, basil, *Ocymum basilicum*
  1. 6. 6-7. root woody; 1. 10. 7. time of leaf-growth; 7. 1. 2-3. time of sowing and germination; 7.2.1. propagation; 7.2.4. grows again when stem is cut; 7. 2. 7-8. root described; 7.3.1. long in flower; 7. 3. 2-3 seeds described; 7. 3. 4. seed very abundant: seed borne at top; 7. 4. 1. only one kind; 7. 5. 2. watered at mid-day; 7. 5. 4. effect of hot weather; 7. 5, 5. seed does not keep well; 7.7. 2. leaf of κόρχορος comp.; 7. 9. 2. flowers borne in succession, *cf.* 7. 3. 1.; 9. 18. 5. leaf of ἀρρενόγονον and of θηλύγονον comp.

ὦχρο;, *Lathyrus Ochrus*
  8. 1. 3. sown early; 8. 3. 1. leaf; 8. 3. 2. stem; 8. 10. 5. infested by 'worms.'

APPENDIX OF UNNAMED PLANTS

The following plants (arranged in the order of mention) which are described or indicated, but not named, in the Enquiry, seem possible to identify:—

1. (ὅμοιον τῷ ἀράκῳ), tine-tare, *Lathyrus tuberosus*
   1. 6. 12. root described.
2. (ὑλήμα ἴδιον τι περὶ Μέμφιν), *Mimosa asperata*
   4. 2. 11. described: 'sensitive.'
3. (δένδρον . . . μεγαλόκαρπον), Jack-fruit, *Artocarpus integrifolia*
   4. 4. 5. used for food by Indian sages.
4. (φύλλον . . . τοῖς τῶν στρουθῶν πτεροῖς ὅμοιον), banana, *Musa sapientum*
   4. 4. 5. described.
5. (καρπὸς σκολιὸς ἐσθιόμενος δὲ γλυκύς), mango, *Mangifera indica*
   4. 4. 5. fruit described: causes dysentery.
6. (καρπὸς ὅμοιος τοῖς κρανέοις), jujube, *Zizyphus Jujuba*
   4. 4. 5.
7. (ὅμοιον τερμίνθῳ), pistachio-nut, *Pistacia vera*
   4. 4. 7. described.
8. (ὅμοιον τῇ ὄψει καὶ τὸ βούκερας), *Phaseolus Mungo*
   4. 4. 9-10. called by Hellenes φακός, and similarly used.
9. (ἄκανθα ἐφ' ἧς γίνεται δάκρυον) (= ἄκανθα (4) ἡ 'Ινδική), *Balsamodendron Mukul*
   4. 4. 12. grows in Aria: described.
10. (ὑλήμα ἥλικον ῥάφανος), Asafoetida, *Scorodosma foetidum*
    4. 4. 12. described: fatal to horses.
11. (ὅμοιον τῇ δάφνῃ φύλλον ἔχον), *Nerium odorum*
    4. 4. 13. effect on animals.
12a. (δένδρα μέγαλα), mangrove, *Bruguiera gymnorhiza*
    4. 7. 4. described.

484

INDEX OF PLANTS

12*b*. (δένδρον φύλλον ἔχον ὅμοιον τῇ δάφνῃ) mangrove, *Rhizophora mucronata*
   4. 7. 4. described (as if identical with 12*a*.).

13. (καρπὸς ὅμοιος τοῖς θέρμοις), *Aegiceras maius*
   4. 7. 5, 6, and 7. described.

14. (δένδρα ὅμοια τῇ ἀιδράχλῃ (= δάφνῃ (6) = ἐλάα (3)), white mangrove, *Avicennia officinalis*
   4. 7. 5. described.

15. (δένδρα τὸ ἄνθος ἔχοντα ὅμοιον τῷ λευκοίῳ) (= (16)), tamarind, *Tamarindus indica*
   4. 7. 8. grows in the island of Tylos.

16. (δένδρον πολύφυλλον) (= (15)), tamarind, *Tamarindus indica*
   4. 7. 8. grows in the island of Tylos: opening and closing of flower.

17. (συκῆ οὐ φυλλοροοῦσα), *Ficus laccifera*
   4. 7. 8. grows in the island of Tylos.

18. (ὅμοιον τοῖς κρίνοις) *Ottelia alismoides*
   4. 8. 6. Egyptian marsh-plant: habitat and leaves: medicinal use.

19. (ἔν τι γένος ἐν ταῖς λίμναις), *Saccharum biflorum*
   4. 8. 13. use for fodder.

20. (γένος παραφυόμενον ἐν τῷ σίτῳ), *Corchorus trilocularis*
   4. 8. 14. treatment as fodder: fruit described.

21. (δένδρον), Sissoo-wood, *Dalbergia Sissoo*
   5. 3. 2. wood described: use for making furniture.

22. (ξύλον), teak, *Tectona grandis*
   5. 4. 7. wood stands sea-water well.

23. (δένδρον), calamander wood, *Diospyros quaesita*
   5. 4. 7. wood described.

24. (ὅμοιον ἀβροτόνῳ), *Artemisia camphorata*
   6.3.6. properties: effect on sheep.

25. (τὸ τῇ νάρδῳ προσεμφερῆ τὴν ὀσμὴν ἔχον), *Valeriana Dioscoridis*
   9. 7. 4. a Thracian ἄρωμα.

26. (ὑποφυόμενον εὐθὺς ἐκ τῆς ῥίζης τῷ κυμίνῳ), broom-rape, *Orobanche versicolor*
   8. 8. 5. parasitic on κύμινον.

27. (ῥίζα θανατηφόρος), Somali arrow-poison, *Acokanthera Schimperi*
   9. 15. 2. Aethiopian: used for poisoning arrows.

KEY TO THE INDEX

I.—LIST OF PLANTS MENTIONED IN THE ENQUIRY UNDER BOTANICAL NAMES

Abies cephalonica	ἐλάτη (1)
—— pectinata	ἐλάτη (2)
Acacia albida	ἄκανθα(1),etc.
—— arabica	ἄκανθα(1),etc.
—— tortilis	ἄκανθα (3)
Acantha arabica	ἄκανθα (7)
Acer campestre	ζυγία
—— creticum	γλεῖνος
—— monspessulanum	σφένδαμνος
—— pseudo-Platanus	κλινότροχος
Acokanthera Schimperi	App. (27)
Aconitum Anthora	ἀκόνιτον, θηλύφονον, μυόφονον, σκορπίος (3)
Acorus Calamus	κάλαμος ὁ εὐώδης
Adiantum Capillus-Veneris	ἀδίαντον, ἀ. τὸ μέλαν
Aegiceras majus	App. (13)
Aegilops ovata	αἰγίλωψ (2)
Ailanthus malabarica	κώμακον
Ajuga Ira	ἐπετίνη
Allium Cepa and vars.	γήθυον, γήτειον, κρόμυον, κ. τὸ σχιστόν, κρομυογήτειον
—— nigrum	μῶλυ
—— Porrum	πράσον (1)
—— sativum	σκόροδον
Alnus glutinosa	κλήθρα
Althaea officinalis	ἀλθαία, μαλάχη ἡ ἀγρία
Amaranthus Blitum	βλίτον
Amelanchier vulgaris	συκῆ ἡ Ἰδαία
Amomum subulatum	ἄμωμον
Anagallis caerulea	κόρχορος
Anchusa tinctoria	ἄγχουσα
Andropogon Ischaemum	ἴσχαιμος
Anemone blanda	ἀνεμώνη ἡ ὀρεία
—— coronaria	ἀ. ἡ λειμωνία
—— pavonina	ἀ. ἡ λειμωνία
—— spp.	ἀνεμώνη
Anethum graveolens	ἄνηθον, ἄννητος
Anthemis chia	ἄνθεμον, ἀ. τὸ φυλλῶδες
Antirrhinum Orontium	ἀντίρρινον
Apium graveolens	ἑλειοσέλινον, σέλινον, σ. τὸ ἕλειον
Arbutus Andrachne	ἀνδράχλη
—— hybrida	ἀφάρκη
—— Unedo	κόμαρος
Aristolochia rotunda	ἀριστολοχία
Artemisia Absinthium	ἀψίνθιον
—— arborescens	ἀβρότονον
—— camphorata	App. (24)
Artocarpus integrifolia	App. (3)
Arum italicum	ἄρον
Arundo Donax	δόναξ, κάλαμος ὁ αὐλητικός etc.
Asparagus acutifolius	ἀσφάραγος
Asphodelus ramosus	ἀσφόδελος, πόθος (2)
Asplenium Ceterach	ἡμίονιον
—— Trichomanes	ἀδίαντον τὸ λευκόν, τριχομανές

487

KEY TO THE INDEX

Aster Amellus	ἀστέρισκος	Carthamus leucocaulos	κνῆκος ἡ ἀγρία
—— Tripolium	τριπόλιον	—— tinctorius	κνῆκος, κ. ἡ ἥμερος, κρόκος ὁ ἀκανθώδης
Astragalus creticus	τραγάκανθα (1)		
—— Parnassi	τραγάκανθα (2)		
Atractylis gummifera	ἄκανθα (8), ἄκανος, ἰξία (2), ἰξίνη, χαμαιλέων ὁ λευκός	Castanea vesca	διοσβάλανος
		—— var.	καρύα ἡ Εὐβοϊκή
		Celtis australis	λωτός (1)
Atriplex Halimus	ἅλιμον	Centaurea calcitrapa	παντάδουσα
—— rosea	ἀδράφαξυς	—— salonitana	κενταύριον
Atropa Belladonna	μανδραγόρας (2)	—— solstitialis	τετράλιξ
		Ceratonia Siliqua	κερωνία
Avena sativa	βρόμος	Cerris Siliquastrum	κερκίς (1) σημύδα
Avicennia officinalis	δάφνη (6), ἐλάα (3), App. (14)		
		Chamaerops humilis	φοῖνιξ (3)
		Cheiranthus Cheiri	φλόγινον, φλόξ
		Chelidonium majus	χελιδόνιον
		Chondrilla juncea	χόνδρυλλα
Balanites aegyptiaca	βάλανος	Cicer arietinum	ἐρέβινθος
Ballota acetabulosa	ψευδοδίκταμνον	Cichorium Intybus	κιχόριον
		Cinnamomum Cassia	κινάμωμον
—— pseudo-Dictamnus	δίκταμνον (ἕτερον)	—— iners	κασία
Balsamodendron Mukul	ἄκανθα (4), App. (9)	Cistus salvifolius	κίσθος ὁ θῆλυς
		—— villosus	κίσθος, κ. ὁ ἄρρην
—— Myrrha	σμύρνα	Citrus medica	μηλέα ἡ Περσική (Μηδική)
—— Opobalsamum	βάλσαμον		
Bambusa arundinacea	κάλαμος ὁ Ἰνδικός	Clematis vitalba	ἀθραγένη
Beta maritima	τεῦτλον	Cnicus Acarna	ἄκορνα
Brassica alba	νᾶπυ	—— benedictus	κνῆκος ἡ ἀγρία (ἑτέρα)
—— cretica	ῥάφανος		
—— Rapa	γογγυλίς	Colchicum parnassicum	ἐφήμερον, σπάλαξ
Bruguiera gymnorhiza	App. (12a)		
Bryonia cretica	ἄμπελος ἡ ἀγρία, μήλωθρον	Colocasia antiquorum	οὔϊγγον
Buxus sempervirens	πύξος	Colutea arborescens	κολυτέα
		Conium maculatum	κώνειον
		Convolvulus Scammonia	σκαμμωνία
Calamintha incana	ἐλένιον		
Calamogrostis Epigeios	κάλαμος (ἐπίγειος)	—— sepium	ἰασιώνη
		Corchorus trilocularis	App. (20)
Callitriche verna	λέμνα	Cordia Myxa	κοκκυμηλέα (ἡ Αἰγυπτία)
Callitris quadrivalvis	θύον		
Callophyllis laciniata	φοῖνιξ (4)	Coriandrum sativum	κορίαννον
Calycotome villosa	ἀσπάλαθος	Cornus Mas	κράνεια
Capparis spinosa	κάππαρις	—— sanguinea	θηλυκράνεια
Cardopatium corymbosum	χαμαιλέων ὁ μέλας	Corydalis densiflora	θήσειον
		Cotoneaster Pyracantha	ὀξυάκανθος
Carduus arvensis	ἄκανθα (2)		
—— acanthoides	πολυάκανθος	Corylus avellana	καρύα
Carex riparia	βούτομος	—— var.	καρύα ἡ Ἡρακλεωτική
Carlina corymbosa	χάλκειος		
Carthamus lanatus	ἀτρακτυλίς, φόνος	Crataegus Heldreichii	κράταιγος
		—— orientalis	μεσπίλη ἡ ἀνθηδών

488

KEY TO THE INDEX

Crataegus oxyacantha	μεσπίλη ἡ ἀνθηδονοειδής	Elettaria Cardamomum	καρδάμωμον
Crepis Columnae	ἀπαργία	Ephedra campylopoda	θραύπαλος
Crocus cancellatus	κ. ὁ λευκός		
—— sativus	κ. ὁ εὔοσμος	Erianthus Ravennae	φλεώ
—— spp.	κρόκος	Erica arborea	ἐρείκη
Cucumis sativus	σίκυος	Eruca sativa	εὔζωμον
Cucurbita maxima	κολοκύντη	Ervum Lens	φακός
Cuminum Cyminum	κύμινον	Eryngium campestre	ἠρύγγιον
Cupressus sempervirens	κυπάριττος	Erythraea Centaurium	κενταυρίς
Cuscuta europaea	ὀροβάγχη	Euonymus europaeus	εὐώνυμος
Cyclamen graecum	κυκλάμινος	—— latifolius	τετραγωνία
Cydonia vulgaris	κυδώνιον, στρουθίον (1)	Euphorbia antiquorum	ἄκανθα (5), (6)
Cymbopogon Schoenanthus	σχοῖνος (2)	—— Apios	ἄπιος (2), ἰσχάς, ῥάφανος ἡ ὀρεία
Cymodocea nodosa	φῦκος (6)		
Cynara Cardunculus	κάκτος (1)	—— Myrsinites	τιθύμαλλος ὁ μυρτίτης
—— Scolymus	κάκτος (2)		
Cynodon Dactylon	ἄγρωστις	—— paralias	τιθύμαλλος ὁ παράλιος
Cyperus auricomus	σάρι		
—— esculentus	μαλιναθάλλη, μνάσιον	—— Peplus	μηκώνιον, τιθύμαλλος
—— longus	κύπειρος	—— Sibthorpii	τιθύμαλλος ὁ ἄρρην
—— Papyrus	πάπυρος		
—— rotundus	κύπειρον	—— spp.	μηκώνιον, ἱπποφέως, τιθύμαλλος
Cystoseira Abies marina	ἐλάτη (3)		
—— ericoides	δρῦς (7)		
—— foeniculosa	φῦκος (4)	Fagonia cretica	τρίβολος (2)
Cytisus aeolicus	κολοιτία (1)	Fagus silvatica	ὀξύη
		Ferula communis	ναρθηκία, νάρθηξ
Dalbergia Sissoo	App. (21)		
Daphne Gnidium	κνέωρον	—— nodosa	πάνακες τὸ Ἀσκληπίειον
—— oleoides	κνέωρος ὁ λευκός		
Datura Stramonium	θρύορον, περιττός, στρύχνος ὁ μανικός	—— tingitana	σίλφιον
		Ferulago galbanifera	πάνακες
		Ficus bengalensis	συκῆ ἡ Ἰνδική
Daucus Carota	δαῦκον (1)	—— Carica	συκῆ (1)
Delphinium Ajacis	ὑάκινθος ἡ σπαρτή	—— —— var.	συκῆ ἡ Ἀραβική
—— orientale	πόθος (1)	—— laccifera	App. (17)
—— Staphisagria	ἀσταφίς	—— Sycamorus	συκάμινος ἡ Αἰγυπτία
Dendrocalamus strictus	κάλαμος ὁ Ἰνδικός		
Diospyros Ebenum	ἐβένη (1)	—— —— var.	συκῆ ἡ Κυπρία
—— Melanoxylon	ἐβένη (2)	Fraxinus excelsior	βουμέλιος
Dracunculus vulgaris	δρακόντιον	—— Ornus	μελία
Drypis spinosa	δρυπίς	Fucus spiralis	ἄμπελος (3)
		Fungi	μύκης
Ecballium Elaterium	σίκυος ὁ ἄγριος		
Echinops spinosus	ῥύτρος	Galanthus nivalis	λευκόϊον (2)
Echium diffusum	ὀνοχειλές	Galium Aparine	ἀπαρίνη
		Genista acanthoclada	σκορπίος (1)

489

KEY TO THE INDEX

Gladiolus segetum	ξίφιον, ξίφος, φάσγανον
Glaucium flavum var Serpierii	μήκων ἡ κερατῖτις
Glycyrrhiza glabra	γλυκεῖα (ρίζα), Σκυθική
Gossypium arboreum	(δένδρον τὸ) ἐριόφορον
Hedera Helix	ἕλιξ, κιττός
Helichrysum siculum	ἐλειόχρυσος
Heliotropium villosum	ἡλιοτρόπιον
Helleborus cyclophyllus	ἐλλέβορος, ἐ. ὁ μέλας
Herniaria glabra	ἐλλεβορίνη
Hippuris vulgaris	ἵπνον
Hordeum sativum and vars.	κριθή
Hyphaene thebaica	κόϊξ, κουκιόφορον
Ilex Aquifolium	κήλαστρος
Imperata arundinacea	θρύον
Inula Helenium	πάνακες τὸ Χειρώνειον
—— graveolens	κόνυζα ἡ θήλεια
—— viscosa	κόνυζα ἡ ἄρρην
—— spp.	κόνυζα
Iris foetidissima	ξίρις
—— pallida etc.	ἶρις
—— Sisyrinchium	σισυρίγχιον
Juglans regia	καρύα ἡ Περσική
Juncus acutus	σχοῖνος ὁ ὀξύς
—— spp.	σχοῖνος
Juniperus communis	κεδρίς
—— excelsa	κέδρος (2)
—— foetidissima	θυία
—— phoenicea	ἄρκευθος, κέδρος (3)
—— Oxycedrus	κέδρος(1), ὀξύκεδρος
Laburnum vulgare	κύτισος (1)
Lactuca graeca	λιβανωτίς
—— sativa	θρίδαξ
—— scariola	θριδακίνη
Lagenaria vulgaris	σικύα
Laminaria saccharina	φῦκος (2)
Lapidium sativum	κάρδαμον
Lathyrus amphicarpus	ἀράχιδνα
—— Ochrus	ὦχρος
—— sativus	λαθύρος
—— tuberosus	App. (1)
Laurus nobilis	δάφνη (1)
Lavandula spica	ἴφυον
Lavatera arborea	μαλάχη (1)
Lecokia cretica	λιβανωτός ἡ κάρπιμος
Lemna minor	ἴκμη
Ligustrum vulgare	σπειραία
Lilium candidum etc.	κρίνον, κρινωνία, λείριον(1)
—— chalcedonicum	κρίνον τὸ πορφυροῦν
—— Martagon	ἡμεροκαλλές
Limnanthemum nymphoides	μήνανθος
Linum usitatissimum	λίνον
Lolium temulentum	αἶρα
Lonicera etrusca	κλύμενον
Loranthus europaeus	ἰξία (1)
Lupinus alba	θέρμος
Lychnis coronaria	λυχνίς
Lycoperdon Bovista	πέζις
—— giganteum	ἄσχιον
Malabaila aurea	δαῦκον (2)
Malva silvestris	μαλάχη (2)
Mandragora officinarum	μανδραγόρας (1)
Mangifera indica	App. (5)
Marrubium peregrinum	πράσιον
Matthiola incana	ἴον τὸ λευκόν, ἰωνία (ἡ λευκή), λευκόϊον (1)
Matricaria Chamomilla	ἄνθεμον τὸ ἀφύλλανθες
Medicago arborea	κύτισος (2)
—— sativa	(πόα) ἡ Μηδική
Melissa officinalis	μελισσόφυλλον
Mentha aquatica	σισύμβριον
—— Pulegium	βληχώ
—— viridis	ἡδύοσμον, μίνθη
Mercurialis perennis	ἀρρενόγονον, θηλύγονον

490

KEY TO THE INDEX

Mespilus germanica etc.	μεσπίλη, μεσπίλη ἡ σατάνειος	Orobanche cruenta	αἱμόδωρον
Mimosa asperata	App. (2)	—— versicolor	App. (26)
Mimusops Schimperi	περσέα, πέρσιον	Oryza sativa	ὄρυζον
		Ostrya carpinifolia	ὀστρύα, ὄστρυς
Musa sapientum	App. (4)	Ottelia alismoides	App. (18)
Muscari comosum etc.	βολβός	Paeonia officinalis	γλυκυσίδη, παιωνία
Myrtus communis	μυρρίνη		
		Paliurus australis	παλίουρος
Nannorhops ritchiana	φοῖνιξ (2)	Pancratium maritimum	βολβὸς ὁ ἐριόφορος
Narcissus poeticus	νάρκισσος (2)	Panicum miliaceum	κέγχρος
—— serotinus	λείριον (2), νάρκισσος(1)	Papaver hybridum	μήκων ἡ ῥοιάς
—— Tazetta	λείριον (2)	—— Rhoeas	μήκων ἡ μέλαινα
—— spp.	λείριον (2)	—— somniferum	μήκων (ἡ ὁπώδης), νηπενθές
Nardostachys Jatamansi	νάρδον		
Nelumbium speciosum	κύαμος ὁ Αἰγύπτιος	—— spp.	μήκων
		Parietaria cretica	ἀλσίνη
Nephrodium Filix-mas	πτερίς	Petroselinum sativum	ὀρεισσέλινον
		Peucedanum officinale	πευκέδανον
Nerium Oleander	δάφνη ἡ ἀγρία, ὀνοθήρας	Phillyrea media	φιλυρέα
—— odorum	App. (11)	Phoenix dactylifera	φοῖνιξ (1)
Neslia paniculata	μελάμπυρον	Phragmites communis	κάλαμος ὁ χαρακίας
Nuphar luteum	μαδωνάϊς, νυμφαία		
		Pinus brutia	πίτυς ἡ φθειρόποιος
Nymphaea alba	σίδη		
—— stellata	λωτός (2)	—— halepensis	πίτυς; see also under πίτυς ἡ ἀγρία
Ocymum basilicum	ὤκιμον	—— Laricio	πεύκη ἡ ἄκαρπος, π. ἡ θήλεια, π. ἡ Ἰδαία
Olea cuspidata	ἐλάα (2)		
—— europaea	ἐλάα (1)		
—— Oleaster	ἀγριέλαιος, κότινος	—— pinea	πεύκη ἡ ἥμερος, π. ἡ κωνόφορος
Ononis antiquorum	ὀνωνίς		
Onopordon illyricum	ὀνόπυξος	—— spp.	πεύκη
Opoponax hispidus	πανάκεια, πάνακες τὸ Ἡράκλειον	Pimpinella Anisum	ἄννησον
		Piper nigrum	πέπερι
Orchis longicruris	ὄρχις	Pistacia Lentiscus	σχῖνος
—— papilionacea	ὄρχις	—— Terebinthus	τέρμινθος
Origanum Dictamnus	δίκταμνον	—— vera	App. (7)
—— heracleoticum	ὀρίγανος ἡ λευκή	Pisum sativum	πισός
		Plantago Coronopus	κορωνόπους
—— Majorana	ἀμάρακον	—— crassifolia	θρυαλλίς
—— viride etc.	ὀρίγανος, ὀρίγανος ἡ μέλαινα	—— Lagopus	ὄρτυξ, στελέφουρος
		—— lanceolata	κύνωψ
Ornithogalum pyrenaicum	σκίλλα ἡ Ἐπιμενίδειος	—— major	ἀρνόγλωσσον
		Platanus orientalis	πλάτανος
—— umbellatum	βολβίνη	Polygonum maritimum	περδίκιον

491

KEY TO THE INDEX

Polygonum Persicaria	κραταίγονος	Ranunculus Ficaria	ἀφία
Polypodium vulgare	πολυπόδιον	Raphanus Raphanistrum	κεραΐς, ῥάφανος ἡ ἀγρία
Polypogon mouspeliensis	ἀλωπέκουρος	—— sativus	ῥαφανίς
Polyporus igniarius	πῖλος	Rhamnus alaternus	φιλύκη
Populus nigra	αἴγειρος	—— graeca	ῥάμνος ἡ λευκή
—— tremula	κερκίς (2)	—— oleoides	ῥάμνος ἡ μέλαινα
Portulaca oleracea	ἀνδράχνη	—— spp.	ῥάμνος
Potentilla reptans	πενταπετές, πεντάφυλλον	Rhizophora mucronata	App. (12*b*)
Poterium spinosum	στοιβή, φεώς	Rhus Coriaria	ῥοῦς
Prangos ferulacea	ἱππομάραθον, μαγύδαρις	—— Cotinus	κοκκυγέα
Prunus Amygdalus	ἀμυγδαλῆ	Ricinus communis	κρότων
—— avium	κέρασος, λακάρη	Rosa canina	κυνόρροδον
—— domestica	κοκκυμηλέα	—— centifolia var.	ῥοδωνία
—— insititia	προύμνη, σποδιάς	—— dumetorum	ῥόδον τὸ ἄγριον
		—— sempervirens	κυνόσβατος
—— Mahaleb	πάδος (πηδός?)	Rubus ulmifolius	βάτος, χαμαίβατος
Pteris aquilina	θηλύπτερις		
Puccinia graminis	ἐρυσίβη	Roccella tinctoria	φῦκος (5)
Punica Granatum	ῥόα	Rumex conglomeratus	λάπαθον τὸ ἄγριον
—— var.	ῥόα ἡ ἀπύρηνος	—— Patientia	λάπαθος
Pyrethrum Parthenium	παρθένιον	Ruscus aculeatus	κεντρομυρρίνη, μυάκανθος
Pyrus amygdaliformis	ἀχράς	—— Hypophyllum	δάφνη ἡ Ἀλεξάνδρεια
—— communis	ἄπιος (1)	Ruta graveolens	πήγανον
—— —— var. Pyraster	ὄγχνη		
—— Malus	μηλέα	Saccharum biflorum	App. (19)
—— —— vars.	μηλέα ἡ γλυκεῖα, μ. ἡ ἐαρινή, μ. ἡ ὀξεῖα	Salix alba	ἰτέα ἡ λευκή
		—— amplexicaulis	ἰτέα ἡ μέλαινα
		—— cinerea	κολοιτία (2)
		—— fragilis	ἑλίκη
		—— spp.	ἰτέα
Quercus Aegilops	δρῦς ἡ ἀγρία, φηγός	Salvia calycina	σφάκος
—— Cerris	αἰγίλωψ (1), ἄσπρις	—— Horminum	ὅρμινον
		—— triloba	ἐλελίσφακος
—— coccifera	πρῖνος	Sambucus nigra	ἀκτέος, ἀκτή
—— Ilex typica	σμῖλαξ (1)	Saponaria officinalis	στρουθίον (2), στρούθος
—— —— var. agrifolia	ἀρία, ἴψος, φελλόδρυς	Sargassum vulgare	δρῦς (8)
—— infectoria	ἡμερίς (1)	Satureia Thymbra	θύμβρα
—— lanuginosa	δρῦς ἡ πλατύφυλλος	Saussurea Lappa	κόστος
		Scandix australis	ἔνθρυσκον
—— Pseudo-Robur	δρῦς ἡ ἀλίφλοιος, δ. ἡ εὐθύφλοιος	—— Pecten-Veneris	σκάνδιξ
		Schoenus Holoschoenus	σχοῖνος ὁ ὁλόσχοινος
—— Robur	δρῦς, δ. ἡ ἥμερος, ἐτυμόδρυς, ἡμερίς (2)	—— nigricans	μελαγκρανίς, σχοῖνος ὁ κάρπιμος
		Scilla autumnalis	τίφυον
—— Suber	φελλός, ἴψος(?)	—— bifolia	ὑάκινθος ἡ ἀγρία

KEY TO THE INDEX

Scolopendrium vulgare	σκολοπένδριον	Trapa natans	τρίβολος (3)
Scolymus hispanicus	λειμωνία σκόλυμος	Tribulus terrestris	τρίβολος (1)
		Trifolium fragiferum	λωτός (3)
Scorodosma foetidum	App. (10)	Trigonella Foenum-graecum	βουκέρας, τῆλις
Securigera Coronilla	πελεκῖνος		
Sedum anopetalum	ἐπίπετρον	—— graeca	μελίλωτος, λωτός (5)
Sempervivum tectorum	ἀείζωον		
		Triticum dicoccum	ζειά, ὀλύρα
Senecio vulgaris	ἠριγέρων	—— monococcum	τίφη
Sesamum indicum	σήσαμον	—— vulgare	πυρός
Setaria italica	ἔλυμος, μέλινος	—— —— vars.	πυρός
		Tuber aestivum	κεραύνιον
Silene venosa	Ἡρακλεία, μήκων ἡ Ἡρακλεία	—— cibarium	ὕδνον
		Typha angustata	τύφη
Silybum marianum	λευκάκανθα		
Smilax aspera	σμῖλαξ (2)	Ulmus glabra	πτελέα
Smyrnium Olusatrum	ἱπποσέλινον	—— montana	ὀρειπτελέα
Solanum nigrum	στρύχνος ὁ ἐδώδιμος	Ulva Lactuca	βρύον
		Urginea maritima	σκίλλα
Sonchus Nymani	σόγκος	Urtica urens	ἀκαλύφη
Sorbus domestica	ὄη	Usnea barbata	φάσκος
Sorghum halepense	κριθαὶ αἱ ἀγρίαι (Indian)		
		Valeriana Dioscoridis	App. (25)
Spartium junceum	λινόσπαρτον	Veratrum album	ἐλλέβορος ὁ λευκός
Spiraea filipendula	οἰνάνθη (1)		
Storax officinalis	στύραξ	Verbascum sinuatum	φλόμος ἡ μέλαινα
Tamarindus indica	App. (15) (16)	Vicia angustifolia	ἀφάκη
Tamarix articulata	μυρίκη (2)	—— Ervilla	ὄροβος
—— tetrandra	μυρίκη (1)	—— Faba	κύαμος
Taraxacum officinale	ἀπάπη	—— Sibthorpii	ἄρακος
Taxus baccata	μίλος	Vigna sinensis	δόλιχος
Tectona grandis	App. (22)	Vinca herbacea	χαμαιδάφνη
Teucrium Polium	πόλιον	Viola odorata	ἴον τὸ μέλαν, ἰωνία ἡ μέλαινα
Thapsia garganica	θαψία		
Thymelaea hirsuta	κνέωρος ὁ μέλας	Viscum album	ὕφεαρ
		Vitex Agnus-castus	ἄγνος, οἶσος
Thymbra capitata	θύμον (1)	Vitis vinifera	ἄμπελος (1)
Thymus atticus	ἕρπυλλος (2)	—— —— var. corinthiaca	ἄμπελος (2)
—— Sibthorpii	ἕρπυλλος (1)		
Tilia platyphyllos	φίλυρα	—— silvestris	οἰνάνθη ἡ ἀγρία
—— tomentosa	φίλυρα, φ. ἡ θήλεια		
		Zizyphus Jujuba	App. (6)
Tordylium apulum	καυκαλίς	—— Lotus	λωτός (4)
—— officinale	σέσελι	—— Spina-Christi	παλίουρος ὁ Αἰγύπτιος
Tragopogon porrifolius	τραγοπώγων		

KEY TO THE INDEX

II.—LIST OF PLANTS MENTIONED IN THE ENQUIRY UNDER POPULAR NAMES

Abele	λεύκη	Bog-rush	μελαγκρανίς, σχοῖνος ὁ κάρπιμος
Acacias	ἄκανθα (1), (3)		
Alaternus	φιλύκη		
Alder	κλήθρα	Bottle-gourd	σικύα
Alexanders	ἱπποσέλινον	Box	πύξος
Alkanet	ἄγχουσα	Bracken	θηλύπτερις
Allheal	πάνακες	Brambles	βάτος, χαμαίβατος
Almond	ἀμυγδαλῆ		
Andrachne	ἀνδράχλη	Broadleaved oak	δρῦς ἡ πλατύφυλλος
Anemones	ἀνεμώνη		
Apples	μηλέα	Broom-rapes	αἱμόδωρον, App. (26)
Arbutus	κόμαρος		
—— hybrid	ἀφάρκη	Brooms	λινόσπαρτον, σκορπίος (1)
Arrow-poison (Somali)	App. (27)		
		Bryony	ἄμπελος ἡ ἀγρία, μήλωθρον
Artichoke	κάκτος (1)		
Asafoetida	App. (10)		
Ashes	βουμέλιος, μελία	Buckthorns	ῥάμνος, φιλύκη
		Bugloss	ὀνοχειλές
Asparagus	ἀσφάραγος	Bullfist	πέζις
Aspen	κερκίς (2)	Bulrush	τύφη
Asphodel	ἀσφόδελος, πόθος (2)	Bush-grass	κάλαμος (ἐπίγειος)
Axe-weed	πελεκῖνος	Butcher's broom	κεντρομυρρίνη, μυάκανθος
Bachelor's buttons	παρθένιον		
Balm	μελισσόφυλλον		
Balsam of Mecca	βάλσαμον	Cabbage	ῥάφανος
Bamboos	κάλαμος ὁ Ἰνδικός	Calamander-wood	App. (23)
		Calamint	ἐλένιον
Banyan	συκῆ ἡ Ἰνδική	Calavance	δόλιχος
Barbary nut	σισυρίγχιον	Caltrop	τρίβολος(1),(2)
Barley	κριθή	Caper	κάππαρις
Basil	ὤκιμον	Cardamom	καρδάμωμον
Bay (sweet)	δάφνη (1)	—— Nepaul	ἄμωμον
Bean	κύαμος	Cardoon	κάκτος (1)
Bedstraw	ἀπαρίνη	Carnation	διόσανθος
Beet	τεῦτλον,	Carob	κερωνία, συκῆ ἡ Αἰγυπτία
Bergamot-mint	σισύμβριον		
Bindweed	ἰασιώνη	Cassia	κασία
Bird-cherry	κέρασος, λακάρη	Castor-oil plant	κρότων
		Cat's ear	ὑποχοιρίς
Birth-wort	ἀριστολοχία	Cedar, odorous	θυία
Bladder-senna	κολυτέα	—— prickly	κέδρος (1), ὀξύκεδρος
Blite	βλίτον	—— Syrian	κέδρος (2)

KEY TO THE INDEX

Celandine, greater	χελιδόνιον	Dittany	δίκταμνον
—— lesser	ἀφία	Dock	λάπαθον τὸ ἄγριον
Celery	σέλινον		
Centaury	κενταύριον	Dodder	ὀροβάγχη
Chamaeleon	χαμαιλέων	Dog-mercury	ἀρρενόγονον, θηλύγονον
Chamomile, wild	ἄνθεμον τὸ ἀφύλλανθες	Dog-rose	κυνόρροδον
Charlock	κεράϊς, ῥαφανὶς ἡ ἀγρία	Dog's tooth grass	ἀγρωστις
		Doum-palm	κόϊξ, κουκιόφορον
Chaste-tree	ἄγνος (οἶσος)		
Cheese-flower	μαλάχη (2)	Dropwort	οἰνάνθη (1)
Chervil	ἔνθρυσκον	Duckweed	ἴκμη
—— wild	σκάνδιξ	Dwarf palm	φοῖνιξ ὁ χαμαιρριφής
Chestnuts	διοσβάλανος, καρύα ἡ Εὐβοϊκή		
		Ebony	ἐβένη
Chick-pea	ἐρέβινθος	Edder-wort	δρακόντιον
Chicory	κιχόριον	Elder	ἀκτέος, ἀκτῆ
Christ's thorn	παλίουρος	Elecampane	πάνακες τὸ Χειρώνειον
Cinnamon	κινάμωμον		
Cinquefoil	πενταπετές, πεντάφυλλον	Elms	πτελέα, ὀρειπτελέα
Citron	μηλέα ἡ Περσική	Eryngo	ἠρύγγιον
Clematis	ἀθραγένη	Fenugreek	βουκέρας, τῆλις
Cork-oak	ἴψος (?), φελλός	Ferns	ἀδίαντον, ἡμιόνιον, θηλυπτερίς, πολυπόδιον, πτερίς, τριχομανές, σκολοπένδριον
Coriander	κορίαννον		
Cornel	θηλυκράνεια		
Cornelian cherry	κράνεια		
Corn-flag	ξίφιον, ξίφος, φάσγανον		
Corn-thistle	ἄκανθα (2)		
Cotoneaster	ὀξυάκανθος	Ferula	ναρθηκία, νάρθηξ
Cotton-plant	(δένδρον τὸ) ἐριόφορον		
		Feverwort	κενταυρίς
Crack willow	ἑλίκη	Fig, wild	ἐρινεός
Cress	κάρδαμον	Figs	συκῆ, σ. ἡ Ἀραβική, σ. ἡ Ἰνδική, App. (17)
Crocus	κρόκος		
Cuckoo-pint	ἄρον		
Cucumber	σίκυος		
—— squirting	σίκυος ὁ ἄγριος	Filbert	καρύα ἡ Ἡρακλεωτική
Cummin	κύμινον		
Currant-vine	ἄμπελος (2)	Firs	πεύκη, ἐλάτη
Cyclamen	κυκλάμινος	Flax	λίνον
Cypress	κυπάριττος	Frankincense-tree	λιβανωτός
		French sparrow-grass	σκίλλα ἡ Ἐπιμενίδειος
Dandelion	ἀπάπη	Fungi	ἄσχιον, μύκης, πέζις
Darnel	αἶρα		
Date-palm	φοῖνιξ (1)		
Dill	ἄνηθον, ἄνηητος	Galingale	κύπειρος
		Gall-oak	ἡμερίς (1)
Distaff-thistle	ἀτρακτυλίς, φόνος	Garden nightshade	στρύχνος ὁ ἐδώδιμος

495

KEY TO THE INDEX

Garlics	μῶλυ, σκόροδον	Junipers	ἄρκευθος, θυία, κεδρίς, κέδρος, ὀξύκεδρος
Germander	χαμαίδρυς		
Gilliflower	ἴον τὸ λευκόν		
Ginger-grass	σχοῖνος ὁ εὐώδης (?)		
		Kermes-oak	πρῖνος
Gladwyn	ξίρις		
Globe-thistle	ῥύτρος	Laburnum	κύτισος (1)
Goat's beard	τραγοπώγων	Larkspurs	ἀσταφίς, πόθος (1), ὑάκινθος ἡ σπαρτή
Goat willow	ἐλαίαγνος		
Gold flower	ἐλειόχρυσος		
Golden thistle	λειμωνία (2), σκόλυμος	Leek	πράσον (1)
		Lentil	φακός
Gourd	κολοκύντη	Leopard's bane	σκορπίος (2)
Grasses	αἰγίλωψ (2), θρύον, πόα	Lettuce	θρίδαξ, θριδακίνη
Groundsel	ἠριγέρων	Lilies	ἡμεροκαλλές, κρίνον, κ. τὸ πορφυροῦν, λείριον (1)
Gum arabic	ἄκανθα (7)		
Hartshorn	κορωνόπους		
Hart's tongue	σκολοπένδριον		
Hartwort	σέσελι	Liquorice	γλυκεῖα (ῥίζα) Σκυθική
Hawk's beard	ἀπαργία		
Hawthorn	μεσπίλη ἡ ἀνθηδονοειδής	Limes	φίλυρα, φ. ἡ θήλεια
Hazel	καρύα	Litmus	φῦκος (5)
Heath	ἐρείκη	Lucerne	(πόα) ἡ Μηδική
Hellebores	ἐλλέβορος	Lupin	θέρμος
Hemlock	κώνειον		
Holly	κήλαστρος		
Holm-oaks	ἀρία, ἶψος, σμῖλαξ (1), φελλόδρυς	Madder	ἐρευθεδανόν
		Madonna lily	κρίνον, λείριον (1)
Honeysuckle (Greek)	κλύμενον	Maiden-hair	ἀδίαντον, ἀ. τὸ μέλαν
Hop-hornbeam	ὀστρύα, ὄστρυς	—— English	ἀδίαντον τὸ λευκόν
Horehounds	πράσιον		
Horned poppy	μήκων ἡ κερατῖτις	Mallow	μαλάχη (1)
Horseradish	ῥαφανὶς ἡ ἀμωρέα (?)	Mandrake	μανδραγόρας (1)
		Mango	App. (5)
House-leek	ἀείζωον	Mangroves	App. (12)
Hulwort	πύλιον	—— white	δάφνη (6), ἐλάα (3), App.(14)
		Manna-ash	μελία
Irises	ἶρις, ξίρις, σισυρίγχιον	Maples	γλεῖνος, ζυγία, κλινότροχος, σφένδαμνος
Ivies	ἕλιξ, κιττός		
Jack-fruit	App. (3)	Marestail	ἵππον
Joint-fir	θραύπαλος	Marjorams	ὀρίγανον
Jujube	App. (6)	—— sweet	ἀμάρακον
Judas-tree	κερκίς (1), σημύδα	Marsh celery	ἑλειοσέλινον
		Marsh mallow	ἀλθαία, μαλάχη ἡ ἀγρία

KEY TO THE INDEX

Martagon lily	ἡμεροκαλλές	Onions	γήθυον, γήτειον, κρόμμυον, κρομμυογήτειον
Mastich	σχῖνος		
Meadow saffron	ἐφήμερον, σπάλαξ		
Medlar	μεσπίλη, μ. ἡ σατάνειος	Opium poppy	μήκων (ἡ ὀπώδης), (νηπενθές)
Michaelmas daisy	ἀστερίσκος, (τριπόλιον)	Orach	ἀδράφαξυς
Milk-thistle	λευκάκανθα	Orchis	ὄρχις
Milt-waste	ἡμιόνιον	Oriental thorn	μεσπίλη ἡ ἀνθηδών
Millet	κέγχρος	Oyster-green	βρύον
—— Italian	ἔλυμος	Ox-tongue	κρηπίς
Mints	ἡδύοσμον, μίνθη, σισύμβριον		
Mistletoes	ἰξία (1), ὕφεαρ	Palms	κόϊξ, κουκιόφόρον, φοῖνιξ
Mock-privet	φίλυρα ἡ ἄρρην	Papyrus	πάπυρος
Moly	μῶλυ	Parsley	ὀρεισέλινον
Monk's rhubarb	λάπαθος	Pea	πισός
Mulberry	συκάμινος	Pear	ἄπιος (1)
Mullein	φλόμος ἡ μέλαινα	—— wild	ἀχράς, ὄγχνη
Mushroom	μύκης	Pennyroyal	βλήχω
Mustard, white	νᾶπυ	Peony	γλυκυσίδη, παιωνία
Myrrh	σμύρνα		
Myrtle	μυρρίνη	Pepper	πέπερι
		Periwinkle	χαμαιδάφνη
		Pimpernel, blue	κόρχορος
Narcissus	λείριον (2), νάρκισσος	Pines	πεύκη, πίτυς
Nepaul cardamom	ἄμωμον	Pine-thistle	ἄκανθα (8), ἄκανος, ἰξία (2), ἰξίνη, χαμαιλέων ὁ λευκός
Nettle	ἀκαλύφη		
—— tree	λωτός (1)		
Nightshade, deadly	μανδραγόρας (2) ?		
—— garden	στρύχνος ὁ ἐδώδιμος	Plane-tree	πλάτανος
		Plantains	ἀρνόγλωσσον, θρυαλλίς, κορωνόπους, κύνωψ, ὄρτυξ, στελέφουρος
Oaks	αἰγίλωψ (1), ἀρία, ἄσπρις, δρῦς, ἐτυμόδρυς, ἡμερίς, ἴψος, πρῖνος, σμῖλαξ (1), φηγός, φελλόδρυς, φελλός	Plums	κοκκυμηλέα, πάδος, προύμνη, σποδιάς
		Polypody	πολυπόδιον
		Pole-reed	δόναξ, κάλαμος ὁ Λακωνικός etc.
Oak-mistletoe	ἰξία (1)	Pomegranate	ῥόα
Oats	βρόμος	Poppies	μήκων
Oleander	δάφνη ἡ ἀγρία. ὀνοθήρας	Poplar, black	αἴγειρος
		—— white (abele)	λεύκη
Olives	ἐλάα (1), (2)	Privet	σπείραια
Olive, wild	ἀγριέλαιος, κότινος	Puff-ball	ἄσχιον
		Purslane	ἀνδράχνη
		Purse-tassels	βολβός

KEY TO THE INDEX

Quince	κυδώνιον, στρούθιον (2)	Soapwort	στρούθιον (2), στρούθος
		Sorb	ὄη
Radish	ῥαφανίς	Southernwood	ἀβρότονον
Reeds	δόναξ, κάλαμος, πάπυρος	Spanish broom	λινόσπαρτον
		Spear-grass	κάλαμος ὁ πλόκιμος, κ. ὁ χαρακίας
Restharrow	ὀνωνίς		
Rib-grass	κύνωψ		
Rice	ὄρυζον	Sow-thistle	σόγκος
Rice-wheat	ζειά, ὄλυρα	Spike-lavender	ἴφυον
Rocket	εὔζωμον	Spikenard	νάρδον
Rock-roses	κίσθος	Spindle-tree	εὐώνυμος (τεταγωνία)
Rose-campion	λυχνίς		
Roses	κυνόρροδον, κυνόσβατος, ῥοδωνία	Spurges	ἄκανθα (5), (6), ἄπιος (2), ἱποφέως, ἰσχάς, μηκώνιον, ῥαφανὶς ἡ ὀρεία, τιθύμαλλος
Rue	πήγανον		
Rupture-wort	ἐλλεβορίνη		
Rushes	μελαγκρανίς, σχοῖνος		
		Squills	σκίλλα, τίφυον, ὑάκινθος ἡ ἀγρία
Safflower	κνῆκος, κρόκος ὁ ἀκανθώδης		
Saffron crocus	κρόκος, κ. ὁ εὐώδης	Star-flower	βολβίνη
Sage	σφάκος	Star-thistle	παντάδουσα
Salvia	ἐλελίσφακος	—— yellow	τετράλιξ
Savory	θύμβρα	Stonecrop	ἐπίπετρον
Scammony	σκαμμωνία	Storax	στύραξ
Scrub oak	δρῦς ἡ πλατύφυλλος	Sulphur-wort	πευκέδανον
		Sumachs	ῥοῦς
Sea-bark oak	δρῦς ἡ ἁλίφλοιος, δ. ἡ εὐθύφλοιος	Sweet bay	δάφνη (1)
		Sweet flag	κάλαμος ὁ εὐώδης
Sea spurge	τιθύμαλλος ὁ παράλιος	Sycamore	συκάμινος ἡ Αἰγυπτία, σ. ἡ Κυπρία
Sea-weeds	ἄμπελος (3), βρύον, δρῦς (7), (8), ἐλάτη (3), φοῖνιξ (4), φῦκος		
		Tamarind	App. (15), (16)
		Tamarisks	μυρίκη
		Tare	ἀφάκη
Sebesten	κοκκυμηλέα ἡ Αἰγυπτία	Teak	App. (22)
		Terebinth	τέρμινθος
Sedge	βούτομος	Thistles	ἄκανθα (2), (8), ἄκανος, ἄκορνα, ἰξία (2), ἰξίνη, κάκτος, λειμωνία (2), λευκάκανθα, πολυάκανθος, ῥύτρος, σκόλυμος, σόγκος, χάλκειος, χαμαιλέων
Sesame	σήσαμον		
Shallot	κρόμμυον τὸ σχιστόν		
Silphium	σίλφιον		
Silver-fir	ἐλάτη		
Silver-lim	φίλυρα, φ. ἡ θήλεια		
Sissoo	App. (21)		
Smilax	σμῖλαξ (2)		
Snowdrop	λευκόϊον (2)		
Snapdragon	ἀντίρρινον		

KEY TO THE INDEX

Thorn, oriental	μεσπίλη ἡ ἀνθηδών	Vine, wild	οἰνάνθη ἡ ἀγρία
Thorn-apple	θρύορον, περιττός, στρυχνος ὁ μανικός	Violet	ἴον τὸ μέλαν
		Wall-flower	φλόγινον, φλόξ
Thyine-wood	θύον	Walnut	καρύα ἡ Περσική
Thyme, Attic	ἕρπυλλος ὁ ἄγριος	Water chestnut	τρίβολος (3)
—— Cretan	θύμον (1)	Water chickweed	λέμνα
—— tufted	ἕρπυλλος (ὁ ἥμερος)	Water-lily, Nile	λωτός (2)
		—— white	σίδη
Tine-tare	App. (1)	—— yellow	μαδωναΐς, νυμφαία
Tragacanth	τραγάκανθα		
Traveller's joy	ἀθραγένη	Wheats	πυρός
Tree-medick	κύτισος (2)	—— one-seeded	τίφη
Tree-moss	φάσκος	Wheat-rust	ἐρυσίβη
Trefoil	λωτός (3)	Wig-tree	κοκκυγέα
Truffles	κεραύνιον, ὕδνον	Willows	ἐλαίαγνος, ἑλίκη, ἰτέα, κολοιτία (2)
Turk's cap lily	κρίνον τὸ πορφυροῦν	Willow-weed	κραταίγονος
Turkey oak	αἰγίλωψ (1), ἄσπρις	Withy	οἴσος
		Wolf's bane	ἀκόνιτον, θηλύφονον, μυόφονον, σκορπίος (3)
Turnip	γογγυλίς		
Valonia oak	δρῦς ἡ ἀγρία, φηγός	Wormwood	ἀψίνθιον
		Wych-elm	ὀρειπτελέα
Vetch, bitter	ὄροβος		
Vine	ἄμπελος (1)	Yew	μίλος

*Printed in Great Britain by
Fletcher & Son Ltd, Norwich*

THE LOEB CLASSICAL LIBRARY

VOLUMES ALREADY PUBLISHED

Latin Authors

AMMIANUS MARCELLINUS. Translated by J. C. Rolfe. 3 Vols.
APULEIUS: THE GOLDEN ASS (METAMORPHOSES). W. Adlington (1566). Revised by S. Gaselee.
ST. AUGUSTINE: CITY OF GOD. 7 Vols. Vol. I. G. E. McCracken. Vols. II and VII. W. M. Green. Vol. III. D. Wiesen. Vol. IV. P. Levine. Vol. V. E. M. Sanford and W. M. Green. Vol. VI. W. C. Greene.
ST. AUGUSTINE, CONFESSIONS OF. W. Watts (1631). 2 Vols.
ST. AUGUSTINE, SELECT LETTERS. J. H. Baxter.
AUSONIUS. H. G. Evelyn White. 2 Vols.
BEDE. J. E. King. 2 Vols.
BOETHIUS: TRACTS and DE CONSOLATIONE PHILOSOPHIAE. Rev. H. F. Stewart and E. K. Rand. Revised by S. J. Tester.
CAESAR: ALEXANDRIAN, AFRICAN and SPANISH WARS. A. G. Way.
CAESAR: CIVIL WARS. A. G. Peskett.
CAESAR: GALLIC WAR. H. J. Edwards.
CATO: DE RE RUSTICA. VARRO: DE RE RUSTICA. H. B. Ash and W. D. Hooper.
CATULLUS. F. W. Cornish. TIBULLUS. J. B. Postgate. PERVIGILIUM VENERIS. J. W. Mackail.
CELSUS: DE MEDICINA. W. G. Spencer. 3 Vols.
CICERO: BRUTUS and ORATOR. G. L. Hendrickson and H. M. Hubbell.
[CICERO]: AD HERENNIUM. H. Caplan.
CICERO: DE ORATORE, etc. 2 Vols. Vol. I. DE ORATORE, Books I and II. E. W. Sutton and H. Rackham. Vol. II. DE ORATORE, Book III. DE FATO; PARADOXA STOICORUM; DE PARTITIONE ORATORIA. H. Rackham.
CICERO: DE FINIBUS. H. Rackham.
CICERO: DE INVENTIONE, etc. H. M. Hubbell.
CICERO: DE NATURA DEORUM and ACADEMICA. H. Rackham.
CICERO: DE OFFICIIS. Walter Miller.
CICERO: DE REPUBLICA and DE LEGIBUS. Clinton W. Keyes.

CICERO: DE SENECTUTE, DE AMICITIA, DE DIVINATIONE. W. A. Falconer.
CICERO: IN CATILINAM, PRO FLACCO, PRO MURENA, PRO SULLA. New version by C. Macdonald.
CICERO: LETTERS TO ATTICUS. E. O. Winstedt. 3 Vols.
CICERO: LETTERS TO HIS FRIENDS. W. Glynn Williams, M. Cary, M. Henderson. 4 Vols.
CICERO: PHILIPPICS. W. C. A. Ker.
CICERO: PRO ARCHIA, POST REDITUM, DE DOMO, DE HARUSPICUM RESPONSIS, PRO PLANCIO. N. H. Watts.
CICERO: PRO CAECINA, PRO LEGE MANILIA, PRO CLUENTIO, PRO RABIRIO. H. Grose Hodge.
CICERO: PRO CAELIO, DE PROVINCIIS CONSULARIBUS, PRO BALBO. R. Gardner.
CICERO: PRO MILONE, IN PISONEM, PRO SCAURO, PRO FONTEIO, PRO RABIRIO POSTUMO, PRO MARCELLO, PRO LIGARIO, PRO REGE DEIOTARO. N. H. Watts.
CICERO: PRO QUINCTIO, PRO ROSCIO AMERINO, PRO ROSCIO COMOEDO, CONTRA RULLUM. J. H. Freese.
CICERO: PRO SESTIO, IN VATINIUM. R. Gardner.
CICERO: TUSCULAN DISPUTATIONS. J. E. King.
CICERO: VERRINE ORATIONS. L. H. G. Greenwood. 2 Vols.
CLAUDIAN. M. Platnauer. 2 Vols.
COLUMELLA: DE RE RUSTICA. DE ARBORIBUS. H. B. Ash, E. S. Forster and E. Heffner. 3 Vols.
CURTIUS, Q.: HISTORY OF ALEXANDER. J. C. Rolfe. 2 Vols.
FLORUS. E. S. Forster. CORNELIUS NEPOS. J. C. Rolfe.
FRONTINUS: STRATAGEMS and AQUEDUCTS. C. E. Bennett and M. B. McElwain.
FRONTO: CORRESPONDENCE. C. R. Haines. 2 Vols.
GELLIUS. J. C. Rolfe. 3 Vols.
HORACE: ODES and EPODES. C. E. Bennett.
HORACE: SATIRES, EPISTLES, ARS POETICA. H. R. Fairclough.
JEROME: SELECTED LETTERS. F. A. Wright.
JUVENAL and PERSIUS. G. G. Ramsay.
LIVY. B. O. Foster, F. G. Moore, Evan T. Sage, and A. C. Schlesinger and R. M. Geer (General Index). 14 Vols.
LUCAN. J. D. Duff.
LUCRETIUS. W. H. D. Rouse. Revised by M. F. Smith.
MARTIAL. W. C. A. Ker. 2 Vols.
MINOR LATIN POETS: from PUBLILIUS SYRUS to RUTILIUS NAMATIANUS, including GRATTIUS, CALPURNIUS SICULUS, NEMESIANUS, AVIANUS and others, with "Aetna" and the "Phoenix." J. Wight Duff and Arnold M. Duff.
OVID: THE ART OF LOVE and OTHER POEMS. J. H. Mozley.
OVID: FASTI. Sir James G. Frazer.

OVID: HEROIDES and AMORES. Grant Showerman.
OVID: METAMORPHOSES. F. J. Miller. 2 Vols.
OVID: TRISTIA and EX PONTO. A. L. Wheeler.
PERSIUS. Cf. JUVENAL.
PETRONIUS. M. Heseltine. SENECA: APOCOLOCYNTOSIS. W. H. D. Rouse.
PHAEDRUS and BABRIUS (Greek). B. E. Perry.
PLAUTUS. Paul Nixon. 5 Vols.
PLINY: LETTERS, PANEGYRICUS. Betty Radice. 2 Vols.
PLINY: NATURAL HISTORY. 10 Vols. Vols. I–V and IX. H. Rackham. VI.–VIII. W. H. S. Jones. X. D. E. Eichholz.
PROPERTIUS. H. E. Butler.
PRUDENTIUS. H. J. Thomson. 2 Vols.
QUINTILIAN. H. E. Butler. 4 Vols.
REMAINS OF OLD LATIN. E. H. Warmington. 4 Vols. Vol. I. (ENNIUS AND CAECILIUS) Vol. II. (LIVIUS, NAEVIUS PACUVIUS, ACCIUS) Vol. III. (LUCILIUS and LAWS OF XII TABLES) Vol. IV. (ARCHAIC INSCRIPTIONS)
SALLUST. J. C. Rolfe.
SCRIPTORES HISTORIAE AUGUSTAE. D. Magie. 3 Vols.
SENECA, THE ELDER: CONTROVERSIAE, SUASORIAE. M. Winterbottom. 2 Vols.
SENECA: APOCOLOCYNTOSIS. Cf. PETRONIUS.
SENECA: EPISTULAE MORALES. R. M. Gummere. 3 Vols.
SENECA: MORAL ESSAYS. J. W. Basore. 3 Vols.
SENECA: TRAGEDIES. F. J. Miller. 2 Vols.
SENECA: NATURALES QUAESTIONES. T. H. Corcoran. 2 Vols.
SIDONIUS: POEMS and LETTERS. W. B. Anderson. 2 Vols.
SILIUS ITALICUS. J. D. Duff. 2 Vols.
STATIUS. J. H. Mozley. 2 Vols.
SUETONIUS. J. C. Rolfe. 2 Vols.
TACITUS: DIALOGUS. Sir Wm. Peterson. AGRICOLA and GERMANIA. Maurice Hutton. Revised by M. Winterbottom, R. M. Ogilvie, E. H. Warmington.
TACITUS: HISTORIES and ANNALS. C. H. Moore and J. Jackson. 4 Vols.
TERENCE. John Sargeaunt. 2 Vols.
TERTULLIAN: APOLOGIA and DE SPECTACULIS. T. R. Glover. MINUCIUS FELIX. G. H. Rendall.
VALERIUS FLACCUS. J. H. Mozley.
VARRO: DE LINGUA LATINA. R. G. Kent. 2 Vols.
VELLEIUS PATERCULUS and RES GESTAE DIVI AUGUSTI. F. W. Shipley.
VIRGIL. H. R. Fairclough. 2 Vols.
VITRUVIUS: DE ARCHITECTURA. F. Granger. 2 Vols.

Greek Authors

ACHILLES TATIUS. S. Gaselee.
AELIAN: ON THE NATURE OF ANIMALS. A. F. Scholfield. 3 Vols.
AENEAS TACTICUS. ASCLEPIODOTUS and ONASANDER. The Illinois Greek Club.
AESCHINES. C. D. Adams.
AESCHYLUS. H. Weir Smyth. 2 Vols.
ALCIPHRON, AELIAN, PHILOSTRATUS: LETTERS. A. R. Benner and F. H. Fobes.
ANDOCIDES, ANTIPHON. Cf. MINOR ATTIC ORATORS.
APOLLODORUS. Sir James G. Frazer. 2 Vols.
APOLLONIUS RHODIUS. R. C. Seaton.
THE APOSTOLIC FATHERS. Kirsopp Lake. 2 Vols.
APPIAN: ROMAN HISTORY. Horace White. 4 Vols.
ARATUS. Cf. CALLIMACHUS.
ARISTIDES: ORATIONS. C. A. Behr. Vol. I.
ARISTOPHANES. Benjamin Bickley Rogers. 3 Vols. Verse trans.
ARISTOTLE: ART OF RHETORIC. J. H. Freese.
ARISTOTLE: ATHENIAN CONSTITUTION, EUDEMIAN ETHICS, VICES AND VIRTUES. H. Rackham.
ARISTOTLE: GENERATION OF ANIMALS. A. L. Peck.
ARISTOTLE: HISTORIA ANIMALIUM. A. L. Peck. Vols. I.–II.
ARISTOTLE: METAPHYSICS. H. Tredennick. 2 Vols.
ARISTOTLE: METEOROLOGICA. H. D. P. Lee.
ARISTOTLE: MINOR WORKS. W. S. Hett. On Colours, On Things Heard, On Physiognomies, On Plants, On Marvellous Things Heard, Mechanical Problems, On Indivisible Lines, On Situations and Names of Winds, On Melissus, Xenophanes, and Gorgias.
ARISTOTLE: NICOMACHEAN ETHICS. H. Rackham.
ARISTOTLE: OECONOMICA and MAGNA MORALIA. G. C. Armstrong (with METAPHYSICS, Vol. II).
ARISTOTLE: ON THE HEAVENS. W. K. C. Guthrie.
ARISTOTLE: ON THE SOUL, PARVA NATURALIA, ON BREATH. W. S. Hett.
ARISTOTLE: CATEGORIES, ON INTERPRETATION, PRIOR ANALYTICS. H. P. Cooke and H. Tredennick.
ARISTOTLE: POSTERIOR ANALYTICS, TOPICS. H. Tredennick and E. S. Forster.
ARISTOTLE: ON SOPHISTICAL REFUTATIONS.
On Coming to be and Passing Away, On the Cosmos. E. S. Forster and D. J. Furley.
ARISTOTLE: PARTS OF ANIMALS. A. L. Peck; MOTION AND PROGRESSION OF ANIMALS. E. S. Forster.

ARISTOTLE: PHYSICS. Rev. P. Wicksteed and F. M. Cornford. 2 Vols.

ARISTOTLE: POETICS and LONGINUS. W. Hamilton Fyfe DEMETRIUS ON STYLE. W. Rhys Roberts.

ARISTOTLE: POLITICS. H. Rackham.

ARISTOTLE: PROBLEMS. W. S. Hett. 2 Vols.

ARISTOTLE: RHETORICA AD ALEXANDRUM (with PROBLEMS. Vol. II). H. Rackham.

ARRIAN: HISTORY OF ALEXANDER and INDICA. New Vol. I by P. A. Brunt. Vol. II by E. Iliffe Robson.

ATHENAEUS: DEIPNOSOPHISTAE. C. B. Gulick. 7 Vols.

BABRIUS and PHAEDRUS (Latin). B. E. Perry.

ST. BASIL: LETTERS. R. J. Deferrari. 4 Vols.

CALLIMACHUS: FRAGMENTS. C. A. Trypanis. MUSAEUS: HERO AND LEANDER. T. Gelzer and C. Whitman.

CALLIMACHUS: HYMNS and EPIGRAMS. LYCOPHRON. A. W. Mair. ARATUS. G. R. Mair.

CLEMENT OF ALEXANDRIA. Rev. G. W. Butterworth.

COLLUTHUS. Cf. OPPIAN.

DAPHNIS AND CHLOE. Thornley's Translation revised by J. M. Edmonds. PARTHENIUS. S. Gaselee.

DEMOSTHENES I: OLYNTHIACS, PHILIPPICS and MINOR ORATIONS I–XVII and XX. J. H. Vince.

DEMOSTHENES II: DE CORONA and DE FALSA LEGATIONE. C. A. Vince and J. H. Vince.

DEMOSTHENES III: MEIDIAS, ANDROTION, ARISTOCRATES, TIMOCRATES and ARISTOGEITON I and II. J. H. Vince.

DEMOSTHENES IV–VI: PRIVATE ORATIONS and IN NEAERAM. A. T. Murray.

DEMOSTHENES VII: FUNERAL SPEECH, EROTIC ESSAY, EXORDIA and LETTERS. N. W. and N. J. DeWitt.

DIO CASSIUS: ROMAN HISTORY. E. Cary. 9 Vols.

DIO CHRYSOSTOM. J. W. Cohoon and H. Lamar Crosby. 5 Vols.

DIODORUS SICULUS. 12 Vols. Vols. I–VI. C. H. Oldfather. Vol. VII. C. L. Sherman. Vol. VIII. C. B. Welles. Vols. IX and X. R. M. Geer. Vol. XI. F. Walton. Vol. XII. F. Walton. General Index. R. M. Geer.

DIOGENES LAERTIUS. R. D. Hicks. 2 Vols. New Introduction by H. S. Long.

DIONYSIUS OF HALICARNASSUS: ROMAN ANTIQUITIES. Spelman's translation revised by E. Cary. 7 Vols.

DIONYSIUS OF HALICARNASSUS: CRITICAL ESSAYS. S. Usher. 2 Vols.

EPICTETUS. W. A. Oldfather. 2 Vols.

EURIPIDES. A. S. Way. 4 Vols. Verse trans.

EUSEBIUS: ECCLESIASTICAL HISTORY. Kirsopp Lake and J. E. L. Oulton. 2 Vols.

GALEN: ON THE NATURAL FACULTIES. A. J. Brock.
THE GREEK ANTHOLOGY. W. R. Paton. 5 Vols.
GREEK ELEGY AND IAMBUS with the ANACREONTEA. J. M. Edmonds. 2 Vols.
THE GREEK BUCOLIC POETS (THEOCRITUS, BION, MOSCHUS). J. M. Edmonds.
GREEK MATHEMATICAL WORKS. Ivor Thomas. 2 Vols.
HERODES. Cf. THEOPHRASTUS: CHARACTERS.
HERODIAN. C. R. Whittaker. 2 Vols.
HERODOTUS. A. D. Godley. 4 Vols.
HESIOD and THE HOMERIC HYMNS. H. G. Evelyn White.
HIPPOCRATES and the FRAGMENTS OF HERACLEITUS. W. H. S. Jones and E. T. Withington. 4 Vols.
HOMER: ILIAD. A. T. Murray. 2 Vols.
HOMER: ODYSSEY. A. T. Murray. 2 Vols.
ISAEUS. E. W. Forster.
ISOCRATES. George Norlin and LaRue Van Hook. 3 Vols.
[ST. JOHN DAMASCENE]: BARLAAM AND IOASAPH. Rev. G. R. Woodward, Harold Mattingly and D. M. Lang.
JOSEPHUS. 9 Vols. Vols. I–IV. H. Thackeray. Vol. V. H. Thackeray and R. Marcus. Vols. VI–VII. R. Marcus. Vol. VIII. R. Marcus and Allen Wikgren. Vol. IX. L. H. Feldman.
JULIAN. Wilmer Cave Wright. 3 Vols.
LIBANIUS. A. F. Norman. Vols. I–II.
LUCIAN. 8 Vols. Vols. I–V. A. M. Harmon. Vol. VI. K. Kilburn. Vols. VII–VIII. M. D. Macleod.
LYCOPHRON. Cf. CALLIMACHUS.
LYRA GRAECA. J. M. Edmonds. 3 Vols.
LYSIAS. W. R. M. Lamb.
MANETHO. W. G. Waddell. PTOLEMY: TETRABIBLOS. F. E. Robbins.
MARCUS AURELIUS. C. R. Haines.
MENANDER. F. G. Allison.
MINOR ATTIC ORATORS (ANTIPHON, ANDOCIDES, LYCURGUS, DEMADES, DINARCHUS, HYPERIDES). K. J. Maidment and J. O. Burtt. 2 Vols.
MUSAEUS: HERO AND LEANDER. Cf. CALLIMACHUS.
NONNOS: DIONYSIACA. W. H. D. Rouse. 3 Vols.
OPPIAN, COLLUTHUS, TRYPHIODORUS. A. W. Mair.
PAPYRI. NON-LITERARY SELECTIONS. A. S. Hunt and C. C. Edgar. 2 Vols. LITERARY SELECTIONS (Poetry). D. L. Page.
PARTHENIUS. Cf. DAPHNIS AND CHLOE.
PAUSANIAS: DESCRIPTION OF GREECE. W. H. S. Jones. 4 Vols. and Companion Vol. arranged by R. E. Wycherley.

PHILO. 10 Vols. Vols. I–V. F. H. Colson and Rev. G. H. Whitaker. Vols. VI–IX. F. H. Colson. Vol. X. F. H. Colson and the Rev. J. W. Earp.

PHILO: two supplementary Vols. (*Translation only.*) Ralph Marcus.

PHILOSTRATUS: THE LIFE OF APOLLONIUS OF TYANA. F. C. Conybeare. 2 Vols.

PHILOSTRATUS: IMAGINES. CALLISTRATUS: DESCRIPTIONS. A. Fairbanks.

PHILOSTRATUS and EUNAPIUS: LIVES OF THE SOPHISTS. Wilmer Cave Wright.

PINDAR. Sir J. E. Sandys.

PLATO: CHARMIDES, ALCIBIADES, HIPPARCHUS, THE LOVERS, THEAGES, MINOS and EPINOMIS. W. R. M. Lamb.

PLATO: CRATYLUS, PARMENIDES, GREATER HIPPIAS, LESSER HIPPIAS. H. N. Fowler.

PLATO: EUTHYPHRO, APOLOGY, CRITO, PHAEDO, PHAEDRUS. H. N. Fowler.

PLATO: LACHES, PROTAGORAS, MENO, EUTHYDEMUS. W. R. M. Lamb.

PLATO: LAWS. Rev. R. G. Bury. 2 Vols.

PLATO: LYSIS, SYMPOSIUM, GORGIAS. W. R. M. Lamb.

PLATO: REPUBLIC. Paul Shorey. 2 Vols.

PLATO: STATESMAN, PHILEBUS. H. N. Fowler. ION. W. R. M. Lamb.

PLATO: THEAETETUS and SOPHIST. H. N. Fowler.

PLATO: TIMAEUS, CRITIAS, CLITOPHO, MENEXENUS, EPISTULAE. Rev. R. G. Bury.

PLOTINUS. A. H. Armstrong. Vols. I–III.

PLUTARCH: MORALIA. 17 Vols. Vols I–V. F. C. Babbitt. Vol. VI. W. C. Helmbold. Vols VII and XIV. P. H. De Lacy and B. Einarson. Vol. VIII. P. A. Clement and H. B. Hoffleit. Vol. IX. E. L. Minar, Jr., F. H. Sandbach, W. C. Helmbold. Vol. X. H. N. Fowler. Vol. XI. L. Pearson and F. H. Sandbach. Vol. XII. H. Cherniss and W. C. Helmbold. Vol. XIII 1–2. H. Cherniss. Vol. XV. F. H. Sandbach.

PLUTARCH: THE PARALLEL LIVES. B. Perrin. 11 Vols.

POLYBIUS. W. R. Paton. 6 Vols.

PROCOPIUS: HISTORY OF THE WARS. H. B. Dewing. 7 Vols.

PTOLEMY: TETRABIBLOS. Cf. MANETHO.

QUINTUS SMYRNAEUS. A. S. Way. Verse trans.

SEXTUS EMPIRICUS. Rev. R. G. Bury. 4 Vols.

SOPHOCLES. F. Storr. 2 Vols. Verse trans.

STRABO: GEOGRAPHY. Horace L. Jones. 8 Vols.

THEOPHRASTUS: CHARACTERS. J. M. Edmonds. HERODES, etc. A. D. Knox.

THEOPHRASTUS: DE CAUSIS PLANTARUM. B. Einarson and G. K. K. Link. Vol. I.
THEOPHRASTUS: ENQUIRY INTO PLANTS. Sir Arthur Hort, Bart. 2 Vols.
THUCYDIDES. C. F. Smith. 4 Vols.
TRYPHIODORUS. Cf. OPPIAN.
XENOPHON: CYROPAEDIA. Walter Miller. 2 Vols.
XENOPHON: HELLENICA. C. L. Brownson. 2 Vols.
XENOPHON: ANABASIS. C. L. Brownson.
XENOPHON: MEMORABILIA and OECONOMICUS. E. C. Marchant. SYMPOSIUM and APOLOGY. O. J. Todd.
XENOPHON: SCRIPTA MINORA. E. C. Marchant and G. W. Bowersock.

IN PREPARATION

Latin Authors

MANILIUS. G. P. Goold.

DESCRIPTIVE PROSPECTUS ON APPLICATION

CAMBRIDGE, MASS.   **HARVARD UNIVERSITY PRESS**
LONDON   **WILLIAM HEINEMANN LTD**

DATE DUE

	MAR 7 1985		
	~~4/17/85 page~~		

35

DEMCO NO. 38-298